Springer Atmospheric Sciences

More information about this series at http://www.springer.com/series/10176

Hajime Akimoto

Atmospheric Reaction Chemistry

Springer

Hajime Akimoto
National Institute for Environmental Studies
Tsukuba
Japan

Taiki Hanno Kagaku by Hajime Akimoto.
Original Japanese language edition published by Asakura Publishing Company, Ltd.

ISSN 2194-5217 ISSN 2194-5225 (electronic)
Springer Atmospheric Sciences
ISBN 978-4-431-55868-2 ISBN 978-4-431-55870-5 (eBook)
DOI 10.1007/978-4-431-55870-5

Library of Congress Control Number: 2015958787

Springer Tokyo Heidelberg New York Dordrecht London

Printed on acid-free paper

Springer Japan KK is part of Springer Science+Business Media (www.springer.com)

Preface

Twenty-five years have passed since atmospheric chemistry was established as one of the fundamental sciences deeply relevant to global change. During this period, atmospheric chemistry has been grappling with many important atmospheric environmental issues such as ozone depletion, ozone/oxidant pollution, aerosol/$PM_{2.5}$ pollution, atmosphere–climate interaction, and others from a fundamental point of view of biogeochemical cycles of atmospheric trace components. Giving a "meaning" to academic studies targeting such environmental issues is an integrated point of view for interpreting a cause–effect relationship including human activities, and atmospheric chemistry has been developed with such a holistic aspect of view. Based on this view, many excellent textbooks on atmospheric chemistry have been published in the last 10 years or so.

Meanwhile, what is necessary to get a feeling that a system composed of various elements is "fully understood" and to reliably predict the future? In the field of atmospheric chemistry, a phenomenon would be felt "fully understood" when the controlling chemistry and physics are resolved in terms of their fundamental principles. Therefore, those who learn atmospheric chemistry have to learn fundamental chemistry and physics, which constructs the basis of the discipline. Reaction chemistry, which is a branch of physical chemistry, is one such area of the fundamentals of atmospheric chemistry. In reaction chemistry, it may be implied that the reaction is "understood" when a chemical reaction is fully explained by spectroscopy, photochemistry, and chemical kinetics that have bases in quantum chemistry. This book specializes in atmospheric reaction chemistry, skipping the vast fertile discussion of comprehensive atmospheric chemistry. For overall atmospheric chemistry, readers can refer to the existing textbooks complementarily.

This book is written with the aim of providing fundamental knowledge, and particularly the fundamental way of thinking, about atmospheric chemical reactions for undergraduate and graduate students who wish to learn atmospheric chemistry. Another aim is for it to be used as reference on individual photochemical and thermal reactions. In particular, it is hoped that students in chemistry majors would more deeply understand the science hidden in each atmospheric reaction as being anchored to physical chemistry and quantum chemistry. It is also hoped that students in physics and meteorology majors would understand the principles and ways of thinking about chemical reactions.

Most small molecules which play an important role in atmospheric chemistry can be described accurately in terms of spectroscopy, chemical kinetics, photochemistry, and thermochemistry. Their photolysis and gas-phase homogeneous reactions have mostly been established systematically and understood fundamentally. On the contrary, among the chemical reactions covered in this book, those which do not fall into this category are the oxidation reactions of complex volatile organic compounds (VOCs) such as anthropogenic aromatics and biogenic hydrocarbons, organic aerosol formation processes from high boiling point oxidation products and surface reactions on atmospheric particles (aerosols). There are many aspects that are not yet resolved based on physical and quantum chemistry for those processes. It may be premature to cover these topics in a textbook in the sense that decisive interpretation and/or reliable values are not necessarily obtained yet. In these fields, the description of this book is not necessarily systematic and remains as an introduction of the results of the present research. These fields, however, are now attracting atmospheric scientists the most, so that a part of them were covered under the assumption that they will need to be revised in the near future.

Because this book originally was written as a textbook, the correct amounts of references are not necessarily cited. Although the best effort was made to cite the most important references, when there were too many relevant papers, references were selected arbitrarily to some extent. Apologies are in order for those authors who have been omitted from the citation list. Also, observed concentrations and their variability of the atmospheric species are taken from cited books without quoting original references. Absorption cross sections and rate constants are fully dependent on evaluations of the NASA/JPL Panel and the IUPAC subcommittee for kinetic and photochemical data for atmospheric chemistry and are updated for important findings thereafter. Among the textbooks on atmospheric chemistry, Finlayson-Pitts and Pitts (2000), Brasseur and Solomon (2005), and Seinfeld and Pandis (2006) are often referred to particularly for tropospheric chemistry, stratospheric chemistry, and both fields, respectively, because they contain more detailed chemistry than other textbooks. Also, the book *Photochemistry of Small Molecules* written by Hideo Okabe in 1978 was considered for constituting the chapters.

This book is dedicated to Prof. Ikuzo Tanaka, who was my Ph.D. supervisor in physical chemistry at the Tokyo Institute of Technology; the late Prof. James N. Pitts, Jr., who was my postdoctoral supervisor in chemistry of the atmosphere at the University of California, Riverside; and Prof. Barbara Finlayson-Pitts, at the University of California, Irvine, a friend of mine with whom I worked together on lab kinetics in the Pitts group at that time. Prof. Tanaka opened my eyes to research on photochemistry and molecular science from an international perspective in the then still economically poor and scientifically isolated country of Japan in the 1960s. Prof. Pitts led me to the study of chemistry on photochemical oxidants and tropospheric ozone, and I have been much affected by his pioneering stance to tackle science and policy. As I gradually changed my research field from physical chemistry to atmospheric chemistry, Barbara has been providing me a fixed coordinate like a polar star as an atmospheric scientist anchored to physical chemistry. Also, I wish to dedicate this book to my wife, Yoko, who has been with me supporting my long career apart from the platform of science. Prof. Pitts passed away in June 2014, just before the publication of this book, and it is a major regret that its publication was not in time.

Furthermore, I would like to express appreciation to Drs. Eiji Akiyoshi, Jun Hirokawa, Takashi Imamura, Yoshizumi Kajii, Yugo Kanaya, Yutaka Matsumi, Fumikazu Taketani, and Hiroshi Tanimoto, who read the manuscript in their respective parts and gave me valuable comments. Particularly, I owe Prof. Hirokawa, who kindly checked all equations relevant to heterogeneous kinetics and proposed an alternative easy-to-understand form. I also would like to express appreciation to the Asia Center for Air Pollution Research and its staff, who provided me an opportunity to write this book besides the task as director general, and helping me in several aspects including obtaining references, drawing of figures, etc. Finally, I would like to express my thanks to the editorial staff members of Asakura Publishing Co., who made tremendous efforts for the completion of this book to be published.

Tsukuba, Ibaraki, Japan Hajime Akimoto
July 2014

Preface to the English Edition

It is my great pleasure that my book *Atmospheric Reaction Chemistry*, which was originally published in Japanese in 2014, is now published in an English edition. As I wrote in the preface of the Japanese edition, in order to "fully understand" system science like atmospheric chemistry, one has to learn underlying chemistry and physics as fundamentals for gaining confidence in interpretation of atmospheric phenomena and particularly in future prediction. Reaction chemistry and kinetics, which is a branch of physical chemistry, is one such fundamental of atmospheric chemistry, and this book specializes in this area, skipping vast fertile discussion of comprehensive atmospheric chemistry. For overall atmospheric chemistry, readers can refer complementarily to the cited textbooks.

This book was written with the aim of providing a fundamental way of thinking about chemical reactions in the atmosphere for undergraduate and graduate students who wish to learn atmospheric chemistry. In particular, it is hoped that students in chemistry majors would understand more deeply the science hidden in each atmospheric reaction which is anchored to physical chemistry and quantum chemistry. It is also hoped that students in physics and meteorology majors would understand the fundamental principles of chemical reactions and kinetics. It is also hoped that it will be useful for research scientists in these fields as a reference book for consulting photochemical and thermal reaction pathways, absorption spectrum, and rate constants.

This book covers homogeneous gas-phase kinetics important in the atmosphere, which has been almost established, and provides the solid scientific bases of oxidation of trace gases and oxidant formation. Nevertheless, unresolved problems remain, for example, unsatisfactory reproduction of observed OH/HO_2 mixing ratio by model simulation under certain conditions, and oxidation mechanisms involving isoprene, terpenes and other biogenic hydrocarbons, and anthropogenic aromatic hydrocarbons. Therefore, descriptions of these topics are not completed in the book. Heterogeneous reaction chemistry is not covered well except for the chemistry on polar stratospheric clouds (PSCs) and reactive uptake coefficients of selected inorganic gases on sea salt and other particle surfaces in the troposphere. Heterogeneous organic chemistry related to secondary organic aerosol is now developing fast, and we hope that it will be established in the next decades.

Most of photochemical and kinetic data useful for atmospheric chemistry have been compiled in excellent evaluations by the NASA Panel for Data Evaluation and IUPAC Subcommittee on Gas Kinetic Data Evaluation for Atmospheric Chemistry, which have been frequently referred to in the book. Meanwhile, when they are referred to, I tried to quote selected original papers to be devoted to individual experimental and quantum chemists who actually obtained the data. This book is also an homage to the books *Photochemistry* by J. G. Calvert and J. N. Pitts, Jr (1966) and *Photochemistry of Small Molecules* by H. Okabe (1978), which were my desk-side books in my early career.

This book is dedicated to the late Prof. Ikuzo Tanaka, who was my Ph.D. supervisor in physical chemistry at the Tokyo Institute of Technology; the late Prof. James N. Pitts, Jr., who was my postdoctoral supervisor in chemistry of the atmosphere then at the University of California, Riverside; and Prof. Barbara Finlayson-Pitts, at the University of California, Irvine, a friend of mine with whom I worked together on lab kinetics in the Pitts' group at the beginning of the 1970s. Prof. Pitts and Prof. Tanaka passed away recently, in June 2014 and February 2015, respectively, just before and after the publication of the Japanese Edition. It is my wish that the fundamental photochemistry and reaction kinetics related to atmospheric chemistry will be inherited by the next generation through this book.

Tsukuba, Ibaraki, Japan | Hajime Akimoto
August 2015

Contents

Chapter 1
Introduction to Atmospheric Chemistry

This chapter serves as an introduction to atmospheric chemistry. It provides an overview of the development of the field, from the dawn of modern chemistry to the birth of atmospheric chemistry. Section 1.3 lists the textbooks on chemistry of the atmosphere and atmospheric chemistry that have been published to date. In addition, the literature that served as reference in the writing of this chapter is also listed at the end of this chapter under the bibliography.

1.1 The Dawn of Modern Chemistry and Chemistry of the Atmosphere

The birth of modern chemistry and chemistry of the atmosphere have an inseparable relationship. Chemistry of the atmosphere has its roots in the second half of the 18th century in England, when it first generated much interest as "air chemistry." This also signified the dawn of modern chemistry. Black discovered carbon dioxide in 1755, while Cavendish discovered hydrogen in 1766. Thereafter, nitrogen was discovered in 1772 by Rutherford, and oxygen in 1774 by Priestley. Furthermore, many of the components of air that are known today as air pollutant gases, such as nitrogen monoxide, nitrogen dioxide, sulfur dioxide, hydrogen chloride, ammonia, and nitrous oxide (dinitrogen monoxide), were discovered successively during the same period.

The second half of the 18th century was also the exact period when the transition from alchemy to modern chemistry took place. Incidentally, although the Japanese word for alchemy (*renkinjutsu*, which originates in Chinese word and translates as "hermetic arts") seems to have occult connotations, the English counterpart "alchemy" takes its "al" from the Arabic article for "the," suggesting a more natural sense of continuity between "alchemy" and "chemistry."

H. Akimoto, *Atmospheric Reaction Chemistry*, Springer Atmospheric Sciences,
DOI 10.1007/978-4-431-55870-5_1

Furthermore, this was also the period when the phlogiston theory was propounded, and eventually refuted. This theory, which took the world by storm, hypothesized that combustion was the result of the loss of phlogiston in combustible materials. Today, we know that combustion is the result of a violent exothermic reaction that arises as a result of binding with oxygen in the air. However, according to the phlogiston theory from that time, combustion was thought to be a phenomenon that occurred when phlogiston rose violently from the combusting substance along with the flames, and was released into the air. The phlogiston theory was the mainstream theory propounded in academic circles then, and, for a relatively long time, was believed to be accurate. The inaccuracy of this theory was only acknowledged when Lavoisier formulated the law of conservation of mass in 1774, and it became known that the ashes that remain after the combustion of a substance are not lighter, but rather heavier, than the original substance. In this way, "air chemistry" is also related to the phlogiston theory, and formed a part of the prologue to modern chemistry. At the same time, this period also became the prologue to chemistry of the atmosphere, although no one had been conscious at that time.

The relationship between the ozone layer and the atmosphere triggered a growing awareness for chemists about chemistry of the atmosphere. Ozone molecules, which would later be recognized as a key species in the chemistry of the Earth's atmosphere, were discovered by the German/Swiss chemist Schönbein in his laboratory in the nineteenth century, in 1839. Ozone was discovered in the atmosphere near the surface of the Earth in 1860, about 20 years after the discovery of ozone in the form of molecules, and the relevance of ozone with the atmosphere was uncovered. After that, it became a trend to measure the amount of ozone present in the atmosphere, and activities to measure the amount of ozone near the surface of the Earth began to take place at numerous locations in Europe. Thanks to this trend, records of ozone density at the end of the nineteenth century were preserved as numerical data. These data have taken on great importance, serving as reference values in the study of the increase in tropospheric ozone in the twentieth century. Incidentally, the ozone measurement method at the time involved exposing filter paper soaked in potassium iodide (KI) solution to the air, and measuring the density of the purplish-brown color that is produced when the solution undergoes oxidation. This was known as the Schönbein method. The wet chemical automatized oxidant analyzer, which had also been used in Japan until recently, operates based on this same principle.

On the other hand, ozone was synthesized in laboratories during this time, and an Irish chemist, Hartley, came up with a method for measuring absorption spectrum. As a result, it was found that ozone absorbs ultraviolet light most strongly at 200–320 nm, and this absorption band was therefore named the "Hartley band" after the chemist. Today, a standard method to measure ambient ozone concentration is to use an UV absorption instrument, which is based on the Hartley band. Data on the absorption spectrum of ozone took on even greater significance for chemistry of the atmosphere when it led to the discovery of the ozone layer in the stratosphere. Prior to this, scientists had measured the solar spectrum and learned that ultraviolet rays with wavelengths shorter than 300 nm do not reach the surface of the Earth at all. As the limits of the wavelengths of these ultraviolet rays were consistent with the rising wavelengths of the ozone's absorption spectrum measured in the laboratory, in around 1880, Hartley concluded that this was definitely the result of the presence of a high density of ozone in the atmosphere. Even further on after that, it was inferred that an even larger amount of ozone was present in the upper air since the measurement data for ozone density near the surface of the Earth at the time could not provide an explanation for this degree of attenuation of solar ultraviolet rays at that density. The existence of the ozone layer in the stratosphere was validated 20–30 years after that in 1913, by two French physicists, Fabry and Buisson.

However, prior to that, the presence of the stratosphere, where temperature increases with altitude, was discovered in 1902 by the French meteorologist Teisserenc de Bort. Partly due to fact that the stratosphere was discovered before the ozone layer, it is common to say that the ozone layer exists in the stratosphere. In actual fact, the ozone layer had formed first in the Earth's atmosphere, and the stratosphere was formed as a consequence of the ozone layer. As ozone absorbs sunlight efficiently, the temperature of the atmosphere rises where there is a high density of ozone, resulting in the temperature reversal phenomenon where temperature rises with higher altitudes, and bringing about the formation of the stratosphere. Against the current concentration of oxygen in the Earth's atmosphere, ozone density is the highest at places between 20 and 25 km above the surface of the Earth. However, as air density falls and heat capacity becomes smaller as the stratum becomes higher, temperature in the stratosphere reaches its peak at an altitude of about 50 km, and not at the center of the ozone layer.

The minds researching chemistry of the atmosphere later turned their attention to chemical reactions in the atmosphere. Research on chemical reactions in the atmosphere stems from two broad streams—one from the perspective of geophysics, and another from the perspective of environmental science and geochemistry. Research from the perspective of geophysics has its roots in research on the chemistry of atmospheric ozone, which seeks to elucidate why the ozone layer was formed in the stratosphere. Chapman, a British geophysicist, was the first to elucidate that the ozone layer is formed through a photochemical reaction when oxygen in the Earth's atmosphere is irradiated by sunlight. This was known as the Chapman mechanism (theory), and was published in 1930. Chapman's theory was also known as "the pure oxygen theory", as it took only oxygen into consideration as the component in the Earth's atmosphere that absorbs sunlight. Oxygen is photolyzed into two oxygen atoms, and these atoms combine with other oxygen molecules to create ozone. If only this simple process operates, all oxygen will eventually be converted to ozone over time. However, this does not happen because the ozone that is produced undergoes photolysis in sunlight and also reacts with oxygen atom, and return to its original form of oxygen molecules. Through this process, oxygen and ozone are repeatedly generated and destroyed in sunlight, resulting in a state of photochemical equilibrium. Based on the absorption spectrum of oxygen and ozone, radiation intensity of sunlight, rate constant of the reaction of oxygen atom with ozone and oxygen, and the altitude distribution of oxygen molecular density in the Earth's atmosphere, Chapman calculated the amount of ozone that was produced in different altitude, and succeeded in explaining the formation of the ozone layer at an altitude of 20–25 km. As for the density of the ozone, that as calculated by Chapman ignored the chemical reactions of trace gases other than oxygen, the values that he obtained were overestimated by about a factor of two as compared to the observed values that we have obtained today. However, the altitude distribution of ozone that he obtained was successfully reproduced in the observed profile.

Thereafter, research on chemical reactions in the stratosphere from the 1960s to the 1970s took into consideration chemical reactions that included trace components in the atmosphere, such as H_2O/CH_4, N_2O, and CH_3Cl. This brought about significant academic advancements, contributing to the schematization of an important chain reaction theory for the atmosphere known as the "HO_x, NO_x, and ClO_x cycle," and at the same time, eliminating the aforementioned inconsistency of results with observations of ozone density in the stratosphere. This developed further in 1974 when Molina and Rowland propounded the theory that predicted the depletion of the ozone layer through chlorofluorocarbon (CFC), which contributed to solving an important environmental problem.

Meanwhile, research on the chemistry of the troposphere from the perspective of environmental science dates back to air pollution arising from coal combustion during the industrial revolution, and has its beginnings in the measurement of the chemical composition of rainfall and snowfall in England in the second half of the 18th century. The term "acid rain" was coined by the 19th century Scottish chemist, Angus Smith, and it was found at the time that rainfall in urban, suburban and unpolluted remote areas typically contained sulfuric acid/ammonium sulfate, ammonium sulfate, and ammonium carbonate, respectively. Such examples of early air pollution research have continued in the academic field of geochemistry in the form of research on precipitation chemistry and gas-aerosol chemistry. However, air chemistry in the field of geochemistry has remained low-key in comparison with other areas of study, such as marine chemistry and mineral chemistry. Furthermore, environmental science and geochemistry research on trace components in the atmosphere focus on research from the perspective of analytical chemistry, and did not contribute significantly to the development of research on chemical reactions in the atmosphere. One of the reasons was that compared to the stratosphere, the troposphere was in general considered to be a chemically static field.

Air pollution research only began to play a significant role in developing research on chemical reactions in the atmosphere during the late of 1940s, when photochemical smog emerged in South California, United States. The Los Angeles basin was shrouded in white smog in summer, and cases of health problems such as eye irritation and respiratory disorders were reported, in addition to reports of crop damage as leaves turned brown and the crops died (necrosis). These problems continued to appear successively into the 1950s, but the cause of the problems remained a mystery for a long period of time. This is because the main air pollutants in this area, which include hydrocarbons and oxides of nitrogen in automobile exhaust gas, do not directly cause health problems and crop damage such as those observed in Southern California. This mystery was solved by Haagen-Smit, then professor at the California Institute of Technology. Haagen-Smit conducted experiments such as exposing automobile exhaust gas to ultraviolet rays, and proved the formation of oxidizing substances (oxidants) including ozone through photochemical reactions in polluted air, which caused damage to human health and plants. Ozone and most of the other oxidants are colorless gases and do not reduce visibility. However, particulate matter (aerosol) is generated as secondary products simultaneously when ozone is generated, and this particulate matter causes the "white smog" that reduces visibility.

Research on chemical reactions in the troposphere devolved by leaps and bounds after that, founded upon the research on the elucidation of mechanism of photochemical smog that was conducted from the 1960s through the 1970s. It began to be addressed in the academic circles as a part of the chemistry of the troposphere, including the free troposphere. The most significant example of that was the OH radical chain reaction theory in the troposphere, which was formulated in the early part of the 1970s. This established the reaction mechanism of photochemical air pollution as an unshakable fact, and at the same time, provided a fundamental theory of troposphere chemistry for the next generation that had greater universality due to its direct relationship with global environmental issues.

1.2 Development into Atmospheric Chemistry

In this way, chemistry of the atmosphere, which has been regarded mainly as chemistry of the ozone layer and chemistry of photochemical smog, entered a significant period of transition in the 1980s. This was its evolution from "chemistry of the atmosphere" to "atmospheric chemistry." Until the 1980s, the chemistry of the stratosphere has been treated as fundamental knowledge constituting a part of geophysics. Conversely, chemistry of the troposphere was regarded as an applied field of chemistry related to air pollution, and was not treated as fundamental science. However, in the mid-1980s, chemistry of the troposphere, which had never historically been treated as a form of science, experienced a dramatic reversal and became widely regarded as a academic research field for the first time. At the same time, atmospheric chemistry, which encompasses chemistry of the troposphere and the stratosphere, came to be established as a new fundamental discipline that constitutes a part of Earth science. While the previous "chemistry of the atmosphere" focused on the analytical and reaction chemistry of atmospheric components, "atmospheric chemistry" sought to elucidate the series of processes related to global material cycle and balance—the process of emissions of atmospheric trace components from anthropogenic and natural sources including the biosphere, the transport processes in atmosphere and the accompanying chemical transformation processes, and the process of the removal of atmospheric components from the atmosphere through wet and dry deposition. By doing so, atmospheric chemistry became a new academic field that had a system science aspect, which sought to elucidate the spatial distribution and temporal variation of atmospheric trace species on a global, regional and urban scale, as well as the biogeochemical cycling of these components. The second half of the 1980s coincided exactly with the dawn of the global environment era, and atmospheric chemistry, together with atmospheric physics, physical and chemical oceanography, as well as terrestrial and marine ecology, gained recognition as fundamental disciplines that were useful for understanding global changes that were brought about by human activities. These global changes included global warming, depletion of the ozone layer, and acid rain.

Atmospheric chemistry is an interdisciplinary academic field that brings together the conventional academic subjects of physical chemistry, analytical chemistry, geophysics, meteorology, and ecosystem science. In order to establish such a holistic academic field in a systematic manner, it is necessary to ensure that the respective elements of science that constitute the field are well organized.

In this context, this textbook focuses on "atmospheric reaction chemistry" as one of key elements of atmospheric chemistry, which is based on gas-phase chemical kinetics and photochemistry that is an aspect of the conventional research field of physical chemistry. A number of excellent textbooks on the subject of atmospheric chemistry as a systems science are listed after this, so please refer to these books for more information on the total subject.

1.3 Textbooks of Atmospheric Chemistry

Two of the textbooks on the chemistry of the atmosphere written in the 1960s are;

- Leighton, P. A., Photochemistry of Air Pollution, Academic Press, 300 pp, 1961, which was born out of photochemical air pollution research, and another,
- Junge, C. E., Air Chemistry and Radioactivity, Academic Press, 382 pp, 1963, which was born from the stream of geochemistry. These books may be called the atmospheric chemist's bibles.

Textbooks written in the 1970s–1980s are:

- Phillips, L. F. and M. J. McEwan, Chemistry of the Atmosphere, 301 pp, Wiley & Sons, 1975.
- Heicklen, J., Atmospheric Chemistry, 406 pp, Academic Press, 1976.
- Shimazaki, T., Minor Constituents in the Middle Atmosphere, 444 pp. D. Reidel, 1985.
- Wayne, R. P., Chemistry of Atmospheres, 355 pp, Clarendon Press, 1985.
- Finlayson-Pitts, B. J., J. N. Pitts, Jr., Atmospheric Chemistry, 1098 pp, John Wiley & Sons, 1986.

- Seinfeld, J. H., Atmospheric Chemistry and Physics of Air Pollution, 738 pp, John Wiley & Sons, 1986.
- Warneck, P., Chemistry of the Natural Atmosphere, 753 pp, Academic Press, 1988.
- Ogawa, T., Physics and Chemistry of the Atmosphere, 224 pp., Tokyodo Shuppan (in Japanese).

These publications served as pioneering textbooks in an era when the study of the chemistry of the atmosphere was not yet so common. Although it is not a textbook on the chemistry of the atmosphere, a textbook on photochemistry that was also useful for a long time for atmospheric chemists, and the style of chapters was referred in writing the present textbook is;

- Okabe, H., Photochemistry of Small Molecules, 431 pp, John Wiley & Sons, 1978.

Textbooks in the era of "atmospheric chemistry" starting in 1990s when a strong consciousness of global environmental issues emerged and a viewpoint of system science was strengthened are as follows;

- Graedel, T. E., P. J. Crutzen, Atmospheric Change, 446 pp, W. H. Freeman and Company, 1993.

A textbook mainly for undergraduate students to understand various phenomena in the atmosphere, geosphere, hydrosphere, and biosphere in the perspective of one earth system.

- Brasseur, G. P., J. J. Orlando, G. S. Tyndall, Eds., Atmospheric Chemistry and Global Change, 654 pp, Oxford University Press, 1999.

From the viewpoint of the relevance of atmospheric chemistry with global environmental issues, processes, chemical compounds, ozone variability, climate change, and so on are discussed in each chapter by experts in each field. It is unique as it includes essays by senior scientists in atmospheric research at the end of each chapter.

- Warneck, P., Chemistry of the Natural Atmosphere, 2nd Edition, 927 pp, Academic Press, 1999.

It is the 2nd edition of the original book published in 1988. It covers chemical reactions, natural and anthropogenic emission sources, removal processes in the gas phase, and in aerosol, clouds and precipitation as well as global distribution and biogeochemical cycles of atmospheric constituents.

- Jacob, D., Introduction to Atmospheric Chemistry, 264 pp, Princeton University Press, 1999.

It is based on a lecture at Harvard University by the author. It is a compact textbook including fundamentals of atmospheric chemistry that can be learned in one semester. The theme of environmental problems which are thought to be important for undergraduate students are selected.

- Finlayson-Pitts, B. J. and J. N. Pitts, Jr., Chemistry of the Upper and Lower Atmosphere, Academic Press, 969 pp, 2000.

Detailed description on the fundamentals of physical chemistry, such as descriptions of spectroscopy, photochemistry, reaction kinetics, homogeneous and heterogeneous reactions are given. The destruction of the ozone layer, photochemical oxidants, acid deposition, hazardous air pollutants, indoor pollution, and so on are widely covered, and their countermeasures are explained based on atmospheric chemistry.

- Wayne, R., Chemistry of Atmospheres, 3rd ed., 775 pp, Oxford University Press, 2000.

In addition to stratospheric and tropospheric chemistry, ions and atmospheric glow in the mesosphere and the chemistry of planetary atmosphere are included. The descriptions on chemical processes in cloud water and heterogeneous reactions as their basis are added in the chapter on photochemistry and reaction kinetics in the third edition.

- Hobbs, P. V., Basic Physical Chemistry for the Atmospheric Sciences, 2nd Edition, 208 pp, Cambridge University Press, 2000.

Principles of chemistry necessary for students learning atmospheric science and planetary science such as chemical equilibrium, thermochemistry, chemical kinetics, photochemistry, and others are concisely described, and it is paired with the following textbook on atmospheric chemistry for students.

- Hobbs, P. V., Introduction to Atmospheric Chemistry, 276 pp, Cambridge University Press, 2000.

A textbook briefly summed up for students on fundamental aspects of atmospheric chemistry including air pollution, ozone hole, global warming. Many problems and answers are provided.

- Akimoto, H, K. Kawamura, K. Nakazawa, and N. Washida eds., Chemistry of the troposphere and Global Change, 223 pp, Gakkai Shuppan Center, 2002. (in Japanese)

A book based on a research project on "tropospheric chemistry global dynamics," which is the first big project in this field funded by Grant-in-Aid in Japan. Greenhouse gases, tropospheric photochemistry and reactive trace constituents, atmospheric homogeneous and heterogeneous reactions, and aerosols and their precursors are covered.

- McElroy, M. B., The Atmospheric Environment: Effects of Human Activities, 326 pp, Princeton University Press, 2002.

A textbook used at Harvard University for a fundamental course for global environmental issues for students of environmental science and social policy science major. Fundamental physics and chemistry of the atmosphere, carbon, nitrogen and sulfur cycles, tropospheric chemistry and rain chemistry, climate change, and other topics are explained in an accessible manner.

- Brasseur, G. P and S. Solomon, Aeronomy of the Middle Atmosphere: Chemistry and Physics of the Stratosphere and Mesosphere, 3rd ed., 644 pp, Springer, 2005.

A unique textbook focused on physics and chemistry in the stratosphere and mesosphere. Dynamics, transport, radiation, and chemical composition are explained in each chapter. Particularly regarding the ozone layer, data from models and observations are explained in detail including gas phase and heterogeneous reactions on PSCs in the chapters of chemical composition and ozone perturbation.

- Seinfeld, J. H. and S. N. Pandis, Atmospheric Chemistry and Physics: From Air Pollution to Climate Change, 2nd ed., 1,203 pp, John Wiley and Sons, 2006.

In addition to homogeneous reactions in the stratosphere and troposphere, characteristics, fluid dynamics, thermodynamics, nucleation, deposition processes, organic aerosols, climate effect regarding aerosols are expounded. In particular, efforts are made to explain each process starting from basic principles.

- Holloway, A. M. and R. P. Wayne, Atmospheric Chemistry, 271 pp, The Royal Society of Chemistry Publishing, Cambridge, 2010.

A textbook for students majoring in chemistry and learning atmospheric chemistry. It is more compact than the above cited "Chemistry of Atmosphere (2000)" although it is written by one of the authors (Wayne). The special feature of this textbook is that the concept of sources and sinks, and atmospheric lifetimes are explained in separate chapters.

1.4 Literatures

General literatures consulted for Chapter 1 are given below:

- Brock, W. H., The Fontana History of Chemistry, 780 pp, Harper Collins Publishers, New York, 1992.
- Cowling, E. B., Acid precipitation in historical perspective, Environ. Sci. Technol., 16, 110A–123A, 1982.
- Ihde, J., A., The Development of Modern Chemistry, 851 pp, Harper & Row Publishers, New York, 1964.
- Warneck, P., Chemistry of the Natural Atmosphere, 2nd Edition, 927 pp, Academic Press, San Diego, 1999.

Chapter 2
Fundamentals of Chemical Reactions

The chemical reaction system in the atmosphere is composed of photolysis and homogeneous reactions in the gas phase, and multiphase processes including heterogeneous reactions on particle surfaces. In this chapter, photochemistry and the chemical kinetics of homogeneous and multiphase reactions are described based on the principles of physical chemistry.

The fundamentals of most chemical reactions are already well established and several textbooks are cited at the end of this chapter. However, the multiphase heterogeneous reactions given in Sect. 2.4 have attracted more interest in recent years, and as fields of research, they are still developing, so it should be noted that their treatment is not necessarily standardized yet.

2.1 Photochemistry and Photolytic Reactions

2.1.1 The First and Second Principle of Photochemistry

The first principle of photochemistry is the Grotthuss-Draper law which states "only the light absorbed by chemical species can cause photochemical reactions," or in other words "without light absorption there is no photochemical reaction." The first principle of photochemistry states that if a molecule does not have absorption spectrum in the wavelength region of light irradiation, photochemical reaction cannot occur even if the strong light is irradiated or the quantum energy of the irradiated light is larger than the bond dissociation energy of the irradiated molecule. This means that the first step to understand photochemical reactions is to grasp the absorption spectrum of molecules, and the starting point of the discussion in atmospheric chemistry should be to understand the overlapping of wavelength range of the absorption spectrum of atmospheric species and solar radiation.

H. Akimoto, *Atmospheric Reaction Chemistry*, Springer Atmospheric Sciences,
DOI 10.1007/978-4-431-55870-5_2

The second principle of photochemistry is called the photochemical equivalence law, or the Stark-Einstein law, which states "the absorption of light occurs in the quantum unit of photon" or "one molecule absorbs one photon, and one or less molecule can be photolyzed accordingly."

Quantum energy of light E is expressed as

$$E = h\nu = \frac{hc}{\lambda} \tag{2.1}$$

where h is the Plank constant, 6.6262×10^{-34} J s, c is the velocity of light in vacuum, 2.9979×10^{8} m s^{-1}, ν is the frequency of light (s^{-1}), and λ is the wavelength of light. The wavelength of visible and ultraviolet light that appears in atmospheric chemistry is usually expressed in nanometers (1 nm = 10^{-9} m), but historically angstroms or ångstroms (1 Å = 10^{-10} m) were also frequently used. Also, in the infrared region, wavenumber ω (cm^{-1}) is in common use, which is the reciprocal of wavelength λ expressed in cm. The meaning of the second principle of photochemistry is that a molecule absorbs the light by the quantized energy corresponding to each wavelength as given in Eq. (2.1).

Instead of molecule, moles (1 mol = 6.022×10^{23} molecules; Avogadro number) are often used in experimental chemistry. The unit of photon energy referenced to one mole is called an einstein. Photon energy per 1 einstein of light with wavelength λ (nm) is expressed as

$$\begin{aligned} E &= \left(6.022 \times 10^{23}\right) \times \frac{hc}{\lambda} \\ &= \frac{1.196 \times 10^{5}}{\lambda} \quad kJ\ einstein^{-1} \\ &= \frac{2.859 \times 10^{4}}{\lambda} \quad kcal\ einstein^{-1} \\ &= \frac{1.240 \times 10^{3}}{\lambda} \quad eV \end{aligned} \tag{2.2}$$

The values of physical constants which appear in this book are given in Table 2.1. Table 2.2 gives energy conversion table for energy units often used in chemistry, kJ (kilojoule), kcal (kilocalorie), and eV (electron volt). Also, conversion table of photon energy of 1 einstein with corresponding wavelengths of visible and ultraviolet light is given in Table 2.3. The energy threshold values for which different photolytic processes are energetically possible are given in the case studies described in Chap. 4.

Table 2.1 Physical constants

Constants	Values
Boltzmann constant (k)	1.3807×10^{-23} J K^{-1}
Plank constants (h)	6.6261×10^{-34} J s^{-1}
Light velocity (in vacuum) (c)	2.9979×10^{8} m s^{-1}
Avogadro number (N)	6.0221×10^{23} mol^{-1}
Gas constant (R)	8.3145 J K^{-1} mol^{-1}
	=0.082058 L atm K^{-1} mol^{-1}

Table 2.2 Energy conversion table

	kJ mol^{-1}	kcal mol^{-1}	eV
kJ mol^{-1} =	1	× 0.2390	× 0.01036
kcal mol^{-1} =	× 4.184	1	× 0.04337
eV =	× 96.49	× 23.06	1

Table 2.3 Photon energy per mole for UV-visible light of typical wavelengths

	Photon energy per mole (E)		
Wavelength (λ) nm	kJ $einstein^{-1}$	kcal $einstein^{-1}$	eV
600	199	47.7	2.07
500	239	57.2	2.48
400	299	71.5	3.10
300	399	95.3	4.13
200	598	143.0	6.20
100	1196	285.9	12.40

λ (nm) = 119600/E (kJ $einstein^{-1}$) = 28590/E (kcal $einstein^{-1}$) = 1240/E(eV)

2.1.2 Photolytic Quantum Yields

Even if light with enough energy for photodissociation is irradiated to a molecule fulfilling the first principle of photochemistry, the molecule is not necessarily photolyzed. That is, fulfilling the first principle is a necessary condition for photolysis, not necessarily a sufficient condition. A molecule that absorbs the light in the visible or ultraviolet region reaches to the electronically excited state. Figures 2.1 and 2.2 illustrate the potential energy curve of the simplest case of a diatomic molecule. The abscissa of each of these figures is interatomic distance in a molecule, and the ordinate is the potential energy. The most stable, lowest energy state of an atom or molecule is called the ground state, and the higher energy state is called the exited state. The case of Fig. 2.1 is an example where the excited states are in a repulsive state that does not have a potential minimum. In this case, an excited molecule AB* (an atom and molecule in their electronically excited state are expressed with an asterisk on the right shoulder) reached by the photo-absorption dissociates into A + B or A* + B immediately. Figure 2.1 illustrates that the dissociation to the channel forming an excited atom, A* + B, generally happens when the

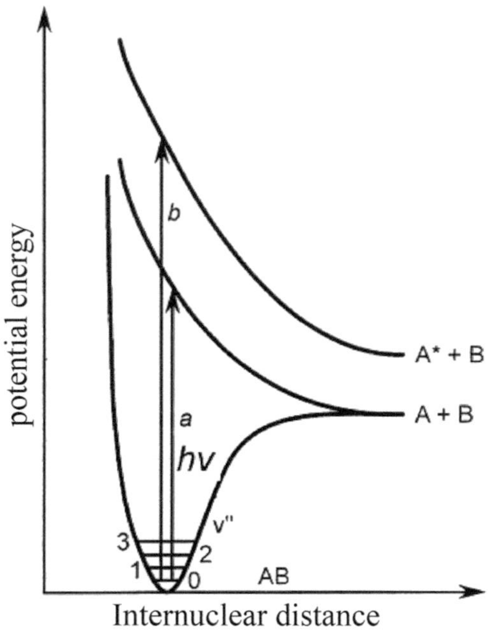

Fig. 2.1 Potential energy curves for a diatomic molecule showing excitations to repulsive electronic states

molecules absorb shorter wavelength light with higher photon energy. The horizontal lines drawn in the energy curve of the ground state designate vibrational energy levels (v″).

Figure 2.2 is an example of a molecule with a bound excited state having a minimum on the potential energy curve. The horizontal lines in the bound excited state designate vibrational levels in the excited state (v′). In this case, an excited molecule cannot dissociate immediately even though the potential energy in the excited state is higher than the dissociation energy, and it stays in the excited state for a certain period of time. During this period, the molecule may emit light and return to the ground state with a natural radiative lifetime. In such cases, photodissociation does not occur. However, an energy curve of the bound excited state is often crossed with a repulsive potential curve that leads to dissociation into A + B as illustrated in Fig. 2.2 by a broken line. When the absorbed photon energy is higher than the crossing point, traversing from the bound excited state to the repulsive state can occur, leading to dissociation into A + B. Dissociation in this form is called "predissociation." Further, when a bound state is reached by an excitation with higher energy than the dissociation energy of A* + B, a molecule dissociates into A* + B immediately.

To which case of Figs. 2.1 or 2.2, the photo-excitation is applicable, can be judged by the appearance of an absorption spectrum whether it is a continuum or has a band structure. In the case of Fig. 2.2, since the absorption occurs to the vibrational energy levels in the bound excited state, absorption spectrum reveals band structure. As an example, absorption spectra of chlorine (Cl_2) (Maric et al. 1993)

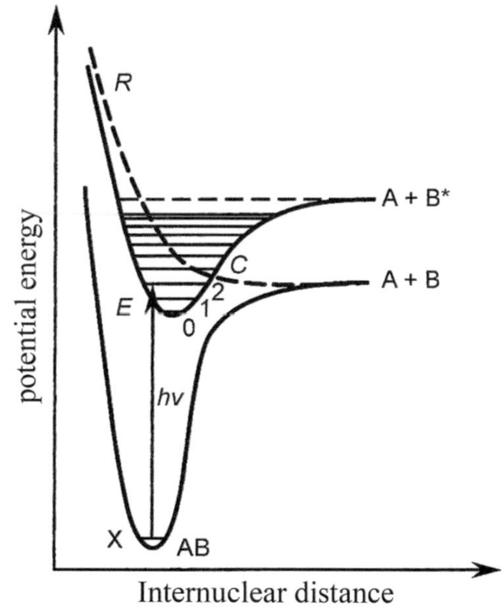

Fig. 2.2 Potential energy curves for a diatomic molecule showing excitation to a bound electronic state that crosses with a repulsive state

and Iodine (I_2) (Saiz-Lopez et al. 2004) are shown in Figs. 2.3 and 2.4, respectively. Absorption spectrum of Cl_2 in the wavelength region of 250–450 nm is a real continuum, which implies that the potential energy curve in the excited state corresponding to the photo-absorption in this region is repulsive. It is known that a Cl_2 molecule photo-excited in this wavelength region dissociates into two ground state chlorine atoms Cl ($^2P_{3/2}$) immediately. On the other hand, the absorption spectrum of I_2 shows a band structure in the 500–650 nm domains overlapping with a continuum spectrum extending from 400 to 500 nm. The band spectrum corresponds to the transition to a quasi-stable state of the potential curve. In general, the probability of photodissociation when a molecule is excited into such a state has to be verified by experiments. The case of Fig. 2.4, showing the band structure overlapping with a continuum, in general corresponds to the predissociation mentioned above. Another example of absorption spectrum corresponding to an excitation to a pure bound potential curve without predissociation is given in Fig. 2.5 for the case of carbon monoxide (CO) (Myer and Samson 1970). In such a case, the spectrum shows perfectly discrete bands, and the excited molecule does not photodissociate, but rather returns to the ground state emitting fluorescence or being deactivated by a collision with another molecules, that is also called quenching.

So far, the explanation was for an example of diatomic molecules. For the polyatomic molecules composed of n-pieces of atoms, the idea of the correspondence of the appearance of the absorption spectrum and the photo-excitation is the same, but the potential surfaces are in $n-1$ dimensions. Since photodissociation can occur with multiple processes such as $ABC \rightarrow AB + C$, $A + BC$, multi-dimensional potential surfaces for each dissociating inter-atomic distance must be considered.

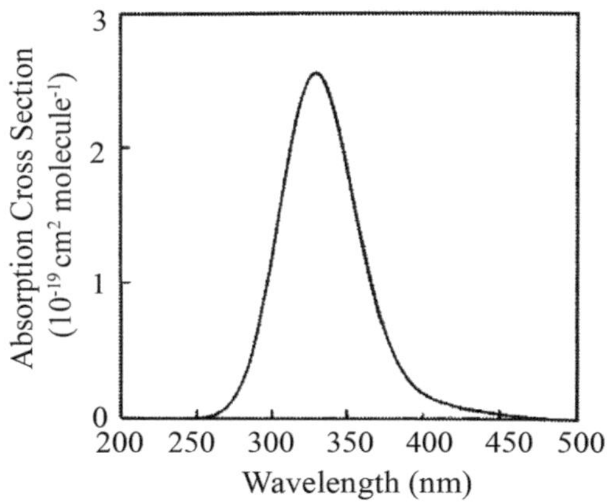

Fig. 2.3 Absorption spectrum of Cl_2 (Adapted from Maric et al. 1993)

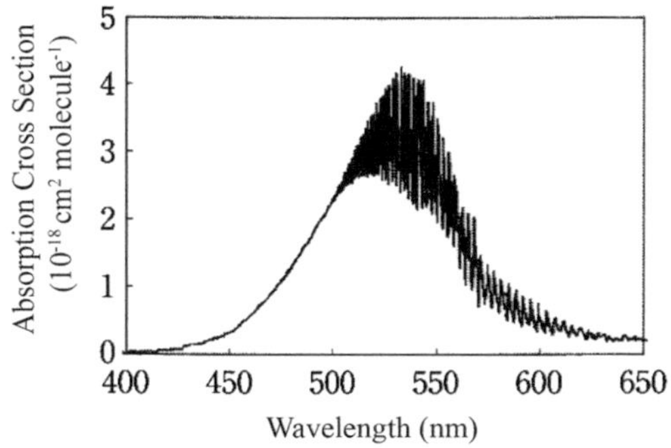

Fig. 2.4 Absorption spectrum of I_2. Resolution: 0.1 nm for 500–630 nm, 1 nm for other region (Adapted from Saiz-Lopez et al. 2004)

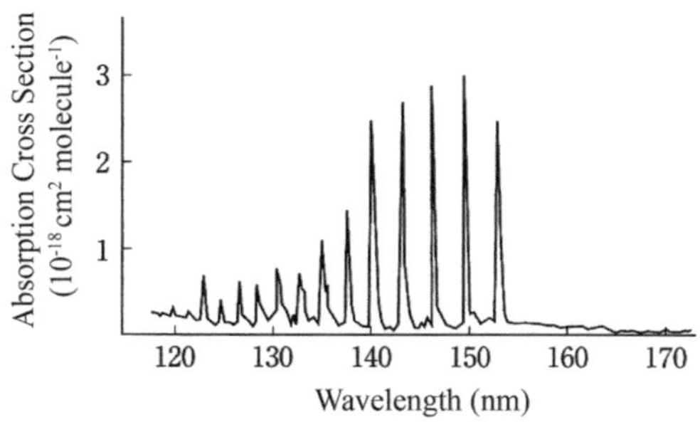

Fig. 2.5 Absorption spectrum of CO. Resolution: 0.025 nm (Adapted from Myer and Samson 1970)

A probability of dissociation of a molecular per absorbed photon is called the photolytic quantum yield. Thus, the photolytic quantum yield Φ is defined by

$$\Phi = \frac{\text{Number of dissociated molecules}}{\text{Number of absorbed photons}}. \tag{2.3}$$

By this definition, the maximum and minimum photolysis quantum yield is 1 and 0, respectively.

From the relationship between the photo-excitation and absorption spectrum, photodissociation quantum yields are in general unity when a molecule is excited to the repulsive potential curve as shown in Fig. 2.1. On the other hand, when an absorption spectrum has a band structure and a molecule is excited to a bound state as shown in Fig. 2.2, photodissociation quantum yields are in general $0 \leq \Phi \leq 1$ and the value has to be determined experimentally.

When a certain wavelength light excites a molecule, photodissociation can occur in energetically permissible multi-pathways, and the photolytic quantum yield for each process has to be determined experimentally. For example, the photodissociation pathways of ozone (O_3) by solar ultraviolet radiation,

$$O_3 + h\nu \rightarrow O_2 + O(^1D) \tag{2.4}$$

$$\rightarrow O_2 + O(^3P) \tag{2.5}$$

are very important in tropospheric chemistry, and many experiments have been done to determine the wavelength dependence of photolytic quantum yields of each process (see Sect. 4.2.1).

2.1.3 *The Beer-Lambert Law*

As shown in Fig. 2.6, when a parallel monochromatic light of wavelength λ with an intensity I_0 (energy or number of photons) irradiates and passes through media with concentration C (mol L^{-1}) and length l (cm), the intensity of the light after the passing can be expressed as

$$\ln\frac{I}{I_0} = -kCl \tag{2.6}$$

$$\frac{I}{I_0} = \exp(-kCl) \tag{2.7}$$

This relationship is called the Beer-Lambert law. The proportionality coefficient k is generally referred to absorption coefficient (L mol^{-1} cm^{-1}).

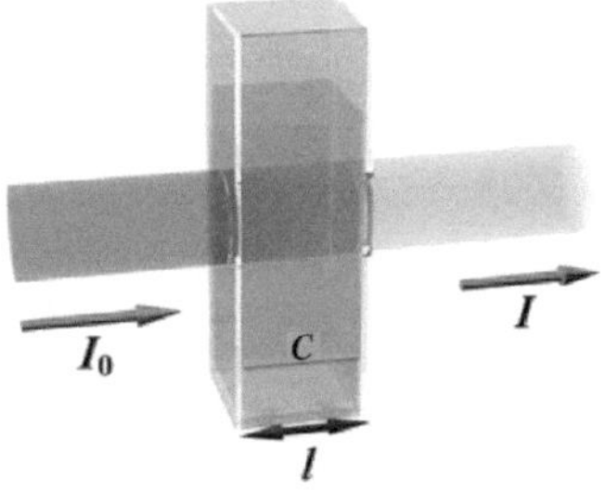

Fig. 2.6 Schematic diagram of Beer-Lambert Law

The Beer-Lambert law is also often expressed taking the base 10 of the logarithm as

$$\log\frac{I}{I_0} = -\varepsilon Cl \tag{2.8}$$

$$\frac{I}{I_0} = 10^{-\varepsilon Cl}, \tag{2.9}$$

where the proportional coefficient ε (L mol^{-1} cm^{-1}) is called the molar extinction coefficient. The log (I/I_0) in Eq. 2.8 is called absorbance A and related to transmittance T with the following relationship,

$$A = -\log\left(\frac{I}{I_0}\right) = -\log T \tag{2.10}$$

The concentration unit of the media is usually taken as molecular number density n (molecules cm^{-3}) in atmospheric chemistry, and in this case the Beer-Lambert law is customarily expressed with the natural logarithm as,

$$\ln\frac{I}{I_0} = -\sigma nl \tag{2.11}$$

$$\frac{I}{I_0} = \exp\ (-\sigma nl) \tag{2.12}$$

Here, the proportionality coefficient, σ (cm^2 molecule^{-1}) has the dimension of area and is called the absorption cross section. Also, the dimensionless number in the above equation σnl is denoted by τ and is called optical depth,

$$\tau = \ln(I_0/I) = \sigma nl. \tag{2.13}$$

Absorption cross sections of many molecules in the atmosphere are given in the tables in Chap. 4. For example, the absorption cross section of the strong Hartley band of O_3 at around 254 nm is $\approx 1 \times 10^{-17}$ cm^2 (c.f. Fig. 3.6, Table 4.1). Cross sections of important absorption of O_3 at 308 nm and NO_2 at 360 nm are $\approx 1 \times 10^{-19}$, $\approx 5 \times 10^{-19}$ cm^2 (c.f. Fig. 3.6, Table 4.1; Fig. 4.9, Table 4.5), respectively, and that of CFC-11 ($CFCl_3$) at around 200 nm which is important for ozone layer depletion, is $\approx 6 \times 10^{-19}$ cm^2 (c.f. Fig. 4.33, Table 4.26). Conceptually, if the molecular diameter is the order of 10^{-8} cm and the light is absorbed 100 % by the geometric surface area, the absorption cross section is in the order of $\approx 10^{-16}$ cm^2. Generally speaking, when absorption cross section σ is larger than 10^{-20} cm^2, it is thought to be strong enough, which can be important in the atmosphere.

When calculating the photolysis rate in atmospheric chemistry, it is desirable to use an absorption cross section as the concentration unit. However, it is sometimes necessary to convert an absorption coefficient in different unit given in literatures to

Table 2.4 Conversion table of absorption coefficients in different units

Absorption cross section / Absorption coefficient	cm^2 molecule^{-1}	L mol^{-1} cm^{-1}	atm^{-1} cm^{-1}, 273 K
cm^2 molecule^{-1} (base e) =	1	$\times 1.66 \times 10^{-21}$	$\times 3.72 \times 10^{-20}$
L mol^{-1} cm^{-1} (base 10) =	$\times 2.62 \times 10^{20}$	1	$\times 2.24 \times 10$
atm^{-1} cm^{-1} (base e, 273 K) =	$\times 2.69 \times 10^{19}$	$\times 4.46 \times 10^{-2}$	1

cross sections. Particularly, absorption spectra of gases in old literatures are often given in the natural logarithm base extinction coefficients in atm unit k (atm^{-1} cm^{-1}). In this case molecular density in the atm unit depends on temperature, and the temperature is usually defined at 273 K or 298 K so that the proper temperature should be used to convert from the extinction coefficient to the cross section. Table 2.4 gives conversion factors between absorption coefficients. In addition, k (atm^{-1} cm^{-1}, 273 K) = 1.09 k (atm^{-1} cm^{-1}, 298 K), and k (base e) = 2.303 × k (base 10).

2.1.4 Photolysis Rate Constants

When a molecule A absorbs light and is photolyzed as

$$A + h\nu \rightarrow B + C, \tag{2.14}$$

the photolysis rate is expressed by the first order reaction rate,

$$d[A]/dt = -k_p[A]. \tag{2.15}$$

Here, h and v in the left-hand side of Eq. (2.14) are the Plank constant and frequency of light appeared in Eq. (2.1) in origin, the [hv] here is customarily used as a symbol designating photon in a photochemical reaction formula. Photolysis rate constant k_p (s^{-1}) is calculated practically by using the irradiating light intensity I (photons cm^{-2} s^{-1}), absorption cross section σ (cm^2 molecule^{-1}, base e), and photolytic quantum yield ϕ. When the irradiation is monochromatic light with wavelength λ, k_p is given as,

$$k_p(\lambda) = \sigma(\lambda)\Phi(\lambda)I(\lambda). \tag{2.16}$$

When the photolysis occurs by the wide range wavelength light as in the atmospheric photochemical reactions, the photolysis rate is obtained by integrating the value at each wavelength,

$$k_p = \int_\lambda \sigma(\lambda)\Phi(\lambda)I(\lambda)d\lambda. \qquad (2.17)$$

In the calculation for atmospheric photodissociation reactions, how to calculate the effective solar intensity is a major issue, because not only direct irradiation from the sun, but light from all directions reflected and scattered by the ground surface, clouds, atmospheric molecules, and aerosols can contribute to photolysis. Furthermore, in the troposphere for example, only solar radiation that has not been absorbed by atmospheric molecules in the higher atmosphere, the stratosphere and above, can cause photolytic reactions. The spherically integrated solar intensity after considering those many atmospheric processes is called the actinic flux $F(\lambda)$ (photons cm^{-2} s^{-1}), which means solar irradiation valid for photochemical effect. In atmospheric chemistry, j_p is often used instead of k_p for representing photolysis rate constant. Photodissociation rate constant in the atmosphere can be expressed using these parameters as

$$j_p = k_p = \int_\lambda \sigma(\lambda)\Phi(\lambda)F(\lambda)d\lambda. \qquad (2.18)$$

Detailed method of calculation of $F(\lambda)$ will be treated later in Chap. 3.

2.1.5 *Spectroscopic Terms and Selection Rules*

In chemical equations for the reactions in the atmosphere, atoms and molecules are sometimes denoted with symbols in a parenthesis, e.g. $O(^3P)$, $O(^1D)$, O_2 $(^3\Sigma_g^-)$, O_2 $(^3\Pi_u)$, etc. These notation are called spectroscopic terms differentiating electronic states by the symmetry of angular momentum wave functions. Here, we do not enter into the theory of quantum chemistry, the meaning and usefulness of the symbols and selection rules which is important for the discussion of photoabsorption and photolysis reactions are remarked.

Spectroscopic Terms Spectroscopic terms for atoms are in general expressed as $^{2S+1}L_J$. Here, S and L is the electron spin and orbital angular momentum, corresponding rotation and revolution of electrons, respectively. J is the total angular momentum, which is the vectorial sum of spin and orbital angular momentum. Orbital angular momentum L is designated by S, P, D, etc. for $L=0, 1, 2$, etc. Spin angular momentum S is denoted by using the values of spin multiplicity, $2S+1$, i.e. 1, 2, 3 for $S = 1/2, 1, 3/2$, respectively. Total angular momentum J is the synthetic vector given by $J = |L+S|, |L+S-1|, |L+S-2|, \ldots, |L-S+2|, |L-S+1|, |L-S|$. Namely, for example, for the ^3P state ($S=1$, $L=1$), J can take the values of 0, 1, 2, giving three different electronic states, $^3P_0$, $^3P_1$, $^3P_2$. When these spin-orbit states are not differentiated, it is denoted simply by ^3P omitting the J values. In atmospheric chemistry, ground and excited states of oxygen atoms are usually denoted as $O(^3P)$,

$O(^1D)$ omitting J values. In the case of halogen atoms, however, discussions are sometimes made for $Cl(^2P_{3/2})$, $Cl(^2P_{1/2})$, $I(^2P_{3/2})$, $I(^2P_{1/2})$, etc. differentiating the electronic states with different J values (Sects. 4.3.8 and 4.4.1).

Spectroscopic symbols for diatomic molecules are in general denoted by $^{2S+1}\Lambda$. Here, the spin angular momentum is the same as in the case of atoms. Orbital angular momentum Λ is the angular momentum around the molecular axis, and designated by Σ, Π, Δ, etc. corresponding to the values of $\Lambda = 0, 1, 2$, etc. Further, for homonuclear diatomic molecules such as O_2 and N_2, another symbol is added representing another molecular symmetry of wave functions. Thus, the initial character, g and u of the German words, gerade and ungerade meaning even and odd, is put after Λ as a subscript to represent a wave function which does not change or change its sign for the inversion at the center of symmetry, respectively, e.g. $O_2\,(^3\Sigma_g^-)$ and $O_2\,(^3\Pi_u)$. Further, for the Σ state, the superscript + or – is put after Σ corresponding to the wave function which does not change or change the sign for the inversion at the plane including the molecular axis. According to these rules, the ground state of oxygen is denoted as $O_2(^3\Sigma_g^-)$, and the excited states are denoted with variety e.g. $O_2\,(^1\Delta_g)$, $O_2\,(^1\Sigma_g^+)$, $O_2\,(^3\Sigma_u^+)$, $O_2\,(^3\Pi_u)$, $O_2\,(^3\Sigma_u^-)$ (see Fig. 3.5). For a molecule, which is not a homonuclear diatomic molecule, g or u symbols are not put since it does not have the center of symmetry. For example, nitric oxide (NO) molecule has the ground state, $NO(^2\Pi)$, and the excited states, e.g. $NO(^2\Sigma^+)$, $NO(^4\Pi)$, $NO(^2\Delta)$ (see Fig. 4.28).

For general non-linear polyatomic molecules, symbols A', A'', A_1, A_2, B_1, B_2 etc. are used depending on whether the sign of wave function is changed or not for the symmetric operation at the symmetric axis or plane. However, they seldom appear in the discussion of atmospheric chemistry.

Furthermore, it is customary to put X in front of the spectroscopic terms representing the ground sate, and put A, B, C., etc., and a, b, c, etc. for the excited states with the same and different spin multiplicity as the ground state, respectively. For example, $O_2\,(X^3\Sigma_g^-)$, $O_2\,(a^1\Delta_g)$, $O_2\,(b^1\Sigma_g^+)$, $O_2\,(A^3\Sigma_u^+)$, $O_2\,(B^3\Pi_u)$ will appear for O_2 (see Fig. 3.5), and NO $(X^2\Pi)$, NO $(A^2\Sigma^+)$, NO $(a^4\Pi)$, NO $(B^2\Pi)$, NO $(C^2\Pi)$ for NO (see Fig. 4.28).

Selection Rules The intensity of photo-absorption for transition from a lower to higher energy state can be determined by the transition probability $|\mathbf{R}|^2$, where $\mathbf{R}$ expresses

$$\mathbf{R} = \int \psi'^* \mu \psi'' dv, \tag{2.19}$$

and $\mathbf{R}$ is called a dipole moment. Here, $\psi''$and ψ' is the wave function, or eigenfunction, for the initial (ground) and final (excited) state, respectively, and μ is the electric dipole moment vector. The * on the right shoulder of ψ'^* represents a so-called conjugated complex function, in which an imaginary part of the wave function, i is replaced by –i. An eigenfunction of a molecule is generally expressed by

$$\psi = \psi_e \psi_v \psi_r, \tag{2.20}$$

where ψ_e, ψ_v, ψ_r are eigenfunctions of electronic, vibrational and rotational motion, respectively. Using an Eq. (2.20), **R** is written as

$$\mathbf{R} = \int \psi_e{}'^{*} \boldsymbol{\mu} \psi_e{}'' dv_e \int \psi_v{}'^{*} \psi_v{}'' dv_v \int \psi_r{}'^{*} \psi_r{}'' \cos \alpha \, dv_r. \tag{2.21}$$

Transition probability is determined only by electronic eigenfunctions for atoms, since there is no vibrational or rotational motion.

In order for the absorption and emission of light to occur, above the transition moment has to have non-zero value, and rules to fulfill this are called selection rules. The selection rule for the electric dipole transition of an atom is

$$\Delta L = 0, \quad \pm 1 \;\; (\text{excluding}\, L = 0 \to 0) \tag{2.22}$$

for the orbital angular moment and

$$\Delta J = 0, \quad \pm 1 \;\; (\text{excluding}\, J = 0 \to 0) \tag{2.23}$$

for the total angular momentum. The case where the total angular momentum **J** is expressed as $J = |L + S|,\ |L + S - 1|,\ |L + S - 2|,\ \ldots,\ |L - S + 2|,\ |L - S + 1|,\ |L - S|$ as previously noted is called Rusell-Sounders coupling, and this approximation can be applied to light atoms. In this case, the selection rule for spin quantum number is

$$\Delta S = 0. \tag{2.24}$$

Transitions with transition moments larger than a certain extent, are called the allowed transitions, and those with near-zero moments are called the forbidden transitions. For example, the transition between $O(^3P) \leftrightarrow O(^1D)$ of O atoms is spin forbidden, and the light emission probability for $O(^1D) \to O(^3P)$ is very small so that the radiative lifetime of $O(^1D)$ is long and the reaction of $O(^1D)$ with other molecule can be important in the atmosphere.

For diatomic molecules, the allowed transitions for the orbital angular momentum are for

$$\Delta \Lambda = 0, \quad \pm 1. \tag{2.25}$$

In the case where g, u, +, and − are applicable for the Σ state,

$$\mathrm{g} \leftrightarrow \mathrm{u}, + \leftrightarrow +, - \leftrightarrow - \tag{2.26}$$

are allowed transitions. For example from the ground state of O_2 molecule, $O_2\ (X^3\Sigma_g{}^-) \leftrightarrow O_2\ (B^3\Pi_u)$ is an allowed transition, and $O_2\ (X^3\Sigma_g{}^-) \leftrightarrow O_2\ (a^1\Delta_g)$, $O_2\ (b^1\Sigma_g{}^+)$, $O_2\ (A^3\Sigma_u{}^+)$ are forbidden transitions. For NO molecule, $NO(X^2\Pi) \leftrightarrow NO(A^2\Sigma^+)$,

$NO(X^2\Pi) \leftrightarrow NO\ (B^2\Pi)$ are allowed transitions, and $NO(X^2\Pi) \leftrightarrow NO\ (a^4\Pi)$ is a spin forbidden transition.

Even for the forbidden transitions, the transition moments are not absolutely zero but in general the absorption and emission of light can be seen with very small probability. This is because, for example, instead of electric dipole transition, magnetic dipole transition with $\Delta\Lambda = \pm 2$ can occur for diatomic molecules, or the molecular symmetry collapses by a coupling between vibrational and electronic motion for polyatomic molecules. Further, the spin forbidden can be applied only when the Rusell-Sounders coupling is applied. For heavy atoms, this rule is broken due to the spin-orbit coupling, and spin forbidden rule cannot be applied. For example, the $Hg(^1S) \leftrightarrow Hg(^3P)$ transition of a mercury atom is spin forbidden, but has a strong absorption and emission of light at 253.7 nm. The ozone meter based on the absorption at this wavelength using the light from a mercury lamp are widely used for monitoring ambient ozone concentrations.

2.2 Bimolecular Reactions

2.2.1 Potential Surface and Transition State

Many of atmospheric reactions are the following type, involving the rearrangement of atoms,

$$AB + C \rightarrow A + BC, \tag{2.27}$$

where AB is a molecule and C is a reactive species such as an atom or a free radical in most cases. Reactions in which rearrangement of atoms occurs between two chemical species are called bimolecular reactions. In view of molecular dynamics, a bimolecular reaction is a process in which AB and C collide to associate, passing through an energy barrier called transition state, and dissociate into A + BC. Figure 2.7 illustrates such a reaction pathway schematically. The abscissa is the reaction coordinate, and the ordinate is the potential energy. The left-hand side of the figure shows the state before the reaction AB + C, called reaction system, and the right-hand side corresponds to the state after the reaction A + BC, called the product system. The energy maximum in-between is called the transition state corresponding to the activated complex $(ABC)^{\ddagger}$. Such a concept is based on the transition state theory, which will be discussed in the following Sect. 2.2.2.

Figure 2.7 schematizes the reaction pathway in one-dimension. However, a reaction pathway is actually represented by an energy surface with two axes, r(A−B) and r(B−C), representing the inter-atomic distances, A−B, and B−C, respectively, as shown in Fig. 2.8. This surface is called the -potential energy surface. The solid lines in Fig. 2.8 are the energy contour connecting the equipotential points. The reaction system AB + C, and the product system A + BC

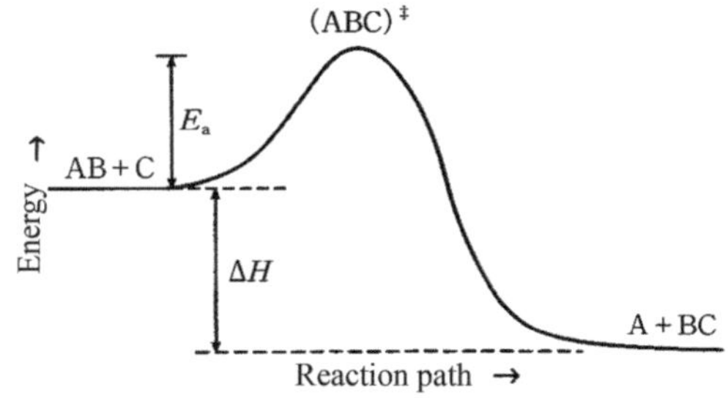

Fig. 2.7 Change of potential energy along the reaction pathway of a bimolecular reaction

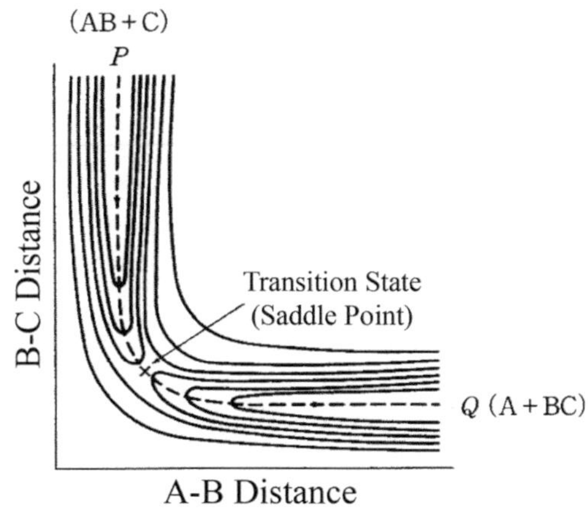

Fig. 2.8 Potential energy surface for a bimolecular reaction

corresponds to the valley of potential energy open toward upper-left, and toward lower right, respectively. The broken line in the figure represents reaction pathway, which reaches from the reaction system to the product system along the lowest energy path passing though the energy maximum point shown by ×. The transition state $(ABC)^{\ddagger}$ shown in Fig. 2.8 corresponds to the energy maximum point × in Fig. 2.8. The point × corresponds to a saddle point of the curved surface of the potential energy, which is the energy maximum along the reaction axis shown by the broken line, but the energy minimum toward the different direction along the surface. A molecule A-B-C in this transition state is called activated complex and represented by $ABC^{\ddagger}$ habitually with a ‡ mark. This maximum point acts as an energy barrier when the reaction proceeds from the reaction system to the product system. The difference of energy between the reaction system and the transition state is called the activation energy and is generally represented by E_a, as seen in Fig. 2.7.

Whether the Reaction (2.27) proceed to the right or, in reverse, to the left, is determined by the difference of the free energy ΔG between the reaction system and the product system. According to the first law of thermodynamics, the free energy change ΔG must be negative for the reaction to proceed. The ΔG is defined by

$$\Delta G = \Delta H - T\Delta S, \tag{2.28}$$

where ΔH is the enthalpy change, ΔS is the entropy change, and T is temperature. For the bimolecular reactions to proceed, the entropy change is negligible so that ΔH has to be negative. The reactions with negative ΔH ($\Delta H < 0$) are called exothermic reactions, and inversely those with positive ΔH ($\Delta H > 0$) are called endothermic reactions. The enthalpy change of the reaction ΔH is obtained as the difference of enthalpy of formation ΔH_f of atoms and molecules in the reaction and product system,

$$\Delta H = \{\Delta H_f(A) + \Delta H_f(BC)\} - \{\Delta H_f(AB) + \Delta H_f(C)\}. \tag{2.29}$$

Enthalpy of formation of typical atoms, molecules and free radicals relevant to atmospheric chemical reactions are cited in Table 2.5. The ΔH shown in Fig. 2.7 is the enthalpy change of reaction given by Eq. (2.29), showing the reaction to proceed from left to right since $\Delta H < 0$ in this figure.

The rate constant of bimolecular reactions represented by Reaction (2.26) is represented as

$$-\frac{d[AB]}{dt} = -\frac{d[C]}{dt} = \frac{d[A]}{dt} = \frac{d[BC]}{dt} = k_r[AB][C]. \tag{2.30}$$

Here, the quantities within the brackets [ ] represent concentration of each species, and the k_r defined above is called a reaction rate constant. The reaction rate constant of a bimolecular reaction has the dimension of $(\text{concentration})^{-1}(\text{time})^{-1}$. In atmospheric chemistry, the concentration of gaseous species is expressed in general by the number density of molecules, molecules cm^{-3}, so that the unit of the rate constant of a bimolecular reaction is commonly expressed as cm^3 $molecule^{-1}$ s^{-1}.

2.2.2 Activation Energy and Reaction Rate Constant

In this paragraph a reaction formula is expressed simply as

$$A + B \rightarrow \text{products}. \tag{2.31}$$

There are two types of molecular theories on reaction rates: collision theory and transition state theory. According to the collision theory, the upper limit of a bimolecular rate constant of a gaseous reaction is the molecular collision frequency obtained by gas kinetics. The molecular collision frequency Z_{AB} between molecules, A and B, is given by

Table 2.5 Enthalpy of formation of molecules, atoms and radicals in the gas phase at 298 and 0 K ($\Delta H^\circ_{f,298}$, $\Delta H^\circ_{f,0}$)

Chemical species	$\Delta H^\circ_{f,298}$ kJ mol^{-1}	$\Delta H^\circ_{f,0}$ kJ mol^{-1}	Chemical species	$\Delta H^\circ_{f,298}$ kJ mol^{-1}	$\Delta H^\circ_{f,0}$ kJ mol^{-1}	Chemical species	$\Delta H^\circ_{f,298}$ kJ mol^{-1}	$\Delta H^\circ_{f,0}$ kJ mol^{-1}
H	218.0	216.0	CH_2OO (CI)[c]	110	118	C_2H_5ONO	−99.4	
H_2	0	0	CH_2O_2 (dioxirane)	5.0	12.6	$C_2H_5ONO_2$	−154.1[a]	
$O(^3P)$	249.2	246.8	CH_3O	21.0	28.4	$C_2H_5OONO_2$	−63.2[a]	
$O(^1D)$	438.9[a]	436.6[a]	CH_3O_2	9.0		$CH_3C(O)O_2NO_2$	−240.1	
O_2	0	0	CH_2OH	−17.0	−10.7	C_3H_6	20.2	34.7[b]
$O_2(^1\Delta)$	94.3[a]	94.3[a]	CH_3OH	−201.0	−190.1	n-C_3H_7	101.3	119.1
$O_2(^1\Sigma)$	156.9[a]	156.9[a]	CH_2OOH	67.2		i-C_3H_7	86.6	107.1[b]
O_3	141.7	144.4	CH_3OOH	−132.2		C_3H_8	−104.7	−82.4[b]
HO	37.4	37.1	HC(O)OH	−378.8[a]	−371.6[a]	C_2H_5CHO	−185.6	−170.6[b]
HO_2	12.3	15.2	$HOCH_2O_2$	−162.1[a]		CH_3COCH_2	−23.9[a]	
H_2O	−241.8	−238.9	CH_3ONO	−64.0		CH_3COCH_3	−217.1	−200.5[b]
H_2O_2	−135.9	−129.9	CH_3ONO_2	−122.2		C_3H_6OH	−74[a]	
N	472.4	470.6	$CH_3O_2NO_2$	−44[a]		$CH_3C(O)CHO$	−271[a]	
N_2	0	0	C_2H_2	227.4	228.0	S	277.2	274.9
NH_2	186.2	189.1	C_2H_4	52.4	61.0	HS	142.9	142.5
NH_3	−45.9	−39.0	C_2H_3	295.4		H_2S	−20.6	−17.7
NO	91.0	90.5	C_2H_5	120.9	131.8	HSO	−6.1	−3.8
NO_2	34.0	36.8	C_2H_6	−84.0	−68.4[b]	SO	4.8	4.7
NO_3	74.7	79.9	CH_2CN	252.6	255.2	SO_2	−296.8	−294.3
N_2O	81.6	85.3	CH_3CN	74.0	81.0	SO_3	−395.9	−390.2
N_2O_4	11.1	20.4	CH_2CO	−49.6	−46.4	HSO_2	−178	
N_2O_5	13.3	22.9	CH_3CO	−10.3	−3.6	$HOSO_2$	−373	

HNO	109.2	112.1	CH_2CHO	10.5		H_2SO_4	−732.7	−720.8
HONO	−78.5	−72.8	CH_3CHO	−166.1	−160.2[b]	CH_3S	124.7	129.9[b]
HNO_3	−134.3	−124.6	C_2H_2OH	121	120[a]	CH_3SO	−70.3	
HO_2NO_2	−54.0		C_2H_5O	−13.6	−0.2	CH_3SOO	75.7[a]	87.9[a]
CH_3	146.7	150.0	C_2H_4OH	−31	−23[a]	CH_3SH	−22.9	−11.9[b]
CH_4	−74.6	−66.6	C_2H_5OH	−234.8	−217.1	CH_3SCH_2	136.3	
CO	−110.5	−113.8	$(CHO)_2$	−212	206.4	CH_3SCH_3	−37.2[a]	−21.0[a]
CO_2	−393.5	−393.1	CH_3CO_2	−207.5[a]		CH_3SSCH_3	−24.7	
HCO	44.2	41.6[b]	$CH_3C(O)OH$	−432.8	−418.1[b]	CS	279.8	276.5
trans-HOCO	−187.9	−183.7	$C_2H_5O_2$	−27.4		CS_2	116.7	115.9
cis-HOCO	−175.7	−171.5	CH_3OOCH_3	−125.5	−106.5	OCS	−141.7	−141.8
CH_2O	−108.7	−104.9	$CH_3C(O)O_2$	−154.4		CS_2OH	108.4	
F	79.4	77.3	ClOO	98.3	99.8	HOBr	−60.5	−50.0
HF	−273.3	−273.3	OClO	99.4	99.0[b]	BrO	123.4	131.0
FO	109	108	ClNO	52.7	54.6	OBrO	163.9	171.1
FO_2	25.4[a]	27.2[a]	$ClNO_2$	12.5	17.9	BrOO	119.8	128.2
FONO	67		cis-ClONO	64.4		BrNO	82.2	91.5
$FONO_2$	15	22	$ClONO_2$	22.9		$BrONO_2$	42.7	
FNO	−65.7	−62.6[b]	CH_3Cl	−82.0[a]	−74.0[a]	CH_2Br	172.8	
FNO_2	−79		$CHClF_2$	−482.6		CH_3Br	−36.4	
FCO	−174.1	−174.5	COFCl	−427[a]	−423[a]	BrCl	14.8	22.2
F_2	0	0	Cl_2	0	0	$Br_2(g)$	30.9	45.7
COF_2	−634.7[a]	−631.6[a]	ClOCl	81.3	83.1	$CHBr_3$	55.4	
CH_3CF_3	−745.6	−732.8	ClOOCl	129.0	132.4	CF_2ClBr	−589.5	
CH_2FCHF_2	−665		Cl_2O_3	139	144	CF_3Br	−641.1	−637.6
CH_2FCF_3	−896	−885	$COCl_2$	−220.1[a]	−218.4[a]	I	106.8	107.2
CHF_2CF_3	−1105	−1095	CH_3CCl_3	−144.6	−131.9[b]	HI	26.5	28.7[b]

(continued)

Table 2.5 (continued)

Chemical species	$\Delta H^\circ_{f,\,298}$ kJ mol^{-1}	$\Delta H^\circ_{f,\,0}$ kJ mol^{-1}	Chemical species	$\Delta H^\circ_{f,\,298}$ kJ mol^{-1}	$\Delta H^\circ_{f,\,0}$ kJ mol^{-1}	Chemical species	$\Delta H^\circ_{f,\,298}$ kJ mol^{-1}	$\Delta H^\circ_{f,\,0}$ kJ mol^{-1}
CF_3	−465.7	−462.8	CF_3Cl	−709.2	−704.2	HOI	−69.6	−64.9
CF_3O	−635		CF_2Cl_2	−493.3[a]	−489.1[a]	IO	125.1	127.2
CF_3OH	−923.4[a]		$CFCl_3$	−284.9[a]	−281.1[a]	OIO	119.7	123.4
CF_3O_2	−614.0[a]		CCl_4	−95.8[a]	−93.3[a]	INO	121.3[a]	124.3[a]
Cl	121.3	119.6	CF_2Cl	−279	–	INO_2	60.2[a]	66.5[a]
HCl	−92.3	−92.1	$CFCl_2$	−89.1		$IONO_2$	37.5	46.1
HOCl	−74.8	−71.5[b]	Br	111.9	117.9	CH_3I	13.2	
ClO	101.7	101.1	HBr	−36.3	−28.4	I_2 (g)	62.4	65.5

Source: NASA/JPL Panel Evaluation No. 17 unless otherwise noted
[a]IUPAC Subcommittee Report Vol. I
[b]CCCBDB List of Species with Enthalpy of Formation at 0 K in Database. http://cccbdb.nist.gov.hf0k.asp
[c]Criegee Intermediate (carbonyl oxide)

$$Z_{AB} = \pi(r_A + r_B)^2 u N_A N_B = \pi(r_A + r_B)^2 \left(\frac{8k_B T}{\pi \mu_{AB}}\right)^{1/2} N_A N_B, \tag{2.32}$$

where r_A, r_B are molecular radii of A and B, respectively, u is the molecular speed, k_B is the Boltzmann constant (Table 2.1), μ_{AB} is the reduced mass of A and B, and N_A and N_B is the number density of each molecule. From the above equation, molecular collision frequency at normal temperature (298 K) is calculated to be 2.5×10^{-10} cm^3 molecule^{-1} s^{-1}, for example, for molecules with a molecular radius of 0.2 nm, and molecular weight of 50 g mol^{-1}. This value is thought to be the upper limit of bimolecular rate constants, and the actual rate constants of many reactions are usually smaller than this.

The most important aspect of bimolecular reactions is that there is an energy barrier to overcome along the reaction path as shown in Figs. 2.7 and 2.8. Considering this factor, the collision theory posits that a molecule with kinetic energy larger than the energy barrier E_0 can react and those with smaller energy than E_0 cannot react. Thus, reflecting such collision probability, exp $(-E_0/kT)$, the reaction rate constant k_r is given by,

$$k_r = P\sigma_{AB} \left(\frac{8k_B T}{\pi \mu_{AB}}\right)^{1/2} exp(-E_0/k_B T), \tag{2.33}$$

where $\sigma_{AB} = \pi(r_A + r_B)^2$ is the collision cross section between molecules A and B, and P is a steric factor discussed later.

On the other hand, for the transition state theory reaction formula is expressed as

$$\mathrm{A} + \mathrm{B} \rightarrow \mathrm{AB}^{\ddagger} \rightarrow \text{products}, \tag{2.34}$$

considering the activated complex described in the previous section. A thermal equilibrium constant $K^{\ddagger}$ between A, B and $AB^{\ddagger}$ is defined by

$$K^{\ddagger} = \frac{[AB^{\ddagger}]}{[A][B]} = \exp(-\Delta G^{\circ\ddagger}/RT) = \exp(-\Delta H^{\circ\ddagger}/RT)\exp(\Delta S^{\circ\ddagger}/R), \tag{2.35}$$

using the free energy in Eq. (2.28). Here, R is the gas constant (Table 2.1). The superscript $^\circ$ attached to ΔG, ΔH, ΔS in the above formula indicates that these values are those referred to the standard state at 1 atm. From this formulation, the reaction rate R_r and reaction rate constant k_r are expressed as,

$$R_r = k_B T/\mathrm{h}[\mathrm{AB}^{\ddagger}] = k_B T/\mathrm{h} \exp(-\Delta G^{\ddagger}/RT)[\mathrm{A}][\mathrm{B}] \tag{2.36}$$

and

$$k_r = k_B T/h \exp\left(-\Delta G^{\ddagger}/RT\right) = k_B T/h \exp\left(\Delta S^{\circ\ddagger}/R\right) \exp\left(-\Delta H^{\circ\ddagger}/RT\right), \quad (2.37)$$

respectively. Replacing the enthalpy difference between the reaction system and transition state, $\Delta H^{\circ\ \ddagger}$ by the activation energy E_a, k_r is given by

$$k_r = k_B T/h \exp\left(\Delta S^{\circ\ddagger}/R\right) \exp(-\mathrm{E_a}/RT). \quad (2.38)$$

Experimentally, in most of the cases bimolecular reaction rate constants are known to be expressed by

$$k_r(T) = A \exp(-E_a/RT). \quad (2.39)$$

This equation is called the Arrhenius expression and is the fundamental equation representing the temperature dependence of reaction rate constants. Comparing the Arrhenius expression Eq. (2.39), with rate constant Eq. (2.33) by the collision theory and (2.38) by the transition state theory, the temperature dependence of the exponential factor is exactly the same as derived by these theories, and E_a of the Arrhenius expression corresponds to the activation energy E_a of the transition state theory. A plot of the logarithm of a reaction rate constant, ln k against $1/RT$, is called an Arrhenius plot, and the experimental value of activation energy can be obtained from the slope of the Arrhenius plot. This linear relationship is known to hold experimentally for numerous reactions, and the activation energy for each reaction has been obtained.

Meanwhile, the pre-exponential factor A in the Arrhenius Eq. (2.39) is the temperature independent factor related to reaction frequency. Comparing the Eq. (2.33) for the collision theory and Eq. (2.38) with the transition state theory, the pre-exponential factors in these theories contain temperature dependences of $\mathrm{T}^{1/2}$ and T^1, respectively. Experimentally, for most of reactions for which the activation energy is not close to zero, the temperature dependence of the reaction rate constants are known to be determined almost solely by exponential factor, and the Arrhenius expression holds as a good approximation. Only for the reaction with near-zero activation energy, the temperature dependence of the pre-exponential factor appears explicitly, and the deviation from the Arrhenius expression can be validated. In this case, an approximated equation modifying the Arrhenius expression can be used,

$$k_r(T) = BT^{\mathrm{n}} \exp(-E_a/RT), \quad (2.40)$$

where B is the temperature independent constant, and the value of n is an experimentally-determined parameter.

The upper limit of the pre-exponential factor A is ca. 2.5×10^{-10} cm^3 molecule^{-1} s^{-1} from the collision theory as mentioned above, but an actual A value of the Arrhenius expression is often smaller than this value. According to the collision theory, it is thought that the molecules need to collide in a specific orientation for the reaction to occur. The P in Eq. (2.33) is called a steric factor reflecting the requirement for the specific orientation. As the steric constraint gets larger, the reaction probability decreases and the value of A gets smaller. In the transition theory, entropy term exp ($\Delta S^{\circ\ddagger}/R$) is included in the pre-exponential factor, and the steric hindrance for the activated complex is interpreted to be reflected in the decrease of entropy leading to smaller reaction rates.

2.3 Termolecular and Unimolecular Reactions

2.3.1 Association Reactions

Among atmospheric reactions, association reactions which are represented as termolecular reactions in the form of,

$$\mathrm{A + B + M \rightarrow AB + M} \tag{2.41}$$

are included, where A and B are atoms, molecules or free radicals. As chain termination reactions, this type of association reaction is often important. M in the above formula is the third body of a reaction. When A and B associate, internal energy is gained ($\Delta H < 0$) corresponding to the bond energy of A-B, and a vibrationally excited molecule $AB^{\dagger}$ is formed. Unless the vibrational energy is taken away, $AB^{\dagger}$ dissociates back into A + B again, and the reaction does not proceed practically. The third body M collides with the vibrationally excited molecule and removes energy to some extent, so that it is a necessary molecule to stabilize the molecule $AB^{\dagger}$ by preventing the dissociation into A + B and complete the reaction substantially. In the atmosphere, N_2 and O_2 molecules play a role of the third body.

Therefore, the termolecularreaction rate constant has pressure dependence, and it is explained by the following scheme, called the Lindemann mechanism. According to the mechanism,

$$\mathrm{A + B \rightarrow AB^{\dagger}} \tag{2.42}$$

$$\mathrm{AB^{\dagger} \rightarrow A + B} \tag{2.43}$$

$$\mathrm{AB^{\dagger} + M \rightarrow AB + M.} \tag{2.44}$$

The vibrationally excited molecule $AB^{\dagger}$ formed by the association of A and B, is assumed to be in equilibrium with the reaction and product system so that,

$$\frac{d[AB^\dagger]}{dt} = k_a[A][B] - k_b[AB^\dagger] - k_c[AB^\dagger][M] = 0 \tag{2.45}$$

$$[AB^\dagger] = \frac{k_a[A][B]}{k_b + k_c[M]} \tag{2.46}$$

$$\frac{d[AB]}{dt} = k_c[AB^\dagger][M] = \frac{k_a k_c[A][B][M]}{k_b + k_c[M]} \tag{2.47}$$

From these equations, the termolecular reaction rate constant k_{ter} is derived as

$$k_{ter} = \frac{k_a k_c[M]}{k_b + k_c[M]} \tag{2.48}$$

where k_a, k_b *and* k_c are a reaction rate constant of reaction (2.42) (2.43) and (2.44), respectively.

When the pressure is low enough, putting [M] = 0 in the denominator of Eq. (2.47), the rate equation is,

$$\frac{d[AB]}{dt} = \frac{k_a k_c}{k_b}[A][B][M]. \tag{2.49}$$

In this case, the termolecular reaction rate constant k_0 is

$$k_0 = \frac{k_a k_c}{k_b} \tag{2.50}$$

and k_0 is called the low-pressure limit rate constant. The dimension of the low-pressure limit rate constant is $(\text{concentration})^{-2}(\text{time})^{-1}$ and the unit used in atmospheric chemistry is $\text{cm}^6\ \text{molecule}^{-2}\ \text{s}^{-1}$.

On the other hand, when pressure is high enough, neglecting k_b in Eq. (2.47) the reaction rate equation is expressed as

$$\frac{d[A]}{dt} = k_a[A][B]. \tag{2.51}$$

In this case, the reaction rate constant is

$$k_\infty = k_a \tag{2.52}$$

and k_∞ is called the high-pressure limit rate constant. Using these k_0 and k_∞, the rate constant Eq. (2.45) is expressed as

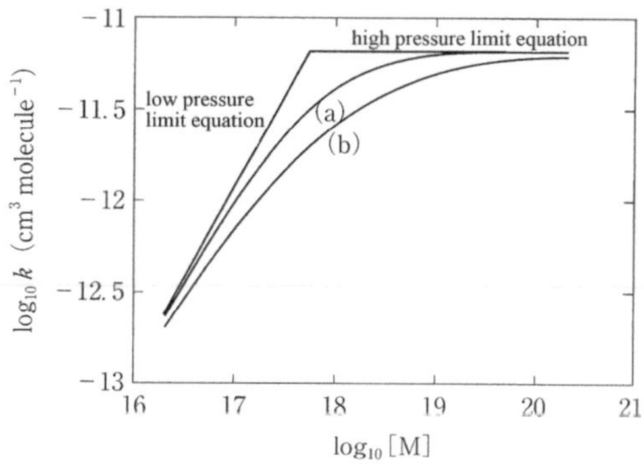

Fig. 2.9 Pressure dependence of termolecular reaction rate constant for (**a**) Hinshelwood-Lindeman mechanism and (**b**) Troe's formula

$$k_{ter} = \frac{k_0 k_\infty [M]}{k_\infty + k_0 [M]}. \tag{2.53}$$

The curve (a) in Fig. 2.9 is the schematic graph of the pressure dependence of a termolecular reaction rate constant according to the Lindemann mechanism. From the figure, it can be seen that the reaction rate constant is proportional to [M] (pressure) in the low-pressure limit, and gets nearly constant independent on the pressure in the high-pressure limit. The intermediate region between these two limits is called the fall-off region.

Although the Lindemann mechanism captures the basic feature of association reactions, and effectively represents the characteristics of pressure dependence of the termolecular reactions, it is not a good enough approximation in quantitative terms. The main reason is that the unimolecular decomposition reaction (see the next section) from $AB^\dagger$ to A + B cannot be represented by a simple first order rate constant. In order the vibrationally excited molecule $AB^\dagger$ to decompose, the vibrationally energy in the molecule has to be localized into the dissociating bond, and the correct unimolecular reaction rate constant has to be determined considering such probability and the energy removed by a single collision. While the unimolecular decomposition theory is explained in the next paragraph, the following Troe's formula proposed by Troe (1979),

$$k_{ter}([M], T) = \left[\frac{k_0(T)[M]}{1 + \frac{k_0(T)[M]}{k_\infty(T)}}\right] F_c^{\left\{1 + \left[\log_{10}\left(\frac{k_0(T)[M]}{k_\infty(T)}\right)^2\right]\right\}^{-1}} \tag{2.54}$$

Is best used for representing the pressure dependence of the termolecular reaction rate constant. This equation is based on the curve fitting to the pressure dependence of the unimolecular decomposition rate by the Kassel theory. In Eq. (2.54), F_c is called a broadening factor, and it is a good experimental fitting for many termolecular reactions in atmospheric chemistry has been obtained by taking e.g. $F_c = 0.6$. The curve (b) in Fig. 2.9 shows the schematic pressure dependence of a termolecular reaction by the Troe's formula. In this case, temperature dependence of k_0 and k_∞ are represented by using parameters, n and m as

$$k_0(T) = k_0^{300}\left(\frac{T}{300}\right)^{-n} cm^6 molecule^{-2} s^{-1} \tag{2.55}$$

$$k_\infty(T) = k_\infty^{300}\left(\frac{T}{300}\right)^{-m} cm^3 molecule^{-1} s^{-1}. \tag{2.56}$$

2.3.2 Unimolecular Decomposition Reactions

The molecule AB formed by the association reaction (2.41) may follow thermal decomposition reaction faster than photolysis or bimolecular reactions when the bond energy is small. In this case, the thermal decomposition reaction of AB is expressed by

$$\mathrm{AB + M \rightarrow A + B + M}. \tag{2.57}$$

This type of reaction is called unimolecular reactions.

Historically, the theory of the termolecular reaction mentioned in the previous paragraph has been developed through the unimolecular reaction theory. This paragraph describes unimolecular decomposition reactions in some detail. The chemical formula for the unimolecular decomposition reactions corresponding to the Lindemann mechanism can be shown as

$$\mathrm{AB + M} \underset{-1}{\overset{1}{\rightleftarrows}} \mathrm{AB^\dagger + M} \tag{2.58}$$

$$\mathrm{AB^\dagger} \xrightarrow{2} \mathrm{A + B} \tag{2.59}$$

Where M is the third body of a reaction previously described. In the above reaction scheme, assuming the stationary state of $[\mathrm{AB}^\dagger]$, the decomposition rate of AB, k_{uni} is given by

$$k_{uni} = \frac{k_1 k_2 [M]}{k_2 + k_{-1}[M]} = \frac{k_2(k_1/k_{-1})}{1 + (k_2/k_{-1}[M])} \tag{2.60}$$

where k_1, k_{-1} are the rate constants of the forward and reverse reaction of Eq. (2.58), k_2 is the rate constant of reaction (2.59). From Eq. (2.60) the rate constant k_∞ in the high-pressure limit, $k_2 \ll k_{-1}[\mathrm{M}]$ is expressed by the first order rate constant (s^{-1}) as

$$k_{\infty} = (k_I/k_{-I})k_2. \tag{2.61}$$

On the other hand, in the low pressure limit $k_2 \gg k_{-I}$[M], the reaction rate is proportional to [M], and the rate constant k_O,

$$k_O = k_I \tag{2.62}$$

is expressed in a bimolecular reaction rate constant (cm^{3}molecule^{-1}s^{-1}). According to the Lindemann mechanism, the formation rate of AB† having energy higher than the dissociation energy E_0 is obtained by a classical solid collision theory using the collision frequency Z, activation energy E_0 and the Boltzmann constant k_B,

$$k_I = \mathrm{Z}\exp(-E_0/k_BT) \tag{2.63}$$

Here, it is assumed that a single collision with M causes deactivation of AB†. The rate constant k_{uni} calculated from Eq. (2.60) was found to qualitatively reproduce the experimental pressure dependence of the unimolecular decomposition, but to produce a large discrepancy for the high-pressure limit rate constants k_∞ with experimental values.

Hinshelwood proposed a formula for the rate of the k_I in which AB† having the energy higher than the dissociation energy E_0, by considering not only the translational energy of a solid ball but also the distribution of vibrational energy. Thus,

$$k_0 = k_1 = \frac{Z}{(s-1)!}\left(\frac{E_0}{k_BT}\right)^{s-1}\exp\left(-\frac{E_0}{k_BT}\right) \tag{2.64}$$

where s is the freedom of normal vibrational modes of a molecule composed of n atoms, s = 2n−1. The statistical fraction the AB† having internal energy $E \frown E + dE$ is

$$\frac{dk_1}{k_{-1}} = \frac{1}{(s-1)!}\left(\frac{E_0}{k_BT}\right)^{s-1}\exp\left(-\frac{E_0}{k_BT}\right)\left(\frac{dE}{k_BT}\right) \tag{2.65}$$

From this equation a formula corresponding to Eq. (2.57) is

$$k_{uni} = \int_{E_0}^{\infty}\frac{k_2(dk_1/k_{-1})}{1 + (k_2/k_{-1}[M])}. \tag{2.66}$$

Energy integration leads to the high-pressure limit equation,

$$k_\infty = \frac{k_2}{(s-1)!}\left(\frac{E_0}{k_B T}\right)^{s-1}\exp\left(-\frac{E_0}{k_B T}\right). \tag{2.67}$$

This treatment is called the Lindemann-Hinshelwood theory. Although the treatment of Hinshelwood was successful in reproducing the experimental values of high-pressure limit rate constants k_∞, the theory still has one defect, that the values of s necessary to explain the experimental values differ largely from the actual number of vibrational freedom and there is a large discrepancy of k_{uni} with experimental values in the fall-off region.

Later development of the unimolecular decomposition theory is on the calculation of the statistical probability related to how the internal energy is distributed within the vibrationally excited molecule $AB^\dagger$ and is localized to a specific chemical bond to break it. The statistical probability of $AB^\dagger$ to have an internal energy $E \frown E + dE$ is replaced by *P(E)*, and k_2 is put k_2 (*E*) considering the energy dependence in Eq. (2.66), k_{uni} can be written as,

$$k_{uni} = \int_{E_0}^{\infty} \frac{k_2(E)P(E)dE}{1 + k_2(E)/k_{-1}[M]}. \tag{2.68}$$

Presently established unimolecular decomposition theory is called the RRKM theory, taking the initials of Rice, Ramsperger, Kassel and Markus. In the RRK theory which preceded the RRKM theory, assuming the molecule is composed of s number of harmonic oscillators with frequency *ν*, and the probability of total energy of the molecule concentrate into a single oscillator. The RRK theory advanced the idea of unimolecular decomposition, it still needed the s and *ν* as adjusting parameters for reproducing experimental values, and their physical meaning was not clear.

Markus improved the idea and the RRKM theory and established the calculation method of the unimolecular decomposition rate constant based on the real vibration-rotation energy levels of reacting molecules. The RRKM theory is a transition state theory, and the equilibrium constant of the transition state is given by

$$K^\ddagger = \frac{W^\ddagger(E^\ddagger)}{\rho(E_v)}. \tag{2.69}$$

Here, $E^\ddagger$ is a transition state energy, W ($E^\ddagger$) is state sum of vibrational-rotational freedom of the transition state molecule, ρ (E_v) is the state density of the reacting molecule with vibrational energy E_v. Thus,

$$k_2(E^\ddagger) = \frac{W^\ddagger(E^\ddagger)}{h\rho(E_v)} \tag{2.70}$$

and the k_2 (T) is obtained by

$$k_2(T) = \int_0^\infty k_2(E^\ddagger)P(E^\ddagger)dE^\ddagger. \tag{2.71}$$

Meanwhile, P($E^\ddagger$) can be written using a partition function Q as

$$P(E^\ddagger) = \frac{\rho(E^\ddagger)\ \exp(-E^\ddagger/k_B T)}{Q} \tag{2.72}$$

Here, the partition function is a state sum defined in general by

$$Q = \int dE\ \exp(-E/k_B T) \tag{2.73}$$

From Eqs. (2.68) and (2.69),

$$k_2(T) = \frac{1}{hQ}\int_0^\infty \rho(E^\ddagger)\ \exp\left(-\frac{E^\ddagger}{k_B T}\right)dE^\ddagger \tag{2.74}$$

and the high-pressure limit equation is

$$k_\infty(T) = \frac{k_\infty T}{h}\frac{Q^\ddagger}{Q}\exp\left(-\frac{E_0}{k_B T}\right) \tag{2.75}$$

which agrees with the equation of the transition state theory. Practical calculation methods for rate constants using these equations should be referred to in the references at end of the chapter.

The RRKM theory reproduces experimental values well,and is now recognized as an established unimolecular reaction theory. The calculation by the RRKM theory has been applied to atmospheric reactions, for example, OH + CO (Sect. 5.2.3), $OH + NO_2 + M$ (Sect. 5.2.4) for obtaining the theoretical rate constants based on quantum chemistry.

Because the termolecular association reaction (2.41) and the unimolecular decomposition are reverse reactions, a thermal equilibrium between A + B and AB,

$$A + B + M \rightleftarrows AB + M \tag{2.76}$$

is established, and each reaction rate constant is interlinked with the thermodynamic equilibrium constant,

$$k_{ter}[A][B] = k_{uni}[AB] \tag{2.77}$$

$$K(T) = \frac{[A][B]}{[AB]} = \exp\left(\frac{-\Delta G^\circ}{RT}\right) = \exp\left(-\frac{\Delta H^\circ}{RT} + \frac{\Delta S^\circ}{R}\right) \tag{2.78}$$

$$k_{uni} = k_{ter}K(T). \tag{2.79}$$

From these relationships, unimolecular decomposition rate constants can be calculated from termolecular reaction rate constants and equilibrium constants (see Table 5.3).

2.4 Multiphase Reactions

A series of a processes including, for example, the uptake of atmospheric molecules into liquid particles such as fog or rain droplets followed by liquid phase reactions is called multiphase reactions, while a process in which molecules react on solid aerosol surface is often called heterogeneous reactions. However, since these terms are often used confusingly, and the processes contain common elements of chemical kinetics, they are treated together in this section.

2.4.1 Accommodation Coefficient and Uptake Coefficient

A fundamental parameter based on quantum chemistry to determine the rate of transfer of a molecule from the gas phase to the liquid or solid phase when a gas phase molecule collides with a particle surface is called a mass accommodation coefficient α and is defined by

$$\alpha = \frac{\text{Number of molecules initially attached to the liquid or solid surface}}{\text{Number of molecule colliding to the liquid or solid surface}}. \tag{2.80}$$

According to the molecular scattering theory, a particle incident to the surface has a certain probability for a certain amount to stay on the surface, and the rest is scattered back into the gas phase. α is an initial attaching probability and is a parameter to be determined through a molecular dynamics simulation based on quantum chemistry. Although in surface chemistry, a thermal accommodation coefficient is also used, in atmospheric chemistry a mass accommodation coefficient is solely used so that here it is simply called an accommodation coefficient. In general α cannot be determined directly by experiments, and the experimentally determined parameter of intake of a gaseous molecule to the particle surface is an uptake coefficient γ defined by

$$\gamma = \frac{\text{Number of molecules lost from the gas phase into the liquid or solid phase}}{\text{Number of molecule colliding to the liquid or solid surface}}. \tag{2.81}$$

γ is defined as a coefficient of a removal reaction rate constant of gaseous molecules as described below.

A flux of a number of molecules colliding on a liquid or solid surface per unit time and unit surface area, J_{col}, is given by

$$J_{col} = \frac{1}{4} u_{av} N_g \tag{2.82}$$

where u_{av} (cm s^{-1}) is the average thermal kinetic velocity, N_g (molecules cm^{-1} s^{-1}) is a molecular density in the gas phase. U_{av} in the above equation is given by gas kinetic theory,

$$u_{av} = \left(\frac{8k_B T}{\pi M}\right)^{1/2} \tag{2.83}$$

Where T is the temperature, M is the mass of a molecule, k_B is the Boltzmann constant. Thus, the net number of molecules taken into a particle surface per unit time and unit surface area, J_{het} (molecules cm^{-2} s^{-1}), can be given by

$$J_{het} = \frac{1}{4} \gamma u_{av} N_g. \tag{2.84}$$

Multiplying with the surface area density A (cm^2 cm^{-3}) of particles contained in the unit volume of gas, the removal rate of molecules in the gas phase by the surface heterogeneous process is given. When the removal rate of gas phase molecules is expressed by the pseudo first order rate equation,

$$\frac{d[N_g]}{dt} = -k_{het}[N_g], \tag{2.85}$$

the heterogeneous reaction rate constant k_{het} (s^{-1}) is given by

$$k_{het} = \frac{1}{4} \gamma u_{av} A. \tag{2.86}$$

From this equation, γ can be obtained experimentally from the experimental value of k_{het} and parameters u_{av} and A.

The surface area A is obtained for a spherical particle such as liquid particle by using size distribution n(r) for a radius in r ~ r + dr,

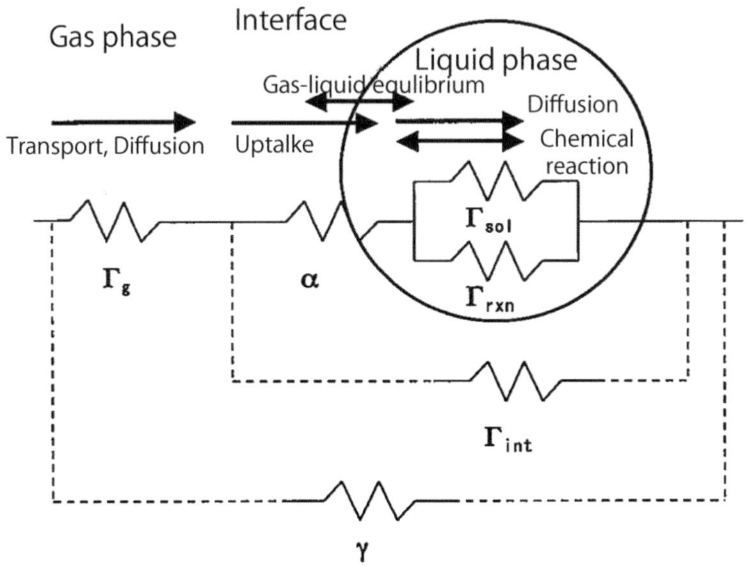

Fig. 2.10 Schematic diagram of resistant model for gas-liquid multiphase reactions

$$A = \int_0^\infty 4\pi r^2 n(r) dr. \tag{2.87}$$

However, solid particles in the atmosphere are mostly non-spherical. Furthermore, the surface is often porous with micropores. In such a case the actual surface area is much larger than the geometrical surface area, and the experimental values of γ differ much depending on which Fig. 2.10 surface area A is used (see Sects. 6.3 and 6.4).

Figure 2.10 illustrates a diagram of schematizing the processes of multiphase reactions on a liquid particle by the resistant model. As shown in the scheme, multiphase reactions include the processes, (1) transport and diffusion of a gaseous molecule to the gas-liquid surface, (2) accommodation at the interface, (3) - gas-liquid equilibrium at the interface, (4) physical dissolution and diffusion into the bulk liquid, and (5) chemical reaction in the bulk liquid.

The resistant model is a method which treats these series of the processes kinetically. In this model, the rate of each process is expressed by conductance Γ, which is the ratio the process to proceed per collision of a gaseous molecule to the interface. The uptake coefficient γ is expressed as the series and parallel coupling of the resistance, which is a reciprocal of Γ.

$$\frac{1}{\gamma} = \frac{1}{\Gamma_g} + \frac{1}{\alpha} + \frac{1}{\Gamma_{sol} + \Gamma_{rxn}} \tag{2.88}$$

where α is the accommodation coefficient defined above, Γ_g is the conductance of diffusion of a gaseous molecule to the interface, Γ_{sol} and Γ_{rxn} are those of diffusion of the taken-up molecule from the interface to the bulk liquid, and of the chemical reaction in the bulk liquid, respectively. In Fig. 2.10, the right side of the Eq. (2.88) lower than the second term is expressed by $1/\Gamma_{int}$. The Γ_{sol} depends not only on the solubility of molecules to the solution, but also on the surface concentration of the accumulated products formed in the surface reaction, so that is in general a time-dependent parameter. Therefore, γ itself is in general dependent on reaction time (see Chap. 6). In this treatment of γ, gas phase diffusion Γ_g can be neglected in the case of $\alpha < 0.01$, but the diffusion process to the interface represented by Γ_g becomes rate determining in the case of α with larger than 0.01.

Since the diffusion rate of gaseous molecules to the spherical particle surface is given by $4\pi r D_g N_g$ (molecules s^{-1}) using the diffusion coefficient D_g, the flux of a number of gaseous molecules transported to the interface by diffusion per unit time and unit volume, J_g (molecules $cm^{-2}\,s^{-1}$), is given by dividing with the surface area of the spherical particles $4\pi r^2$,

$$J_g = \frac{4\pi r D_g}{4\pi r^2} N_g = \frac{D_g}{r} N_g. \tag{2.89}$$

Normalizing J_g with the collision flux to the surface J_{col} given by Eq. (2.82), diffusion conductance Γ_g near the interface is given by

$$\Gamma_g = \frac{J_g}{J_{col}} = \frac{4D_g}{r u_{av}} = \frac{4D_g}{r}\left(\frac{\pi M}{8 k_B T}\right). \tag{2.90}$$

2.4.2 Gas-Liquid Equilibrium and Henry's Law Coefficients

The gas-liquid equilibrium constants determine the solubility of gaseous molecule into the liquid phase so that they are important parameters for the kinetic analysis of multiphase reactions. Liquid reactions are important in the atmosphere for fog and cloud water droplets with the particle diameter of 1–100 μm. When those water droplets coexist in the atmosphere, water soluble molecules X in the gas phase is absorbed by the droplet, and the gas-liquid equilibrium,

$$\mathrm{X(g)} \rightleftarrows \mathrm{X(aq)} \tag{2.91}$$

is established. Here, X(g) and X(aq) are the chemical species X in the gas and aqueous phase, respectively. This gas-liquid equilibrium is expressed by using the equilibrium constant called the Henry's law coefficient, K_H,

$$\frac{[X(aq)]}{p_x} = K_H. \tag{2.92}$$

Table 2.6 Henry's law coefficients of atmospheric molecules for water[a] (298 K)

Chemical Species	K_H (M atm^{-1})	Chemical species	K_H (M atm^{-1})
O_2	1.3×10^{-3}	CO	9.8×10^{-4}
O_3	1.0×10^{-2}	CO_2	3.4×10^{-2}
OH	39	CH_3Cl	0.13
HO_2	690	CH_3Br	0.17
H_2O_2	8.4×10^4	CH_3I	0.20
NH_3	60	HCHO	3.2×10^3
NO	1.9×10^{-3}	CH_3CHO	13
NO_2	1.2×10^{-2}	CH_3OH	200
NO_3	3.8×10^{-2}	CH_3OOH	300
HNO_2 (HONO)	49[b]	$HOCH_2OOH$	1.7×10^6
HNO_3 ($HONO_2$)	2.1×10^5[b]	$CH_3C(O)OOH$	840
SO_2	1.4	CH_3COCH_3	28
H_2S	0.10	HCOOH	8.9×10^3
CH_3SCH_3	0.54	$CH_3C(O)OH$	4.1×10^3
HCl	1.1[b]	CH_3ONO_2	2.0[b]
HOBr	$>1.3 \times 10^2$	$CH_3C(O)O_2NO_2$ (PAN)	2.9[c]

[a]Unless otherwise specified, NASA/JPL Evaluation No.17 (Sander et al. 2011)
[b]Seinfeld and Pandis (2006)
[c]Pandis and Seinfeld (1989)

The [X(aq)] and p_X in the above equation are the concentration of X in the aqueous solution and the partial pressure of X in the gas phase, respectively. Usually, the Henry′s law coefficient is expressed in the unit of [mole l^{-1} atm^{-1}]. If the molar concentration mole l^{-1} is expressed by M, the unit is [M atm^{-1}], and Table 2.6 tabulates the Henry′s law coefficients in this unit for important molecules in the atmosphere. When the concentration of the molecules in the atmosphere is expressed not by partial pressure but by molar concentration [X(g)] in the same unit in the liquid phase, the dimensionless Henry′s law coefficient, $\hat{K}_H$ is defined by

$$\frac{[X(aq)]}{[X(g)]} = \frac{N_{aq}}{N_g} = K_H RT = \hat{K}_H. \tag{2.93}$$

The conversion from K_H to $\hat{K}_H$ is obtained by multiplying RT to K_H as shown in the above equation where R is the gas constant (Table 2.1) and T(K) is the temperature.

The Henry's law coefficient is temperature dependent, and the dependence is given by van't Hoff equation, which is as follows,

$$\frac{d\ln K_H}{dT} = \frac{\Delta H_{A,298}}{RT^2} \tag{2.94}$$

where $\Delta H_{A,298}$ (kJ mole^{-1}) is the enthalpy change (the heat of dissolution) when the dissolution process (2.91) proceeds from left to right. Table 2.7 tabulates the values

Table 2.7 Heat of dissolution of atmospheric molecules for water (298 K)

Chemical species	ΔH_A (kJ mol^{-1})	Chemical species	ΔH_A (kJ mol^{-1})
O_3	−21.1	CO_2	−20.3
H_2O_2	−60.7	HCHO	−53.6
NH_3	−34.2	CH_3OH	−40.6
NO	−12.1	CH_3OOH	−46.4
NO_2	−20.9	$CH_3C(O)OOH$	−51.0
HNO_2 (HONO)	−39.7	HCOOH	−47.7
SO_2	−26.2	$CH_3C(O)O_2NO_2$ (PAN)	−49.0
HCl	−16.7		

Source:Pandis and Seinfeld (1989)

of $\Delta H_{A,298}$ for typical atmospheric molecules. As shown in the table, since enthalpy change associated with dissolution is negative in general, the Henry′s law coefficients increases with the decrease of temperature. Namely, the solubility of gaseous molecule to aqueous solution increases with the decrease of temperature. Since the temperature dependence of ΔH_A is negligible if the temperature change is not very large, the Henry′s law coefficients at different temperatures can be obtained from Eq. 2.94 by

$$\ln \frac{K_H(T_2)}{K_H(T_1)} = \frac{\Delta H_{A,298}}{R} \left(\frac{1}{T_1} - \frac{1}{T_2} \right) \tag{2.95}$$

using $\Delta H_{A,298}$ values given in Table 2.7.

2.4.3 *Diffusion and Reactions in the Liquid Phase*

Here, the diffusion process is considered after a molecule X is taken into the liquid phase. Assuming the liquid phase is the aqueous solution, and the diffusion of dissolved molecules occurs in one-dimension, the process is expressed by the one-dimensional diffusion equation,

$$\frac{\partial N_{aq}}{\partial t} = D_{aq} \frac{\partial^2 N_{aq}}{\partial x^2}, \tag{2.96}$$

where x is the axis distance from the interface along the depth of solution, N_{aq} is the concentration of molecules X (molecules cm^{-3}), and D_{aq} is the diffusion coefficient of the molecules in the aqueous solution. Since the diffusion equation is the partial differential equation of the first order with time and second order with space, one initial condition with regard to N_{aq} and two boundary conditions with regard to N_{aq} in a certain point of the space is necessary. Here, we take $N_{aq} = N_{aq,bulk}$ as an initial condition at $t = 0$ and $x > 0$, and $N_{aq} = N_{aq,int}$ at x = 0 (interface) regardless of time

t as the first boundary condition (gas-liquid equilibrium is established very fast and the liquid interface concentration is always at the equilibrium concentration determined by the Henry's law coefficients), and $N_{aq} = N_{aq,bulk}$ at $x = \infty$ regardless of time t as the second boundary condition (deep inner part of the liquid droplet, the concentration of X does not change from the initial concentration). Solving the diffusion Eq. (2.96) under these conditions, the rate of molecules passing through the unit area of liquid phase after time t has elapsed, $J_{sol}(t)$ (molecules cm^{-2} s^{-1}), can be obtained as

$$J_{sol}(t) = \left(N_{aq,\mathrm{int}} - N_{aq,bulk}\right)\sqrt{\frac{D_{aq}}{\pi t}}. \tag{2.97}$$

As expected, the diffusion rate is dependent on the difference of the concentrations near the gas-liquid interface and in the bulk liquid. Also, the diffusion rate decreases in inverse proportion to the square root of time. This is because the number of re-evaporating molecules from gas-liquid surface to the gas phase increases with time.

Here, putting $N_{aq,bulk} = 0$ at $t = 0$, the above equation becomes

$$J_{sol}(t) = N_{aq,\mathrm{int}}\sqrt{\frac{D_{aq}}{\pi t}}. \tag{2.98}$$

Assuming the gas-liquid equilibrium,

$$N_{aq,\mathrm{int}} = N_g \hat{K}_H, \tag{2.99}$$

Eqs. (2.98) and (2.99) gives,

$$J_{sol}(t) = N_g \hat{K}_H \sqrt{\frac{D_{aq}}{\pi t}}. \tag{2.100}$$

By normalizing $J_{sol}(t)$ with the flux J_{col}, the gaseous molecules collide at the interface in a unit surface area and unit time (Eq. 2.79), the diffusion conductance Γ_{sol} in the liquid phase in Eq. 2.84 is given by

$$\Gamma_{sol}(t) = \frac{J_{sol}(t)}{J_{col}} = \frac{4\hat{K}_H}{u_{av}}\sqrt{\frac{D_{aq}}{\pi t}}. \tag{2.101}$$

As shown in the above equation, Γ_{sol} decreases with time, reflecting the re-evaporation process from the liquid to gas phase. Therefore, after enough time has elapsed ($t \to \infty$), uptake rate and re-evaporation rate gets equal to reach the gas-liquid equilibrium and $\Gamma_{sol} \to 0$. Meanwhile, when the liquid particles are very small, and the interface layer forms bulk layer, gas-liquid equilibrium is completed instantaneously, and Eq. (2.98) does not apply.

When the molecule X is consumed irreversibly by the chemical reactions in the liquid phase, putting the pseudo-first order reaction rate constant is k_{aq} (s^{-1}), the diffusion equation, Eq. (2.93) becomes,

$$\frac{\partial N_{aq}}{\partial t} = D_{aq}\frac{\partial^2 N_{aq}}{\partial t^2} - k_{aq}N_{aq}. \tag{2.102}$$

Solving this equation under the same boundary conditions as mentioned above, assuming that the Henry's law equilibrium is established and that $N_{aq,bulk} = 0$ and kt $\gg$ 1, the flux of molecules dissipating by the reaction in a unit time and unit area (molecules $cm^{-2}\ s^{-1}$), is given by

$$J_{rxn} = N_{aq,\mathrm{int}}\sqrt{D_{aq}k_{aq}}. \tag{2.103}$$

Assuming the Henry equilibrium at the gas-liquid interface,

$$J_{rxn} = N_g\hat{K}_H\sqrt{D_{aq}k_{aq}}. \tag{2.104}$$

By normalizing with J_{col} given in Eq. (2.88), the conductance of the reactions in the liquid phase is

$$\Gamma_{rxn} = \frac{J_{rxn}}{J_{col}} = \frac{4\hat{K}_H}{u_{av}}\sqrt{D_{aq}k_{aq}}. \tag{2.105}$$

This equation is applicable for the irreversible reactions and when the solubility is large enough.

2.5 Literatures

Among the textbooks on atmospheric chemistry, relatively detailed descriptions on chemical reactions are given in the followings:

- Brasseur, G. P and S. Solomon, Aeronomy of the Middle Atmosphere: Chemistry and Physics of the Stratosphere and Mesosphere, 3rd ed., 644 pp, Springer, Dordrecht, the Netherland 2005.
- Finlayson-Pitts, B. J. and J. N. Pitts, Jr., Chemistry of the Upper and Lower Atmosphere, 969 pp, Academic Press, San Diego 2000.
- Wayne, R., Chemistry of Atmospheres, 3rd ed., 775 pp, Oxford University Press, New York 2000.

For deeper understanding on chemical reactions, the following textbooks would be useful:

on photochemistry,

- Turro, N. J., V. Ramamurthy and J. C. Scaiano, Principles of Molecular Photochemistry: An Introduction, 530 pp, University Science Books, Herndon, VA 2008, Wardle, B., Principles and Applications of Photochemistry, 250 pp., John Wiley and Sons, Sussex, UK 2009;

on molecular spectroscopy,

- Harris, D. C., Symmetry and Spectroscopy: An Introduction to Vibrational and Electronic Spectroscopy, 576 pp, Dover Books on Chemistry, New York 1989,
- Obi, K., Fundamentals of Spectroscopic Measurements, 176 pp, Kodansha, Tokyo 2009 (in Japanese);

on reaction kinetics,

- Leidler, K. J., Chemical Kinetics (3rd Edition), 531 pp, Harper Collins Publishers, New York 1987,
- Steinfeld, J. I., J. S. Francisco and W. L. Hase, Chemical Kinetics and Dynamics, 548 pp, Prentice Hall, Upper Saddle River, NJ 1989.
- Houston, P. L., Chemical Kinetics and Reaction Dynamics, 352 pp, Dover Books on Chemistry, New York 2006,
- Hirata Y. and M. Kawasaki, Chemical Reactions, 224 pp, Iwanami Shoten, Tokyo 2007 (in Japanese),
- Koda S., M. M. Kotani, K. Someno, and K. Awaga, Eds., Graduate Lecture, Physical Chemistry II, Reaction Kinetics and Dynamics, 285 pp, Tokyo Kagaku Dojin, Tokyo 2011 (in Japanese),

on multiphase chemistry,

- Pöschl, U., Y. Rudich and M. Ammann, Kinetic model framework for aerosol and cloud surface chemistry and gas-particle interactions – Part 1: General equations, parameters, and terminology, Atmos. Chem. Phys., 7, 5989–6023, 2007.
- Seinfeld, J. H. and S. N. Pandis, Atmospheric Chemistry and Physics: From Air Pollution to Climate Change, 2nd ed., 1203 pp, John Wiley and Sons, Hoboken, NJ 2006

References

Maric, D., Burrows, J.P., Meller, R., Moortgat, G.K.: A study of the UV—visible absorption spectrum of molecular chlorine. J. Photochem. Photobiol. A Chem. **70**, 205–214 (1993)

Myer, J.A., Samson, J.A.R.: Vacuum ultraviolet absorption cross sections of CO, HCl, and ICN between 1050 and 2100 Å. J. Chem. Phys. **52**, 266–271 (1970)

Saiz-Lopez, A., Saunders, R.W., Joseph, D.M., Ashworth, S.H., Plane, J.M.C.: Absolute absorption cross-section and photolysis rate of I_2. Atmos. Chem. Phys. **4**, 1443–1450 (2004)

Sander, S.P., Baker, R., Golden, D.M., Kurylo, M.J., Wine, P.H., Abatt, J.P.D., Burkholder, J.B., Kolb, C.E., Moortgat, G.K., Huie, R.E., Orkin, V.L.: Chemical Kinetics and Photochemical Data for Use in Atmospheric Studies, Evaluation Number 17, JPL Publication 10–6, Pasadena (2011)

Troe, J. Predictive possibilities of unimolecular rate theory. J. Phys. Chem. **83**, 114–126 (1979)

Chapter 3
Fundamentals of Atmospheric Photochemistry

The prime driver of the chemical system of the earth's atmosphere is photochemical reactions caused by solar radiation. The atmosphere of the earth is divided into levels called the troposphere, stratosphere, mesosphere and thermosphere from nearest the ground to farthest according to the characteristics of the temperature gradient as shown in Fig. 3.1. The cause of the temperature inversion in the stratosphere, which characterizes the earth's atmosphere, is the formation of an ozone layer by the photolysis of oxygen, one of the major components of the atmosphere. In this chapter, the spectrum of solar radiation, actinic flux, and so on, that is necessary to calculate the photolysis rate of atmospheric molecules are explained.

3.1 Extraterrestrial Solar Spectrum

In order to discuss the photochemical reactions in the earth's atmosphere, it is necessary to calculate quantitatively how the solar spectrum changes in reaching from outer space to the Earth's surface. For this purpose, let us first look at the extraterrestrial solar spectrum before it is affected by chemical species in the earth's atmosphere. The measurements of the solar radiant intensity outside of the earth's atmosphere are made by satellites for the shorter wavelength of the ultraviolet region, and by ground-based observations for longer wavelengths of the ultraviolet and visible region. Recent data is compiled by 2000 ASTM E490 database (Standard Extraterrestrial Spectrum Reference) by ASTM (American Society for Testing and Materials) (2006). This database covers from 120 nm in the far ultraviolet region to the infrared region with a wavelength longer than 10 μm. It is based on the compilation of the observational data by several satellite sensors by Woods et al. (1996) for the wavelengths shorter than 330 nm, and those by ground-based observational data by Neckel and Labs (1984) for the longer wavelengths. Further, Wehrli Standard Extraterrestrial Solar Irradiance Spectrum by WMO/WRDC

H. Akimoto, *Atmospheric Reaction Chemistry*, Springer Atmospheric Sciences,
DOI 10.1007/978-4-431-55870-5_3

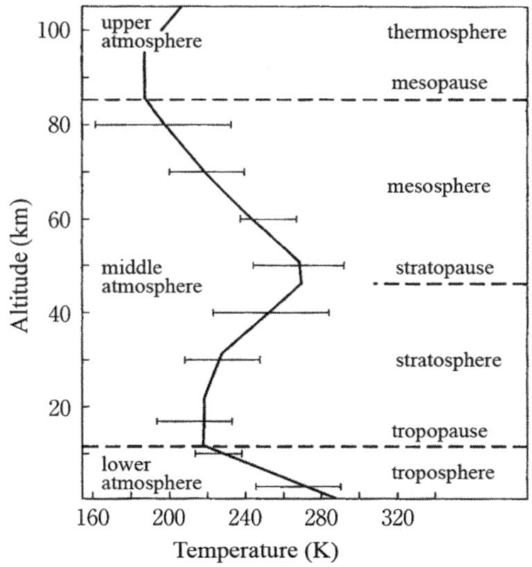

Fig. 3.1 Vertical temperature profile of the atmosphere base on US Standard Atmosphere (Adapted from Goody 1995 based on NOAA, 1976)

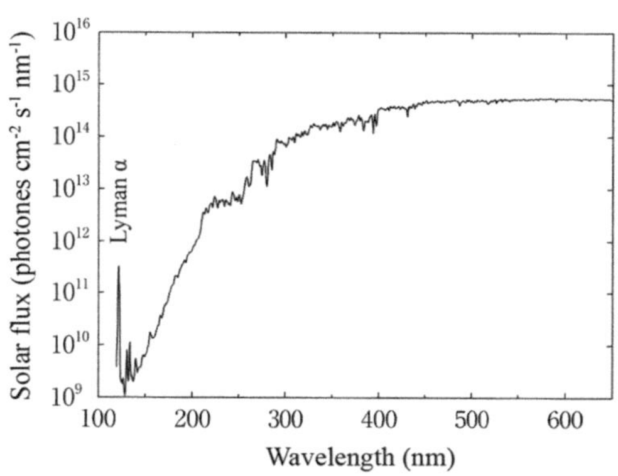

Fig. 3.2 Extraterrestrial solar flux (irradiance) (base on ASTM International, 2006)

(World Radiation Data Center) (1985) has been widely used, which covers a wavelength region of 200 nm–10 μm.

The spectral distribution of extraterrestrial standard solar irradiance (photons $cm^{-2}\ s^{-1}\ nm^{-1}$) at the perpendicular surface to the sun is given in Fig. 3.2 with spectral resolution of 1 nm, and the numerical data in 5 nm intervals is given in Table 3.1. Here, the unit of the solar irradiance given in the figure and the table is photon per unit area, unit time and per given wavelength interval. Although the absorption lines called Fraunhofer lines originating from various atoms can be seen, solar irradiation can be basically treated as a continuous spectrum at wavelengths longer than 200 nm from the viewpoint of atmospheric photochemistry. At wavelengths shorter than 200 nm, line spectrum from the sun cannot be neglected, and the most important is the Lyman-α line at 121.6 nm corresponding to the $2P \rightarrow 1S$ transition of a hydrogen atom, which has the intensity of ca. 3×10^{11}

Table 3.1 Extraterrestrial solar radiation flux [a, b] (average sun-earth distance) (per 5 nm, base e)

Wavelength (nm)	Radiation intensity (photons cm^{-2} s^{-1} 5 nm^{-1})	Wavelength (nm)	Radiation intensity (photons cm^{-2} s^{-1} 5 nm^{-1})	Wavelength (nm)	Radiation intensity (photons cm^{-2} s^{-1} 5 nm^{-1})
120–125[c]	4.15 E+11	290–295	4.05 E+14	460–465	2.38 E+15
125–130	9.15 E+09	295–300	3.77 E+14	465–470	2.35 E+15
130–135	2.77 E+10	300–305	3.94 E+14	470–475	2.37 E+15
135–140	1.48 E+10	305–310	4.45 E+14	485–480	2.44 E+15
140–145	1.81 E+10	310–315	5.30 E+14	480–485	2.46 E+15
145–150	2.49 E+10	315–320	5.46 E+14	485–490	2.25 E+15
150–155	4.85 E+10	320–325	5.87 E+14	490–495	2.42 E+15
155–160	7.25 E+10	325–330	7.89 E+14	495–500	2.45 E+15
160–165	1.02 E+11	330–335	8.07 E+14	500–505	2.37 E+15
165–170	1.91 E+11	335–340	7.60 E+14	505–510	2.48 E+15
170–175	3.25 E+11	340–345	8.06 E+14	510–515	2.46 E+15
175–180	6.09 E+11	345–350	7.97 E+14	515–520	2.29 E+15
180–185	9.76 E+11	350–355	9.02 E+14	520–525	2.48 E+15
185–190	1.41 E+12	355–360	7.99 E+14	525–530	2.46 E+15
190–195	2.04 E+12	360–365	8.87 E+14	530–535	2.54 E+15
195–200	3.03 E+12	365–370	1.09 E+15	535–540	2.57 E+15
200–205	4.39 E+12	370–375	9.73 E+14	540–545	2.52 E+15
205–210	7.39 E+12	375–380	1.11 E+15	545–550	2.58 E+15
210–215	1.78 E+13	380–385	9.09 E+14	550–555	2.60 E+15
215–220	2.14 E+13	285–390	1.01 E+15	555–560	2.57 E+15
220–225	2.90 E+13	390–395	1.03 E+15	560–565	2.62 E+15
225–230	2.65 E+13	395–400	1.26 E+15	565–570	2.62 E+15
230–235	2.81 E+13	400–405	1.73 E+15	570–575	2.66 E+15
235–240	2.81 E+13	405–410	1.71 E+15	575–580	2.66 E+15
240–245	3.53 E+13	410–415	1.79 E+15	580–585	2.72 E+15
245–250	3.21 E+13	415–420	1.81 E+15	585–590	2.61 E+15
250–255	3.29 E+13	420–425	1.85 E+15	590–595	2.68 E+15
255–260	6.95 E+13	425–430	1.73 E+15	595–600	2.68 E+15
260–265	9.18 E+13	430–435	1.71 E+15	600–605	2.66 E+15
265–270	1.66 E+14	435–440	1.95 E+15	605–610	2.68 E+15
270–275	1.40 E+14	440–445	2.12 E+15	610–615	2.64 E+15
275–280	1.25 E+14	445–450	2.21 E+15	615–620	2.63 E+15
280–285	1.63 E+14	450–455	2.31 E+15	620–625	2.65 E+15
285–290	2.34 E+14	455–460	2.35 E+15	625–630	2.65 E+15

[a]Calculated based on ASTM International, Standard Solar Constant and Zero Air Mass Solar Spectral Irradiance Tables, E490-001, 2006

[b]Eleven-year cycle variation (<20 %) due to the solar activity is observed for the radiation intensity in the wavelength range of 120–300 nm (Brasseur and Simon 1981; Brasseur et al. 1999)

[c]Mainly contribution of 121.6 nm Lyman-α line (±30 % fluctuation). (Timothy and Timothy 1970)

(photons $cm^{-2}\ s^{-1}$) as seen in Fig. 3.2. Solar radiation can be approximated by blackbody radiation at 5900 K at the wavelength longer than 400 nm, but the corresponding black body temperature gets lower at wavelengths shorter than 400 nm.

3.2 Attenuation of Solar Irradiance in the Atmosphere by N_2, O_2, and O_3

After entering into the earth's atmosphere, solar radiation is absorbed by N_2 and O_2, the major component of the atmosphere. The bond energy of N_2, D_0 (N-N) is large, 9.76 eV corresponding to the photon energy at the wavelength of 127.9 nm. Figure 3.3 shows the potential energy diagram of molecular N_2. The N_2 molecule has Lyman-Birge-Hopfield bands (L-B-H bands) corresponding to the $a^1\Pi_g - X^1\Sigma_g^+$ transition in the wavelength region of 100–150 nm, which has a vibrational structure. The transition of L-B-H bands with peaks at 135.4 nm is electric dipole forbidden and the absorption cross section is smaller than 4×10^{-21} cm^2 (Lofthus and Krupenie 1977). Because a much stronger absorption spectrum of O_2 exists in this wavelength region, the absorption of N_2 does not actually affect the solar irradiance. The absorption to the lower energy level of N_2 $A^3\Sigma_u^+ - X^1\Sigma_g^+$ is the spin forbidden transition, and the corresponding Vegard-Kaplan bands (V-K bands) is observed in aurora, but can be neglected when calculating solar irradiance. The strong absorption bands of N_2 is in the wavelength region shorter than 100 nm, and a band spectrum and continuum exist in regions from 66 to 100 nm, and shorter than 66 nm, respectively. These are important for photochemistry in the thermosphere,

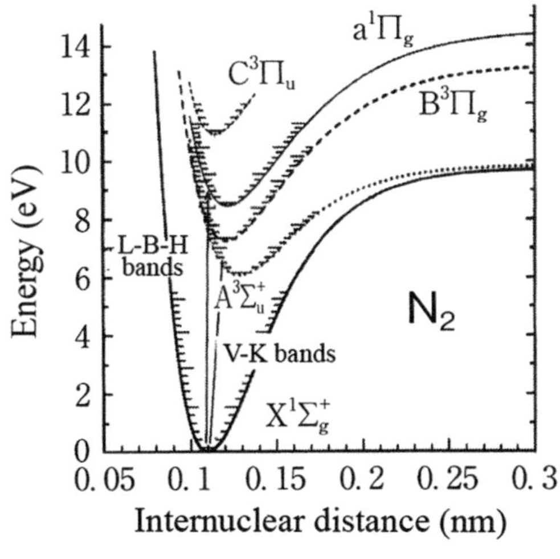

Fig. 3.3 Potential energy curves of N_2 (Adapted from Lofthus and Krupenie 1977)

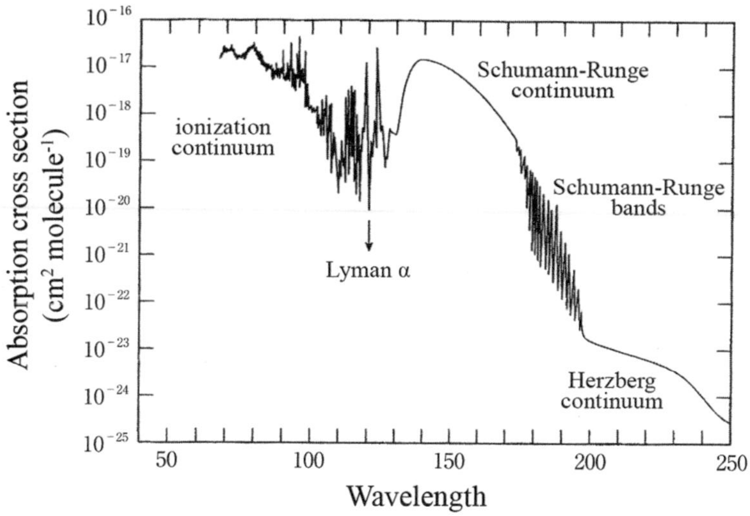

Fig. 3.4 Absorption spectrum of O_2 (Adapted from Goody 1995)

100–300 km above the ground, but the absorption and photolysis of N_2 will not be discussed in this book since it focuses on the troposphere and the stratosphere.

Absorption spectrum, cross sections and energy diagram of molecular O_2, the second major component of the earth's atmosphere, are shown in Fig. 3.4, Table 3.2, and Fig. 3.5, respectively. As shown in Fig. 3.4, O_2 has a strong continuum called the Schumann-Runge Continuum (S-R Continuum) in the 130–175 nm region, which corresponds to the allowed $B^3\Sigma_u^- - X^3\Sigma_g^-$ transition in Fig. 3.5. This continuum merges to the banded spectrum of the Schumann-Runge bands (S-R bands), in the wavelength range longer than 175 nm. This absorption band corresponds to the vibrational levels of the same $B^3\Sigma_u^- - X^3\Sigma_g^-$ transition lower than the dissociation energy to $O_2 \rightarrow O(^3P) + O(^1D)$. In a strict sense, the Schumann-Runge bands overlaps with the continuum, which corresponds to the $^3\Pi_u - X^3\Sigma_g^-$ transition to the repulsive potential curve in Fig. 3.5. As shown in Fig. 3.4, a very weak continuum called Herzberg bands are extended to the further longer wavelengths side of the Schumann-Runge bands in the 190–242 nm range, which corresponds to the forbidden transition $A^3\Sigma_u^+ - X^3\Sigma_g^-$ (H-bands in Fig. 3.5). This continuum is thought to be the transition to the $A^3\Sigma_u^+$ state at higher than the energy of dissociation to $O_2 \rightarrow O(^3P) + O(^3P)$. Absorption of solar radiation by these bands of O_2 is very important when considering the irradiance and chemical reaction system in the stratosphere. The solar radiation in the absorption spectrum range of O_2 is totally absorbed in the stratosphere and does not reach to the troposphere. Further transitions to the lower energy states of O_2, $a^1\Delta_g$ and $b^1\Sigma_g^+$, gives absorption in the near infrared and visible region at 1270, and 762 nm (0–0

Table 3.2 Absorption cross sections of O_2 (205–245 nm) (base e)

Wavelength (nm)	10^{20} σ (cm^2 $molecule^{-1}$)	Wavelength (nm)	10^{20} σ (cm^2 $molecule^{-1}$)	Wavelength (nm)	10^{20} σ (cm^2 $molecule^{-1}$)
205	7.35	220	4.46	235	1.63
206	7.13	221	4.26	236	1.48
207	7.05	222	4.09	237	1.34.
208	6.86	223	3.89	238	1.22
209	6.68	224	3.67	239	1.10
210	6.51	225	3.45	240	1.01
211	6.24	226	3.21	241	0.88
212	6.05	227	2.98	242	0.81
213	5.89	228	2.77	243	0.39
214	5.72	229	2.63	244	0.13
215	5.59	230	2.43.	245	0.05
216	5.35	231	2.25		
217	5.13	232	2.10		
218	4.88	233	1.94		
219	4.64	234	1.78		

Source: NASA/JPL Panel Evaluation No. 17 (Sander et al. 2011)

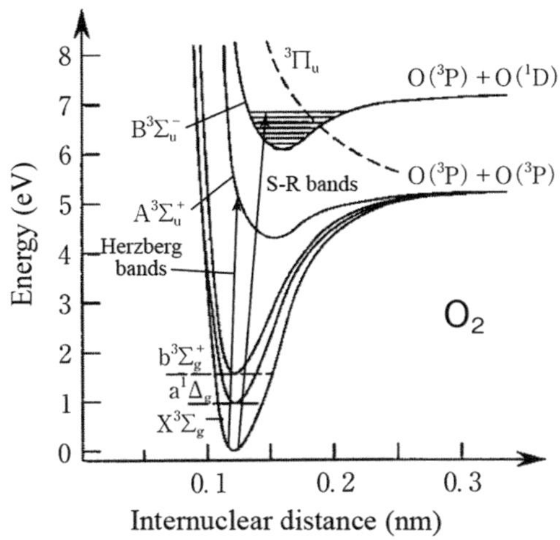

Fig. 3.5 Potential energy curves of O_2 (Adapted from Finlayson-Pitts and Pitts 2000 based on Gaydon 1968)

bands from v" = 0 to v' = 0), respectively. They are both forbidden transitions and very weak.

The photolysis of O_2 at the Schumann-Runge system is an important chemical process to form the ozone layer in the stratosphere. The chemistry of the stratosphere will be discussed in Chap. 8. The absorption spectrum of O_3 formed there plays an important role to determine the vertical distribution of the solar spectrum at different altitude and also the shorter wavelength edge in the troposphere.

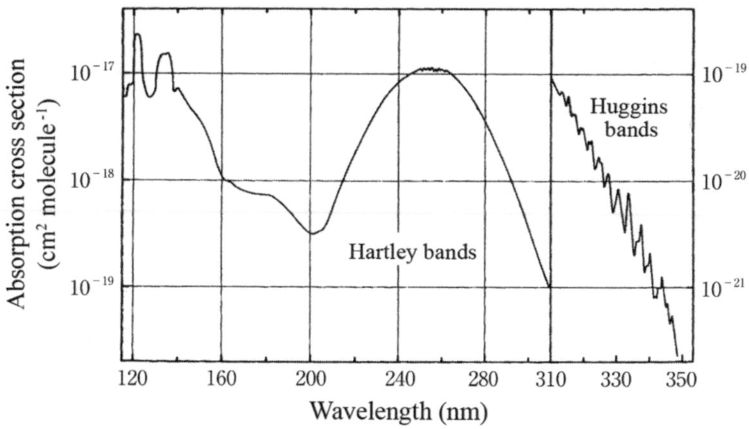

Fig. 3.6 Absorption spectrum of O_2 in the ultraviolet region (Adapted from Warneck 1988)

The absorption of O_3 in the stratosphere is very important to calculate the solar irradiance in the troposphere. Figure 3.6 and Table 4.1 give the absorption spectrum and absorption cross sections of O_3. As shown in Fig. 3.6, O_3 has strong absorption bands in 100–200 and 200–310 nm ranges. Since the stronger bands of O_2 completely absorb the solar radiation in the 100–200 nm region, as an element for determining the solar irradiance spectrum, the absorption of O_3 in this region is not important. In contrast, the absorption bands in the 200–310 nm range called the Hartley bands correspond to the spectral region where the absorption of O_2 becomes weak so that the absorption by O_3 plays a very important role in determining the solar spectrum in the lower stratosphere and the troposphere. This absorption corresponds to the $^1B_2$-X^1A_2 transition which is the transition to the higher energy level than the dissociation into $O_3 \rightarrow O(^1D) + O_2(^1\Delta_g)$ to essentially give a continuum (see Fig. 4.8).

As seen in Fig. 3.6, the long wavelength edge of Hartley bands longer than 310 nm merges to a band spectrum called Huggins bands with a vibrational structure. The photolysis of O_3 in the 310–350 nm range giving an exited oxygen atom $O(^1D)$ is extremely important in the troposphere as the source of OH radicals, and will be discussed in detail in Sect. 4.2.1. Furthermore, O_3 has absorption bands called the Chappuis bands in visible region (440–850 nm) that is a weak continuum with an absorption cross section of 5×10^{-21} cm^2 molecule^{-1} (see Fig. 4.2), but the effect on the solar spectral irradiance is almost negligible.

The intensity of the solar radiation reached to the earth's atmosphere is attenuated according to altitude by the absorption of N_2, O_2 and O_3 as mentioned above. Figure 3.7 depicts the plot of the altitude at which the intensity of the solar radiation by the overhead sun is attenuated 90 % by the absorption of these species against wavelength. As seen in the figure, the solar radiation shorter than 100 nm and 130–180 nm are absorbed by N_2 and O_2 at higher than 80 km above the ground in the thermosphere, and the Lyman α line (121.6 nm) and the surroundings and the longer wavelength radiation reaches to the mesosphere, 45–80 km above the

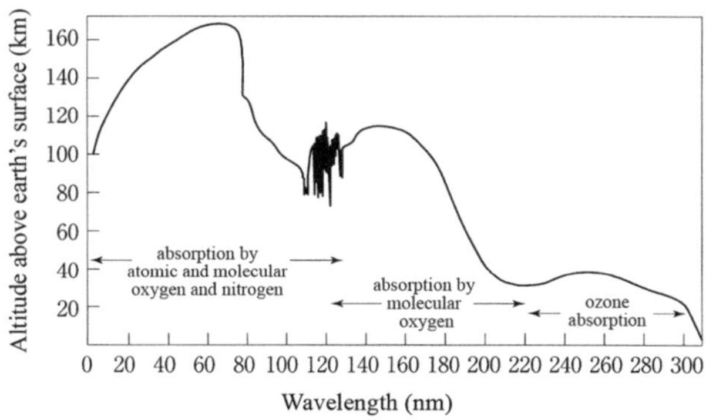

Fig. 3.7 Penetrating Depth of solar ultraviolet light into the earth's atmosphere. The height where the absorption of the solar light perpendicular to the earth surface by each atmospheric constituents is in the maximum) (Adapted from Finlayson-Pitts and Pitts 2000 based on Friedman 1960)

ground. Solar radiation longer than 185 nm reached to the stratosphere (15–45 km). Especially, solar radiation at around 200 nm reaches deep into the stratosphere, and this fact plays an important role in ozone depletion by anthropogenic chlorofluorocarbons (CFC) as will be discussed in Chap. 8. Due to the absorption by O_3 in the stratosphere, the solar radiation at wavelengths shorter than 310 nm are almost attenuated before reaching the troposphere. However, a small fraction of 295–310 nm radiation still reaches to the troposphere, and the formation of electronically excited $O(^1D)$ atoms in the photolysis of O_3 in this wavelength's region, and its subsequent reaction with water vapor give an important effect on the production of OH radicals in the troposphere.

3.3 Solar Zenith Angle and Air Mass

When the attenuation of solar radiation penetrating to the earth's atmosphere is caused by the absorption of O_2 and O_3, the solar irradiance at the altitude z_0 and wavelength λ on the surface perpendicular to the sun is expressed according to the Beer-Lambert law,

$$I(\lambda, z_0) = I_0(\lambda)\exp\left\{-\int_{z_0}^{\infty}\left[\sum_k \sigma_k(\lambda)n_k(z)\right]dz\right\} \tag{3.1}$$

$$\sum_k \sigma_k(\lambda)n_k(z) = \sigma_{o_2}(\lambda)n_{o_2}(z) + \sigma_{o_3}(\lambda)n_{o3}(z). \tag{3.2}$$

Here, I_0 is the extraterrestrial solar irradiance, σ_k and n_k are absorption cross section and molecular density of each molecular species, respectively.

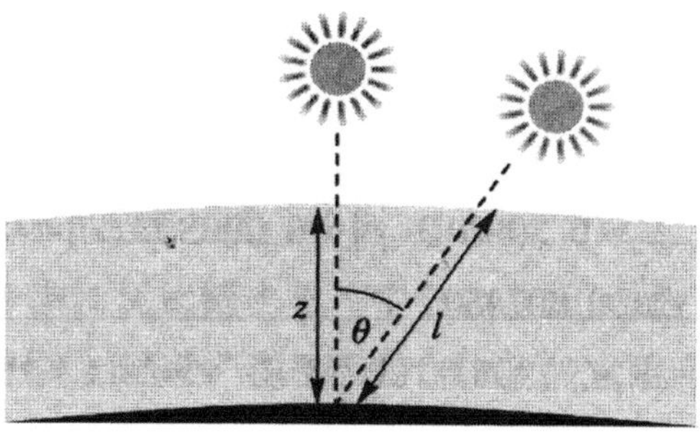

Fig. 3.8 The definition of solar zenith angle and air mass

Table 3.3 Air mass for different solar zenith angle

Solar Zenith Angle θ (°)	m = sec θ	Air mass
0	1.00	1.00
10	1.02	1.02
20	1.06	1.06
30	1.15	1.15
40	1.31	1.31
50	1.56	1.56
60	2.00	2.00
70	2.92	2.90
78	4.81	4.72
86	14.3	12.4

Source: Demerjian et al. (1980)

The distance l of a certain point on the earth surface from the sun gets longer with the solar zenith angle θ than the vertical distance z, as shown in Fig. 3.8, and more photons are absorbed by the molecules. Considering the horizontal earth surface, the relationship between l andθ is expressed as,

$$l = z/\cos\theta = \mathrm{z}\sec\theta. \tag{3.3}$$

The ratio of the distance l, the direct radiation of the sun passes through the atmosphere, and the vertical distance z from the sun is called air mass m. The air mass can be closely approximated for $\theta < 60°$, by

$$m = l/z = \sec\theta. \tag{3.4}$$

Table 3.3 shows the values of air mass comparing with those of sec θ. As seen in Table 3.3, the differences between them are not negligible only for the large solar zenith angle $\theta > 60°$, due to the curvature and refraction of the atmosphere. The solar irradiance for air mass m at the altitude z_0 from the earth's surface is given by

$$I(\lambda, z_0) = I_0(\lambda)\exp\left\{-m\int_{z_0}^{\infty}\left[\sum_k \sigma_k(\lambda)n_k(z)\right]dz\right\} \tag{3.5}$$

from Eq. (3.1).

The solar zenith angle θ necessary to calculate air mass in Eq. (3.4) can be obtained from the latitude φ (degree), solar declination δ (radian), and hour angle h (radian),

$$\cos\theta = \sin\delta\sin\varphi + \cos\delta\cos\varphi\cos h \tag{3.6}$$

where the solar declination δ is given (Vermote et al. 1997) by

$$\delta(\text{radian}) = \beta_1 - \beta_2\cos N + \beta_3\sin N - \beta_4\cos 2N + \beta_5\sin 2N - \beta_6\cos 3N + \beta_7\sin 3N \tag{3.7}$$

$$N(radian) = \frac{2\pi d_h}{365} \tag{3.8}$$

$\beta_1 = 0.006918, \beta_2 = 0.399912, \beta_3 = 0.070257, \beta_4 = 0.006758, \beta_5 = 0.000907, \beta_6 = 0.002697, \beta_7 = 0.001480.$

The d_n in Eq. (3.8) is the day number of a year with 0 on January 1st and 364 on December 31st. The hour angle h in Eq. (3.6) is obtained from Greenwich mean time GMT (hour), longitude λ (degree), and equation of time EQT (Vermote et al. 1997),

$$h(\text{radian}) = \pi[(\text{GMT}/12) - 1 + (\lambda/180)] + \text{EQT} \tag{3.9}$$

$$\text{EQT} = \alpha_1 + \alpha_2\cos N - \alpha_3\sin N - \alpha_4\cos 2N - \alpha_5\sin 2N \tag{3.10}$$

$\alpha_1 = 0.000075, \alpha_2 = 0.001868, \alpha_3 = 0.032077, \alpha_4 = 0.014615, \alpha_5 = 0.040849$

The equation of time is the difference of right ascension between the average and apparent sun, and caused by the fact that the movement of the sun in a day shifts east and west since the revolution angular velocity of the earth is different by season due to the elliptical orbit and the declination of the earth's axis from the celestial equator by 23°27′.

3.4 Scatter by Atmospheric Molecules and Particles, and Surface Albedo

The solar radiation penetrating the atmosphere changes its intensity and spectral distribution as affected by the light scattering by atmospheric molecules and particulate matters (aerosols) besides the absorption by N_2, O_2, and O_3. Particularly, for the calculation of the actinic flux (see next section) in the troposphere, these effects have to be considered.

Taking in to consideration light scattering in addition to absorption, the Lambert-Beer law can be expressed by

$$\ln\frac{I}{I_0} = -sm \tag{3.11}$$

$$s = s_{sm} + s_{am} + s_{sp} + s_{ap}, \tag{3.12}$$

where, s_{sm}, s_{am}, s_{sp}, and s_{ap} are optical attenuation coefficients by the scattering and absorption by gaseous molecules, and the scatter and absorption by atmospheric particulates, respectively, and m is the air mass mentioned in the previous section.

The scattering by gaseous molecules is called Rayleigh scattering, and its scattering coefficient is given by

$$s_{sm} = \frac{24\pi^3}{\lambda^4 N^2}\left(\frac{n^2 - 1}{n^2 + 1}\right) \tag{3.13}$$

where n is refractive index and N (molecules cm^{-3}) is the molecular density. Since the refractive index of air is close to unity (1.000278 at 530 nm), the equation can be approximated by

$$s_{sm} = \frac{32\pi^3}{3\lambda^4}\left(\frac{n - 1}{N}\right), \tag{3.14}$$

showing the Rayleigh scattering coefficient is inversely proportional to the fourth power of wavelength (Bohren and Huffman 1981; Ahrens 2007). In this way, the shorter wavelength's light is more easily scattered, and it is well known that the blue light of the solar radiation reaching to the troposphere is more strongly scattered, causing the blue sky. Using the atmospheric molecular density $N_O = 2.687 \times 10^{19}$ cm^{-3} at the scale height $H = 7.996 \times 10^5$ cm, approximately 8 km (the altitude where the atmospheric pressure is reduced to 1/e of the ocean surface; e is the base of natural logarithm), the Rayleigh scattering coefficient is

$$(s_{sm})_0 = \frac{1.044 \times 10^5 (m_{0,\lambda} - 1)^2}{\lambda^4} \tag{3.15}$$

where $m_{0,\ \lambda}$ is the refractive index of air for the wavelength λ at this altitude (Leighton 1961; Finlayson-Pitts and Pitts 2000). From Eq. (3.15), light attenuation coefficient by the molecular scattering s_{sm} in the atmosphere for each altitude can be given by

$$s_{sm} = \frac{(s_{sm})_0 N}{N_0} = (s_{sm})_0 \frac{P}{P_0} \tag{3.16}$$

where P_0 is the atmospheric pressure corresponding to the scale height, i.e. 1/e atm.

Detailed treatments of scattering and absorption by particles are given by Bohren and Huffman (1981). The scattering by particles is a function of particle diameter and light wavelength, and the light attenuation coefficient by particle scattering is in general expressed as

$$s_{sp} = \frac{b}{\lambda^a} \tag{3.17}$$

Here, if the parameter of particle diameter α is defined by

$$\alpha = \frac{2\pi r}{\lambda}, \tag{3.18}$$

a in Eq. (3.18) is $a = 4$ for small particle diameter ($\alpha \ll 1$), the same as the molecular scattering, and a decreases with the increase of the diameter reaching to $a = 0$ for a large diameter compare to the wavelength ($\alpha \gg 1$), i.e. the scatter is expressed by geometric optical approximation (Leighton 1961). The scattering by particles in the intermediate range is called Mie scattering.

In the case of scattering by molecules and particles, being different from the absorption, solar radiation is not lost but the scattered light is also utilized effectively for the photolysis of atmospheric molecules. Thus, although the direct radiation intensity is attenuated, multiple scattered lights contribute to actinic flux (Sect. 3.5).

Figure 3.9 shows plots of transmittance of light for each process against wavelengths in the troposphere considering molecular and particle scattering and absorption by ozone. The transmitting coefficients T in the figure are

$$T = \frac{I}{I_0} \tag{3.19}$$

$$T_{\text{total}} = T_a T_m T_p \tag{3.20}$$

$$T_s = T_m T_p \tag{3.21}$$

where T_a, T_m, T_p, T_s, T_{total} are the transmittance for the molecular absorption, molecular scattering, particle scattering, total scattering and total transmittance, respectively. In the calculation of Fig. 3.9, solar zenith angle 45°, ozone column density 0.300 atm-cm (=300 D.U.; D.U. means Dobson unit), aerosol optical density 0.295 (at 500 nm) are used. The general characteristics of wavelength dependence of each process is well demonstrated; particle scattering is important for wavelengths longer than 450 nm, molecular scattering (Rayleigh scattering) is

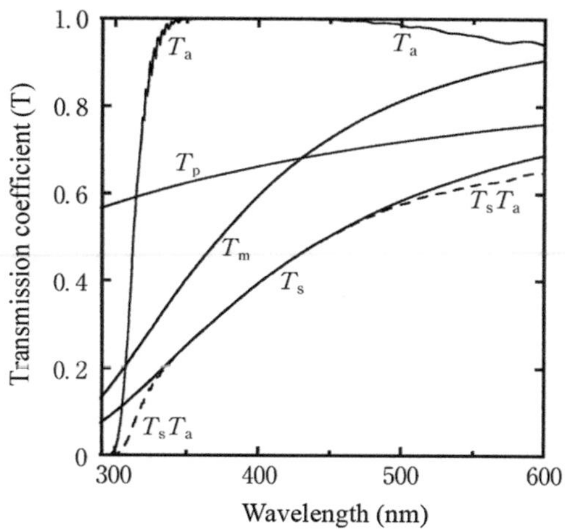

Fig. 3.9 Atmospheric transmittance at the solar zenith angle 45°. T_a: Absorption by ozone, T_m: Rayleigh scattering by molecules, T_p: Mie scattering by particles, $T_s = T_m + T_p$, $T_{total} = T_aT_s$ (Courtesy of H. Irie)

Table 3.4 Typical albedos for various earth surfaces

Surface	Albedo
Snow (old snow ~ new snow)	0.40–0.95
Ocean ice	0.30–0.40
Water surface ($\theta < 25°$)	0.03–0.10
Water surface ($\theta > 25°$)	0.10–1.00
Tundra	0.18–0.25
Sands, Desert	0.15–0.45
Soil (dark wet soil ~ bright dry soil)	0.05–0.40
Grassland	0.16–0.26
Cultivated land (agricultural crops)	0.05–0.20
Forest (broadleaf trees)	0.15–0.20
Forest (coniferous trees)	0.05–0.15
Thick cloud	0.60–0.90
Thin cloud	0.30–0.50

Source: Budikova (2010)

important for the shorter wavelengths and absorption by ozone predominates in the shorter wavelength than 320 nm.

In addition, since a portion of the solar radiation which reaches the earth surface is reflected back into the atmosphere, it is necessary to take the surface reflected light into consideration when calculating the actinic flux, in addition to the direct and scattered radiation. The reflectance at the earth surface is called Albedo and differs depending on the nature of the surface. Table 3.4 summarizes the reported values of Albedo for various earth surfaces. The highest Albedo is for the fresh snow, 0.75–0.95 close to unity. Meanwhile, the Albedos for water surfaces, such as the ocean, is very small, 0.1 for the small solar zenith angle but increasing with the zenith angle. Further Albedos of surfaces covered by vegetation, such as forest, grassland, and agricultural fields are in general middle ranged values of 0.1–0.3 and naturally changes by season. Although the values of Albedo depend on wavelength, details have not been studied much (McLinden et al. 1997).

3.5 Actinic Flux and Photolysis Rate Constants

Photolysis rate constants of atmospheric molecules can be calculated by the formula given in Chap. 2,

$$j_p = k_p = \int_\lambda \sigma(\lambda)\Phi(\lambda)F(\lambda)d\lambda \tag{2.18}$$

where $\sigma(\lambda)$ (cm^2) is the absorption cross section of a molecule to be photolyzed, $\Phi(\lambda)$ is the quantum yield for giving a specific photolysis product, and $F(\lambda)$ (photons cm^{-2} s^{-1} nm^{-1}) is the actinic flux integrated spherically. Actinic flux $F(\lambda)$ is the total number of photons incident to a small spherical surface in the atmosphere, which is obtained by integrating the solar radiance $L(\lambda,\theta,\phi)$ (photons cm^{-2} s^{-1} nm^{-1} sr^{-1}) for all angles (solar zenith angle θ, and azimuth angle ϕ) (Madronich 1987),

$$F(\lambda) = \int_0^{2\pi}\int_{-\pi/2}^{\pi/2} L(\lambda,\theta,\phi)\sin\theta d\theta d\phi, \tag{3.22}$$

where $\sin\theta$ is a coefficient appearing in the conversion of the solid angle to the spherical coordinate. It should be noted that the intervals of integral is for -π/2⁀π/2, which implies not only the radiation from the upper side but also the light from the lower side is also effective for photolysis.

Figure 3.10 illustrates the schematic diagram of various radiation types contributing to the actinic flux. Thus, for the calculation of $F(\lambda)$, all of the processes, such as absorption by stratospheric ozone, Rayleigh scattering by atmospheric molecules, scattering and absorption by aerosols and clouds, and reflection at the earth's surface have to be considered, and not only the direct radiation from the sun but also the light from all directions reflected and scattered by aerosols must also be accounted for.

Meanwhile, the solar irradiance $E(\lambda)$ (photons cm^{-2} s^{-1} nm^{-1}) mentioned in Sect. 3.1 is defined by

$$E(\lambda) = \int_0^{2\pi}\int_0^{\pi/2} L(\lambda,\theta,\phi)\cos\theta\sin\theta d\theta d\phi \tag{3.23}$$

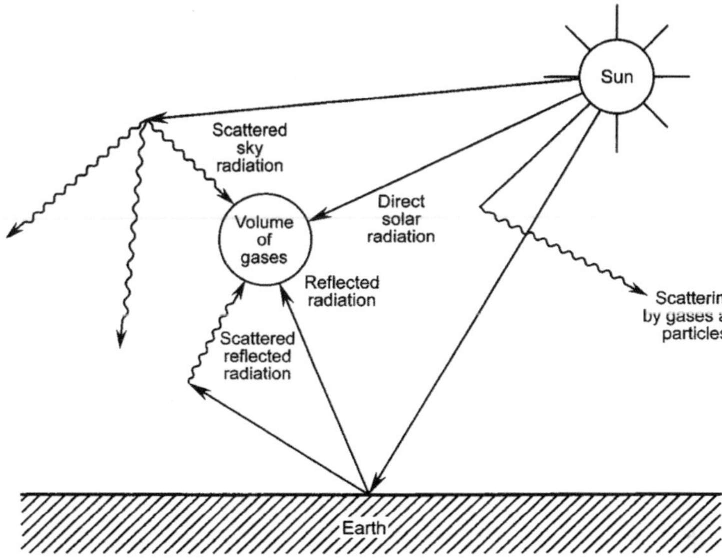

Fig. 3.10 Various sources of radiation contributing to the actinic flux for a small volume in the atmosphere (Adapted from Finlayson-Pitts and Pitts 2000)

which is the radiation flux incident upon a plane in the atmosphere. That is, $E(\lambda)$ is the hemispherically integrated direct and scattered solar radiation incident upon a plane in a fixed space orientation from the upper side, and differentiated from the actinic flux defined by Eq. (3.22) in the respect that it contains cos θ which depends on the direction of the irradiation. $E(\lambda)$ is often given in the energy unit, W cm^{-2} nm^{-1}, and it can be converted to photon per unit time by using Eq. (2.2). Incidentally, in the discussion of the extraterrestrial solar spectrum, the irradiance $E(\lambda)$ at the plane perpendicular to the sun (cos $\theta = 1$) is equal to the actinic flux $F(\lambda)$.

The actinic flux $F(\lambda)$ is an important parameter necessary for calculations of photolysis rate constants in the atmosphere, and there are two methods to obtain the values; one is the method obtained by the theoretical calculation using the radiative transfer model, and the other is the method to measure the irradiance $E(\lambda)$ by a radiometer and convert it to $F(\lambda)$ by using a theoretical formula. Here, we shall see the method of obtaining actinic flux by using the radiative transfer model equation first. The pioneering work on this treatment was undertaken by Leighton (1961), and he calculated the actinic flux applicable to photochemical air pollution by using simplified radiative transfer model for different solar zenith angles in the wavelength region of 290–800 nm. Although the work of Leighton (1961) is already more than 50 years old, with all the factors mentioned in this chapter taken into consideration, it can be concluded that the basic view on photolysis in the atmosphere was established at this time.

Peterson (1976) developed the work of Leighton (1961) using new input parameters, and was further succeeded by Demerjian (1980) and Madronich (1987) giving the values of actinic flux with higher precision. Peterson (1976) calculated the

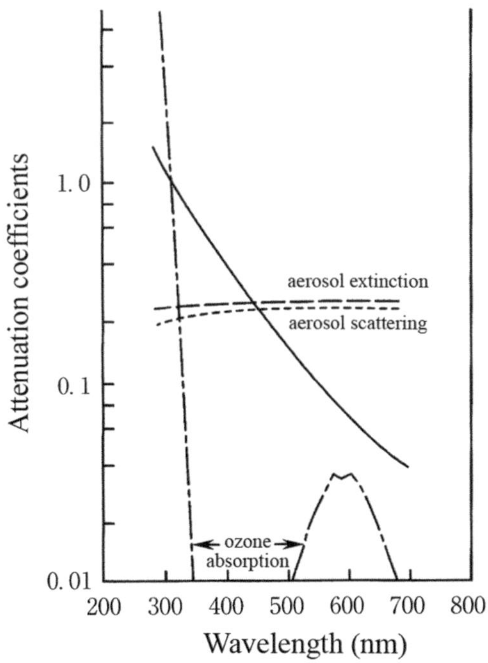

Fig. 3.11 Attenuation coefficients of solar radiation due to absorption and scattering by atmospheric molecules and particles (Adapted from Peterson 1976)

actinic flux at the zenith angle 0° dividing the spectral region by 5 nm (290–420 nm), 10 nm (420–580 nm) and 20 nm (580–700 nm), and assuming the amount of stratospheric ozone of 0.285 atm-cm (285 D.U.) and the boundary layer ozone of 100 ppb in the 0–1 km layer from the ground. As for the calculation of radiation characteristics of aerosols, assuming the distribution of particle diameters given in Eq. (3.18), column density of 4.99×10^7 particles cm^{-2} was used. The optical parameter of aerosols complex index of refraction, $n = 1.5\text{-}0.01i$, was used assuming partial absorption. Figure 3.11 shows the wavelength dependence of radiative extinction coefficient s defined by Eqs. (3.11) and (3.12) for absorption by ozone, scatter and absorption by particles, and Rayleigh scattering by atmospheric molecules using the above parameters. In Fig. 3.11, it is shown that the extinction by aerosol is mostly by scattering and the contribution of absorption is about 9 %, and the absorption by the Chappuis Bands at around 600 nm is less than 3 %.

It is known that in addition to these parameters, the surface albedo has a large effect on the actinic flux (Luther and Gelinas 1976). The contribution ratios of albedo are 5 % (290–400 nm), 6 % (400–450 nm), 8 % (450–500 nm), 10 % (500–550 nm), 11 % (550–600 nm), 12 % (600–640 nm), 13.5 % (640–660 nm), and 15 % (660–700 nm) according to Peterson (1976). Table 3.5 gives the values of actinic flux calculated by Peterson (1976). Figure 3.12 depicts the solar zenith angle dependence of the actinic flux for several wavelengths. As shown in Fig. 3.12, the decrease of the actinic flux is relatively small in the zenith angle range of 0–50°, but

Table 3.5 Actinic flux at the earth surface for different solar zenith angles (photons cm^{-2} s^{-1} 5 nm^{-1})

Wavelength (nm)	Power	Actinic Flux: Solar Zenith Angle (°)									
		0	10	20	30	40	50	60	70	78	86
290–295	14	0.001	0.001	–	–	–	–	–	–	–	–
295–300	14	0.041	0.038	0.030	0.019	0.009	0.003	–	–	–	–
300–305	14	0.398	0.381	0.331	0.255	0.167	0.084	0.027	0.004	0.001	–
305–310	14	1.41	1.37	1.25	1.05	0.800	0.513	0.244	0.064	0.011	0.002
310–315	14	3.14	3.10	2.91	2.58	2.13	1.56	0.922	0.357	0.090	0.009
315–320	14	4.35	4.31	4.10	3.74	3.21	2.52	1.67	0.793	0.264	0.030
320–325	14	5.48	5.41	5.19	4.80	4.23	3.43	2.43	1.29	0.502	0.073
325–330	14	7.89	7.79	7.51	7.01	6.27	5.21	3.83	2.17	0.928	0.167
330–335	14	8.35	8.25	7.98	7.50	6.76	5.72	4.30	2.54	1.15	0.241
335–340	14	8.24	8.16	7.91	7.46	6.78	5.79	4.43	2.69	1.25	0.282
340–345	14	8.89	8.80	8.54	8.09	7.38	6.36	4.93	3.04	1.44	0.333
345–350	14	8.87	8.79	8.54	8.11	7.43	6.44	5.04	3.15	1.51	0.352
350–355	14	10.05	9.96	9.70	9.22	8.48	7.39	5.83	3.69	1.77	0.414
355–360	14	9.26	9.18	8.94	8.52	7.86	6.88	5.47	3.50	1.69	0.391
360–365	14	10.25	10.16	9.91	9.46	8.76	7.71	6.17	3.99	1.94	0.444
365–370	15	1.26	1.25	1.22	1.17	1.08	0.958	0.772	0.505	0.247	0.055
370–375	15	1.14	1.13	1.10	1.06	0.983	0.873	0.708	0.467	0.230	0.051
375–380	15	1.27	1.26	1.23	1.18	1.10	0.983	0.802	0.535	0.265	0.058
380–385	15	1.05	1.04	1.02	0.980	0.917	0.820	0.673	0.453	0.226	0.049
385–390	15	1.15	1.15	1.12	1.08	1.01	0.909	0.750	0.510	0.257	0.054
390–395	15	1.19	1.18	1.16	1.11	1.05	0.943	0.783	0.537	0.273	0.057
395–400	15	1.44	1.43	1.40	1.35	1.28	1.15	0.962	0.666	0.341	0.070
400–405	15	1.73	1.72	1.69	1.63	1.53	1.39	1.16	0.809	0.418	0.085
405–410	15	1.94	1.93	1.90	1.83	1.17	1.57	1.32	0.926	0.482	0.097

(continued)

Table 3.5 (continued)

Wavelength (nm)	Power	Actinic Flux: Solar Zenith Angle (°)									
		0	10	20	30	40	50	60	70	78	86
410–415	15	2.05	2.04	2.00	1.93	1.83	1.66	1.41	0.993	0.522	0.104
415–420	15	2.08	2.07	2.03	1.96	1.86	1.70	1.44	1.03	0.543	0.107
420–430	15	4.08	4.06	3.99	3.87	3.67	3.36	2.87	2.07	1.11	0.216
430–440	15	4.20	4.18	4.11	3.99	3.80	3.49	3.01	2.19	1.20	0.229
440–450	15	4.87	4.85	4.77	4.64	4.43	4.09	3.54	2.61	1.45	0.272
450–460	15	5.55	5.51	5.43	5.27	5.03	4.64	4.02	2.99	1.67	0.312
460–470	15	5.68	5.65	5.57	5.42	5.17	4.79	4.17	3.12	1.77	0.325
470–480	15	5.82	5.79	5.70	5.55	5.31	4.91	4.32	3.26	1.87	0.341
480–490	15	5.78	5.75	5.67	5.53	5.29	4.93	4.33	3.29	1.90	0.339
490–500	15	5.79	5.76	5.68	5.54	5.31	4.96	4.37	3.34	1.95	0.344
500–510	15	5.99	5.96	5.87	5.71	5.47	5.09	4.47	3.41	1.99	0.340
510–520	15	5.88	5.86	5.77	5.62	5.38	5.02	4.43	3.40	2.00	0.340
520–530	15	5.98	5.95	5.87	5.72	5.48	5.11	4.52	3.47	2.04	0.336
530–540	15	5.98	5.95	5.87	5.72	5.48	5.12	4.52	3.48	2.05	0.326
540–550	15	5.88	5.85	5.77	5.62	5.40	5.04	4.46	3.44	2.03	0.317

Source: Peterson (1976)

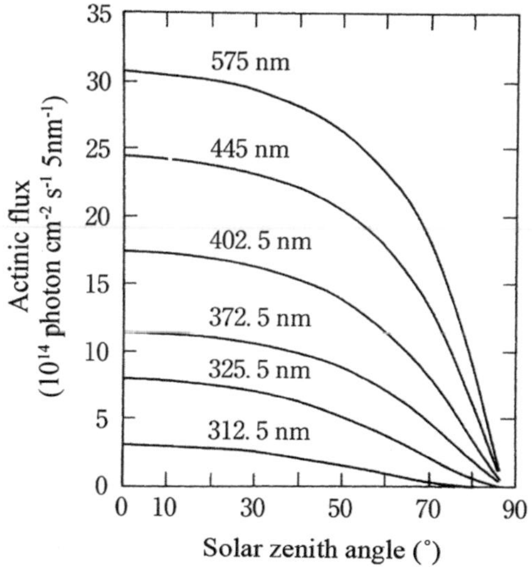

Fig. 3.12 Solar zenith angle dependence of the actinic flux at the earth surface (Adapted from Peterson 1976)

it deceases rapidly for 50–90°. This is because the air mass increases rapidly when the zenith angle increases over 50° as seen in Table 3.3. Further, the rapid increase of the actinic flux at the earth's surface in the spectral range of 300–400 nm corresponds to the rapid decrease of absorption of ozone with the increase of wavelengths. The values of actinic flux for higher altitudes at 15, 25, 40 km are given by Finlayson-Pitts and Pitts (2000).

The direct physical measurement of the spectral actinic flux $F(\lambda)$ is not easy, although attempts have been made (Shetter and Müller 1999; Hofzumahaus et al. 1999). Generally, irradiance $E(\lambda)$ (radiation flux per unit area, W m^{-2} nm^{-1}) is measured by radiometers, and experiments to compare solar spectral intensity in the field with radiative transfer models have been made in order to convert the spectral irradiance $E(\lambda)$ to $F(\lambda)$. In these analyses, downward actinic flux $F_d(\lambda)$ is obtained by upper-hemispherical integration of observed spectral radiance $L(\lambda,\theta,\phi)$ (radiation flux per solid angle, W sr^{-1} m^{-2} nm^{-1}), and $F_d(\lambda)$ is expressed as the sum of the flux of direct radiation $F_0(\lambda)$ and downward diffusive flux $F_\downarrow(\lambda)$,

$$F_d(\lambda) = \int_0^{2\pi} \int_0^{\pi/2} L(\theta, \phi) \sin\theta d\theta d\phi = F_\downarrow(\lambda) + F_0(\lambda). \tag{3.24}$$

Similarly, the irradiance $E(\lambda)$ is expressed as a sum of the direct component, $E_0(\lambda)$ ($= \cos\theta\, F_0(\lambda)$) and downward diffusive component, $E_\downarrow(\lambda)$,

$$E(\lambda) = \int_0^{2\pi} \int_0^{\pi/2} L(\theta, \phi) \cos\theta \sin\theta d\theta d\phi = E_\downarrow(\lambda) + E_0(\lambda). \tag{3.25}$$

Using the approximation by van Weele et al. (1995), the ratio of $F_d(\lambda)$ and $E(\lambda)$ is given by

$$\frac{F_d(\lambda)}{E(\lambda)} = \frac{F_\downarrow(\lambda)}{E_\downarrow(\lambda)} + \left(\frac{1}{\cos\theta} - \frac{F_\downarrow(\lambda)}{E_\downarrow(\lambda)}\right) \frac{E_0(\lambda)}{E(\lambda)}. \tag{3.26}$$

The ratio of the downward actinic flux F_d and irradiance E have been obtained from several field observations, and comparisons have been made with model calculations (Kazandzis et al. 2000; Webb et al. 2002a, b; McKenzie et al. 2002; Bais et al. 2003; Kylling et al. 2005; Palancer et al. 2011). The ratio of $F_\downarrow(\lambda)/E_\downarrow(\lambda)$ is expected to be 2 for the completely isotropic diffusive radiation, and the observed values are in general 1.4 to 2.6, depending on wavelength, solar zenith angle, and aerosol density (Kazandzis et al. 2000; Webb et al. 2002a). These values agree with the theoretical model in the case of no cloud, and the actinic flux can be obtained from the observed irradiance using the validated ratio.

On the contrary, if there are clouds, the discrepancy between the observation and the model is large in general, and the cause of uncertainty is thought to be the contribution of albedo of clouds. When the actinic flux F_{tot} is divided by direct radiation component F_0, and downward and upward diffusive radiation component $F_\downarrow$, and $F_\uparrow$, respectively, assuming a Lambertian surface i.e. a virtual completely diffusive surface for which radiance is constant being independent of the direction of observation (isotropic scattering),

$$F_\uparrow = A\left(2\cos\theta_0 F_0 + \mathrm{F}_\downarrow\right) \tag{3.27}$$

where A is the surface albedo, and

$$F_{\text{tot}} = F_0 + F_\downarrow + F_\uparrow = F_0(1 + 2A\cos\theta_0) + F_\downarrow(1 + A) \tag{3.28}$$

is derived (Madronich 1987). Here, θ_0 is the solar zenith angle. Equation (3.28), in the limiting case of $\mathrm{A} = 1$, $\theta_0 = 0$, and $\mathrm{F}_\downarrow = 0$, gives $F_{\text{tot}} = F_0 + F_\downarrow = 3F_0$, showing that the actinic flux can be larger than the direct radiation component by a factor of 3. In general, direct radiation is reflected isotropically at the surface, and actinic flux increases by a factor of $2\mathrm{A}\cos\theta_0$.

When the actinic flux is obtained, the photolysis rate constant $j_p(\mathrm{s}^{-1})$ can be calculated by Eq. (2.17). Practically, it is often calculated from the equation,

$$j_p\left(s^{-1}\right) = \sum_\lambda \sigma_{av}(\lambda)\phi_{av}(\lambda)F_{av}(\lambda) \tag{3.29}$$

by substituting the integral to the sectional mensuration. Here, $\sigma_{av}(\lambda)$, ϕ_{av}, and F_{av} are absorption cross section, photolysis quantum yield, and actinic flux, respectively, each averaged over wavelength interval $\Delta\lambda$ around the central wavelength λ.

The photolysis rate constants calculated by Eq. (3.29) and actually measured values for example for NO_2 are sometimes agree well and sometimes have large discrepancy (Kraus and Hofzumahaus 1998). Particularly in the case of the photolysis rate above cloud, in cloud and above snow the discrepancy between the calculated and observed values are large. It is discussed that there are uncertainties in the treatments of albedo effects (Van Weele and Duynkerke 1993; Junkermann 1994; Wild et al. 2000; Lee-Taylor and Madronich 2002; Simpson et al. 2002; Brasseur et al. 2002; Thiel et al. 2008).

References

Ahrens, C.D.: Meteorology Today, an Introduction to Weather, Climate and the Environment, 8th edn. Brooks Cole, Publishing Co., Monterey (2007)

ASTM (American Society for Testing and Materials): ASTM Standard Extraterrestrial Spectrum Reference E-490-00. http://www.astm.org/Standards/E490.htm (2006)

Bais, A., Madronich, S., Crawford, J., Hall, S.R., Mayer, B., Van-Weele, M., Lenoble, J., Calvert, J.G., Cantrell, C.A., Shetter, R.E., Hofzumahaus, A., Koepke, P., Monks, P.S., Frost, G., McKenzie, R., Krotkov, N., Kylling, A., Lloyd, S., Swartz, W.H., Pfister, G., Martin, T.J., Roeth, E.P., Griffioen, E., Ruggaber, A., Krol, M., Kraus, A., Edwards, G.D., Mueller, M., Lefer, B.L., Johnston, P., Schwander, H., Flittner, D., Gardiner, B.G., Barrick, J., Schmitt, R.: International photolysis frequency measurement and model intercomparison (IPMMI) spectral actinic solar flux measurements and modeling. J. Geophys. Res. **108**(D16), 8543 (2003). doi:10.1029/2002/JD002891

Bohren, C., Huffman, D.: Scattering and Absorption of Light by Small Particles. Wiley, New York (1981)

Brasseur, G., Simon, P.C.: Stratospheric chemical and thermal response to long-term variability in solar uv irradiation. J. Geophys. Res. **86**, 7343–7368 (1981)

Brasseur, G.P., Orlando, J.L., Tyndall, G.S. (eds.): Atmospheric Chemistry and Global Change. Oxford University Press, New York (1999)

Brasseur, A.L., Ramaroson, R., Delannoy, A., Skamarock, W., Barth, M.: Three-dimensional calculation of photolysis frequencies in the presence of clouds and impact on photochemistry. J. Atmos. Chem. **41**, 211–237 (2002)

Budikova, D.: The Encyclopedia of Earth, Albedo (Pidwiny, M. ed.). http://www.eoearth.org (2010)

Demerjian, K.L., Shere, K.L., Peterson, J.T.: Theoretical estimate of actinic (spherically integrated) flux and photolytic rate constants, of atmospheric species in the lower troposphere. Adv. Environ. Sci. Technol. **10**, 369–459 (1980)

Finlayson-Pitts, B.J., Pitts Jr., J.N.: Chemistry of the Upper and Lower Atmosphere. Academic (2000)

Friedman, H.: Physics of the Upper Atmosphere (Ratcliffe, J.A. ed.). Academic, New York (1960)

Gaydon, A.G.: Dissociation Energies and Spectrum of Diatomic Molecules, 3rd edn. Chapman and Hall, London (1968)

Goody, R.: Principles of Atmospheric Physics and Chemistry. Oxford University Press (1995)

Hofzumahaus, A., Kraus, A., Muller, M.: Solar actinic flux spectroradiometry: a new technique to measure photolysis frequencies in the atmosphere. Appl. Opt. **38**, 4443–4460 (1999)

Junkermann, W.: Measurements of the $J(O^1D)$ actinic flux within and above stratiform clouds and above snow surfaces. Geophys. Res. Lett. **21**, 793–796 (1994)

Kazandzis, S., Bais, A.F., Balis, D., Zerefos, C., Blumthaler, M.: Retrieval of downwelling UV actinic flux density spectra from spectral measurements of global and direct solar UV irradiance. J. Geophys. Res. **105**, 4857–4864 (2000)

Kraus, A., Hofzumahaus, A.: Field measurements of atmospheric photolysis frequencies for O_3, NO_2, HCHO, CH_3CHO, H_2O_2 and HONO by UV spectroradiometry. J. Atom. Chem. **31**, 161–180 (1998)

Kylling, A., Webb, A.R., Kift, R., Gobbi, G.P., Ammannato, L., Barnaba, F., Bais, A., Kazandzidis, S., Wendisch, M., Jäkel, E., Schmidt, S., Kniffka, A., Thiel, S., Junkermann, W., Blumthaler, M., Silbernagl, R., Schallhart, B., Schmitt, R., Kjeldstad, B., Thorseth, T.M., Scheirer, R., Mayer, B.: Spectral actinic flux in the lower troposphere: measurement and 1-D simulations for cloudless, broken cloud and overcast situations. Atmos. Chem. Phys. **5**, 1975–1997 (2005)

Lee-Taylor, J., Madronich, S.: Calculation of actinic fluxes with a coupled atmosphere–snow radiative transfer model. J. Geophys. Res. **107**(D24), 4796 (2002). doi:10.1029/2002JD002084

Leighton, P.A.: Photochemistry of Air Pollution. Academic Press, New York (1961)

Lofthus, A., Krupenie, P.H.: The spectrum of molecular nitrogen. J. Phys. Chem. Ref. Data **6**, 113–307 (1977)

Luther, F.M., Gelinas, R.J.: Effect of molecular multiple scattering and surface albedo on atmospheric photodissociation rates. J. Geophys. Res. **81**, 1125–1132 (1976)

Madronich, S.: Photodissociation in the atmosphere 1. Actinic flux and the effect of ground reflections and clouds. J. Geophys. Res. **92**, 9740–9752 (1987)

McKenzie, R.L., Johnston, P.V., Hofzumahaus, A., Kraus, A., Madronich, S., Cantrell, C., Calvert, J., Shetter, R.: Relationship between photolysis frequencies derived from spectroscopic measurements of actinic fluxes and irradiances during the IPMMI campaign. J. Geophys. Res. **107**, 4042 (2002). doi:10.1029/2001JD000601

McLinden, C.A., McConnell, J.C., Griffoen, E., McElroy, C.T., Pfister, L.: Estimating the wavelength-dependent ocean Albedo under clear-sky conditions using NASA ER2 spectrometer measurements. J. Geophys. Res. **102**, 18801–18811 (1997)

Neckel, H., Labs, D.: The solar spectrum between 3300 and 12500. Solar. Phys. **90**, 205–258 (1984)

NOAA: U.S. Standard Atmosphere, Publication NOAA-S/T76-1562. U.S. Government Printing Office, Washington, DC (1976)

Palancer, G.G., Shetter, R.E., Hall, S.R., Toselli, B.M., Madronich, S.: Ultraviolet actinic flux in clear and cloudy atmospheres: model calculations and aircraft-based measurements. Atmos. Chem. Phys. **11**, 5457–5469 (2011)

Peterson, J.T.: Calculated Actinic Fluxes (290–700 nm) for Air Pollution Photochemistry Applications. U.S. Environmental Protection Agency report no. EPA-600/4-76-025, June (1976)

Sander, S.P., Baker, R., Golden, D.M., Kurylo, M.J., Wine, P.H., Abatt, J.P.D., Burkholder, J.B., Kolb, C.E., Moortgat, G.K., Huie, R.E., Orkin, V.L.: Chemical Kinetics and Photochemical Data for Use in Atmospheric Studies, Evaluation Number 17, JPL Publication 10–6, Pasadena, California. Website: http://jpldataeval.jpl.nasa.gov/ (2011)

Shetter, R., Müller, M.: Photolysis frequency measurements using actinic flux spectroradiometry during the PEM-tropics mission: instrumentation description and some results. J. Geophys. Res. **104**, 5647–5661 (1999)

Simpson, W.R., Kinga, M.D., Beineb, H.J., Honrath, R.E., Zhou, X.: Radiation-transfer modeling of snow-pack photochemical processes during ALERT 2000. Atmos. Environ. **36**, 2663–2670 (2002)

Thiel, S., Ammannato, L., Bais, A., Bandy, B., Blumthaler, M., Bohn, B., Engelsen, O., Gobbi, G. P., Gröbner, J., Jäkel, E., Junkermann, W., Kazadzis, S., Kift, R., Kjeldstad, B., Kouremeti, N., Kylling, A., Mayer, B., Monks, P.S., Reeves, C.E., Schallhart, B., Scheirer, R., Schmidt, S., Schmitt, R., Schreder, J., Silbernagl, R., Topaloglou, C., Thorseth, T.M., Webb, A.R.,

Wendisch, M., Werle, P.: Influence of clouds on the spectral actinic flux density in the lower troposphere (INSPECTRO): overview of the field campaigns. Atmos. Chem. Phys. **8**, 1789–1812 (2008)

Timothy, A.F., Timothy, J.G.: Long term variations in the solar helium II Ly-alpha line. J. Geophys. Res. **75**, 6950 (1970)

Van Weele, M., Duynkerke, P.G.: Effect of clouds on the photodissociation of NO_2: observations and modelling. J. Atmos. Chem. **16**, 231–255 (1993)

Van Weele, M., de Arrelano, J.V.-G., Kuik, F.: Combined measurements of UV–A actinic flux, UV–A irradiance and global radiation in relation to photodissociation rates. Tellus **47B**, 353–364 (1995)

Vermote, E.F., Tanre, D., Deuze, J.L., Herman, M., Morcette, J.J.: Second simulation of the satellite signal in the solar spectrum, 6S: an overview. IEEE Trans. Geosci. Remote Sens. **35**, 675–686 (1997)

Warneck, P.: Chemistry of the Natural Atmosphere. 2nd Ed., Academic (1999)

Webb, A.R., Bais, A.F., Blumthaler, M., Gobbi, G.-P., Kylling, A., Schmitt, R., Thiel, S., Barnaba, F., Danielsen, T., Junkermann, W., Kazantzidis, A., Kelly, P., Kift, R., Liberti, G.L., Misslbeck, M., Schallhart, B., Schreder, J., Topaloglou, C.: Measuring spectral actinic flux and irradiance: experimental results from the Actinic Flux Determination from Measurements of Irradiance (ADMIRA) project. J. Atmos. Ocean. Technol. **19**, 1049–1062 (2002a)

Webb, A.R., Kift, R., Thiel, S., Blumthaler, M.: An empirical method for the conversion of spectral UV irradiance measurements to actinic flux data. Atmos. Environ. **36**, 4044–4397 (2002b)

Wild, O., Zhu, Z., Prather, M.J.: Fast-J: accurate simulation of in- and below-cloud photolysis in tropospheric chemical models. J. Atmos. Chem. **37**, 245–282 (2000)

WMO: Wehrli Standard Extraterrestrial Solar Irradiance Spectrum. World Meteorological Organization, Geneva (1985)

Woods, T.N., Prinz, D.K., Rottman, G.J., London, J., Crane, P.C., Cebula, R.P., Hilsenrath, E., Brueckner, G.E., Andrews, M.D., White, O.R., Van Hoosier, M.E., Floyd, L.E., Herring, L.C.: Validation of the UARS solar ultraviolet irradiances: comparison with the ATLAS I and 2 measurements. J. Geophys. Res. **101**, 9541–9569 (1996)

Chapter 4
Spectra and Photolytic Reactions of Atmospheric Molecules

Photochemical reactions of atmospheric molecules are the primary driving force of chemical reaction systems in the troposphere and stratosphere. Therefore it is very important to understand the photolytic processes of each species for comprehending atmospheric chemistry. In this chapter, the absorption spectra and cross sections, photodissociation pathways, and their quantum yields (see Sect. 2.1.2) are described for atmospheric molecules that are photolyzed under the actinic flux in the troposphere and stratosphere. For atmospheric molecules at which altitudes they are photolyzed are determined by the overlap of the spectrum of the actinic flux and absorption spectrum, and by the photodissociation quantum yield at each wavelength. The photolysis of some molecules is important both in the troposphere and stratosphere. In particular, O_3 is mainly discussed in the section of the photolysis in the troposphere (Sect. 4.2.1), and supplemental description is given in the Sect. 4.3.2 for the processes which occur only in the stratosphere. Since the photolysis of organic halogen species including chlorofluorocarbons is particularly important in the stratosphere, they are compiled in the section for the photolysis in the stratosphere including several species that are photolyzed also in the troposphere. Further, since most of inorganic halogen species are relevant both in the troposphere and stratosphere, they are consolidated separately in Sect. 4.3 as the photolysis of inorganic halogens.

The absorption cross sections and photolytic processes of atmospheric molecules are evaluated and compiled by the NASA Panel for Data Evaluation and IUPAC Subcommittee on Gas Kinetic Data Evaluation for Atmospheric Chemistry. The descriptions in this chapter refer to the NASA/JLP Panel Evaluation No. 17 (Sander et al. 2011) mainly for absorption cross sections, and to the IUPAC Subcommittee Reports Vol. I, II, III and IV (Atkinson et al. 2004, 2006, 2007, 2008, respectively) mainly for photolytic processes. Numerical values of absorption cross sections for each chemical species cited in this chapter are given at the end of this chapter as appended tables.

After each photolytic reaction formula, the reaction enthalpy ΔH° and threshold wavelength below which the photodissociation is energetically possible is given. It

H. Akimoto, *Atmospheric Reaction Chemistry*, Springer Atmospheric Sciences,
DOI 10.1007/978-4-431-55870-5_4

should be noted that the values shown in this chapter is for the values of ΔH_0° calculated from heat of formation $\Delta H_{f,0}^\circ$ at 0 K given in Table 2.5. For the heat of reaction, the values at 298 K ΔH_{298}° are generally used for gas phase reactions and they are given in Chap. 5. However, for the photolytic reactions the values at 0 K, the values of ΔH_0° are useful for the discussion of contribution of vibrational and rotational energies and they are employed in this chapter.

4.1 Solar Spectra in the Troposphere and Stratosphere

Figure 4.1 shows the solar actinic flux in the ultraviolet region for different altitudes at the solar zenith angle 60°, that are calculated considering the processes described in Chap. 3. As shown in the figure, the shortest wavelength of solar radiation reaching to the troposphere is ca. 295 nm, and the molecules that have absorption spectra in the ultraviolet and visible region longer than this wavelength are the subject of discussion. On the other hand, the solar radiation with wavelengths longer than ca. 185 nm reaches to the upper stratosphere, and the spectrum is very much dependent on the altitude due to the absorption of ozone present in the stratosphere. In particular, the ultraviolet radiation at 190–230 nm reaches to the middle and lower stratosphere, and this wavelength region is called "atmospheric window" of the stratosphere. This wavelength region happens to overlap with the absorption spectrum of chlorofluorocarbons, whose photolysis through the atmospheric window triggers ozone layer depletion.

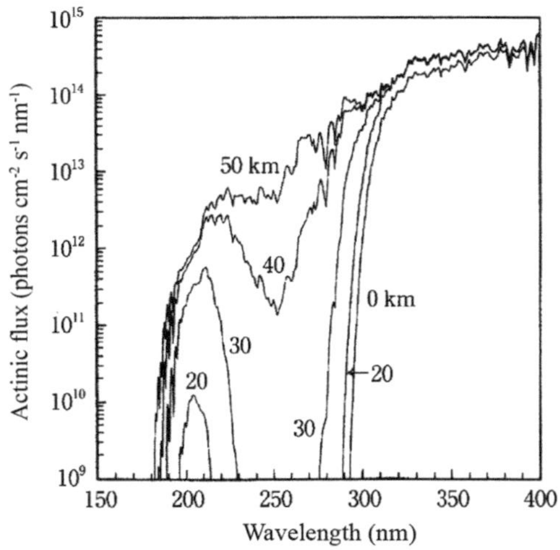

Fig. 4.1 Solar actinic flux for different altitudes. Solar zenith angle 30°, surface albedo 0.3 (Adapted from Demore et al. 1997)

4.2 Photolysis in the Troposphere

In this section, among the chemical species that are photodecomposed in the troposphere, important molecules in atmospheric chemistry are discussed, with the exception of inorganic halogens, which are described in Sect. 4.4.

4.2.1 Ozone (O_3)

The production of OH radicals by the reaction of H_2O with the electronically excited oxygen atoms $O(^1D)$ formed in the photolysis of ozone (O_3) is the most important reaction in the natural atmosphere to trigger tropospheric photochemistry. Here, the absorption spectrum and the $O(^1D)$ production quantum yields in the photolysis in the troposphere are described in detail.

Absorption Spectrum and Absorption Cross Sections Although the absorption spectrum of O_3 in the ultraviolet region has already been given in Fig. 3.6, Fig. 4.2 depicts the spectrum including the visible region, and Fig. 4.3 illustrates the UV absorption spectrum with the axis of ordinate in a linear scale. The strong absorption in the wavelength region of 200–310 nm is called the Hartley bands, and corresponds to the allowed transition from the ground X^1A_1 state to the electronically excited $^1B_2$ state (see Fig. 4.8 shown later). The absorption of Hartley bands is almost a continuum, and in most cases, a molecule excited to this electronic state is thought to dissociate. However, as seen in Fig. 4.3, a weak vibrational structure is discernible at the peak of the Hartley bands near 250 nm. This implies that the $^1B_2$ state is a bound state crossed with a repulsive potential curve, which leads the most of the molecule to dissociate, but the transition to the higher energy than the crossing point still reveals vibrational structure due to the nuclear motion. The absorption cross

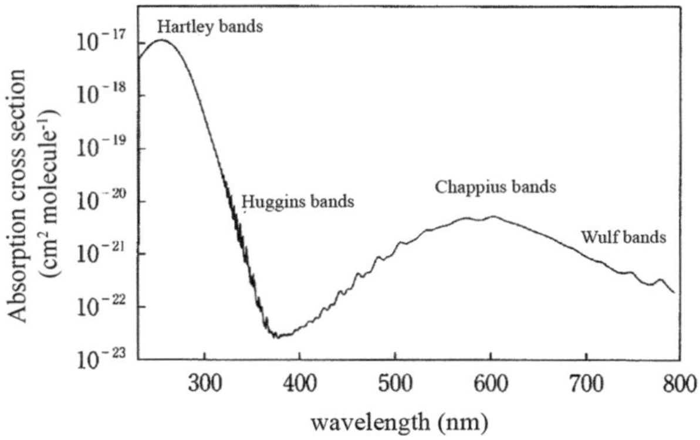

Fig. 4.2 Absorption spectrum of O_3 in the tropospheric actinic flux region (Adapted from Orphal 2003)

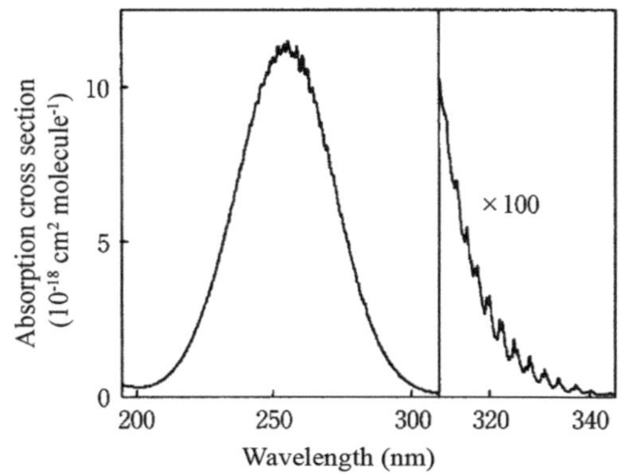

Fig. 4.3 Ultraviolet absorption spectrum of O_3 in a linear scale (Adapted from Matsumi and Kawasaki 2003)

sections (293–298 K) in the wavelength region of 186–390 nm based on the NASA/JPL Evaluation No. 17 (Sander et al. 2011) are tabulated in Table 4.1.

The absorption bands with vibrational structure in the region of wavelength longer than 310 nm are called the Huggins bands. Although the bands are thought to cause the overlapping of the transition to the energy state $^1B_2$ lower than the crossing point with the dissociative potential curve, and to the bound electronically exited $^1A_1$ state, the assignment of the excited states of the Huggins bands has not been established theoretically (Matsumi and Kawasaki 2003).

In addition, O_3 has weak absorption bands called the Chappuis bands in the visible range and the Wulf bands in further longer wavelengths as shown in Fig. 4.2. These bands corresponds to the forbidden transitions to the lower electronically excited states that cross with repulsive potential curve dissociating into the ground states of an O atom and O_2 molecule.

The important process in the troposphere is the photo-absorption in the Hartley and Huggins bands in the wavelength range from 295 nm to around 360 nm, the solar actinic flux region. It is known that the absorption cross sections in this region has strong temperature dependence and that they decrease with temperature. Figure 4.4 shows the absorption spectrum of O_3 in this wavelength range at 293 and 202 K. As shown in the figure, the temperature dependence gets stronger with wavelength, and it can also be seen that the temperature dependence is very different for the peaks and valleys of spectrum in the Huggins bands. The absorption cross sections of O_3 in the 186–390 nm has been given in Table 4.1, here those of Huggins bands (310–345 nm) at different temperatures at 298, 263, and 226 K (Molina and Molina 1986) are cited in Table 4.2. The temperature dependence of the O_3 absorption cross sections are approximated by the empirical formula of a quadratic function (Orphal 2003),

Table 4.1 Absorption cross sections of O_3 (186–390 nm, T = 293 ~ 298 K) (Base e)

Wavelength (nm)	10^{20} σ (cm^2 $molecule^{-1}$)	Wavelength λ (nm)	10^{20} σ (cm^2 $molecule^{-1}$)	Wavelength λ (nm)	10^{20} σ (cm^2 $molecule^{-1}$)
186	61.9	242	897	298	51.2
188	56.6	244	972	300	39.2
190	51.1	246	1033	302	30.3
192	46.1	248	1071	304	23.4
194	40.7	250	1124	306	17.9
196	36.7	252	1155	308	13.5
198	33.5	254	1159	310	10.2
200	31.5	256	1154	312	7.95
202	31.8	258	1124	314	6.25
204	33.7	260	1080	316	4.77
206	38.6	262	1057	318	3.72
208	46.4	264	1006	320	2.99
210	57.2	266	949	322	2.05
212	71.9	268	875	324	1.41
214	91.0	270	798	326	1.01
216	115	272	715	330	0.697
218	144	274	614	335	0.320
220	179	276	545	340	0.146
222	220	278	467	345	0.0779
224	268	280	400	350	0.0306
226	323	282	325	355	0.0136
228	383	284	271	360	0.0694
230	448	286	224	365	0.00305
232	518	288	175	370	0.00130
234	589	290	142	375	0.000850
236	672	292	111	380	0.000572
238	749	294	87.1	385	0.000542
240	831	296	67.3	390	0.000668

Source: 186–298 nm (298 K): Molina and Molina 1986. 300–390 nm (293–298 K): NASA/JPL Panel Evaluation No. 17

$$\sigma(\lambda, T) = a(\lambda) + b(\lambda)(T - 230) + c(\lambda)(T - 230)^2, \tag{4.1}$$

and the values of a(λ), b(λ) and c(λ) in the 280–320 nm range are given in the literatures (Molina and Molina 1986; Finlayson-Pitts and Pitts 2000).

Photodissociation Quantum Yields The wavelength thresholds below which photodissociation is energetically possible in the photolysis of O_3 are shown in Table 4.3 (Okabe 1978). From the table, energetically possible photodissociation processes by the solar flux reaching to the troposphere are the following five processes,

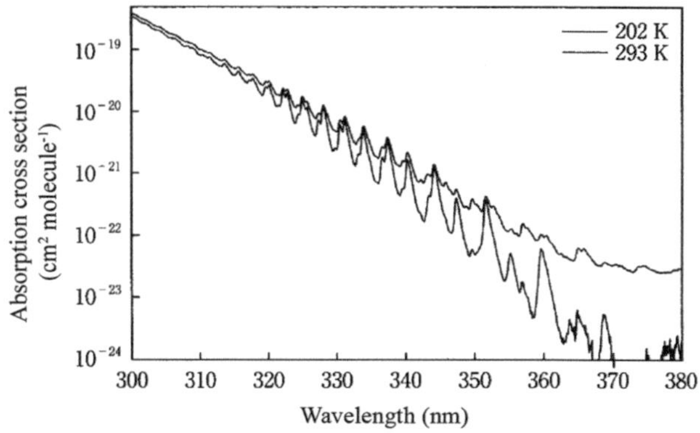

Fig. 4.4 Absorption spectrum of O_3 Huggins bands at 201 and 293 K (Adapted from Orphal et al. 2003)

Table 4.2 Absorption cross sections of O_3 in the 280–350 nm region[a] at 298, 263, and 226 K

	Absorption Cross Sections (10^{-20} cm^2 molecule^{-1})		
Wavelength (nm)	298 K	263 K	226 K
310[b]	10.5	9.66	9.14
315	5.55	4.92	4.56
320	2.80	2.46	2.21
325	1.38	1.18	1.01
330	0.706	0.599	0.506
335	0.329	0.263	0.214
340	0.149	0.112	0.0832
345	0.0781	0.0586	0.0442

[a]Average absorption cross sections for λ-2.5 nm–λ + 2.5 nm unless otherwise noted
[b]307.69 nm < λ < 312.5 nm
Source: Molina and Molina (1986)

Table 4.3 Wavelength thresholds (nm) below which the production of each pair of O and O_2 is energetically possible in the photolysis of O_3 (0 K)

O / O_2	$X^3\Sigma^-{}_g$	$a^1\Delta_g$	$b^1\Sigma^+{}_g$	$A^3\Sigma^+{}_u$	$B^3\Sigma^-{}_u$
^3P	1180	611	463	230	173
^1D	411	310	266	168	136
^1S	237	200	180	129	109

Source: Okabe (1978)

$$O_3 + h\nu(\lambda < 310\ \text{nm}) \rightarrow O(^1D) + O_2(a^1\Delta_g) \quad \Delta H_0^\circ = 386\ \text{kJ mol}^{-1} \tag{4.2}$$

$$+ h\nu\ (\lambda<411\ \text{nm}) \rightarrow O(^1D) + O_2\,(X^3\Sigma^-{}_g) \quad \Delta H_0^\circ = 291\ \text{kJ mol}^{-1} \tag{4.3}$$

$$+ h\nu\ (\lambda<463\ \text{nm}) \rightarrow O(^3P) + O_2\,(b^1\Sigma^+{}_g) \quad \Delta H_0^\circ = 258\ \text{kJ mol}^{-1} \tag{4.4}$$

$$+ h\nu\ (\lambda<611\ \text{nm}) \rightarrow O(^3P) + O_2\,(a^1\Delta_g) \quad \Delta H_0^\circ = 195\ \text{kJ mol}^{-1} \tag{4.5}$$

$$+ h\nu\ (\lambda<1180\ \text{nm}) \rightarrow O(^3P) + O_2\,(X^3\Sigma^-{}_g). \quad \Delta H_0^\circ = 101\ \text{kJ mol}^{-1} \tag{4.6}$$

The $O(^1D)$ atoms formed in the photolysis of O_3 reacts with water vapor,

$$O(^1D) + H_2O \rightarrow 2OH \tag{4.7}$$

to produce OH radicals, that is very important in the tropospheric chemistry. Although it has been known that the allowed transition for the Hartley bands corresponds to the photolytic process, Eq. (4.2) to yield $O(^1D)$ atoms, the quantum yield of $O(^1D)$ production has not been measured accurately until recently. Particularly, the quantum yields of $O(^1D)$ production in the Huggins bands gives large impact on tropospheric chemistry since the solar flux gets stronger at the wavelengths longer than 310 nm.

Matsumi's group (Matsumi and Kawasaki 2003) determined the $O(^1D)$ production quantum yields in the region of 305–330 nm accurately by detecting $O(^3P)$ and $O(^1D)$ directly in the spectroscopic methods and their data superseded the past one (Ravishankara et al. 1998). Accordingly, the international evaluation panel was formed, and the results (Matsumi et al. 2002) were adopted in the NASA/JPL Evaluation 14 (Sander et al. 2003), which is succeeded in the later evaluation (Sander et al. 2011).

Absolute quantum yields at 308 nm measured by the spectroscopic method in the past (Greenblatt and Wiesenfeld 1983; Talukdar et al. 1997b, 1998) agrees well, and the value $\Phi(^1D) = 0.79$ has been adopted at 298 K. Based on this value, re-normalized values of wavelength dependence (298 K) of $O(^1D)$ production quantum yields are shown in Fig. 4.5(a) (Matsumi et al. 2002).

As shown in Fig. 4.5(a) the quantum yields of $O(^1D)$ production decreases with wavelength from the value of 0.79 at 308 nm, and levels-off at 0.1 in wavelengths longer than 325 nm. As seen in Table 4.3, the energy threshold of 310 nm for the allowed transition corresponding to reaction (4.2)

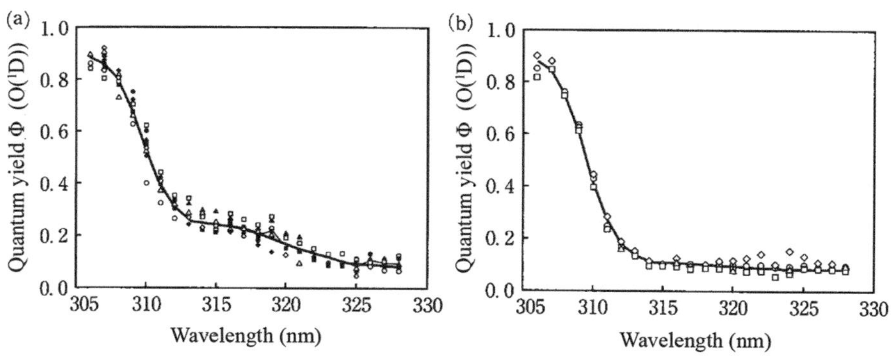

Fig. 4.5 Wavelength dependence of $O(^1D)$ quantum yields in the photolysis of O_3 at >305 nm: (**a**) 298 K, (**b**) 227 K (Adapted from Matsumi et al. 2002). Symbols are experimental values (Refer to the source for each literature), *Solid lines* are the recommended curve fitting

$$O_3 + h\nu(\lambda < 310\ \text{nm}) \rightarrow O(^1D) + O_2(a^1\Delta_g) \quad \Delta H_0^\circ = 386\ \text{kJ mol}^{-1} \tag{4.2}$$

is 310 nm, Fig. 4.5(a) shows the quantum yields of $O(^1D)$ production has substantial values even at wavelengths longer than this value. This means that a process other than reaction (4.2) is operating in the photolysis of O_3 in this wavelength range, which has a large implication for tropospheric photochemistry.

The hint of which processes are involved in the photolysis of O_3 at the Huggins bands is given by the temperature dependence of the quantum yields of $O(^1D)$ production. The temperature dependence has been obtained in the 200–320 K, and the quantum yields Φ(308 nm, T) at 308 nm at temperature T when normalized at 0.79 at 298 K is expressed (Matsumi et al. 2002) as

$$\Phi(308\ \text{nm}, \text{T}) = (6.10 \times 10^{-4})\text{T} + 0.608. \tag{4.8}$$

The wavelength dependence of the quantum yields of $O(^1D)$ production at 277 K obtained by several studies are shown Fig. 4.5(b). As shown in Fig. 4.5(b), the $O(^1D)$ production quantum yields at 227 K decreases more rapidly from the value at 0.75 at 308 nm than at 298 K and levels-off at ca. 0.1 for wavelengths longer than 315 nm. The wavelength dependence of the quantum yields of $O(^1D)$ production obtained from these results are given in Table 4.4 for 306–328 nm at 321, 298, 273, 253, 223, and 203 K (Matsumi et al. 2002).

The production quantum yields of $O(^1D)$ at 290–305 nm for wavelengths shorter than 308 nm, is recommended as 0.90 ± 0.09 by NASA/JPL Evaluation (Matsumi et al. 2002; Sander et al. 2011) based on the results of recent measurements (Talukdar et al. 1998; Taniguchi et al. 2000). The quantum yields of $O(^1D)$ production for further shorter wavelength have been reported as 0.85–0.95. Including these values the data for the $O(^1D)$ production quantum yields in the Hartley bands in 220–290 nm are shown in Fig. 4.6. In this wavelength range, temperature dependence is not observed, and NASA/JPL Evaluation No. 17 recommends the value of 0.90 at the wavelength shorter than 306 nm (Sander et al. 2011).

Meanwhile, the production quantum yields of $O(^1D)$ is not 0 even for wavelengths longer than 328 nm, and the corresponding photolytic process is thought to be $O(^1D) + O_2\ (X^3\Sigma^-{}_g)$ as will be discussed later. Since the threshold wavelength for this process is 411 nm, the production of $O(^1D)$ is thought to continue to this wavelength. NASA/JPL Evaluation No. 17 (Sander et al. 2011) recommends the $O(^1D)$ production quantum yield of 0.08 ± 0.04 for wavelengths longer than 340 nm being independent of temperature, although the photolysis in this region is not so important for the tropospheric photochemistry since the absorption cross section of O_3 is very small at wavelengths longer than 340 nm.

Here again, let us see what photochemical processes are participated in the photolysis of O_3 based on the wavelength and temperature dependence of the $O(^1D)$ production quantum yields shown in Figs. 4.5 and 4.6. Figure 4.7 depicts wavelength and temperature dependence of the $O(^1D)$ production quantum yields in 305–330 nm region calculated by Matsumi et al. (2002). Also, a schematic

Table 4.4 Quantum yields of O(^{1}D) production in the photolysis of O_3 at each temperature (321, 298, 273, 253, 223, 203 K)

	O(^{1}D) Production Quantum Yields					
Wavelength (nm)	321 K	298 K	273 K	253 K	223 K	203 K
306	0.893	0.884	0.878	0.875	0.872	0.872
307	0.879	0.862	0.850	0.844	0.838	0.835
308	0.821	0.793	0.772	0.760	0.748	0.744
309	0.714	0.671	0.636	0.616	0.595	0.585
310	0.582	0.523	0.473	0.443	0.411	0.396
311	0.467	0.394	0.334	0.298	0.259	0.241
312	0.390	0.310	0.246	0.208	0.169	0.152
313	0.349	0.265	0.200	0.162	0.126	0.112
314	0.332	0.246	0.180	0.143	0.108	0.095
315	0.325	0.239	0.173	0.136	0.102	0.090
316	0.317	0.233	0.168	0.133	0.100	0.088
317	0.300	0.222	0.162	0.129	0.098	0.087
318	0.275	0.206	0.152	0.123	0.096	0.086
319	0.246	0.187	0.141	0.116	0.093	0.085
320	0.214	0.166	0.129	0.109	0.090	0.083
321	0.183	0.146	0.117	0.101	0.087	0.082
322	0.155	0.128	0.107	0.095	0.084	0.080
323	0.132	0.113	0.098	0.089	0.082	0.079
324	0.114	0.101	0.091	0.085	0.080	0.078
325	0.101	0.092	0.086	0.082	0.079	0.078
326	0.091	0.086	0.082	0.080	0.078	0.077
327	0.085	0.082	0.080	0.079	0.077	0.077
328	0.081	0.080	0.078	0.078	0.077	0.077

Source: Matsumi et al. (2002)

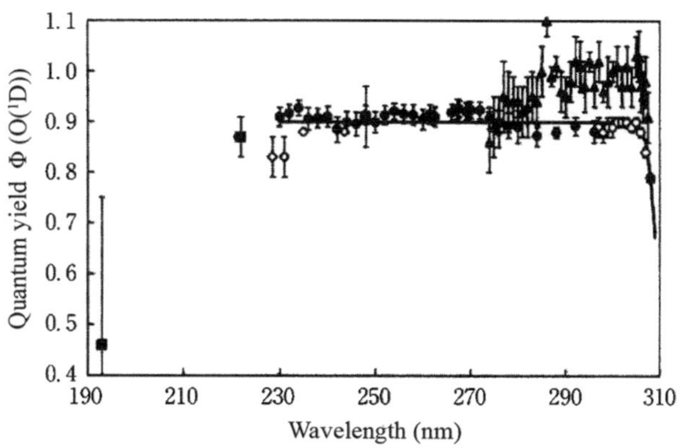

Fig. 4.6 Wavelength dependence of O(^{1}D) quantum yields in the photolysis of O_3 at <310 nm (Adapted from Matsumi et al. 2002)

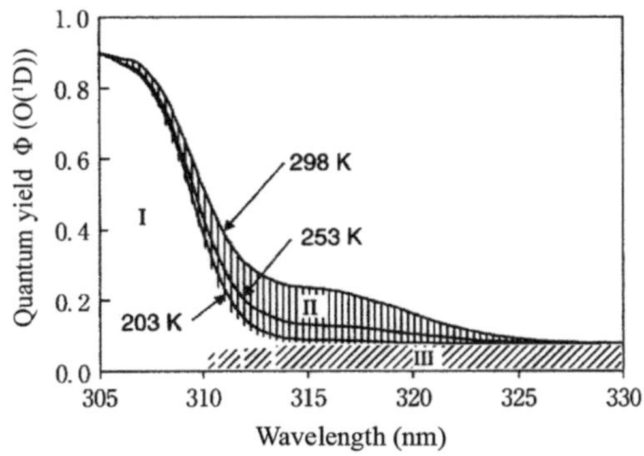

Fig. 4.7 Calculated recommended values of $O(^1D)$ quantum yields in the photolysis of O_3 at >305 nm for 203, 253 and 298 K (Adapted from Matsumi et al. 2002). Region I: $O(^1D) + O_2(a^1\Delta_g)$ via reaction (4.2a), Region II: $O(^1D) + O_2(a^1\Delta_g)$ by the hot band excitation via reaction (4.2a), Region III: $O(^1D) + O_2(X^3\Sigma_g^-)$ via reaction (4.2b)

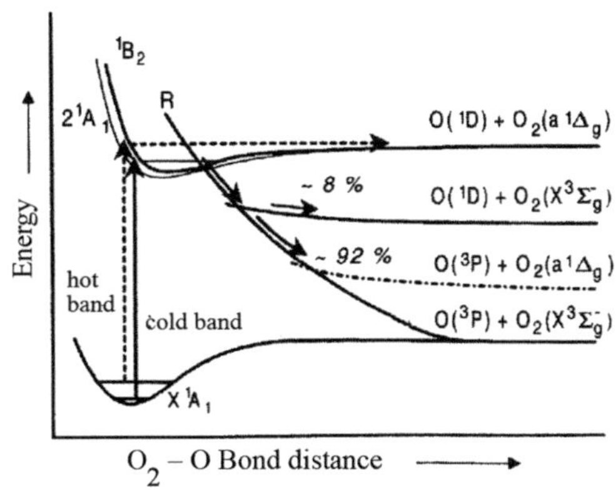

Fig. 4.8 Schematic potential energy curves of O_3 (Adapted from Matsumi and Kawasaki 2003)

potential diagram of O_3 (Matsumi and Kawasaki 2003) taking a dissociating bond length as an axis of abscissa is shown in Fig. 4.8.

In the Hartley bands corresponding the $^1B_2 - X^1A_1$ transition shown in Fig. 4.8,

$$O_3 + h\nu \rightarrow O(^1D) + O_2(a^1\Delta_g) \tag{4.2}$$

$$+ h\nu \rightarrow O(^3P) + O_2(X^3\Sigma^-_g) \tag{4.6}$$

are the main pathways, and the production of $O(^1D)$ and $O(^3P)$ occurs in the ratios of 0.9 and 0.1, respectively (Adler-Golden et al. 1982). According to the measurement of accurate O-O bond energy by Taniguchi et al. (1999), $\Delta H_f°(O_3)$ is -144.31 ± 0.14 kJ mol^{-1} and the wavelength threshold for reaction (4.2) is 309.44 ± 0.02 nm. However, the production of $O(^1D)$ is discernible at the wavelength longer than this value. From the fact that the temperature dependence is observed for the $O(^1D)$ production quantum yields in this wavelength region, the formation of $O(^1D)$ here is thought to be due to the photoabsorption form the vibrationally excited ground state molecule of O_3 to the Hartley bands (hot bands). Although the ratio of vibrationally excited molecules is not large at the temperature range of 200–320 K, the transition probability from the ground state with the excited anti-symmetric stretching vibration (ν_3) to the $^1B_2$ state is large, and the absorption cross sections of the hot bands are sizable, since the O_3 in the $^1B_2$ state is an asymmetric molecule with unequal O-O bond lengths in contrast to in the ground state which is symmetric molecule with equal O-O bond distances. Theoretical calculation also shows that the absorption cross sections for the long–wavelength edge of Harley bands get larger with the excitation of anti-symmetric stretching vibration ν_3 (Adler-Golden 1983). In Fig. 4.7, the $O(^1D)$ production quantum yields for the transition from the vibrational ground state in the Hartley bands correspond to region I, and those for the transition from the ν_3 vibrationally excited states correspond to region II at 298, 253, and 203 K.

For the $O(^1D)$ formation pathways in the Huggins bands, it is known that the reaction pathway,

$$O_3 + h\nu(\lambda < 411 \text{ nm}) \rightarrow O(^1D) + O_2(X^3\Sigma^-{}_g) \tag{4.3}$$

is involved in addition to reaction (4.2), which is proved from the measurement of translational energy of formed $O(^1D)$ (Takahashi et al. 1996; Denzer et al. 1998). The temperature independent production of $O(^1D)$ seen in Figs. 4.5 and 4.7 assigned to the spin forbidden process (4.3) and corresponds to region III in Fig. 4.7. The $O(^1D)$ production quantum yields by this spin forbidden process are nearly constant at about 0.08 in the wavelength longer than 310 nm for the absorption by the Huggins bands, and the wavelength dependence corresponding to the vibrational structure of the absorption spectrum in 313–320 nm is not observed (Takahashi et al. 1996, 1998). Although the assignment of the upper state of the spin forbidden transition for the Huggins bands has not been established distinctively, Takahashi et al. (1997) proposed the 2^1A_1 state as shown in Fig. 4.8 from the recent measurement of rotational constants.

Using these new data of the quantum yields, the contributions to the $O(^1D)$ formation in the tropospheric chemistry by the above mentioned vibrationally excited O_3 molecule in the Hartley bands, and the spin forbidden transition process are estimated as 25–40 %, and 30 % at the solar zenith angle of 40–80°, and 80°, respectively (Matsumi et al. 2002).

4.2.2 Nitrogen Dioxide (NO_2)

The formation of oxygen atom $O(^3P)$ in the photolysis of nitrogen dioxide (NO_2) is the fundamental reaction that causes direct production of O_3 in the troposphere. In this section, absorption spectrum and $O(^3P)$ production quantum yields relevant to the tropospheric photochemistry are described.

Absorption Spectrum and Cross Section The absorption spectrum of NO_2 in 240–800 nm region is shown in Fig. 4.9 (Orphal 2003). As shown in the figure, NO_2 has a continuous absorption spectrum from ultraviolet to a whole range of visible and near infrared with a maximum at around 400 nm. This spectrum has been assigned to the D^2B_2-X^2A_1 transition for the wavelength shorter than 240 nm, and to the B^2B_1-X^2A_1, and A^2B_2-X^2A_1 mixed with the C^2A_2-X^2A_1 (forbidden transition) transitions for the 300–790 nm region (Douglas 1966; Stevens et al. 1973). Therefore, the absorption spectrum of NO_2 in the near ultraviolet and visible region is very complex. Figure 4.10 shows the high resolution absorption spectrum in the

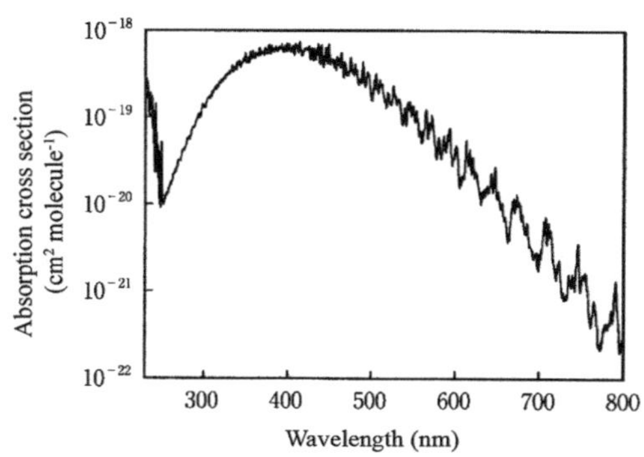

Fig. 4.9 Absorption spectrum of NO_2 in the ultraviolet and visible region (Adapted from Orphal 2003)

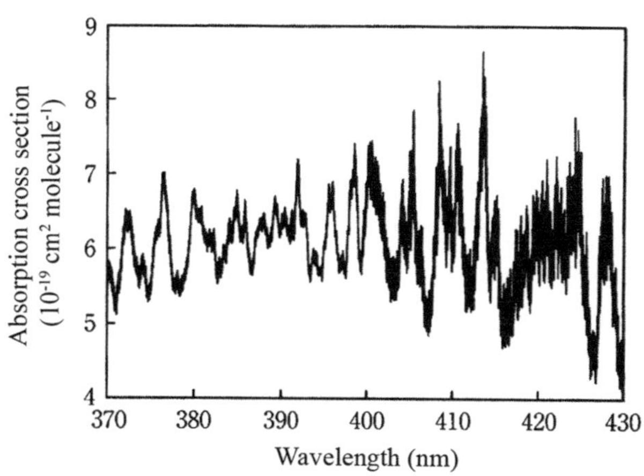

Fig. 4.10 High-resolution absorption spectrum of NO_2 near the dissociation wavelength threshold (398 nm) for $NO_2 \rightarrow NO + O(^3P)$. Resolution 0.03 nm at 445 nm (Adapted from Orphal et al. 2003)

region of 370–430 nm (Orphal 2003). The rotational lines are very sharp and dense particularly near the dissociation limit of 398 nm, and the line widths are determined by the pressure broadening. Since the lifetime of the excited state is shorter at wavelengths shorter than the dissociation limit, the line widths are broader according to the uncertainty principle as seen in the figure. It is known that the absorption spectrum of NO_2 has temperature dependence, and the differences between peaks and valleys of absorption lines are larger at lower temperature which makes the absorption lines more distinct (Harder et al. 1997; Vandaele et al. 2002). The temperature and pressure dependence of the high resolution spectrum of NO_2 are important for the spectral analysis of remote sensing data of satellites.

Table 4.5 tabulates the absorption cross sections of NO_2 at 298 K recommended by IUPAC Subcommittee Report Vol. I (Atkinson et al. 2004) based on Vandaele et al. (1998) and others.

Photolysis Quantum Yields. Table 4.6 shows the wavelength threshold below which the production of the pairs of either the ground or excited states of NO($X^2\Pi$, $A^2\Sigma^+$) and O(^3P, ^1D, ^1S) in the photolysis of NO_2 (Okabe 1978). From Table 4.6 it

Table 4.5 Absorption cross sections of NO_2(298 K)

Wavelength (nm)	10^{20} σ (cm^2 molecule^{-1})	Wavelength λ (nm)	10^{20} σ (cm^2 molecule^{-1})	Wavelength λ (nm)	10^{20} σ (cm^2 molecule^{-1})
205	33.8	305	16.0	405	57.7
210	44.5	310	18.8	410	61.5
215	48.9	315	21.6	415	58.9
220	46.7	320	25.4	420	59.2
225	39.0	325	28.8	425	56.7
230	27.7	330	31.9	430	54.0
235	16.5	335	35.9	435	55.5
240	8.30	340	40.2	440	48.4
245	3.75	345	41.8	445	48.8
250	1.46	350	46.1	450	48.1
255	1.09	355	49.8	455	41.2
260	1.54	360	50.8	460	43.0
265	2.18	365	55.0	465	40.9
270	2.92	370	56.1	470	33.6
275	4.06	375	58.9	475	38.5
280	5.27	380	59.2	480	33.4
285	6.82	385	59.4	485	25.2
290	8.64	390	62.0	490	30.7
295	10.6	395	59.2	495	29.3
300	13.0	400	63.9		

Source: IUPAC Subcommittee Report Vol. I (Atkinson et al. 2004)

Table 4.6 Wavelength thresholds (nm) below which the production of each pair of NO and O is energetically possible in the photolysis of NO_2 (0 K)

NO / O	^{3}P	^{1}D	^{1}S
$X^2\Pi$,	397.8	243.9	169.7
$A^2\Sigma^+$	144.2	117.4	97.0

Source: Okabe (1978)

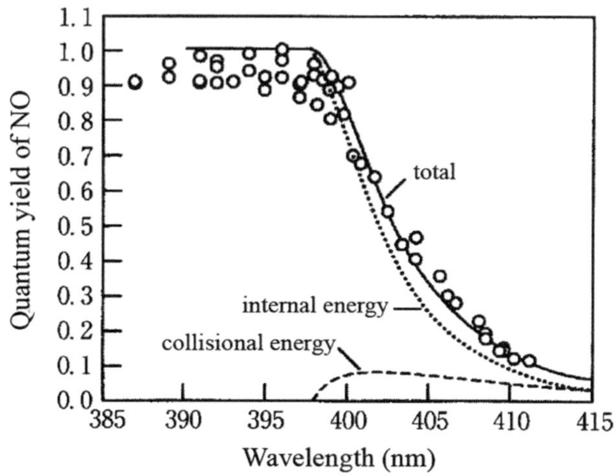

Fig. 4.11 Quantum yields of NO formation in the photolysis of NO_2. The dotted and broken lines show the contribution of the internal energy of vibration and rotation, and the collisional energy to the photolysis beyond the dissociation limit, respectively (Adapted from Roehl et al. 1994)

can be seen that only the photolytic process which is possible by the actinic flux in the troposphere, is

$$NO_2 + h\nu(\lambda < 398 \text{ nm}) \rightarrow NO(X^2\Pi) + O(^3P), \quad \Delta H_0^\circ = 301 \text{ kJ mol}^{-1} \quad (4.9)$$

and the threshold wavelength to give the ground states of NO and O is 397.8 nm at 0 K. The NO production quantum yields in the photolytic process (4.9) near the threshold wavelength at 298 K is shown in Fig. 4.11 (Roehl et al. 1994; Troe 2000). Also, the quantum yields of the photodissociation of NO_2 in the 300–420 nm range recommended by NASA/JPL Evaluation No. 17 based on Roehl et al. (1994) and Troe (2000) are shown in Table 4.7. As clearly shown in Fig. 4.11, Although the quantum yields of NO production falls rapidly at wavelengths longer than the dissociation limit of 398 nm, it is not zero even beyond the threshold wavelength and the production of NO extends to around 420 nm. The production of NO beyond the wavelength range of dissociation limit is interpreted as it is mainly due to the addition of internal energy from the vibronic excited states of ground state molecules and also to the further supplement of the translational energy by a collision of another molecule to the excited NO_2 molecule after photoabsorption. Figure 4.11 depicts the calculated contributions of these processes by dotted and broken lines, respectively.

Table 4.7 Quantum yields for the NO_2 photolysis in the 300–420 nm region (298, 248 K)

Wavelength (nm)	Quantum Yields Φ		Wavelength (nm)	Quantum Yields Φ		Wavelength (nm)	Quantum Yields Φ	
	298 K	248 K		298 K	248 K		298 K	248 K
300–398	1.00	1.00	406	0.30	0.22	414	0.08	0.04
399	0.95	0.94	407	0.26	0.18	415	0.06	0.03
400	0.88	0.86	408	0.22	0.14	416	0.05	0.02
401	0.75	0.69	409	0.18	0.12	417	0.04	0.02
402	0.62	0.56	410	0.15	0.10	418	0.03	0.02
403	0.53	0.44	411	0.13	0.08	419	0.02	0.01
404	0.44	0.34	412	0.11	0.07	420	0.02	0.01
405	0.37	0.28	413	0.09	0.06			

As seen in Fig. 4.9, the absorption spectrum of NO_2 extends over the whole range of the visible region, and the formation rate of excited NO_2 molecules by the absorption of solar radiation beyond the wavelength longer than the dissociation threshold is large. Most of the non-dissociative electronic excited NO_2 molecules undergo the processes,

$$NO_2^* \rightarrow NO_2 + h\nu' \tag{4.10}$$

$$NO_2^* + M \rightarrow NO_2 + M \tag{4.11}$$

and return to the ground state either by emitting fluorescence or being deactivated (quenched) by other atmospheric molecules. The radiative rate of electronically excited NO_2 is ~1.5×10^4 s^{-1} and the fluorescence lifetime is ~70 μs (Donnelly and Kaufman 1978). Meanwhile, the rate constants of quenching by N_2 and O_2 molecules are ~5×10^{-11} cm^3 $molecule^{-1}$ s^{-1} (Donnelly et al. 1979), and the fluorescence quantum yield is about 10^{-5} under the pressure of the lower troposphere, so that most of the excited NO_2 formed by the absorption of the solar radiation are quenched by N_2 and O_2.

However, if the rate constants of the reaction or energy transfer between the electronically excited NO_2 and other atmospheric molecules are large enough, the possibility of such processes cannot be denied. An energy transfer process,

$$NO_2^* + O_2(X^3\Sigma^-{}_g) \rightarrow NO_2 + O_2(a^1\Delta_g) \tag{4.12}$$

to form O_2 $(a^1\Delta_g)$ has been studied by Jones and Bayes (1973), but the role of the electronically excited O_2 molecules formed in reaction (4.12) are thought to be unimportant in the lower atmosphere. On the other hand, it has recently been reported that OH radicals are formed when NO_2 molecules are excited by visible light at 565, 590, 613 nm in the presence of water vapor, and the reaction,

$$NO_2^* + H_2O \rightarrow HONO + OH \tag{4.13}$$

has been proposed (Li et al. 2008). If the reaction has the rate constant $k_{4.13} = 1.7 \times 10^{-13}$ cm^3 $molecule^{-1}$ s^{-1} as reported by Li et al. (2008), this reaction could have the importance in the lower troposphere comparable to the reaction of $O(^1D) + H_2O$ succeeding to the photolysis of O_3. Replication experiments (Carr et al. 2009) and theoretical studies are attempted (Fang et al. 2010).

4.2.3 *Nitrous Acid (HONO)*

Nitrous acid (HONO) is formed not only by the homogeneous reaction of OH + NO, but by the heterogeneous reactions of NO_2 with ground surface, which is further enhanced by photo-irradiation (see Sect. 6.4.2), it exists in the polluted air

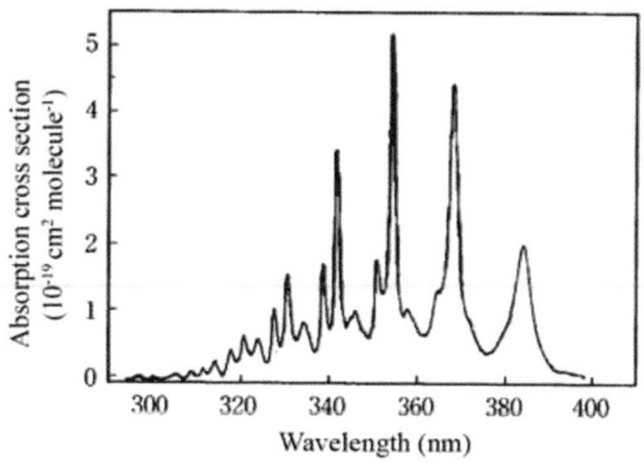

Fig. 4.12 Absorption spectrum of HONO (Adapted from Stutz et al. 2000)

at relatively high concentration even in the daytime when the photolysis rate is fast. Therefore, the photolytic reaction of HONO is very important as a source of the OH radicals in the polluted atmosphere.

Absorption Spectrum and Cross Sections The absorption spectrum of HONO is shown in Fig. 4.12. Since pure HONO cannot be obtained in the laboratory, and a trace amount of NO_2 is always present in the sample, it has been a big problem about how to remove the effect of NO_2 in the measurements of absorption spectrum and cross sections of HONO (Stockwell and Calvert 1978). The absorption spectrum shown in Fig. 4.12 is based on Stutz et al. (2000) with the resolution 0.08 nm, which agrees well with the previous data (Vasudev 1990; Bongartz et al. 1991). The absorption bands of HONO in the wavelength region of 300–400 nm is assigned to the A $A'' \leftarrow$ X A' transition, and the spectrum reveals a clear vibrational structure. This vibrational structure corresponds to the -N = O stretching frequency of the excited states, and the bands at 369, 355 and 342 nm are assigned to 1–0, 2–0 and 3–0, respectively (Vasudev et al. 1984). As for the absolute values of absorption cross sections, the uncertainty was large due to the effect of the impurity NO_2. The absolute values of absorption cross sections are usually compared at the maximum peak of the absorption bands at 354 nm, and relative values normalized to this are given. It should be noted, however, that the recent values are for the measurements with 0.1 nm or higher resolution, but the previous values are for the measurements with lower resolution, care should be taken for the comparison. The absorption cross section at 354 nm by Stuts et al. (2000) is $51.9 \pm 0.03 \times 10^{-20}$ cm^2 (resolution 0.08 nm), that agrees well with the values by Stockwell and Calvert (1978), Bongartz et al. (1994), and Pagsberg et al. (1997) within 5 %. Table 4.8 gives absorption cross sections recommended by NASA/JPL Evaluation No. 17 (Sander et al. 2011). These values are based on Stuts et al. (2000) and averaged over 1 nm intervals.

Photolysis Quantum Yields The broad bands without rotational structure in the spectrum shown in Fig. 4.12 implies that the dissociation lifetime of the excited states is short. Indeed, the HONO molecules absorbing the radiation in this wavelength region is known to dissociate in the pathway,

Table 4.8 Absorption cross section of HONO (298 K)

Wave-length (nm)	10^{20} σ (cm^2 $molecule^{-1}$)	Wave-length (nm)	10^{20} σ (cm^2 $molecule^{-1}$)	Wave-length (nm)	10^{20} σ (cm^2 $molecule^{-1}$)	Wave-length (nm)	10^{20} σ (cm^2 $molecule^{-1}$)	Wave-length (nm)	10^{20} σ (cm^2 $molecule^{-1}$)
300	0.617	320	4.66	340	7.79	360	6.87	380	7.21
301	0.690	321	5.96	341	16.1	361	6.05	381	9.13
302	0.579	322	4.05	342	29.4	362	5.98	382	12.4
303	0.925	323	4.56	343	11.4	363	7.39	383	17.0
304	1.04	324	5.89	344	7.79	364	11.5	384	19.5
305	1.57	325	4.05	345	8.77	365	12.8	385	16.1
306	1.29	326	2.65	346	9.64	366	14.8	386	10.5
307	0.916	327	6.44	347	7.80	367	25.1	387	6.59
308	1.45	328	9.22	348	6.63	368	43.6	388	4.30
309	2.01	329	5.20	349	6.00	369	31.5	389	2.81
310	1.51	330	9.92	350	9.06	370	15.1	390	1.71
311	2.07	331	14.3	351	16.9	371	9.49	391	0.992
312	2.42	332	6.94	352	12.4	372	7.96	392	0.731
313	2.25	333	6.31	353	16.3	373	6.30	393	0.597
314	3.35	334	8.35	354	48.7	374	4.59	394	0.528
315	2.54	335	7.71	355	27.6	375	3.55	395	0.403
316	1.61	336	5.33	356	11.1	376	3.36	396	0.237
317	3.21	337	4.23	357	9.45	377	3.66		
318	4.49	338	9.38	358	9.84	378	4.33		
319	3.19	339	14.3	359	8.37	379	5.66		

Source: NASA/JPL Panel Evaluation No. 17 (Sander et al. 2011)

$$HONO + h\nu(\lambda < 598\ \text{nm}) \rightarrow OH + NO \quad \Delta H_0^\circ = 200\ \text{kJ mol}^{-1} \tag{4.14}$$

with quantum yields of unity (Cox and Derwent 1976). The ratio of another conceived photolytic pathway,

$$HONO + h\nu(\lambda < 367\ \text{nm}) \rightarrow H + NO_2 \quad \Delta H_0^\circ = 326\ \text{kJ mol}^{-1} \tag{4.15}$$

to produce H atoms has been reported to be less than 0.01 (Wollenhaupt et al. 2000).

4.2.4 Nitrogen Trioxide (NO_3), Dinitrogen Pentoxide (N_2O_5)

Nitrogen trioxide (NO_3) is also called nitrate radical, and is formed by the reaction of $O_3 + NO_2$. Dinitrogen pentoxide (N_2O_5) is then formed from the NO_3 and NO_2 in the equilibrium reaction of $NO_3 + NO_2 \rightleftarrows N_2O_5$. The NO_3 and N_2O_5 are important intermediates in nighttime chemistry, and their daytime concentrations are very low since NO_3 absorbs solar radiation strongly and is easily photolyzed.

Absorption Spectrum and Cross Sections of NO_3 NO_3 is a free radical with unpaired electron, and many studies have been performed from spectroscopic interest (Wayne et al. 1991). Figure 4.13 shows the absorption spectrum of NO_3 in the visible region (Sander et al. 2011). As seen in the figure, NO_3 has very strong absorption bands with vibrational structures in the broad range of the visible region 400–700 nm, particularly in the red region of 600–700 nm. These absorption bands correspond to the B-X transition, and the strongest peaks at 662 and 623 nm are assigned to 0–0 and 1–0 bands. The spectrum shown in Fig. 4.13 is those averaged over 1 nm interval. These bands consist of many rotational lines and the high-resolution spectrum separating them are also obtained (Orphal et al. 2003; Osthoff et al. 2007).

The absorption cross section at 662 nm has been used for obtaining the concentration of NO_3 in its detection in the laboratory and ambient air, and many laboratory studies have been conducted to obtain the value. The recommended

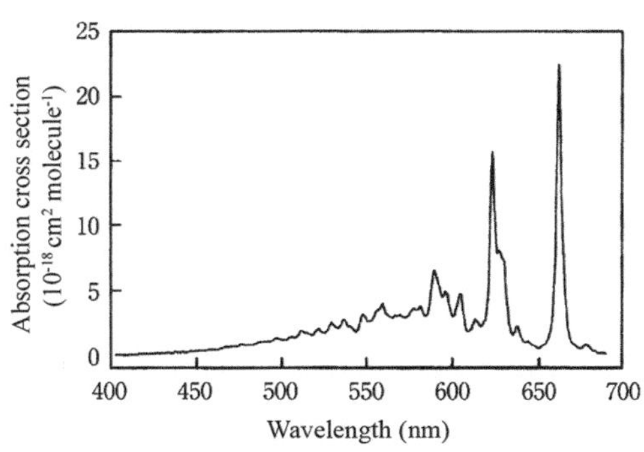

Fig. 4.13 Absorption spectrum of NO_3 (Adapted from Sander et al. 2011)

value by Wayne et al. (1991) is $(2.10 \pm 0.20) \times 10^{-17}$ cm^2 molecule^{-1} and agrees well with the recommended value of $(2.25 \pm 0.15) \times 10^{-17}$ cm^2 molecule-^{1} by the NASA/JPL Evaluation No. 17 (Sander et al. 2011). The absorption cross section of NO_3 at 662 nm peak is known to have temperature dependence, and Osthoff et al. (2007) presented the formula

$$\sigma(662\ \text{nm}, \text{T}) = (4.582 \pm 0.096) \\ = [(0.00796 \pm 0.0031) \times \text{T}] \times 10^{-17}\text{cm}^2\ \text{molecule}^{-1} \quad (4.16)$$

modifying the parameters of the experimental formula by Yokelson et al. (1994). The cause of the temperature dependence has been ascribed to the temperature change in the Boltzmann distribution of the vibration-rotation levels (Orphal et al. 2003).

The absorption cross sections of NO_3 averaged over 1 nm intervals recommended by NASA/JPL Evaluation No. 17 (Sander et al. 2011) are given in Table 4.9. These values are based on the data by Sander (1986) that were normalized by the value at 662 nm mentioned above.

Photolysis Quantum Yields of NO_3 The two reaction pathways of the photolysis of NO_3 in the visible region can be,

$$NO_3 + h\nu(\lambda < 1031\ \text{nm}) \rightarrow NO + O_2 \quad \Delta H_0^\circ = 11.6\ \text{kJ mol}^{-1} \quad (4.17)$$

$$+ h\nu\ (\lambda < 587\ \text{nm}) \rightarrow NO_2 + O(^3P). \quad \Delta H_0^\circ = 205\ \text{kJ mol}^{-1} \quad (4.18)$$

Furthermore, as judged from the high resolution spectrum, the dissociation lifetime of the excited state of NO_3 for the bands at wavelengths longer than 600 nm where sharp vibration-rotation structure is observed, is long and the observation of fluorescence has been known from the laboratory studies,

$$NO_3 + h\nu \rightarrow NO_3{}^* + h\nu'. \quad (4.19)$$

Reaction (4.17) is nearly thermo-neutral and there is no energy restriction, but it has large energy barrier since it passes the three-centered O-N-O transition state. The barrier height has been obtained as 198 kJ mol^{-1} from the photofragment experiment (Davis et al. 1993). On the other hand, the wavelength threshold for reaction (4.18) has been determined as 587 ± 3 nm, from which the new values of heat of formation of NO_3 at 0 K and 298 K are obtained as 79.0 ± 1.4, 73.7 ± 1.4 kJ mole^{-1}, respectively.

Figure 4.14 depicts the plot of quantum yields of Φ ($NO + O_2$) and Φ ($NO_2 + O$) for the reaction (4.17) and (4.18) as a function of wavelength together with the fluorescence quantum yields. As shown in Fig. 4.14, Φ ($NO_2 + O$) is nearly unity at the threshold wavelength of 587 nm. Φ ($NO_2 + O$) decreases with the wavelength in the longer wavelength region, but still has the value of ca. 0.1 at around 635 nm. In the photolysis of reaction (4.18) in the longer wavelength range than 587 nm, the

Table 4.9 Absorption cross sections of NO_3 (298 K)

Wave-length (nm)	10^{20} σ (cm^2 molecule^{-1})	Wave-length (nm)	10^{20} σ (cm^2 molecule^{-1})	Wave-length (nm)	10^{20} σ (cm^2 molecule^{-1})	Wave-length (nm)	10^{20} σ(cm^2 molecule^{-1})	Wave-length (nm)	10^{20} σ (cm^2 molecule^{-1})
420	9	470	63	520	180	570	299	620	350
422	10	472	69	522	206	572	294	622	1090
424	10	474	66	524	176	574	306	624	1290
426	15	476	84	526	175	576	350	626	783
428	13	478	78	528	225	578	354	628	789
430	18	480	75	530	239	580	358	630	724
432	16	482	76	532	216	582	351	632	350
434	20	484	83	534	218	584	302	634	176
436	16	486	98	536	275	586	355	636	181
438	23	488	102	538	251	588	540	638	217
440	21	490	111	540	225	590	638	640	132
442	23	492	107	542	201	592	548	642	99
444	21	494	109	544	183	594	449	644	102
446	26	496	129	546	260	596	495	646	80
448	26	498	128	548	320	598	393	648	66
450	31	500	121	550	265	600	296	650	53
452	36	502	118	552	264	602	355	652	65
454	38	504	135	554	298	604	468	654	88
456	38	506	143	556	349	606	355	656	142
458	39	508	136	558	376	608	198	658	260
460	42	510	162	560	355	610	189	660	798
462	42	512	189	562	311	612	239	662	2250

(continued)

Table 4.9 (continued)

Wave-length (nm)	10^{20} σ (cm^2 $molecule^{-1}$)	Wave-length (nm)	10^{20} σ (cm^2 $molecule^{-1}$)	Wave-length (nm)	10^{20} σ (cm^2 $molecule^{-1}$)	Wave-length (nm)	10^{20} σ(cm^2 $molecule^{-1}$)	Wave-length (nm)	10^{20} σ (cm^2 $molecule^{-1}$)
464	51	514	169	564	291	614	273	664	1210
466	58	516	167	566	305	616	224	666	532
468	60	518	154	568	305	618	256	668	203

Source: NASA/JPL Panel Evaluation No.17 (Sander et al. 2011)

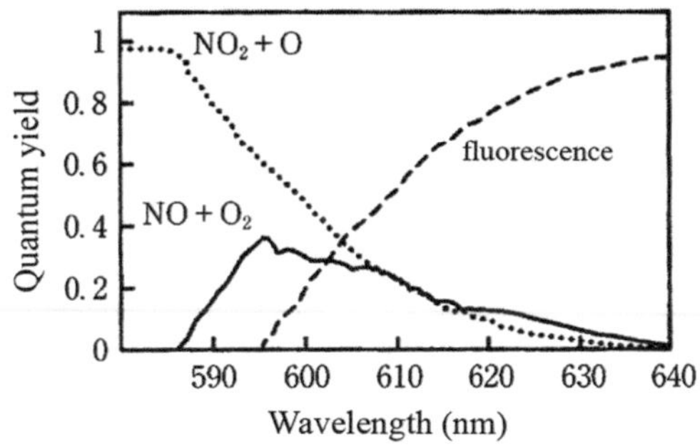

Fig. 4.14 Photolysis and fluorescence quantum yields of NO_3 (Adapted from Johnston et al. 1996)

energy is supplemented by the vibronic energy in the ground state molecule as in the case of NO_2 mentioned before. The values of Φ ($NO + O_2$) and Φ ($NO_2 + O$) obtained by Johnston et al. (1996) agrees well with those obtained by different experimental method by Orlando et al. (1993) except a part of wavelength region (605–620 nm). As seen in Fig. 4.14, the values of Φ ($NO + O_2$) is 0 at the shorter wavelength where Φ_2 ($NO_2 + O$) is unity, it increases with wavelength and take a maximum value of 0.35 at 595 nm. The Φ ($NO + O_2$) decreases gradually at the longer wavelength, and is ca. 0.1 at around 630 nm. In the longer wavelength than 587 nm, the photodissociation quantum yields of NO_3, the sum of Φ ($NO + O_2$) and Φ ($NO_2 + O$) is smaller than unity, and the fluorescence quantum yields of the pathway (4.13) increases. Photolysis quantum yields of NO_3 in the wavelength region longer than 587 nm shows large temperature dependence and decreases with temperature (Johnston et al. 1996). Particularly, since the reaction (4.18) in this region is due to the hot bands the temperature dependence of Φ ($NO_2 + O$) is much larger than Φ ($NO + O_2$). Table 4.10 gives photolytic quantum yields of NO_3 at 298, 230 and 190 K recommended by NASA/JPL Evaluation No. 17 (Sander et al. 2011).

Absorption Spectrum and Cross Sections of N_2O_5 Figure 4.15 shows absorption spectrum of N_2O_5 (Harwood et al. 1998). The absorption spectrum of N_2O_5 has a maximum at 160 nm, decreases monotonically toward longer wavelength, and extends to near ultraviolet region of tropospheric solar actinic flux. Table 4.11 tabulates absorption cross sections of N_2O_5 recommended by IUPAC Subcommittee Report Vol. I (Atkinson et al. 2004). The recommendation is based on the values of Harwood et al. (1993, 1998) in the wavelength region longer than 240 nm, and of Yao et al. (1982), Osborne et al. (2000), etc. in regions shorter than this.

The temperature dependence is also found in the absorption cross sections of N_2O_5. The dependence is particularly large in wavelength regions longer than 280 nm where the photolysis occurs in the troposphere, and the cross sections decreases with the decrease of temperature.

Photolysis Quantum Yields of N_2O_5 The photolytic process of N_2O_5 of the interest in tropospheric chemistry is

Table 4.10 Quantum yields of the photolysis of NO_3 (298, 230, 190 K)

	Φ ($NO + O_2$)			Φ ($NO_2 + O$)		
λ (nm)	298 K	230 K	190 K	298 K	230 K	190 K
586	0.015	0.026	0.038	0.97	0.97	0.96
588	0.097	0.16	0.22	0.89	0.84	0.78
590	0.19	0.30	0.40	0.79	0.70	0.60
592	0.25	0.38	0.50	0.73	0.61	0.51
594	0.33	0.49	0.61	0.65	0.51	0.39
596	0.36	0.50	0.60	0.59	0.43	0.31
598	0.32	0.42	0.47	0.53	0.37	0.25
600	0.29	0.35	0.36	0.47	0.31	0.20
602	0.29	0.32	0.31	0.42	0.25	0.15
604	0.28	0.28	0.50	0.35	0.20	0.11
606	0.27	0.25	0.21	0.30	0.16	0.080
608	0.25	0.22	0.17	0.26	0.13	0.062
610	0.24	0.19	0.14	0.23	0.11	0.048
612	0.20	0.15	0.10	0.21	0.10	0.042
614	0.17	0.11	0.071	0.17	0.068	0.028
616	0.16	0.10	0.060	0.14	0.053	0.020
618	0.14	0.084	0.045	0.11	0.039	0.014
620	0.13	0.072	0.036	0.090	0.030	0.010
622	0.12	0.062	0.029	0.070	0.022	0.0070
624	0.11	0.050	0.022	0.055	0.016	0.0048
626	0.092	0.041	0.017	0.044	0.012	0.0034
628	0.074	0.030	0.012	0.034	0.0087	0.0023
630	0.065	0.025	0.0090	0.026	0.0063	0.0015
632	0.051	0.018	0.0060	0.020	0.0043	0.0010
634	0.043	0.014	0.0045	0.016	0.0034	0.0007
636	0.032	0.0099	0.0029	0.012	0.0023	0.0005
638	0.027	0.0077	0.0022	0.0096	0.0018	0.0003
640	0.020	0.0054	0.0014	0.0072	0.0012	0.0002

Source: Extracted from NASA/JPL Evaluation No.17 (Sander et al. 2011)

$$N_2O_5 + h\nu\ h\nu(\lambda < 1289\ \text{nm}) \rightarrow NO_3 + NO_2. \quad \Delta H_0 = 94.8\ \text{kJ mol}^{-1} \quad (4.20)$$

The photolytic quantum yields of N_2O_5 is reported to be close to unity (Harwood et al. 1993, 1998).

4.2.5 *Formaldehyde (HCHO), Acetaldehyde (CH_3CHO)*

Formaldehyde (HCHO) exists globally in the natural atmosphere as an oxidation product of CH_4. It is also an oxidation product of biogenic hydrocarbons of plant

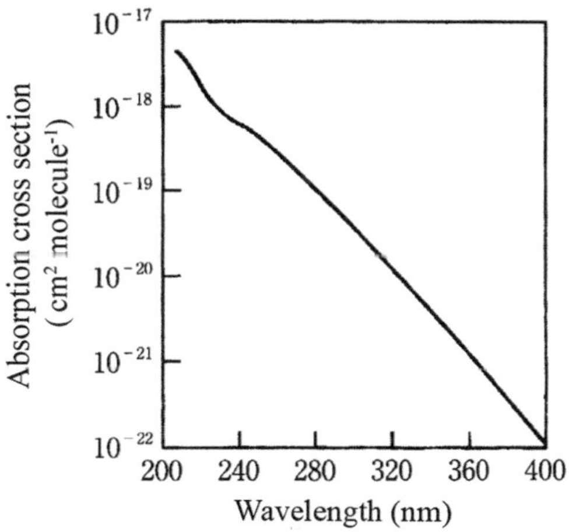

Fig. 4.15 Absorption spectrum of N_2O_5 (Adapted from Harwood et al. 1998)

Table 4.11 Absorption cross sections of N_2O_5 (298 K)

Wavelength (nm)	10^{20} σ (cm^2 molecule^{-1})	Wavelength (nm)	10^{20} σ (cm^2 molecule^{-1})	Wavelength λ (nm)	10^{20} σ (cm^2 molecule^{-1})
210	470	280	11	350	0.22
215	316	285	8.6	355	0.16
220	193	290	6.7	360	0.12
225	128	295	5.1	365	0.091
230	91	300	3.9	370	0.072
235	73	305	2.9	375	0.053
240	60	310	2.2	380	0.041
245	51	315	1.6	385	0.032
250	40	320	1.2	390	0.023
255	32	325	0.89	395	0.017
260	26	330	0.67	400	0.014
265	20	335	0.50	405	0.010
270	16	340	0.38	410	0.008
275	13	345	0.28		

Source: IUPAC Subcommittee Report Vol. I (Atkinson et al. 2004)

origin over lands. On the other hand, HCHO is a secondary pollutant formed by the oxidation of anthropogenic hydrocarbons in the polluted atmosphere, and also a primary pollutant emitted from auto-exhaust and biomass burning. HCHO is the aldehyde that exists usually in the highest concentration in polluted atmospheres. Photolysis of HCHO gives H atoms HCO radicals which transform to HO_2 radicals in the atmosphere, and has a large impact on the photochemical ozone production in the polluted atmosphere.

Acetaldehyde (CH_3CHO) is an important aldehyde next to HCHO in polluted atmospheres. As with HCHO, CH_3CHO is formed as a secondary pollutant from

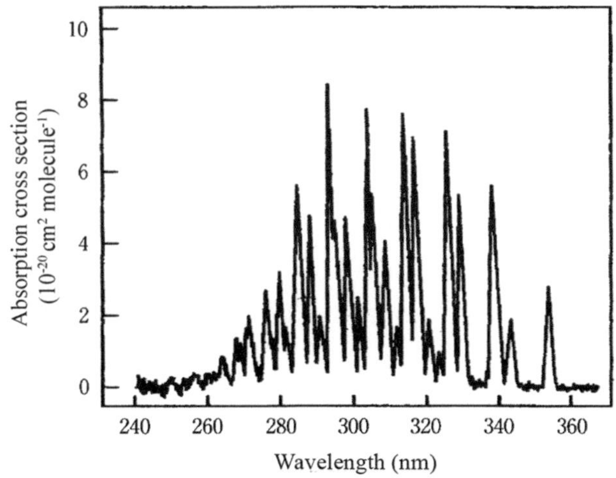

Fig. 4.16 Absorption spectrum of HCHO. Resolution 0.1 nm (Adapted from Rogers 1990)

anthropogenic hydrocarbons and also emitted as a primary pollutant from auto-exhaust and biomass burning. The photolysis of CH_3CHO is also important for the photochemical ozone production as a source of radicals.

Absorption Spectrum and Cross Sections of HCHO For carbonyl compounds such as aldehydes and ketones, absorption bands due to the electronic transition called $n - \pi^*$ transition, in which the isolated pair of lone-pair electrons on the O atom of carbonyl group ($-C=O$) is excited to the excited π orbital of the double bond, appears around 300 nm. Since this transition is a forbidden transition, the absorption cross sections are not very large (~10^{-20} cm^2 molecule^{-1}) in general. However, since the absorption bands extend to near 350 nm where solar actinic flux grows, their photolyses are very important in the troposphere.

Figure 4.16 shows absorption spectrum of HCHO (Rogers 1990). As shown in Fig. 4.16, the absorption spectrum of HCHO spreads over 260–360 nm and it characterized by many vibrational structures. Since the absorption has the banded structure, its spectral shape is very much dependent on the spectral resolution. The spectrum with the resolution less than 0.1 nm has been obtained by Rogers(1990), Cantrell et al. (1990), Meller and Moortgat (2000), etc. with good agreement to each other. Higher resolution spectrum (0.001 nm, 0.1 cm^{-1}) separates the rotational lines (Pope et al. 2005a; Co et al. 2005; Smith et al. 2006).

Temperature dependence is known for the absorption cross sections of HCHO (Cantrell et al. 1990; Meller and Moortgat 2000; Smith et al. 2006), and a linear approximation formula,

$$\sigma(\lambda, T) = \sigma(\lambda, 298\ \text{K}) + \Gamma(\lambda) \times (T - 298), \tag{4.21}$$

has been proposed by Meller and Moortgat (2000) in the 223–323 nm region.

Table 4.12 gives absorption cross sections of HCHO taken form the NASA/JPL Evaluation No. 17 (Sander et al. 2011) and IUPAC Subcommittee Report Vol. II (Atkinson et al. 2006). These values are based on the high-resolution absorption

Table 4.12 Absorption cross sections of HCHO (298 K) (Averaged over 1 nm interval)

Wave-length (nm)	10^{20} σ (cm^2 molecule^{-1})	Wave-length (nm)	10^{20} σ (cm^2 molecule^{-1})	Wave-length (nm)	10^{20} σ (cm^2 molecule^{-1})	Wave-length (nm)	10^{20} σ (cm^2 molecule^{-1})
240	0.078	270	0.963	300	0.964	330	3.87
241	0.078	271	1.94	301	1.62	331	1.41
242	0.123	272	1.43	302	0.854	332	0.347
243	0.159	273	0.811	303	3.02	333	0.214
244	0.110	274	0.658	304	7.22	334	0.159
245	0.131	275	2.14	305	4.75	335	0.097
246	0.163	276	2.58	306	4.29	336	0.126
247	0.151	277	1.57	307	1.78	337	0.383
248	0.234	278	1.03	308	1.38	338	1.92
249	0.318	279	2.45	309	3.25	339	5.38
250	0.257	280	2.34	310	1.74	340	3.15
251	0.204	281	1.56	311	0.462	341	0.978
252	0.337	282	0.973	312	1.19	342	0.509
253	0.289	283	0.722	313	0.906	343	1.92
254	0.342	284	4.26	314	5.64	344	1.27
255	0.450	285	4.05	315	5.57	345	0.437
256	0.628	286	2.10	316	2.56	346	0.119
257	0.443	287	1.15	317	5.78	347	0.044
258	0.307	288	3.17	318	3.15	348	0.075
259	0.617	289	3.22	319	0.978	349	0.038
260	0.605	290	1.17	320	1.19	350	0.036
261	0.659	291	1.84	321	1.60	351	0.089
262	0.603	292	0.797	322	0.722	352	0.729
263	1.08	293	3.12	323	0.328	353	2.27

(continued)

Table 4.12 (continued)

Wave-length (nm)	10^{20} σ (cm^2 $molecule^{-1}$)	Wave-length (nm)	10^{20} σ (cm^2 $molecule^{-1}$)	Wave-length (nm)	10^{20} σ (cm^2 $molecule^{-1}$)	Wave-length (nm)	10^{20} σ (cm^2 $molecule^{-1}$)
264	0.947	294	7.15	324	0.858	354	1.64
265	0.531	295	4.05	325	1.58	355	0.696
266	0.539	296	2.47	326	6.88	356	0.148
267	1.36	297	1.37	327	4.37	357	0.035
268	1.24	298	4.22	328	1.22	358	0.019
269	0.991	299	3.17	329	3.12	359	0.011

Source: IUPAC Subcommittee Report Vol. II (Atkinson et al. 2006)

coefficients by Meller and Moortgat (2000), and are averaged over the wavelength intervals for the use of atmospheric model calculations.

Photolysis Quantum Yields of HCHO Two reaction pathways,

$$\mathrm{HCHO} + h\nu(\lambda < 330\ \mathrm{nm}) \rightarrow \mathrm{H} + \mathrm{HCO} \quad \Delta H_0^\circ = 363\ \mathrm{kJ\ mol^{-1}} \tag{4.22}$$

$$+ h\nu\ (\text{all wavelength}) \rightarrow \mathrm{H_2} + \mathrm{CO} \quad \Delta H_0^\circ = -8.9\ \mathrm{kJ\ mol^{-1}} \tag{4.23}$$

are energetically possible in the photolysis of HCHO in the near ultraviolet region. The energy threshold of reaction (4.22) is 363 kJ mole^{-1},and the corresponding wavelength threshold is 330 nm. On the other hand, the reaction enthalpy of the process (4.23) is negative and there is no thermochemical threshold restriction. Among these processes, reaction (4.23) produces only stable molecules whereas HO_2 radicals are produced from H atoms and HCO radicals formed in the reaction (4.22) by,

$$\mathrm{H} + \mathrm{O_2} + \mathrm{M} \rightarrow \mathrm{HO_2} + \mathrm{M} \tag{4.24}$$

$$\mathrm{HCO} + \mathrm{O_2} \rightarrow \mathrm{HO_2} + \mathrm{CO}, \tag{4.25}$$

and affects largely photochemical processes in the troposphere. Therefore, obtaining the accurate value of quantum yield of reaction (4.22) is very important form the atmospheric chemistry point of view.

Photolytic quantum yields Φ (H + HCO) for reaction (4.22) have been measured with the middle-resolution (0.62 nm) (Smith et al. 2002), and the high-resolution (0.0035 nm) (Carbajo et al. 2008). Troe (2007) reviewed the literature values of Φ (H + HCO) and Φ (H_2 + CO), and summarized their wavelength dependence at 300 K and 1 atm as illustrated in Fig. 4.17. The recent values of photolytic quantum

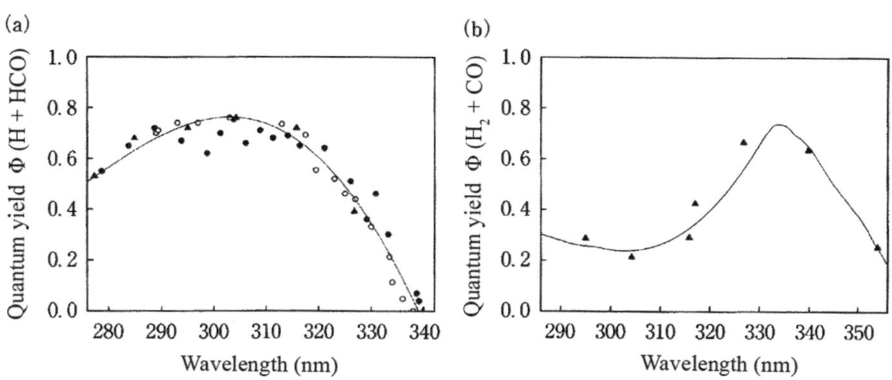

Fig. 4.17 Photolytic quantum yields of HCHO: (**a**) Φ(H + HCO), (**b**) Φ(H_2 + CO) (Adapted from Troe 2007)

Table 4.13 Quantum yields for the photolysis of HCHO (300 K, 1 atm)

λ (nm)	Φ (H + HCO)	Φ (H_2 + CO)	λ (nm)	Φ (H + HCO)	Φ (H_2 + CO)
250	0.310	0.490	310	0.737	0.263
255	0.304	0.496	315	0.685	0.315
260	0.307	0.493	320	0.603	0.397
265	0.343	0.477	325	0.489	0.511
270	0.404	0.441	330	0.343	0.657
275	0.479	0.391	335	0.165	0.735
280	0.560	0.347	340	0.0	0.645
285	0.633	0.307	345	0.0	0.505
290	0.690	0.278	350	0.0	0.375
295	0.734	0.256	355	0.0	0.220
300	0.758	0.242	360	0.0	0.04
305	0.760	0.240			

Source: NASA/JPL Evaluation No.17 (Sander et al. 2011)

yields Φ (H + HCO) agree well with the previous one when normalized by the value of 0.753 at 303.75 nm. However, the values of Φ (H + HCO) obtained in the high-resolution showed banded structure and gave considerably smaller values particularly around 305 nm (Carbajo, et al. 2008).

Table 4.13 gives the values of Φ (H + HCO) and Φ (H_2 + CO) recommended by NASA/JPL Evaluation No. 17. These values are obtained by integrating the previous data by Horowitz and Calvert (1978), Moortgat and Warneck (1979), Moortgat et al. (1983) and others, and the recent one by Smith et al. (2002), Pope et al. (2005b) and Carbajo et al. (2008). The quantum yields of the formation of CO combining reaction (4.22) and (4.23) is unity for the range of 290–350 nm being independent on temperature and pressure (Moortgat and Warneck 1979; Moortgat et al. 1983). In the wavelength region longer than 350 nm, Φ (H + HCO) does not show any temperature and pressure dependence, but large temperature and pressure dependences are seen for Φ (H_2 + CO) (Moortgat et al. 1983).

Absorption Spectrum and Cross Sections of CH_3CHO Figure 4.18 shows the absorption spectrum of CH_3CHO together with those for other open-chain aldehydes such as propanal CH_3CH_2CHO, butanal CH_3 (CH_2) $_2CHO$, and iso-butanal (CH_3) $_2CHCHO$ (Martinez et al. 1992). For these aldehydes, absorption peaks corresponding to the $n - \pi^*$ transition appear near 290 nm similar to formaldehyde. The absorption spectra of these aldehydes are characterized by diffuse vibrational structures near the peak of absorption, that are different from the sharp bands of formaldehyde, and also shifted to a shorter wavelength than formaldehyde. The wavelength range of the absorption spectra does not change, but the absorption cross section in the longer wavelength side increases as the carbon number increase for these open-chain aldehydes.

Table 4.14 gives the absorption cross sections of CH_3CHO recommended by NASA/JPL Evaluation No. 17 (Sander et al. 2011), which is based on Martinez et al. (1992) and Libuda and Zabel (1995).

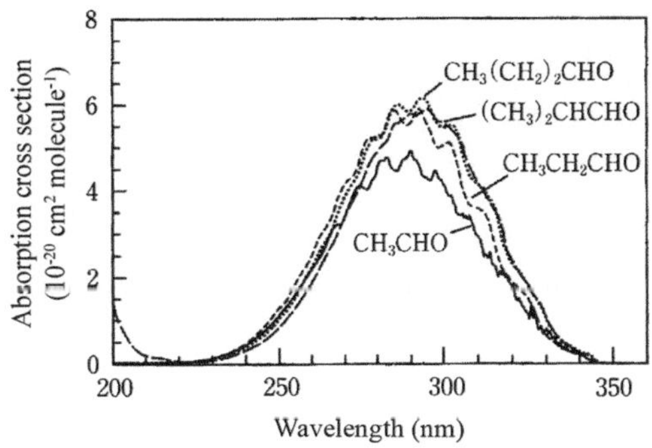

Fig. 4.18 Absorption spectra of CH_3CHO and other aliphatic aldehydes (Adapted from Martinez et al. 1992)

Table 4.14 Absorption cross sections of CH_3CHO (298 K)

Wavelength (nm)	10^{20} σ (cm^2 molecule^{-1})	Wavelength (nm)	10^{20} σ (cm^2 molecule^{-1})	Wavelength (nm)	10^{20} σ (cm^2 molecule^{-1})
230	0.151	290	4.86	326	1.09
234	0.241	292	4.66	328	0.715
238	0.375	294	4.31	330	0.699
242	0.639	296	4.24	332	0.496
246	0.887	298	4.41	334	0.333
250	1.18	300	4.15	336	0.227
254	1.57	302	3.87	338	0.212
258	2.03	304	3.46	340	0.135
262	2.45	306	3.41	342	0.042
266	3.06	308	3.31	344	0.027
270	3.38	310	2.92	346	0.020
274	4.03	312	2.52	348	0.016
278	4.15	314	2.38	350	0.008
280	4.48	316	2.07	352	0.005
282	4.66	318	1.98	354	0.004
284	4.58	320	1.70	356	0.005
286	4.41	322	1.38	358	0.004
288	4.69	324	1.06	360	0.003

Source: NASA/JPL Panel Evaluation No. 17 (Sander et al. 2011)

Photolysis Quantum Yields of CH_3CHO: Three reaction pathways,

$$CH_3CHO + h\nu(\lambda < 340\ \text{nm}) \rightarrow CH_3 + HCO \quad \Delta H_0^\circ = 352\ \text{kJ mole}^{-1} \tag{4.26}$$

$$+ h\nu\ (\text{all wavelength}) \rightarrow CH_4 + CO \quad \Delta H_0^\circ = -20\ \text{kJ mole}^{-1} \tag{4.27}$$

$$+ h\nu\ (\lambda < 321\ \text{nm}) \rightarrow CH_3CO + H \quad \Delta H_0^\circ = 373\ \text{kJ mole}^{-1} \tag{4.28}$$

are known for the photolytic processes of CH_3CHO in the wavelength range of solar actinic flux in the troposphere. Among these processes, the energy thresholds (0 K)

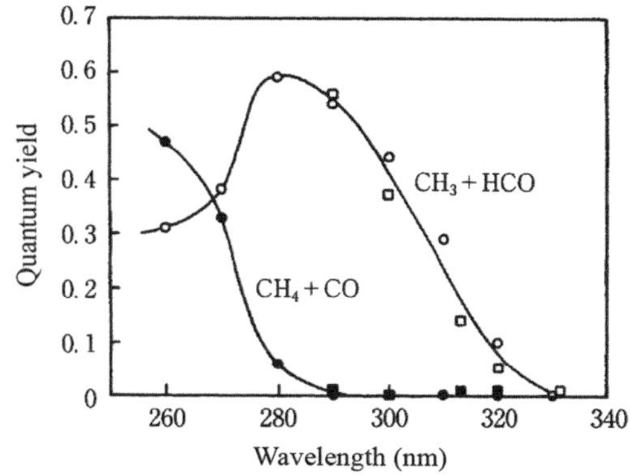

Fig. 4.19 Photolytic quantum yields of CH_3CHO. ○, □: Φ(CH_3+HCO), ●, ■: Φ(CH_4+CO) (Adapted from Atkinson and Lloyd 1984)

of reaction (4.26) and (4.28) are 352, 373 kJ mole^{-1} corresponding to the wavelengths thresholds of 340, 321 nm, respectively. Meanwhile, reaction (4.27) is a exothermic process, so there exists no thermochemical threshold.

The measurements of photolytic quantum yields of CH_3CHO is reported in the 250–330 nm region at 1 atm. as a function of wavelength by Meyrahn et al. (1981), Horowitz and Calvert (1982) based on the product yield analysis. Figure 4.19 depicts the wavelength dependence of Φ (CH_3+HCO) and Φ (CH_4+CO) of the photolysis of CH_3CHO at 1 atm. given by Atkinson and Lloyd (1984). Also, the recommended numerical values by the NASA/JPL Evaluation No. 17 (Sander et al. 2011) for Φ (CH_3+HCO) and Φ (CH_4+CO) for 256–332 nm at 1 atm. are shown in Table 4.15. These values are based on the Meyrahn et al. (1981), Horowitz and Calvert (1982), and the review by Atkinson and Lloyd (1984). According to these results, Φ (CH_4+CO) is 0.48 at 256 nm, and decreases with the increase of wavelengths to 0 at 294 nm. On the contrary, the peak value of Φ (CH_3+HCO) is 0.59 at around 283 nm, decreases with the increase of wavelength, but maintains a value larger than 0.01 until 330 nm. The values of Φ (CH_3CO+H), which are not shown in the figure and table is 0.025 at 300 nm and decreases with wavelength reaching to 0 at 320 nm. Therefore, reaction (4.27) does not occur in the photolysis in the troposphere, and the quantum yields of reaction (4.28) are very small so that only reaction (4.26) should be considered in the tropospheric chemistry. Temperature and pressure dependences are known for the photolysis quantum yields of CH_3CHO, and they decrease with pressure. This suggests that the dissociative lifetime of the excited state of CH_3CHO is long enough to be subjected to molecular collisions and the quenching of the excited state occurs as the pressure increases.

4.2.6 Acetone (CH_3COCH_3)

Acetone CH_3COCH_3 is formed secondarily in the atmosphere by the oxidation reactions of non-methane hydrocarbons in addition to direct emissions from

Table 4.15 Quantum yields for the photolysis of CH_3CHO (298 K, 1 atm)

λ (nm)	Φ ($CH_3 + HCO$)	Φ ($CH_4 + CO$)	λ (nm)	Φ ($CH_3 + HCO$)
256	0.29	0.48	296	0.47
258	0.30	0.47	298	0.45
260	0.31	0.45	300	0.43
262	0.32	0.43	302	0.40
264	0.34	0.40	304	0.38
266	0.36	0.37	306	0.35
268	0.38	0.33	308	0.31
270	0.41	0.29	310	0.28
272	0.44	0.25	312	0.24
274	0.48	0.20	314	0.19
276	0.53	0.16	316	0.15
278	0.56	0.09	318	0.12
280	0.58	0.06	320	0.10
282	0.59	0.04	322	0.07
284	0.59	0.03	324	0.05
286	0.58	0.02	326	0.03
288	0.56	0.01	328	0.02
290	0.54	0.01	330	0.01
292	0.52	0.005	332	0.00
294	0.50	0.00		

Source: NASA/JPL Evaluation No.17 (Sander et al. 2011)

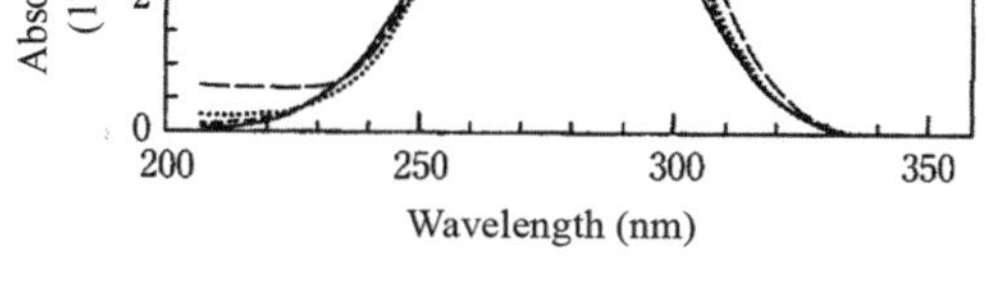

Fig. 4.20 Absorption spectra of CH_3COCH_3 and other aliphatic ketones (Adapted from Martinez et al. 1992)

biogenic and anthropogenic sources. Since the atmospheric lifetime of acetone is relatively long, it exists over the whole range of the troposphere at the concentration of ca. 1 ppbv. The photolysis of acetone in the troposphere is important as a source of HO_x radicals in the free troposphere as well as the major loss process of acetone itself.

Absorption Spectrum and Cross Sections Figure 4.20 shows the absorption spectrum of CH_3COCH_3 together with other homologue ketones (Martinez et al. 1992). The ketones show absorption in 200–350 nm due to the n-π* transitions similar to the aldehydes, but the absorption shifts to a little shorter wavelength and vibrational structures are inconspicuous compared to aldehydes.

The absorption spectrum of CH_3COCH_3 has temperature dependence and the absorption cross sections decrease as the temperature decreases (Hynes et al. 1992; Gierczak et al. 1998). Table 4.16 shows the extracts of absorption cross sections (1 nm resolution) recommended by NASA/JPL Evaluation No. 17 (Sander et al. 2011) based on Gierczak et al. (1998).

Photolytic Quantum Yields: Two reaction pathways,

$$CH_3COCH_3 + h\nu(\lambda < 338\ \text{nm}) \rightarrow CH_3CO + CH_3 \quad \Delta H_0^\circ = 354\ \text{kJ mole}^{-1} \tag{4.29}$$

$$CH_3COCH_3 + h\nu(\lambda < 299\ \text{nm}) \rightarrow 2CH_3 + CO \quad \Delta H_0^\circ = 400\ \text{kJ mol}^{-1} \tag{4.30}$$

are known for the photolytic processes of CH_3COCH_3, and the wavelength thresholds for reactions (4.29) and (4.30) are 338, 299 nm, respectively (Atkinson et al. 2002). From these wavelength thresholds, it can be conceived that the reaction (4.29) is the more important photolytic process in the troposphere. It is also know that the photolytic quantum yields of CH_3COCH_3 has strong pressure dependence as well as temperature dependence (Meyrahn et al. 1986; Gierczak et al. 1998; Emrich and Warneck 2000; Blitz et al. 2004). From these results, the photolysis of CH_3COCH_3 is thought to actually proceed in the following processes,

$$CH_3COCH_3 + h\nu \rightarrow {}^1[CH_3COCH_3]^* \tag{4.31}$$

$${}^1[CH_3COCH_3]^* \rightarrow {}^3[CH_3COCH_3]^* \tag{4.32}$$

$${}^1[CH_3COCH_3]^* \rightarrow 2CH_3 + CO \tag{4.33}$$

$${}^1[CH_3COCH_3]^* + M \rightarrow CH_3COCH_3 + M \tag{4.34}$$

$${}^3[CH_3COCH_3]^* \rightarrow CH_3CO + CH_3 \tag{4.35}$$

$${}^3[CH_3COCH_3]^* + M \rightarrow CH_3COCH_3 + M \tag{4.36}$$

Here, ${}^1[CH_3COCH_3]^*$ is the excited singlet state molecule reached directly by the photo-absorption, ${}^3[CH_3COCH_3]^*$ is the excited triplet state molecule formed by intersystem crossing (non-radiative transition between the different spin multiplicity such as between the singlet and triplet states). It has been shown that the reaction processes (4.29) and (4.30) occur through the ${}^3[CH_3COCH_3]^*$ and ${}^1[CH_3COCH_3]^*$ molecule, respectively (Emrich and Warneck 2000). Both ${}^1[CH_3COCH_3]^*$ and ${}^3[CH_3COCH_3]^*$ have relatively long lifetimes and are subject to collisional deactivation such as (4.34) and (4.36). This is the cause of the strong pressure dependence of Φ (CH_3CO+ CH_3) and Φ ($2CH_3 + CO$) (Meyrahn et al. 1986; Emrich and Warneck 2000; Blitz et al. 2004).

Table 4.16 Absorption cross sections of CH_3COCH_3 (298 K)

Wave-length (nm)	10^{20} σ (cm^2 molecule^{-1})	Wave-length (nm)	10^{20} σ (cm^2 molecule^{-1})	Wave-length (nm)	10^{20} σ (cm^2 molecule^{-1})	Wave-length (nm)	10^{20} σ (cm^2 molecule^{-1})
220	0.246	250	2.47	280	4.91	310	1.36
222	0.294	252	2.74	282	4.79	312	1.14
224	0.346	254	3.01	284	4.62	314	0.944
226	0.419	256	3.30	286	4.44	316	0.760
228	0.492	258	3.57	288	4.28	318	0.598
230	0.584	260	3.81	290	4.06	320	0.455
232	0.693	262	4.07	292	3.82	322	0.348
234	0.815	264	4.32	294	3.57	324	0.248
236	0.956	266	4.49	296	3.26	326	0.174
238	1.11	268	4.64	298	2.98	328	0.113
240	1.30	270	4.79	300	2.67	330	0.0740
242	1.50	272	4.91	302	2.45	332	0.0465
244	1.72	274	4.94	304	2.18	334	0.0311
246	1.95	276	4.93	306	1.89	336	0.0199
248	2.20	278	4.94	308	1.61	338	0.0135

Source: NASA/JPL Panel Evaluation No. 17 (Sander et al. 2011)

Table 4.17 Quantum yields for the photolysis of CH_3COCH_3 (295, 273, 218 K)

	Φ				Φ		
λ (nm)	295 K	273 K	218 K	λ (nm)	295 K	273 K	218 K
280	0.60	0.55	0.66	304	0.18	0.087	0.056
282	0.56	0.51	0.62	306	0.14	0.061	0.032
284	0.52	0.47	0.57	308	0.10	0.043	0.018
286	0.48	0.43	0.50	310	0.077	0.030	0.010
288	0.45	0.39	0.43	312	0.058	0.022	0.0057
290	0.41	0.35	0.36	314	0.045	0.016	0.0033
292	0.38	0.31	0.29	316	0.035	0.012	0.0019
294	0.35	0.28	0.23	318	0.028	0.0088	0.0011
296	0.32	0.25	0.18	320	0.022	0.0068	0.0007
298	0.27	0.20	0.16	322	0.018	0.0053	0.0004
300	0.26	0.20	0.11	324	0.015	0.0041	0.0002
302	0.24	0.13	0.079	326	0.012	0.0033	0.0001

Source: NASA/JPL Evaluation No. 17 (Sander et al. 2011)

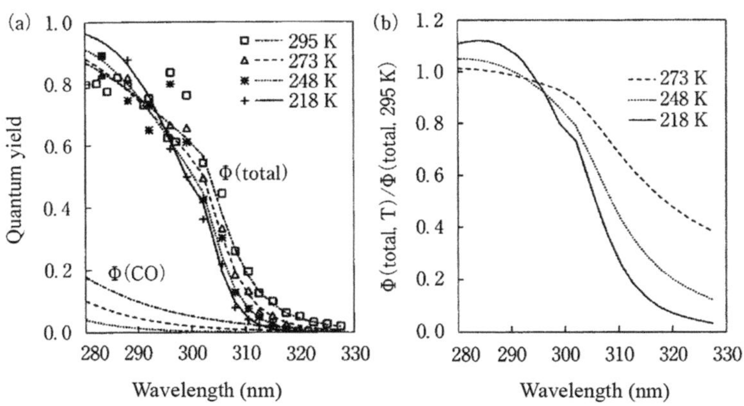

Fig. 4.21 Quantum yields of the photolysis of CH_3CCH_3: (**a**) Φ(total) and Φ(CO), symbols are experimental values and lines are calculation by parameterized formula $[M] = 5 \times 10^{18}$ molecules cm^{-3}, (**b**) Φ(total, T)/Φ(total, 295 K) (Adapted from Blitz et al. 2004)

As for the temperature dependence of the photolytic quantum yields of CH_3COCH_3, the total quantum yields increase with temperature in the wavelength region shorter than 295 nm, while they decrease with temperature in the wavelength region longer than 295 nm. NASA/JPL Evaluation No. 17 (Sander et al. 2011) gives the approximation formula of the quantum yields for the pressure and the temperature dependence based on Blitz et al. (2004). Table 4.17 shows the temperature dependent photolysis quantum yields and Fig. 4.21 depicts the photolytic quantum yields calculated by the approximation compared with experimental values (Blitz et al. 2004).

4.2.7 *Hydrogen Peroxide (H_2O_2), Methyl Hydroperoxide (CH_3OOH)*

Hydrogen peroxide H_2O_2 is formed by the radical termination reaction $HO_2 + HO_2$, and exists in the troposphere generally at the mixing ratio in order of ppbv. Since H_2O_2 is water soluble, it is removed by the dissolution into cloud and fog water, while photolytic reaction is another important removal process. Methyl hydroperoxide CH_3OOH also exists in the whole region of the troposphere in natural atmosphere as an oxidation product of methane. Its photolytic reaction is important as its removal process, and also as a radical source in the upper troposphere.

The absorption spectrum of H_2O_2 is a continuum decreasing monotonically from vacuum ultraviolet region (wavelength shorter than 200 nm) of 190 nm toward the longer wavelength as shown in Fig. 4.22 (Vaghjiani and Ravishankara 1989). Although the absorption cross sections are relatively small $\sigma < 1 \times 10^{-20}$ cm^2 molcule^{-1}, the spectrum extends to 350 nm and the photolysis in the troposphere is important. The absorption cross sections obtained after the latter half of 1970s agree well each other (Lin et al. 1978; Molina and Molina 1981; Nicovich and Wine 1988; Vaghjiani and Ravishankara 1989). Table 4.18 cites the extracted values from the NASA/JPL Evaluation No. 17 (Sander et al. 2011), based on the average of these studies.

The absorption cross sections of H_2O_2 are known to have temperature dependence due to the effect of transition from vibronic excited states in the electronic ground state (Nicovich and Wine 1988; Knight et al. 2002),and their temperature dependence at 260 nm is given by Nicovich and Wine (1988) as a function of wavelength.

The absorption spectrum of CH_3OOH bears a resemblance to H_2O_2, and a continuum in which the absorption cross sections decreases monotonically from ultraviolet to near ultraviolet region as shown in Fig. 4.22. There has been considerable uncertainty in the absorption cross sections of CH_3OOH due to the effect of impurities, and the recommended values of NASA/JPL Evaluation No. 17 based on Vaghjiani and Ravishankara (1989) are cited in Table 4.18.

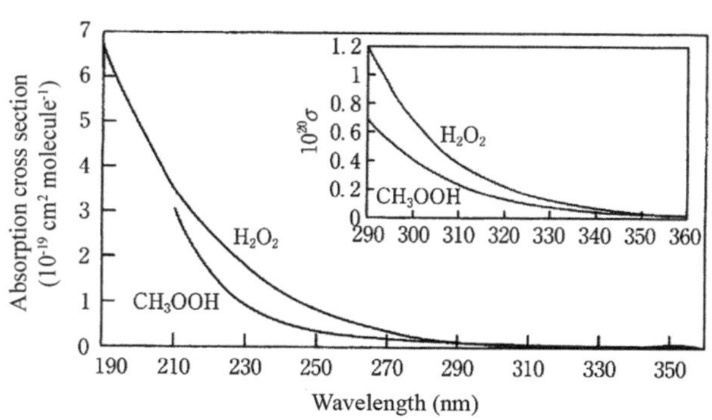

Fig. 4.22 Absorption spectra of H_2O_2 and CH_3OOH (Adapted from Finlayson-Pitts and Pitts, 2000 based on Vaghjiani and Ravishankara, 1989)

Table 4.18 Absorption cross sections of H_2O_2 and CH_3OOH (298 K)

	$10^{20}\sigma$ (cm^2 molecule^{-1})			$10^{20}\sigma$ (cm^2 molecule^{-1})	
wavelength (nm)	H_2O_2	CH_3OOH	wavelength (nm)	H_2O_2	CH_3OOH
200	47.5	–	280	2.0	1.09
205	40.8	–	285	1.5	0.863
210	35.7	31.2	290	1.2	0.691
215	30.7	20.9	295	0.90	0.551
220	25.8	15.4	300	0.68	0.413
225	21.7	12.2	305	0.51	0.313
230	18.2	9.62	310	0.39	0.239
235	15.0	7.61	315	0.29	0.182
240	12.4	6.05	320	0.22	0.137
245	10.2	4.88	325	0.16	0.105
250	8.3	3.98	330	0.13	0.079
255	6.7	3.23	335	0.10	0.061
260	5.3	2.56	340	0.07	0.047
265	4.2	2.11	345	0.05	0.035
270	3.3	1.70	350	0.04	0.027
275	2.6	1.39	355	–	0.021

Source: NASA/JPL Panel Evaluation No. 17 (Sander et al. 2011)

The photolytic processes of H_2O_2 and CH_3OOH in near ultraviolet region are

$$H_2O_2 + h\nu(\lambda < 587\ \text{nm}) \rightarrow OH + OH \quad \Delta H_0^\circ = 204\ \text{kJ mol}^{-1} \tag{4.37}$$

$$CH_3OOH + h\nu(\lambda < 640\ \text{nm}) \rightarrow CH_3O + OH. \quad \Delta H_0^\circ = 187\ \text{kJ mole}^{-1} \tag{4.38}$$

The photolytic quantum yield for each process is unity, as expected from the complete continuum of the absorption spectrum (Vaghjiani and Ravishankara 1989).

4.2.8 *Peroxynitric Acid (HO_2NO_2)*

Peroxynitric acid HO_2NO_2 is a molecule formed in the recombination reaction between HO_2 radicals and NO_2. Although it is not detected in the atmosphere, it is thought to be an important atmospheric species to be included in the tropospheric chemistry models. In the lower troposphere thermal decomposition reaction is as a major loss process of HO_2NO_2, but in the middle and upper troposphere where the temperature is lower, the regeneration of HO_2 radicals by the photolysis becomes important.

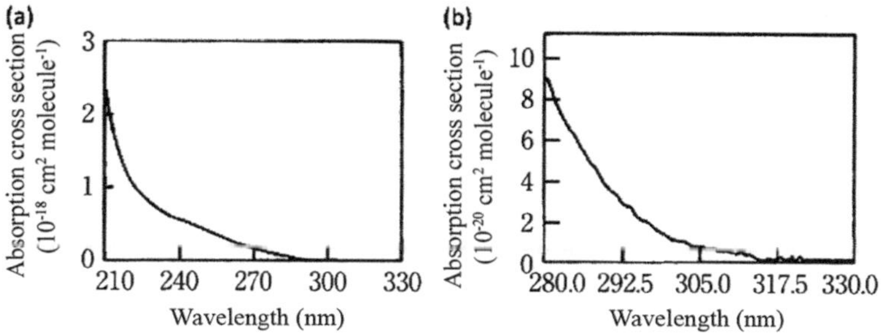

Fig. 4.23 Absorption spectrum of HO_2NO_2: (**a**) 210–330 nm, (**b**) enlarged for 280–330 nm (Adapted from Singer et al. 1989)

Absorption Spectrum and Cross Sections Figure 4.23 depicts absorption spectrum of HO_2NO_2. Similar to H_2O_2, the absorption spectrum of HO_2NO_2 has a maximum at wavelengths shorter than 200 nm and decreases toward longer wavelengths monotonically and extends to around 320 nm (Molina and Molina 1981; Singer et al. 1989; Knight et al. 2002). The absorption cross sections at 298 K recommended by the NASA/JPL Evaluation No. 17 (Sander et al. 2011) are given in Table 4.19. The absorption cross sections of HO_2NO_2 are known to be temperature dependent (Knight et al. 2002).

Photolytic Quantum Yields The major photolytic processes for HO_2NO_2 are thought to be,

$$HO_2NO_2 + h\nu(\lambda < 1184\ \text{nm}) \rightarrow HO_2 + NO_2. \quad \Delta H_0^\circ = 101\ \text{kJ mol}^{-1} \tag{4.39}$$

$$+ h\nu\ (\lambda < 724\ \text{nm}) \rightarrow OH + NO_3. \quad \Delta H_0^\circ = 165\ \text{kJ mol}^{-1} \tag{4.40}$$

Other reaction pathway which split into three product species, e.g.,
are also energetically possible within the wavelength range of tropospheric solar actinic flux. The photolytic quantum yields of HO_2NO_2 are unity at the wavelength longer than 200 nm measured by MacLeod et al. (1988), Roehl et al. (2001), and Jimenez et al. (2005). Since the quantum yields of HO_2 and NO_2 are 0.8 and those of OH and NO_3 are 0.2 (Sander et al. 2011), $\Phi\ (HO_2 + NO_2) = 0.8$, and $\Phi\ (OH + NO_3) = 0.2$, if they are ascribed to reactions (4.39) and (4.40), respectively.

4.2.9 *Nitric Acid (HNO_3) and Methyl Nitrate (CH_3ONO_2)*

Nitric acid HNO_3 ($HONO_2$) is mainly formed by the chain termination reaction $OH + NO_2$, and ubiquitously exists in the troposphere. The photolysis rate of HNO_3 in the troposphere is not very large, and the formation of ammonium nitrate aerosols

Table 4.19 Absorption cross sections of HO_2NO_2 (298 K)

Wave-length (nm)	10^{20} σ (cm^2 $molecule^{-1}$)	Wave-length (nm)	10^{20} σ (cm^2 $molecule^{-1}$)	Wave-length (nm)	10^{20} σ (cm^2 $molecule^{-1}$)	Wave-length (nm)	10^{20} σ (cm^2 $molecule^{-1}$)
200	563	240	58.1	280	9.29	312	0.465
205	367	245	49.0	284	6.93	316	0.313
210	239	250	41.3	288	4.91	320	0.216
215	161	255	35.0	292	3.37	324	0.152
220	118	260	28.5	296	2.30	328	0.110
225	93.5	265	23.0	300	1.52	332	0.079
230	79.2	270	18.1	304	1.05	336	0.054
235	68.3	275	13.4	308	0.702	340	0.037

Source: NASA/JPL Panel Evaluation No. 17 (Sander et al. 2011)

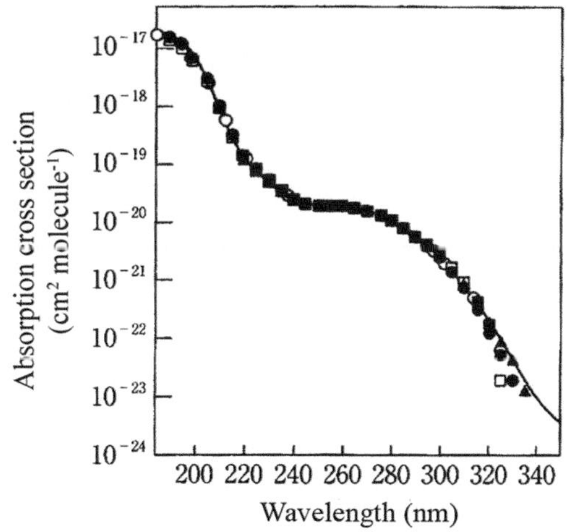

Fig. 4.24 Absorption spectrum of HNO_3 (Adapted from Burkholder et al. 1993)

and wet and dry deposition occurs preferentially to remove the species in the lower troposphere. However, the photolysis is important both as the loss process of HNO_3 and the OH radical source in the upper troposphere. Similarly, methyl nitrate CH_3ONO_2 is formed by the reaction of NO_2 with CH_3O radicals. The CH_3O is formed in the oxidation processes of CH_4 and other hydrocarbons. As well as the reaction with OH radicals, photolysis is an important removal process of CH_3ONO_2.

Absorption Spectrum and Cross Sections As shown in Fig. 4.24, the absorption spectrum of HNO_3 consists of a very strong band with a maximum at around 180 nm, and a second continuum with a shoulder peak near 270 nm overlapping to the skirt of the former band (Rattigan et al. 1992; Burkholder et al. 1993). The absorption cross sections are known to have strong temperature dependence, and Burkholder et al. (1993) presented an approximate expression,

$$\sigma(\lambda, T) = \sigma(\lambda, 298\ K)\exp[B(\lambda)(T - 298)]. \tag{4.41}$$

Table 4.20 carries absorption cross sections at 298 K for 200–345 nm recommended by IUPAC Subcommittee Report Vol. I (Atkinson et al. 2004) based on the data of Burkholder et al. (1993).

The absorption spectrum of CH_3ONO_2 and other alkyl nitrates are similar to HNO_3, and consist of a strong continuum in 190–250 nm which is followed by monotonically decreasing continuum in the 250–340 nm region as shown in Fig. 4.25 (Roberts and Fajer 1989). The absorption cross sections of CH_3ONO_2 are also known to have temperature dependence (Talukdar et al. 1997a), and the absorption cross sections at 298 K are given in Table 4.20 together with HNO_3 (Sander et al. 2011; Atkinson et al. 2006) based on the data of Talukdar et al. (1997a) and Taylor et al. (1980).

Table 4.20 Absorption cross sections of HNO_3[a], and CH_3ONO_2[b] (298 K)

	10^{20} σ (cm^2 $molecule^{-1}$)			10^{20} σ (cm^2 $molecule^{-1}$)			10^{20} σ (cm^2 $molecule^{-1}$)	
Wave-Length (nm)	HNO_3	CH_3ONO_2	Wave-length (nm)	HNO_3	CH_3ONO_2	Wave-length (nm)	HNO_3	CH_3ONO_2
200	588	1180	250	1.97	3.59	300	0.263	0.360
205	280	700	255	1.95	3.30	305	0.150	0.214
210	104	360	260	1.91	3.06	310	0.081	0.134
215	36.5	145	265	1.80	2.77	315	0.041	0.063
220	14.9	70	270	1.62	2.39	320	0.020	0.032
225	8.81	33	275	1.38	2.00	325	0.0095	0.014
230	5.78	18	280	1.12	1.58	330	0.0043	0.0066
235	3.75	10	285	0.858	1.19	335	0.0022	0.0027
240	2.58	5.88	290	0.615	0.850	340	0.0010	0.0012
245	2.11	4.19	295	0.412	0.568	345	0.0006	

Source: [a]IUPAC Subcommittee Report Vol. I (Atkinson et al. 2004), [b]200–235 nm: NASA/JPL Panel Evaluation No.17 (Sander et al. 2011); 240–345 nm: IUPAC Subcommittee Report Vol. II (Atkinson et al. 2006)

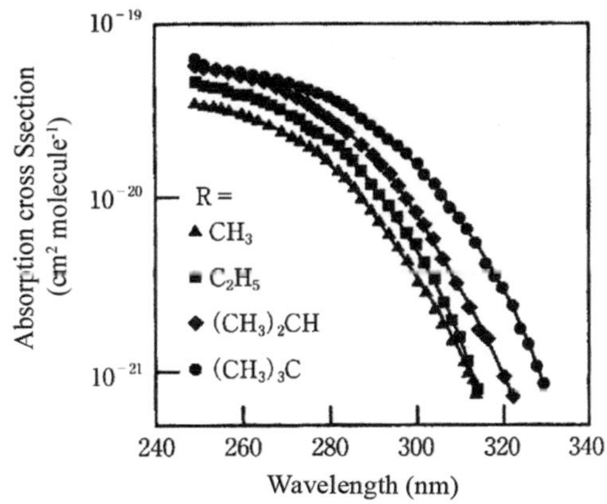

Fig. 4.25 Absorption spectra of CH_3ONO_2 and other alkyl nitrates (Adapted from Roberts and Fajer 1989)

Photolysis Quantum Yield The photolytic process of HNO_3 is known to occur via

$$HNO_3 + h\nu(\lambda < 604\ \text{nm}) \rightarrow OH + NO_2 \quad \Delta H_0^\circ = 198\ \text{kJ mol}^{-1} \tag{4.42}$$

with a quantum yield of unity for the wavelength range of 200–350 nm (Johnston et al. 1974). Only for the wavelength shorter than 200 nm, the reaction

$$HNO_3 + h\nu(\lambda < 401\ \text{nm}) \rightarrow HONO + O(^3P), \quad \Delta H_0^\circ = 298\ \text{kJ mol}^{-1} \tag{4.43}$$

is reported to be important (Sander et al. 2011).

The photolytic process of CH_3ONO_2 analogues to reaction (4.42) for HNO_3 is thought to occur at wavelengths longer than 200 nm,

$$CH_3ONO_2 + h\nu(\lambda < 703\ \text{nm}) \rightarrow CH_3O + NO_2 \quad \Delta H_0^\circ = 170\ \text{kJ mol}^{-1} \tag{4.44}$$

with a quantum yield of unity. At 193 nm the production of $O(^3P)$ atoms by the process analogous to the reaction (4.43),

$$CH_3ONO_2 + h\nu(\lambda < 394\ \text{nm}) \rightarrow CH_3ONO + O(^3P) \quad \Delta H_0^\circ = 304\ \text{kJ mol}^{-1} \tag{4.45}$$

is important (Sander et al. 2011).

4.2.10 *Peroxyacetyl Nitrate ($CH_3C(O)OONO_2$)*

Peroxyacetyl nitrate $CH_3C(O)OONO_2$ usually called PAN is a unique compound produced in the polluted atmosphere, and is important as a reservoir to carry NO_x to clean free troposphere. The atmospheric concentration of $CH_3C(O)OONO_2$ is

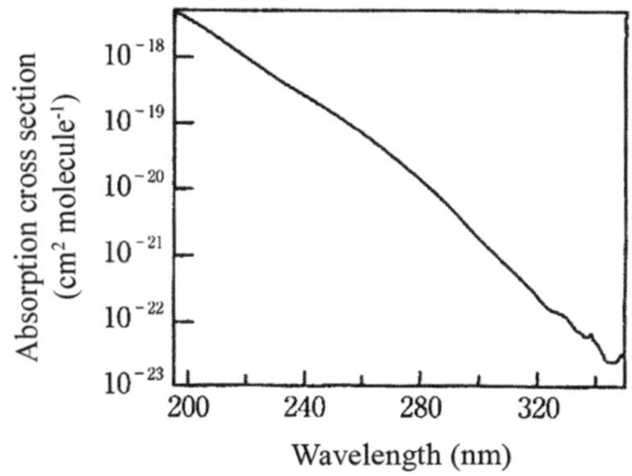

Fig. 4.26 Absorption spectrum of $CH_3C(O)OONO_2$ (Adapted from Talukdar et al. 1995)

maintained in equilibrium with the thermal decomposition reaction. The thermal decomposition is preferential in the lower troposphere where temperature is relatively high, but the photolysis becomes important as a removal process in the upper troposphere together with the OH radical reaction (Talukdar et al. 1995).

Absorption Spectrum and Cross Sections As shown in Fig. 4.26, the absorption spectrum of $CH_3C(O)OONO_2$ is similar to H_2O_2 and HNO_3, and consists of continuum with a peak at wavelengths shorter than 200 nm and decreases monotonically toward longer wavelengths in the 200–340 nm range (Talukdar et al. 1995). Absorption cross sections reported by Harwood et al. (2003) are given in Table 4.21 (Sander et al. 2011).

Photolysis Quantum Yields The photolytic processes of $CH_3C(O)OONO_2$ can be conceived as,

$$CH_3C(O)OONO_2 + h\nu(\lambda < 1004\ \text{nm}) \rightarrow CH_3C(O)OO + NO_2 \Delta H^\circ_{298} = 119\ \text{kJ mol}^{-1} \tag{4.46}$$

$$h\nu\ (\lambda<1004\ \text{nm}) \rightarrow CH_3C(O)O + NO_3.\ \ \Delta H^\circ_{298}=124\ \text{kJmol}^{-1} \tag{4.47}$$

Harwood et al. (2003) reported the quantum yields of NO_3 formation to be 0.22 ± 0.04, 0.39 ± 0.04 at 248 and 308 nm, respectively. These values are thought to correspond to Φ ($CH_3C(O)O + NO_3$), so that Φ ($CH_3C(O)OO + NO_2$) $= 1 - \Phi$ ($CH_3C(O)O + NO_3$) since the total photolytic quantum yield is thought to be unity.

4.3 Photolysis in the Stratosphere

In this section, absorption spectra and cross sections, and photolytic processes of atmospheric molecules that are not photolyzed by the solar actinic flux in the troposphere and photolyzed only in the stratosphere are described. Photolyses of many inorganic halogen molecules, which are important in the stratosphere, are

Table 4.21 Absorption cross section of $CH_3C(O)OONO_2$ (PAN) (298 K)

Wave-length (nm)	10^{20} σ (cm^2 $molecule^{-1}$)	Wave-length (nm)	10^{20} σ (cm^2 $molecule^{-1}$)	Wave-length (nm)	10^{20} σ (cm^2 $molecule^{-1}$)	Wave-length (nm)	10^{20} σ (cm^2 $molecule^{-1}$)
200	361	240	24.4	280	1.46	320	0.0252
204	292	244	18.8	284	1.01	324	0.0166
208	226	248	14.6	288	0.648	328	0.0117
212	168	252	11.4	292	0.447	332	0.0086
216	122	256	8.86	296	0.297	336	0.0061
220	89.7	260	6.85	300	0.189	340	0.0042
224	67.6	264	5.23	304	0.125	344	0.0029
228	52.0	268	3.94	308	0.0816	348	0.0020
232	40.4	272	2.87	312	0.0538		
236	31.4	276	2.07	316	0.0363		

Source: NASA/JPL Panel Evaluation No. 17 (Sander et al. 2011) (Data for 336–348 nm are smoothed)

summarized separately in the next Sect. 4.4 since most of them play a role also in the troposphere.

4.3.1 Oxygen (O_2)

Photolytic reaction of oxygen O_2 is the most fundamental reaction of stratosphere chemistry. In the first place, the formation of stratosphere in the earth's atmosphere has been brought on by the temperature inversion of the atmosphere due to the absorption of solar radiation by O_3 molecules that were produced by the reaction of $O(^3P)$ atoms from the photolysis of O_2 with another O_2 molecule. In this sense the photolytic reaction of O_2 is the basic of basics of stratospheric chemistry.

As the absorption spectrum, absorption cross sections, and potential curves has already been described in Chap. 3 (Sect. 3.2, Fig. 3.4, 3.5, Table 3.2), only photolytic processes will be discussed here. In the photolysis of O_2, three different electronic states of oxygen atoms can be formed as follows,

$$O_2 + h\nu(\lambda < 242\ \text{nm}) \rightarrow O(^3P) + O(^3P) = 494\ \text{kJ mol}^{-1} \tag{4.48}$$

$$+ h\nu\ (\lambda < 175\ \text{nm}) \rightarrow O(^3P) + O(^1D) \qquad = 683\ \text{kJ mol}^{-1} \tag{4.49}$$

$$+ h\nu\ (\lambda < 133\ \text{nm}) \rightarrow O(^3P) + O(^1S). \qquad = 906\ \text{kJ mol}^{-1} \tag{4.50}$$

Since the wavelengths of the solar radiation to reach to the stratosphere is $\lambda \geq \sim 190$ nm as seen in Fig. 4.1, only reaction (4.48) to form $O(^3P)$ is energetically possible as a photolytic process of O_2 in the stratosphere. The absorption bands of O_2 in this wavelength region are the Herzberg bands in $200 < \lambda < 250$ nm and the Schuman-Runge (S-R) bands in $175 < \lambda < 200$ nm with a vibrational structure as shown in Fig. 3.4. As seen in the potential energy curves of Fig. 3.5, O_2 molecules which are excited to the Herzberg bands with shorter wavelength radiation than the dissociation threshold of 242 nm are dissociated to $O(^3P) + O(^3P)$ along with the $A^3\Sigma_u^+$ potential curve with a quantum yield of unity. On the other hand, O_2 molecules excited to the Schuman-Runge bands reaches to the bound state of $B^3\Sigma_u^-$ then crosses to the repulsive $^3\Pi_u$ state and predissociate also into $O(^3P) + O(^3P)$ with a quantum yield of unity. Incidentally, O_2 excited to the Schuman-Runge continuum at wavelength shorter than 175 nm dissociates into $O(^3P) + O(^1D)$ with a quantum yield of unity, which occurs only in the higher altitude than the mesosphere.

Furthermore, O_2 has forbidden transitions to two low lying energy levels $a^1\Delta_g$ (94 kJ mol^{-1} higher than the ground state) and $b^1\Sigma_g^+$ (157 kJ mol^{-1} higher than the ground state), and their 0–0 bands are observed at 1270 nm and 762 nm, respectively. The radiative lifetimes of the $O_2(a^1\Delta_g)$ and $O_2\left(b^1\Sigma_g^+\right)$ are long, 67 min and 12 s, respectively (Wallace and Hunten 1968; Slanger and Cosby 1988). Thus, the

O_2 molecules excited into these states are mostly deactivated by collisions with atmospheric molecules and returns to the ground state O_2 as,

$$O_2(a^1\Delta_g) + M \rightarrow O_2\left(X^3\Sigma_g^-\right) + M \tag{4.51}$$

$$O_2\left(b^1\Sigma_g^+\right) + M \rightarrow O_2\left(X^3\Sigma_g^-\right) + M. \tag{4.52}$$

4.3.2 *Ozone (O_3)*

The photolysis of O_3 has an unambiguous importance in the stratospheric chemistry together with the photolysis of O_2 mentioned above. In the stratosphere, ozone layer has been formed by the equilibrium between the formation of O_3 by the reaction of $O(^3P)$ atoms generated by the photolysis of O_2 with O_2 molecules and the loss of O_3 by the photolysis of O_3 itself and the reaction with $O(^3P)$ atoms (refer to Sect. 8.1).

Since the absorption spectrum, absorption cross sections, and photolytic processes in the tropospheric solar actinic flux region, and photolytic quantum yields have already been described in detail in the Sects. 3.2 (Fig. 3.6) and 4.2.1 (Fig. 4.2, 4.3, 4.4, 4.5, 4.6, 4.7, and 4.8, Table 4.1, 4.2, 4.3, and 4.4), only photolytic processes in the stratosphere will be described here. As seen in Fig. 3.6, the absorption spectrum of the Hartley bands of O_3 extend over broad range of 200–300 nm that is the most important in the stratospheric solar actinic flux. The photolytic process of O_3 molecules reached by the absorption of photons in the Hartley bands is thought to be

$$O_3 + h\nu(\lambda < 310\ \text{nm}) \rightarrow O(^1D) + O_2(a^1\Delta_g), \tag{4.2}$$

as mentioned in the Sect. 4.2.1, and the quantum yields of $O(^1D)$ formation have been obtained in laboratory studies (see Fig. 4.6) (Cooper et al. 1993; Takahashi et al. 2002; Matsumi and Kawasaki 2003). Based on these data, NASA/JPL Evaluation No. 17 recommended 0.90 as the quantum yield of $O(^1D)$ formation in this wavelength region (Sander et al. 2011). Since the total quantum yield of the O_3 photolysis in the Hartley bands is thought to be unity, the quantum yield of $O(^1D)$ formation less than unity implies the formation of $O(^3P)$ by some other processes.

Furthermore, the quantum yields of $O(^3P)$ and $O(^1D)$ formation at 193 nm have been reported to be 0.57 ± 0.14 and 0.46 ± 0.2, respectively, thus showing the quantum yields of $O(^1D)$ formation in this region is much smaller than in the Hartley bands as shown in Fig. 4.6 (Turnipseed et al. 1991). Also, at this wavelength the formation of O_2 () has been measured with the quantum yield 0.50 ± 0.38, and the occurrence of the process,

$$O_3 + h\nu(\lambda < 260\ \text{nm}) \rightarrow O(^1D) + O_2\left(b^1\Sigma_g^+\right) \tag{4.53}$$

has been suggested (Turnipseed et al. 1991).

4.3.3 Nitric Oxide (NO)

Nitric oxide NO is a major oxides of nitrogen formed by the photolysis of N_2O in the stratosphere, but it was not paid much attention in the past since the photolysis rate is not very large. However, the photolysis of NO forms N atoms that leads to the loss of odd nitrogen by the reaction $N + NO \rightarrow N_2 + O$, so that it is now recognized to be important in the stratospheric chemistry.

The absorption spectrum of NO has long been measured and shows band structure at wavelengths shorter than 250 nm, as shown in Fig. 4.27. The assignment of electronic transitions and vibrational levels are also shown in the figure, and Fig. 4.28 shows the potential energy curves of NO (Okabe 1978). The γ-, β-, δ- and ε-bands in the 196–227 nm range shown in Fig. 4.27 correspond to the $A^2\Sigma$-$X^2\Pi$, B $^2\Pi$-$X^2\Pi$, $C^2\Pi$-$X^2\Pi$, and $D^2\Sigma$-$X^2\Pi$ transition in Fig. 4.28, respectively (Callear and Pilling 1970).

As the photolytic process of NO,

$$NO + h\nu(\lambda < 191 \text{ nm}) \rightarrow N(^4S) + O(^3P) \quad \Delta H_0 = 627 \text{ kJ mol}^{-1} \tag{4.54}$$

is energetically possible at wavelengths shorter than 191 nm. It is known from laboratory studies that the excitation to the β- and γ-band (Fig. 4.28) in the stratospheric solar actinic flux region, yields the fluorescence of NO as follows,

$$NO + h\nu \rightarrow NO^* \tag{4.55}$$

$$NO^* \rightarrow NO + h\nu'. \tag{4.56}$$

In contrast, the fluorescence of NO is not observed for the excitation by the δ(0–0) and δ(1–0) bands, and the photolytic process (4.54) occurs with the quantum yield

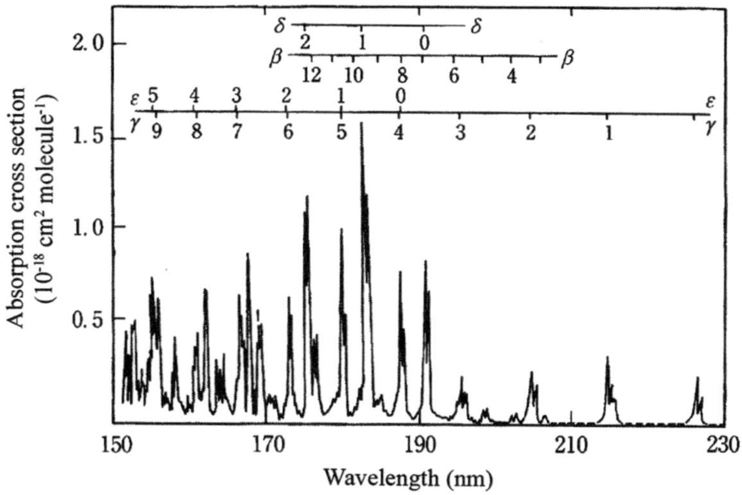

Fig. 4.27 Absorption spectrum of NO (Adapted from Okabe 1978)

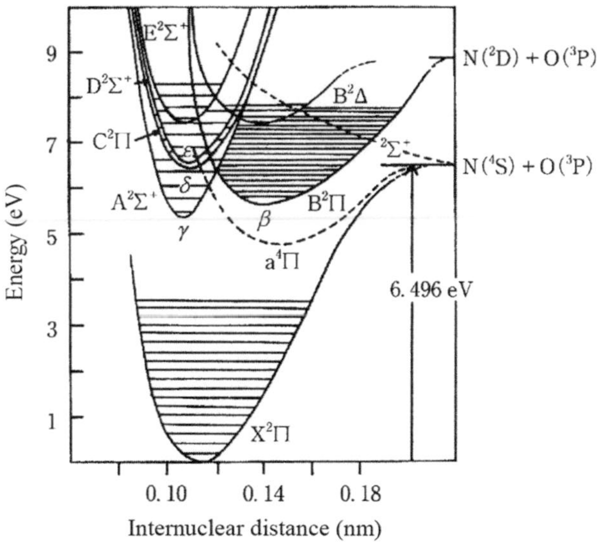

Fig. 4.28 Potential energy curves of NO (Adapted from Okabe 1978)

of unity. Therefore, it is enough to consider only the absorption by these δ(0–0) and δ(1–0) bands for the photolysis rate of NO in the stratosphere.

Since the wavelength region of the absorption bands of the δ(0–0) (189.4–191.6 nm) and δ(1–0) (181.3–183.5 nm) overlaps with those of the Schuman-Runge bands of O_2, the accurate wavelengths of rotational lines of both NO and O_2 and oscillator strengths data which are the bases of absorption cross sections for individual rotational lines of NO is necessary. The oscillator strengths is a non-dimensional number to express the ratio of contribution of a single electron to the absorption. Although the oscillator strength of electronic transition is in general smaller than unity, it is close to unity for the very strong allowed transition. For the oscillator strength of NO, the values of 5.78×10^{-3} (δ(1–0) + β(10–0)) and $2{,}49 \times 10^{-3}$ (δ(0–0) + β(7–0)) by Bethke (1959) have long been used, but about 50 % smaller values than these were reported and the J_{NO} by using these values were reported later (Fredlick and Hudson 1979; Nicolet and Cieslik 1980). However, Minschwaner and Siskind (1993) made a line-by-line calculation of the NO photolysis rate J_{NO} using the high-resolution spectrum of O_2 in which rotational lines are separated (Yoshino et al. 1983; Lewis et al. 1986) and newly observed high resolution spectrum and cross sections of NO δ(0–0) and δ(1–0) bands. The absorption cross sections used here coincide with those of Bethke (1959), and recent values by Imajo et al. (2000) (oscillator strength of 5.4×10^{-3} and absorption cross section of 4.80×10^{-15} cm^2 cm^{-1} for δ(1–0)). The calculated photolysis rate of NO $J_{NO} \approx 10^{-7}$ s^{-1} (the sum of δ(0–0) and δ(1–0)) is not negligible as the loss process of odd nitrogen in the upper stratosphere (Minschwaner and Siskind 1993; Mayor et al. 2007).

4.3.4 Dinitrogen Monoxide (N_2O)

Dinitrogen monoxide N_2O is also called nitrous oxide, and emitted from natural and anthropogenic sources from the ground. Since N_2O does not dissipate in the troposphere, it reaches to the stratosphere and subjects photolysis to provide reactive oxides of nitrogen (odd nitrogen). For this reason, the photolysis of N_2O in the stratosphere is very important.

As shown in Fig. 4.29, absorption spectrum of N_2O is a broad continuum with a peak at around 180 nm, decreases monotonically toward longer wavelengths and extends to near 240 nm (Johnston and Selwyn 1975). Table 4.22 cites the absorption cross sections of N_2O in the 160–240 nm region as recommended by NASA/JPL Evaluation No. 17 (Sander et al. 2011). These values are based on the data by Hubrich and Stuhl (1980) (160, 165, 170 nm), and Selwyn et al. (1977) (173–240 nm). The absorption cross sections are temperature dependent and decrease with the increase of temperature for which the formula for approximation is given by Selwyn et al. (1977) (Sander et al. 2011).

The following photolytic process has been known to occur with a quantum yield of unity for 140–230 nm covering the stratospheric photolysis region,

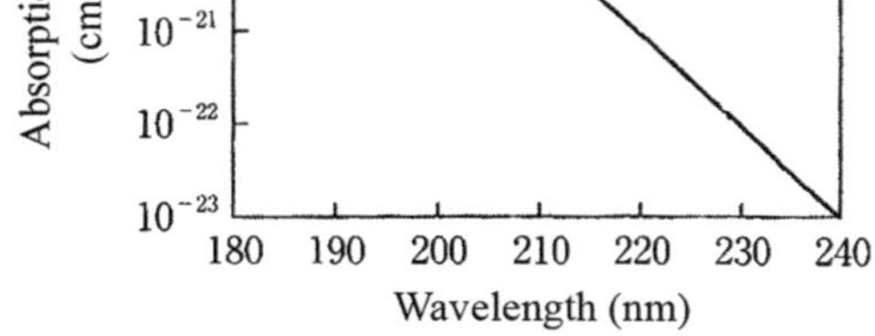

Fig. 4.29 Absorption spectrum of N_2O (Adapted from Carlon et al. 2010)

Table 4.22 Absorption cross sections of N_2O (298 K)

Wavelength (nm)	10^{20} σ (cm^2 molecule^{-1})	Wavelength (nm)	10^{20} σ (cm^2 molecule^{-1})	Wavelength (nm)	10^{20} σ (cm^2 molecule^{-1})
160	4.30	190	11.1	220	0.922
165	5.61	195	7.57	225	0.030
170	8.30	200	4.09	230	0.0096
175	12.6	205	1.95	235	0.0030
180	14.6	210	0.755	240	0.0010
185	14.3	215	0.276		

Source: NASA/JPL Panel Evaluation No. 17 (Sander et al. 2011)

$$N_2O + h\nu(\lambda < 340\ \text{nm}) \rightarrow N_2 + O(^1D) \quad \Delta H_0 = 352\ \text{kJ mol}^{-1} \tag{4.57}$$

by the studies before 1970s (Paraskevopoulos and Cvetanovic 1969; Preston and Barr 1971). The result is confirmed by recent studies, and the formations of $O(^3P)$ and $N(^4S)$ by the spin forbidden processes,

$$N_2O + h\nu(\lambda < 739\ \text{nm}) \rightarrow N_2 + O(^3P) \quad \Delta H_0 = 162\ \text{kJ mol}^{-1} \tag{4.58}$$

$$+ h\nu\ (\lambda < 251\text{nm}) \rightarrow N(^4S) + NO(^2\Pi) \quad \Delta H_0 = 476\ \text{kJ mol}^{-1} \tag{4.59}$$

are less than 1 %, respectively (Nishida et al. 2004; Greenblatt and Ravishankara 1990).

4.3.5 *Other Oxides of Nitrogen (NO_2, NO_3, N_2O_5, HNO_3, HO_2NO_2), Hydrogen peroxide (H_2O_2), Formaldehyde (HCHO)*

Other than NO and N_2O, NO_2, NO_3, N_2O_5, HNO_3, HO_2NO_2 are also important oxides of nitrogen in the stratosphere. Since their photolyses are also all important in the troposphere, their absorption spectra, cross sections, and photolytic processes have been described in the previous section including the stratospheric actinic flus region. H_2O_2 and CH_4 also exist in the stratosphere in considerable concentration as they are formed by the mutual chain termination reaction of HO_2, and in the oxidation process of CH_4, respectively, as in the troposphere. Their absorption spectra, cross sections, and photolytic processes have already been described in the previous section to be referred.

4.3.6 *Carbonyl Sulfide (COS)*

Carbonyl sulfide COS is emitted from terrestrial soil, ocean and biomass burning into the atmosphere, but since their loss rate in the troposphere is very small, most of them reach to the stratosphere. The photolysis of COS in the stratosphere is a very crucial reaction as it provides sulfur into the atmosphere forming sulfuric aerosol layer (the Junge Layer) in the stratosphere. Incidentally, although COS is often described as OCS in the textbooks and literature of atmospheric chemistry, the notation of COS is used in this book according to the recommendation of IUPAC (International Union of Pure and Applied Chemistry).

As shown in Fig. 4.30, the absorption spectrum of COS is a broad continuum with a peak at ca. 222 nm extending to around 300 nm, and a weak vibrational structure can be seen near the maximum (Okabe 1978; Molina et al. 1981; Rudolph and Inn 1981). The absorption cross sections of COS are known to have

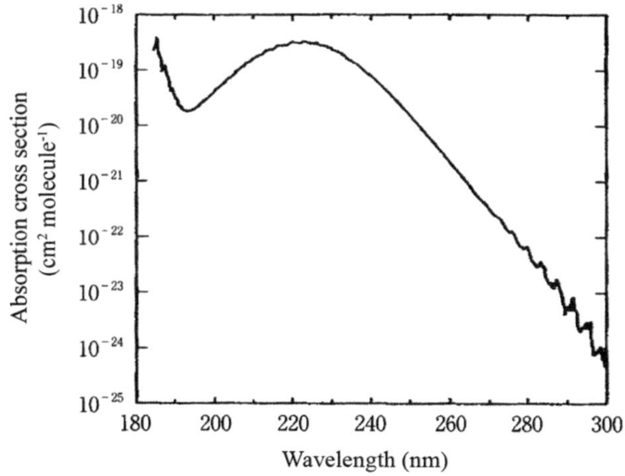

Fig. 4.30 Absorption spectrum of COS (Adapted from Molina et al. 1981)

Table 4.23 Absorption cross sections of COS (295 K)

Wavelength (nm)	10^{20} σ (cm^2 molecule^{-1})	Wavelength (nm)	10^{20} σ (cm^2 molecule^{-1})	Wavelength (nm)	10^{20} σ (cm^2 molecule^{-1})
185	19.0	225	31.0	265	0.096
190	3.97	230	24.3	270	0.038
195	2.02	235	15.4	275	0.015
200	3.93	240	8.13	280	0.0054
205	8.20	245	3.82	285	0.0022
210	15.1	250	1.65	290	0.0008
215	24.2	255	0.664	295	0.0002
220	30.5	260	0.252	300	0.0001

Source: IUPAC Subcommittee Report Vol. I (Atkinson et al. 2004)

temperature dependence, and Wu et al. (1999) reported from the high resolution measurement (0.06 nm) that hot bands can be seen in the 215–260 nm region, and the temperature dependence is particularly large at wavelengths longer than 224 nm.

Table 4.23 gives absorption cross sections of COS recommended by NASA/JPL Evaluation No. 17 (Sander et al. 2011) based on Molina et al. (1981).

The photolytic processes of COS,

$$COS + h\nu(\lambda < 395\ \text{nm}) \rightarrow CO + S(^3P) = 303\ \text{kJ mol}^{-1} \tag{4.60}$$

$$+ h\nu\ (\lambda < 290\ \text{nm}) \rightarrow CO + S(^1D) \quad = 413\ \text{kJ mol}^{-1} \tag{4.61}$$

$$+ h\nu\ (\lambda < 180\ \text{nm}) \rightarrow CS + O(^3P) \quad = 665\ \text{kJ mol}^{-1} \tag{4.62}$$

have threshold wavelengths of 395, 290 and 180 nm, respectively. Therefore, the formation of CO and S atoms by reactions (4.60) and (4.61) are possible in the stratosphere referring to the absorption spectrum depicted in Fig. 4.30. Experimentally, the formation of CO has been confirmed to be a main product in the photolysis of 214–254 nm region, and the CO formation quantum yield of >0.95 at 248 nm has been reported (Zhao et al. 1995). Based on these results, NASA/JPL Evaluation No. 17 recommends the quantum yield of unity for the photolysis of COS in the wavelength range of 220–254 nm. There are two possibilities for the formation of S atoms in the photolysis of COS; the formation of $S(^1D)$ by the direct dissociation from the excited singlet state reached by the photoexcitation, and the formation of $S(^3P)$ by the dissociation from to the excited triplet state reached by intersystem crossing. From the recent research of molecular dynamics, S atoms formed by the photoabsorption at 228 nm are found to be mostly $S(^3P)$, so that at least at this wavelength the reaction (4.60) via the excited triplet state is the major photolytic process of COS (Zhao et al. 1995; Katayanagi et al. 1995).

4.3.7 *Sulfur Dioxide (SO_2)*

Sulfur dioxide SO_2 is one of the most typical air pollutants, as is NO_x. Although SO_2 has an absorption spectrum in the tropospheric actinic flux region, the SO_2 molecules absorbing the radiation in this region reach an excited state which does not dissociate. Meanwhile, the molecule that absorbs radiation shorter than 219 nm in the stratospheric actinic flux region can be photolyzed, but since the SO_2 molecules emitted in the troposphere are mostly removed by the reaction with OH radicals or uptake into cloud and fog water, they do not reach to the stratosphere. Therefore, it is necessary to consider the photolysis of SO_2 only for those formed by the photochemical reactions of COS in the stratosphere, and for those introduced directly into the stratosphere by large eruptions of volcanoes or by the emissions of aircrafts. Therefore, although the photolysis of SO_2 is not in general very important in atmospheric chemistry, it is described here in the interest of fundamental photochemistry of the atmospheric molecule.

As shown in Fig. 4.31 (Manatt and Lane 1993), the absorption spectrum of SO_2 consists of a very strong band which extends to 230 nm with a peak at around 200 nm, medium strength bands in the region of 230–340 nm with a peak at 290 nm, and a very weak band in the 340–400 nm region. These absorption bands are assigned to the transitions from the ground state to the C^1B_2, the mixed state of B^1B_1 and A^1A_2, and a^3B_1 state, respectively (Okabe 1978). Among these bands, the C^1B_2-X^1A_1 and B^1B_1-X^1A are allowed transitions, and A^1A_2-X^1A_1 and a^3B_1-X^1A_1 are forbidden transitions. Theoretical calculations on photochemical processes and potential energy surfaces have been studied by Katagiri et al. (1997) and Li et al. (2006).

The data of absorption cross sections of SO_2 prior to 1993 have been compiled by Manatt and Lane (1993) for the 106–403 nm, and later the measurements in high-resolution (Vandaele et al. 1994; Rufus et al. 2003) and the temperature dependence have been reported (Prahlad and Kumar 1997; Bogumil et al. 2003).

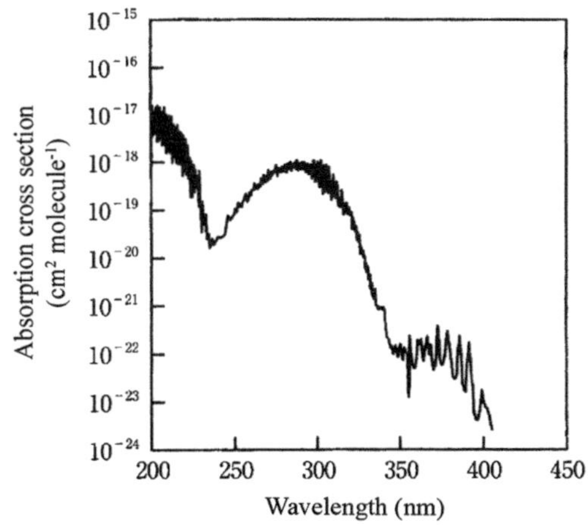

Fig. 4.31 Absorption spectrum of SO_2 (Adapted from Manatt and Lane 1993)

The photochemical processes of SO_2 are known (Okabe 1978) to be,

$$SO_2 + h\nu(\lambda < 219\ \text{nm}) \rightarrow SO(X^3\Sigma^-) + O(^3P) = 545\ \text{kJ mol}^{-1} \tag{4.63}$$

$$+ h\nu\ (\lambda > 217\ \text{nm}) \rightarrow SO_2^*(C^1B_2, B^1B_1, A^1A_2, a^3B_1) \tag{4.64}$$

$$SO_2^*(C^1B_2, B^1B_1, A^1A_2, a^3B_1) \rightarrow SO_2(X^1A_1) + h\nu' \tag{4.65}$$

$$+ M \rightarrow SO_2\ (X^1A_1) + M. \tag{4.66}$$

The energy threshold for the photolytic reaction (4.63) to form ground states, SO ($X^3\Sigma^-$) and O (^3P) is 545 kJ mole^{-1} corresponding to the wavelength threshold of 219 nm. Therefore, even though there are lower lying excited states reached by the absorption of the tropospheric and stratospheric actinic flux, the high dissociation energy is the reason that the photolysis of SO_2 is not very important in the atmosphere. Thus, the photolysis of SO_2 is possible only at the wavelengths shorter than 219 nm in the stratosphere. Indeed, Okabe (1971) found that the fluorescence of SO_2 from the C^1B_2 state is not observed with the divide of 219 nm suggesting that the photolysis occurs at shorter wavelengths than this. The absolute quantum yield of photolytic reaction (4.63) has not been known until fairly recently, and Abu-Bajeh et al. (2002) reported the formation quantum yield of O (^3P) as $\Phi\ (O(^3P)) = 0.13 \pm 0.05$ at 222.4 nm where photo-dissociation can occur by the absorption from the rotationally excited levels in the electronic ground state. As expected from the vibrational structure in the absorption spectrum near 200 nm, the SO_2 molecule in the C^1B_2 state is in the bound state and predissociation is suggested to occur by the transfer to another dissociating state.

The absorption of the solar radiation at wavelengths longer than 219 nm forms the excited SO_2 molecules following the reaction (4.64), and photo-emission is observed in laboratory studies. The radiative lifetime of the excited state reached by the absorption of photons at 340–400 nm is rather long, 8.1 ± 2.5 ms (Su et al. 1977), which is thought to correspond to phosphorescence (light-emission between the electronic state with different spin multiplicity) from the a^3B_1 state. On the other hand, fluorescence (light-emission between the electronic state with same spin multiplicity) lifetime in the 230–340 nm region consists of a shorter component of ~50 μs which is independent of the wavelength, and the longer one, 80–600 μs, which is longer at the longer wavelengths. Each of them are assigned to the emissions from B^1B_1 and A^1A_2 state, respectively (Brus and McDonald 1974). Furthermore, the fluorescence lifetime from the C^1B_2 state in the longer wavelengths than 219 nm is very short ~50 ns (Hui and Rice 1972). In any case, the SO_2 molecules excited to the non-dissociative state are returned to the electronic ground state mostly through quenching by the atmospheric species and partially through light-emission, and the photolysis of SO_2 does not generally play an important role in atmospheric chemistry.

4.3.8 Methyl Chloride (CH_3Cl), Methyl Bromide (CH_3Br), Methyl Iodide (CH_3I)

Methyl halides such as methyl chloride CH_3Cl, methyl bromide CH_3Br, and methyl iodide CH_3I are natural origin species emitted from terrestrial and oceanic sources, but anthropogenic emissions are also important for CH_3Br. Among these, CH_3I are photolyzed mostly by the actinic flux in the troposphere.

Figure 4.32 depicts absorption spectra of CH_3Cl, CH_3Br, and CH_3I in the 180–360 nm region drawn according to the absorption cross sections recommended by the NASA/JPL Evaluation No. 17 (Sander et al. 2011). As seen in the figure, the absorption spectra of methyl halides are all continuum without vibrational structure. CH_3Cl has a broad absorption band with a strong peak at around 170 nm and an extended tail to 230 nm (Hubrich and Stuhl 1980). The shape of the spectrum of CH_3Br is similar to CH_3Cl but shifted to longer wavelength with a peak at around 180 nm and a tail is extended to 280 nm (Robbins 1976; Molina et al. 1982). The peak and the

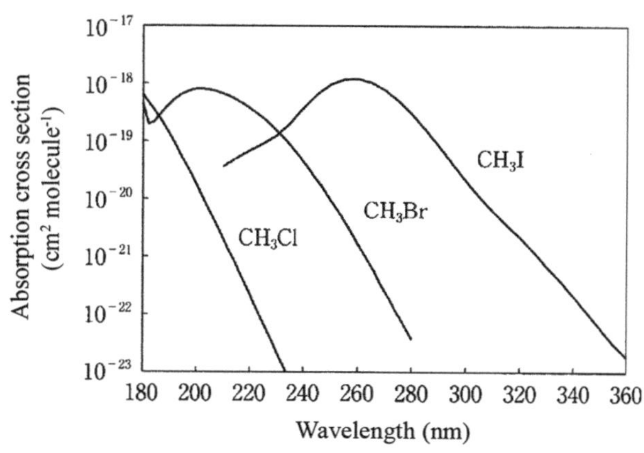

Fig. 4.32 Absorption spectra of CH_3Cl, CH_3Br and CH_3I (Constructed based on NASA/JPL Evaluation No. 17, Sander et al. 2011)

Table 4.24 Absorption cross sections of CH_3Cl and $CH3_3Br$ (298 K)

Wavelength (nm)	10^{20} σ (cm^2 $molecule^{-1}$)		Wavelength (nm)	10^{20} σ (cm^2 $molecule^{-1}$)		Wavelength (nm)	10^{20} σ (cm^2 $molecule^{-1}$)	
	CH_3Cl	CH_3Br		CH_3Cl	CH_3Br		CH_3Cl	CH_3Br
180	63.6	44.6	214	0.0860	54.2	248	–	1.31
182	46.5	19.8	216	0.0534	47.9	250	–	0.921
184	35.0	21.0	218	0.0345	42.3	252	–	0.683
186	25.8	27.8	220	0.0220	36.6	254	–	0.484
188	18.4	35.2	222	0.0135	31.1	256	–	0.340
190	12.8	44.2	224	0.0086	26.6	258	–	0.240
192	8.84	53.8	226	0.0055	22.2	260	–	0.162
194	5.83	62.6	228	0.0035	18.1	262	–	0.115
196	3.96	69.7	230	0.0022	14.7	264	–	0.0795
198	2.68	76.1	232	0.0014	11.9	266	–	0.0551
200	1.77	79.0	234	0.0009	9.41	268	–	0.0356
202	1.13	79.2	236	0.0006	7.38	270	–	0.0246
204	0.731	78.0	238	–	5.73	272	–	0.0172
206	0.482	75.2	240	–	4.32	274	–	0.0114
208	0.313	70.4	242	–	3.27	276	–	0.0081
210	0.200	65.5	244	–	2.37	278	–	0.0055
212	0.127	59.9	246	–	1.81	280	–	0.0038

Source: NASA/JPL Panel Evaluation No. 17 (Sander et al. 2011)

Table 4.25 Absorption cross sections of CH_3I (298 K)

λ (nm)	10^{20} σ (cm^2)	λ (nm)	10^{20} σ (cm^2)	λ (nm)	10^{20} σ (cm^2)	λ (nm)	10^{20} σ (cm^2)
210	3.62	250	96.3	290	8.04	330	0.0684
215	5.08	255	117.7	295	4.00	335	0.0388
220	6.90	260	119.7	300	2.06	340	0.0212
225	9.11	265	102.9	305	1.10	345	0.0114
230	12.6	270	75.9	310	0.621	350	0.0061
235	20.5	275	49.6	315	0.359	355	0.0032
240	38.1	280	29.2	320	0.221	360	0.0019
245	65.6	285	15.6	325	0.126	365	0.0009

Source: NASA/JPL Panel Evaluation No. 17 (Sander et al. 2011)

extended tail edge of the absorption bands of CH_3I are further shifted to the longer wavelengths to 260, and 360 nm, respectively (Fahr et al. 1995; Roehl et al. 1997).

Absorption cross sections recommended by the NASA/JPL Evaluation No. 17 (Sander et al. 2011) are the averaged values based on above and other studies. Tables 4.24, and 4.25 give the absorption cross sections of CH_3Cl and CH_3Br, and CH_3I extracted from the evaluation, respectively. The approximation formula of the temperature dependences of absorption cross sections are also given by Sander et al. (2011).

Photolytic processes of CH_3Cl and CH_3Br in the above spectral regions are thought to be

$$CH_3Cl + h\nu(\lambda < 342\ \text{nm}) \rightarrow CH_3 + Cl \quad \Delta H^\circ_{298} = 350\ \text{kJ mol}^{-1} \tag{4.67}$$

$$CH_3Br + h\nu(\lambda < 396\ \text{nm}) \rightarrow CH_3 + Br, \quad \Delta H^\circ_{298} = 302\ \text{kJ mol}^{-1} \tag{4.68}$$

whose quantum yields are unity (Takacs and Willard 1977; Talukdar et al. 1992). In the case of CH_3I processes forming two different electronic states of I atom $I(^2P_{3/2})$ (ground state) and $I^*(^2P_{1/2})$ are known to occur,

$$CH_3I + h\nu(\lambda < 500\ \text{nm}) \rightarrow CH_3 + I(^2P_{3/2}), \quad \Delta H^\circ_{298} = 239\ \text{kJ mol}^{-1} \tag{4.69}$$

$$CH_3I + h\nu(\lambda < 362\ \text{nm}) \rightarrow CH_3 + I^*(^2P_{1/2}). \quad \Delta H^\circ_{298} = 330\ \text{kJ mol}^{-1} \tag{4.70}$$

The quantum yields of each process are reported at several wavelengths, but the total photolytic quantum yield summing the reactions (4.69) and (4.70) is thought to be unity (Kang et al. 1996).

4.3.9 *Chlorofluorocarbons (CFCs), Hydrochlorofluorocarbons (HCFCs)*

Chlorofluorocarbons CFCs and hydrochlorofluorocarbons HCFCs are all anthropogenic species and are the causative agents of ozone layer destruction as well as greenhouse gasses. CFCs are the molecules in which all the hydrogen atoms of hydrocarbons are substituted by chlorine and fluorine atoms. They do not have absorption bands in the tropospheric actinic flux region and also do not react with OH radicals. Therefore, they do not have any dissipation process in the troposphere, and can be photolyzed only after they reached to the stratosphere. On the other hand, HCFCs is molecules in which at least one of chlorine or fluorine atom of CFCs is substituted by hydrogen atom. Since HCFCs react with OH radicals, they are removed in the troposphere, but a portion of them reach the stratosphere and photolyzed, similar to CFCs.

Among organic chlorinated compounds which do not contain H atoms in molecules, five CFCs; CFC-11 ($CFCl_3$), CFC-12 (CF_2Cl_2), CFC-113 ($CF_2ClCFCl_2$), CFC-114 (CF_2ClCF_2Cl) and CFC-115 (CF_3CF_2Cl) and carbon tetrachloride (CCl_4), have relatively high concentrations in the stratosphere. Figure 4.33 depicts absorption spectra of these compound (Hubrich and Stuhl 1980). As can be seen in the figure, all the spectra have peaks around 180–200 nm in common, and have extended tails toward longer wavelength side, whose shape is similar to CH_3Cl shown in Fig. 4.32. In the figure, it can be seen that the absorption cross sections are lager and the absorption shift to longer wavelengths as the number of chlorine atoms in a molecule increases.

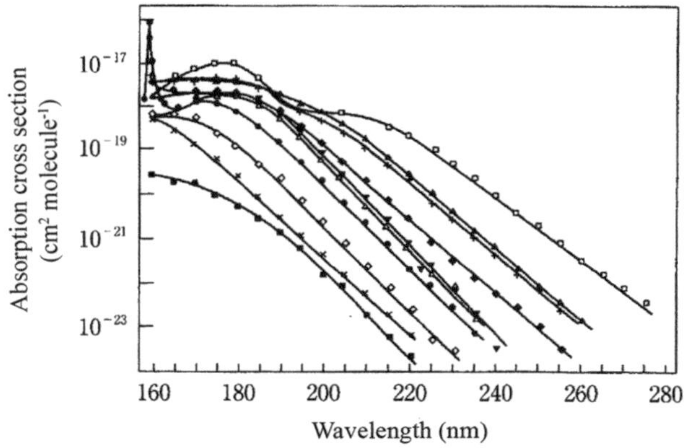

Fig. 4.33 Absorption spectra of CFCs. 298 K: + $CHCl_3$; △ $CHFCl_2$; × CHF_2Cl; ◆ CH_2Cl_2; ◇ CH_2FCl; ● CH_3Cl; □ CCl_4; ▲ $CFCl_3$ (CFC-11); ▼ CF_2Cl_2 (CFC-12); ■ CF_3Cl (Adapted from Hubrich and Stuhl 1980)

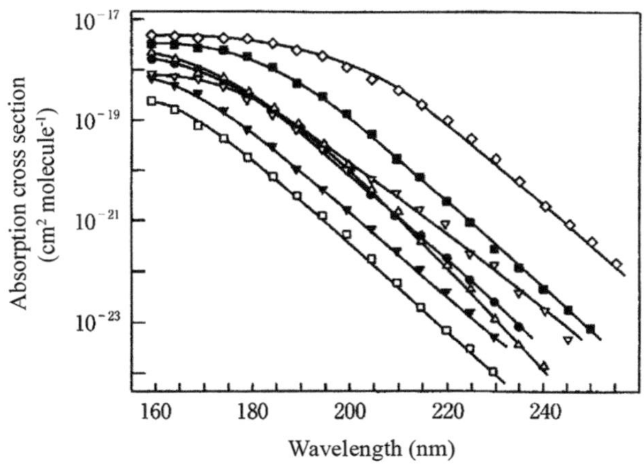

Fig. 4.34 Absorption spectra of HCFCs. 298 K: △ CH_3CCl_3; ▽ CF_3CH_2Cl; △ CH_3CH_2Cl; □ CF_3CF_2Cl; ▼ CH_3CF_2Cl; ● CF_2ClCF_2Cl; ■ $CFCl_2CF_2Cl$ (Adapted from Hubrich and Stuhl 1980)

Meanwhile, among the HCFCs, HCFC-22 (CHF_2Cl), HCFC-141b (CH_3CFCCl_2), and HCFC142b (CH_3CF_2Cl) have relatively high concentrations in the stratosphere. Figure 4.34 shows the absorption spectra of these molecules (Hubrich and Stuhl 1980). The absorption spectra of HCFCs also have peaks near 180–200 nm with tails extending toward longer wavelengths, similar to those of CFCs.

As seen in Fig. 4.1, the 190–220 nm region is the valley between the strong absorption of O_2 in the shorter wavelength side and that of O_3 in the longer wavelength side. The actinic flux in this range reaches the middle stratosphere. The absorption peaks of CFCs and HCFCs shown in Figs. 4.33 and 4.34 incidentally coincide with this range, they are efficiently photolyzed in the stratosphere, which is the direct cause of ozone layer destruction by the anthropogenic species.

NASA/JPL Evaluation No. 17 (Sander et al. 2011) gives recommended absorption cross sections and their temperature dependence for these compounds based on

Table 4.26 Absorption cross sections of CCl_4 and CFCs (295–298 K)

Wavelength (nm)	$10^{20}\sigma$ (cm^2 $molecule^{-1}$) CCl_4	$CFCl_3$ (CFC-11)	CF_2Cl_2 (CFC-12)	$CF_2ClCFCl_2$ (CFC-113)	CF_2ClCF_2Cl (CFC-114)	CF_3CF_2Cl (CFC-115)
176	1010	324	186	(192)[a]	43.0	3.08
180	806	314	179	155	26.2	1.58
184	479	272	134	123	15.0	0.790
188	227	213	82.8	83.5	7.80	0.403
192	99.6	154	45.5	48.8	3.70	0.203
196	69.5	99.1	21.1	26.0	1.75	0.0985
200	66.0	63.2	8.71	12.5	0.800	0.0474
204	61.0	37.3	3.37	5.80	0.370	0.0218
208	52.5	20.4	1.26	2.65	0.160	(0.0187)[b]
212	41.0	10.7	0.458	1.15	0.0680	(0.0070)[c]
216	27.8	5.25	0.163	0.505	0.0290	(0.0027)[d]
220	17.5	2.51	0.062	0.220	0.0122	0.0011
224	10.2	1.17	0.023	0.0950	0.0053	–
228	5.65	0.532	0.0090	0.0410	0.0023	–
232	3.04	–	0.0034	0.0188	0.0010	–
236	1.60	(0.132)[e]	0.0013	0.008		–
240	0.830	0.047	–	0.0036	–	–
244	0.413	(0.017)[f]	–	0.0016	–	–
248	0.210	–	–	0.0007	–	–
250	0.148	0.0066	–	0.0005	–	–
260	0.025	0.0015	–	–	–	–

Source: NASA/JPL Panel Evaluation No. 17 (Sander et al. 2011), [a]175 nm, [b]205 nm, [c]210 nm, [d]215 nm, [e]235 nm, [f]245 nm

measurements by Hubrich and Stuhl (1980), Simon et al. (1988), Mérienne et al. (1990), Gillotay and Simon (1991), and Fahr et al. (1993) and others. Table 4.26 gives absorption cross sections of CCl_4, $CFCl_3$, CF_2Cl_2, $CF_2ClCFCl_2$, CF_2ClCF_2Cl, and CF_3CF_2Cl, and Table 4.27 cites those of CHF_2Cl, CH_3CFCCl_2, and CH_3CF_2Cl extracted from the evaluation.

The photolytic processes of CFCs and HCFCs in the region of stratospheric actinic flux are in general known as,

$$CFCl_3 + h\nu(\lambda < 377\ \text{nm}) \rightarrow CFCl_2 + Cl \quad \Delta H^\circ_{298} = 317\ \text{kJ mol}^{-1} \tag{4.71}$$

$$CF_2Cl_2 + h\nu(\lambda < 346\ \text{nm}) \rightarrow CF_2Cl + Cl \quad \Delta H^\circ_{298} = 346\ \text{kJ mol}^{-1} \tag{4.72}$$

$$CH_3CF_2Cl + h\nu(\lambda < 360\ \text{nm}) \rightarrow CF_2Cl + Cl, \quad \Delta H^\circ_{298} = 335\ \text{kJ mol}^{-1} \tag{4.73}$$

and Cl atoms are released (Clark and Husain 1984; Brownsword et al. 1999; Hanf et al. 2003). Recently, Taketani et al. (2005) detected $Cl(^2P_{3/2})$ and $Cl(^2P_{1/2})$

Table 4.27 Absorption cross sections of HCFCs (298 K)

	10^{20} σ (cm^2 $molecule^{-1}$)				10^{20} σ (cm^2 $molecule^{-1}$)		
Wave-length (nm)	CHF_2Cl (HCFC-22)	CH_3CFCl_2 (HCFC-141b)	CH_3CF_2Cl (HCFC-142b)	Wave-length (nm)	CHF_2Cl (HCFC-22)	CH_3CFCl_2 (HCFC-141b)	CH_3CF_2Cl (HCFC-142b)
176	4.04	163	(14.0)[a]	212	0.0029	1.40	0.0105
180	1.91	172	6.38	216	0.0013	0.589	0.0040
184	0.842	146	(2.73)[b]	220	0.0006	0.248	0.0015
188	0.372	104	–	224	–	0.105	0.0005
192	0.156	63.6	0.706	228	–	0.0444	0.0001
196	0.072	34.1	0.324	232	–	0.0189	–
200	0.032	16.6	0.145	236	–	0.0080	–
204	0.0142	7.56	0.0622	240	–	0.0033	–
208	0.00636	3.30	0.0256		–	–	–

Source: NASA/JPL Panel Evaluation No. 17 (Sander et al. 2011), [a]175 nm, [b]185 nm

spectroscopically in the photolysis of CFCs, CCl_4 and HCFCs, and determined the formation quantum yields of Cl atoms. The obtained quantum yields are 1.03 ± 0.09, 1.01 ± 0.08, 1.41 ± 0.14, 1.02 ± 0.08 for CF_2Cl_2, $CFCl_3$, CCl_4 and $CHFCl_2$, respectively, which implies that single C-Cl bond rupture occurs with the quantum yields of unity for CFCs and HCFCs except CCl_4. It has been revealed that the process releasing two Cl atoms such as,

$$CCl_4 + h\nu(\lambda < 207\ \text{nm}) \rightarrow CCl_2 + 2Cl \quad \Delta H^\circ_{298} = 577\ \text{kJ mol}^{-1} \tag{4.74}$$

occurs partially in the photolysis of CCl_4 at 193 nm.

4.3.10 Bromochlorofluorocarbons (Halons)

Among halocarbons (halogenated hydrocarbons), bromochlorofluorocarbons in which at least one chlorine or fluorine atom of chlorofluorocarbons is substituted by bromine is called halons. Halons, as with CFCs, do not have any dissipating process in the troposphere, and reach to the stratosphere and are photolyzed, which affects the ozone layer destruction. Among halons, CF_2ClBr (halon–1211) and CF_3Br (halon–1301) have the highest atmospheric concentrations, and these two compounds are taken up here.

Figure 4.35 depicts absorption spectra of halons and bromohydrofluorovcarbons (Orkin and Kasimovskaya 1995). As seen in the figure the absorption bands of halons have peaks at around 200–210 nm, and extend abroad tails to 300, and 280 nm for CF_2ClBr and CF_3Br, respectively. Although its absorption band overlaps with the tropospheric actinic flux region particularly for CF_2ClB, the absorption cross sections are very small so that the photolysis in the troposphere is almost negligible. Since halons has large absorption cross sections near 200 nm, they are easily photolyzed in the stratosphere.

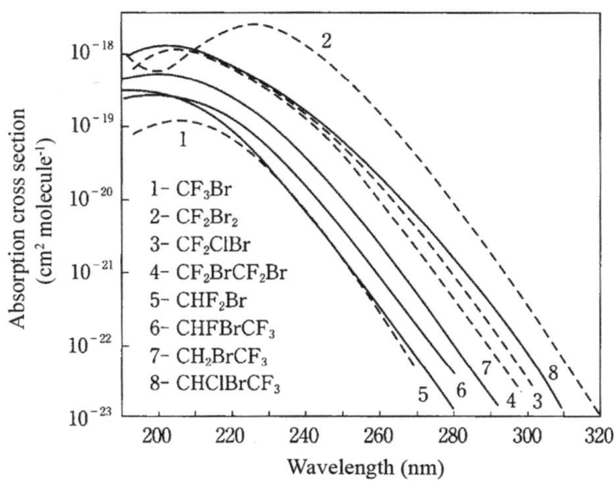

Fig. 4.35 Absorption spectra of halons (Adapted from Orkin and Kasimovskaya 1995)

Table 4.28 Absorption cross sections of CF_2ClBr and CF_3Br (298 K)

Wave-length (nm)	$10^{20}\sigma$ (cm^2 $molecule^{-1}$)		Wave-length (nm)	$10^{20}\sigma$ (cm^2 $molecule^{-1}$)		Wave-length (nm)	$10^{20}\sigma$ (cm^2 $molecule^{-1}$)	
	CF_2ClBr Halon-1211	CF_3Br Halon-1301		CF_2ClBr Halon-1211	CF_3Br Halon-1301		CF_2ClBr Halon-1211	CF_3Br Halon-1301
176	121	1.60	228	45.7	3.69	280	0.0991	0.0006
180	58.1	2.61	232	33.8	2.32	284	0.0527	(0.0002)[a]
184	35.0	4.02	236	24.4	1.39	288	0.0282	–
188	38.9	5.82	240	16.9	0.766	292	0.0148	–
192	57.0	7.58	244	11.4	0.414	296	0.0076	–
196	81.4	9.61	248	7.50	0.212	300	0.0039	–
200	106	11.3	252	4.76	0.107	304	0.0021	–
204	117	12.4	256	2.94	0.0516	308	0.0011	–
208	118	12.4	260	1.76	0.0248	312	0.0006	–
212	109	11.4	264	1.03	0.0118	316	0.0003	–
216	93.6	9.71	268	0.593	0.0058	320	0.0002	
220	76.8	7.56	272	0.336	0.0027	–	–	
224	60.4	5.47	276	0.184	0.0013	–	–	

Source: NASA/JPL Panel Evaluation No. 17 (Sander et al. 2011), [a]285 nm

The absorption cross sections of CF_2ClBr and CF_3Br are cited in Table 4.28. These absorption cross sections are extracted from NASA JPL Evaluation No. 17 (Sander et al. 2011) based on Gillotay and Simons (1989), Burkholder et al. (1991), Orkin and Kasimovskaya (1995), etc.

As the photolytic processes of CF_2ClBr,

$$CF_2ClBr + h\nu(\lambda < 441\ \text{nm}) \rightarrow CF_2Cl + Br \quad \Delta H^\circ_{298} = 271\ \text{kJ mol}^{-1} \tag{4.75}$$

$$+ h\nu\ (\lambda < 344\ \text{nm}) \rightarrow CF_2Br + Cl \quad \Delta H^\circ_{298} = 348\ \text{kJ mol}^{-1} \tag{4.76}$$

$$+ h\nu\ (\lambda < 245\ \text{nm}) \rightarrow CF_2 + Cl + Br \quad \Delta H^\circ_{298} = 488\ \text{kJ mol}^{-1} \tag{4.77}$$

can be conceived and the measurements of Talukdar et al. (1996) gave the quantum yields for the production of Cl and Br atoms at 193, 222, and 248 nm as $\Phi(Cl) = 1.03 \pm 0.14$, 0.27 ± 0.04, and 0.18 ± 003, $\Phi(Br) = 1.04 \pm 0.13$, 0.86 ± 0.11, and 0.75 ± 0.13, respectively. It is suggested that the simultaneous release of Cl and Br atoms by reaction (4.77) occurs at 193 nm, and reactions (4.75) and (4.76) take place with the total quantum yield of unity at the longer wavelengths than 200 nm.

As for CF_3Br, the quantum yields of Br formation, $(Br) = 1.12 \pm 0.16$, 0.92 ± 0.15 at 193, and 222 nm, respectively, are reported by Talukdar et al. (1992), and the reaction,

$$CF_3Br + h\nu(\lambda < 404\ \text{nm}) \rightarrow CF_3 + Br \quad \Delta H^\circ_{298} = 296\ \text{kJ mol}^{-1} \tag{4.53}$$

is thought to take place with the quantum yield of unity.

4.4 Photolysis of Inorganic Halogens

As a result of the photolyses of halogenated methane, CFCs, HCFCs, and halons, large amounts of Cl and Br atoms are released into the stratosphere, and chain reactions giving rise to ozone layer destruction are formed. Many inorganic halogen compounds formed as the chain carrier or chain terminating species in these chain reactions are photolyzed again in the stratosphere to regenerate halogen atoms or radicals. The calculations of the photolysis rates of these species are very important for determining the efficiency of the chain reactions. On the other hand, the same kinds of inorganic halogen compounds and radicals are formed in the troposphere from the photolysis or OH radical reactions of biogenic organic halogenated molecules and inorganic halogen molecules formed from heterogeneous reactions on sea salts. In this section, photolytic reactions of these inorganic halogen compounds which appear in common in the stratospheric and tropospheric chemistry will be summarized.

4.4.1 Chlorine (Cl_2), Bromine Monochloride (BrCl), Bromine (Br_2), Iodine (I_2)

Chlorine Cl_2, and bromine monochloride BrCl are formed in the reactions of $ClONO_2$, $BrONO_2$, HCl, HBr, HOCl, HOBr in the heterogeneous reaction in the polar stratospheric clouds (see Sect. 6.5), and their photolyses play an important role in the chain reactions of the ozone hole formation. In the troposphere, Cl_2 is known to be produced in the heterogeneous reactions on sea salts, but observational data is still limited. Bromine Br_2 is known to be produced by the heterogeneous chain reactions in the tropospheric ozone destruction in the arctic region. Meanwhile, iodine I_2 is released from sea weeds in coastal regions.

Figure 4.36 illustrates absorptio006E spectra of Cl_2, BrCl, Br_2, and I_2. As shown, the spectrum of Cl_2 is a broad band extending from the ultraviolet region of 260 nm to visible region of 500 nm with an maximum at around 330 nm. Therefore, Cl_2 can be photolyzed in the lower stratosphere at the altitude of ca. 20 km and in the troposphere by the photoabsorption in the near ultraviolet and visible light region at wavelengths longer than 290 nm. The absorption spectrum of Br_2 consists of relatively weak bands in the 190–300 nm with an maximum at around 225 nm, and strong bands in the 300–650 nm with an maximum near 415 nm. Furthermore, the second and third bands can be seen as shoulders of the latter band at around 480 and 550 nm. Also, weak vibrational bands appear overlapping with the continuum at wavelengths longer than 510 nm, although it is not discernible in the logarithmic scale chart of Fig. 4.36. Br_2 is photolyzed by visible light with large rate both in the stratosphere and troposphere. The spectrum of BrCl is similar to Br_2 and consists of the 190–290 nm bands with a peak at around 230 nm and 290–600 nm bands with a peak at around 375 nm. Therefore, BrCl can be photolyzed in the

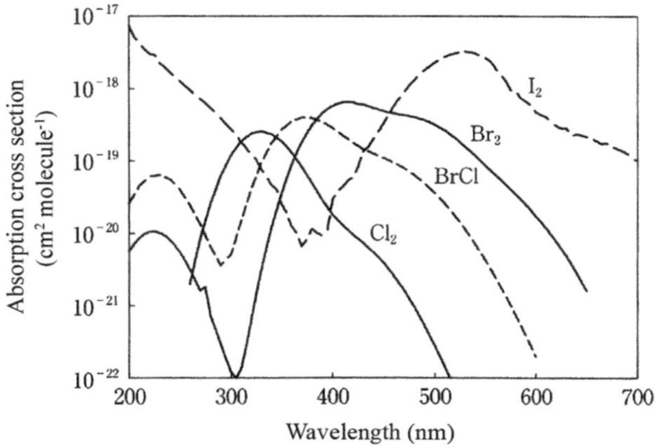

Fig. 4.36 Absorption spectra of Cl_2, BrCl, Br_2 and I_2 (Constructed based on NASA/JPL Evaluation No. 17, Sander et al. 2011, and IUPAC Subcommittee Report Vol. III, Atkinson et al. 2007)

lower stratosphere at the atmospheric window region near 230 nm, but the photolysis rate at the visible radiation is much larger. I_2 has a very strong absorption bands in the ultraviolet region of 200–300 nm and in the visible region of 450 – >700 nm, and easily photolyzed in the troposphere. Further distinct vibrational band structure can be seen (see Fig. 2.4) at the wavelengths longer than 500 nm in the visible spectrum although it is not shown in Fig. 4.36 with the vertical axis in the logarithmic sale.

Absorption cross sections of Cl_2, BrCl, Br_2, and I_2 extracted from NASA/JPL Evaluation No. 17 (Sander et al. 2011) are cited in Table 4.29 that are based on Maric et al. (1993) for Cl_2,Maric et al. (1994) for BrCl and Br_2, and Saiz-Lopez et al. (2004) for I_2.

The absorption spectrum of Cl_2 in the 250–450 nm is assigned to the transition from the ground $X^1\Sigma_g$ state to the dissociative excited states $^1\Pi_u$ and $^3\Pi_u$. From the $^1\Pi_u$ state $Cl(^2P_{3/2}) + Cl(^2P_{3/2})$ and from the $^3\Pi_u$ state $Cl(^2P_{3/2}) + Cl^*(^2P_{1/2})$ are thought to be formed (Matsumi et al. 1992) following the reactions,

$$Cl_2 + h\nu(\lambda < 500\ \text{nm}) \rightarrow Cl(^2P_{3/2}) + Cl(^2P_{3/2}) \quad \Delta H_0^\circ = 239\ \text{kJ mol}^{-1} \tag{4.79}$$

$$+ h\nu\ (\lambda < 480\text{nm}) \rightarrow Cl(^2P_{3/2}) + Cl^*(^2P_{1/2}). \quad \Delta H_0^\circ = 249\ \text{kJ mol}^{-1} \tag{4.80}$$

The formation ratio of the excited chlorine atom $Cl^*(^2P_{1/2})$ is small 0.01 at wavelengths shorter than 350 nm, and increases to 0.47 at 475 nm, near the dissociation limit of reaction (4.80) (Park et al. 1991; Matsumi et al. 1992; Samartzis et al. 1997). The total photolysis quantum yields $\Phi(Cl(^2P_{3/2})) + \Phi(Cl^*(^2P_{1/2}))$ are thought to be unity at the wavelengths shorter than 450 nm.

The photolytic processes of Br_2 is thought to be similar to Cl_2 (Lindeman and Wiesenfeld 1979).

Table 4.29 Absorption cross sections of Cl_2, BrCl, Br_2

	10^{20} σ (cm^2 molecule^{-1})				10^{20} σ (cm^2 molecule^{-1})		
Wavelength (nm)	Cl_2	BrCl	Br_2	Wavelength (nm)	Cl_2	BrCl	Br_2
200	–	2.64	0.562	430	0.732	14.6	60.1
210	–	4.59	0.870	440	0.546	12.6	54.0
220	–	6.13	1.05	450	0.387	11.0	48.8
230	–	6.48	1.01	460	0.258	9.52	45.2
240	–	5.60	0.808	470	0.162	8.02	42.8
250	–	4.05	0.544	480	0.0957	6.47	40.3
260	0.198	2.50	0.316	490	0.0534	4.99	36.6
270	0.824	1.35	0.161	500	0.0283	3.68	31.8
280	2.58	0.653	0.0728	510	0.0142	2.59	26.2
290	6.22	0.357	0.0299	520	0.0068	1.74	20.6
300	11.9	0.504	0.0122	530	0.0031	1.13	15.7
310	18.5	1.47	0.0135	540	0.0014	0.700	11.7
320	23.7	4.08	0.0626	550	0.0006	0.419	8.68
330	25.6	9.25	0.300	560	–	0.243	6.43
340	23.5	17.2	1.14	570	–	0.136	4.77
350	18.8	26.7	3.49	580	–	0.0739	3.50
360	13.2	35.0	8.66	590	–	0.0390	2.52
370	8.41	39.6	17.8	600	–	0.0200	1.76
380	5.00	39.3	30.7	610	–	–	1.19
390	2.94	34.9	45.1	620	–	–	0.767
400	1.84	28.6	57.4	630	–	–	0.475
410	1.28	22.5	64.2	640	–	–	0.282
420	0.956	17.8	64.5	650	–	–	0.161

Source: NASA/JPL Panel Evaluation No. 17 (Sander et al. 2011)

$$Br_2 + h\nu(\lambda < 629\ \text{nm}) \rightarrow Br(^2P_{3/2}) + Br(^2P_{3/2}) \quad \Delta H_0^\circ = 190\ \text{kJ mol}^{-1} \tag{4.81}$$

$$+ h\nu\ (\lambda < 511\ \text{nm}) \rightarrow Br\ (^2P_{3/2}) + Br^*(^2P_{1/2}) \quad \Delta H_0^\circ = 234\ \text{kJ mol}^{-1} \tag{4.82}$$

The formation quantum yields of the excited bromine atoms $Br(^3P_{1/2})$ increases from 0.4 at 444 nm to 0.89 at 510 nm close to the dissociation limit of reaction (4.82), and then decreases (Peterson and Smith 1978). As for the photolytic processes of Br_2, several other studies have also been made (Haugen et al. 1985; Cooper et al. 1998), but no absolute photolytic quantum yields has been measured. For the purpose of atmospheric chemistry total photolytic quantum yields of Br_2 can be approximated to be unity for the wavelength region of 200–510 nm.

As for BrCl, the quantum yields for each of the processes,

$$BrCl + h\nu(\lambda < 559\ \text{nm}) \rightarrow Br(^2P_{3/2}) + Cl(^2P_{3/2}) \quad \Delta H_0^\circ = 214\ \text{kJ mol}^{-1} \tag{4.83}$$

$$+ h\nu\ (\lambda < 534\ \text{nm}) \rightarrow Br(^2P_{3/2}) + Cl^*(^2P_{1/2}) \quad \Delta H_0^\circ = 224\ \text{kJ mol}^{-1} \tag{4.84}$$

$$+ h\nu\ (\lambda < 464\ \text{nm}) \rightarrow Br^*(^2P_{1/2}) + Cl(^2P_{3/2}) \quad = 258\ \text{kJmol}^{-1} \tag{4.85}$$

$$+ h\nu\ (\lambda < 446\ \text{nm}) \rightarrow Br^*(^3P_{1/2}) + Cl^*(^2P_{1/2}) \quad = 268\ \text{kJ mol}^{-1} \tag{4.86}$$

are reported at 235 nm as $\Phi[Br(^2P_{3/2}) + Cl(^2P_{3/2})] = 0.26 \pm 0.05$, $\Phi[Br(^2P_{3/2}) + Cl^*(^2P_{1/2})] = 0.16 \pm 0.05$, and $\Phi[Br^*(^2P_{1/2}) + Cl(^2P_{3/2})] = 0.58 \pm 0.05$ (Park et al. 2000). The total quantum yields of photodissociation may be approximated to be unity.

The photolysis of I_2 is similar to those of Cl_2 and Br_2, and the reactions,

$$I_2 + h\nu(\lambda < 803\ \text{nm}) \rightarrow I(^2P_{3/2}) + I(^2P_{3/2}) = 149\ \text{kJ mol}^{-1} \tag{4.87}$$

$$+ h\nu\ (\lambda < 498\ \text{nm}) \rightarrow I(^2P_{3/2}) + I^*(^2P_{1/2}) \quad = 240\ \text{kJ mol}^{-1} \tag{4.88}$$

with the total photolytic quantum yields of 0.33–0.9 depending on the wavelengths at the 501–624 nm region, and of unity in the continuum at <500 nm are reported (Brewer and Tellinghuisen 1972).

4.4.2 Chlorine Nitrate ($ClONO_2$), Bromine Nitrate ($BrONO_2$), Iodine Nitrate ($IONO_2$)

Chlorine nitrate $ClONO_2$ and bromine nitrate $BrONO_2$ are important reservoir molecules formed by the chain termination reactions, $ClO + NO_2$ and $BrO + NO_2$, in the ClO_x and BrO_x cycles in the stratosphere, respectively. Iodine nitrate $IONO_2$ plays a similar role in the in the iodine chemistry in the troposphere.

As shown in Fig. 4.37, the absorption spectra of $ClONO_2$ and $BrONO_2$ have peaks in common in the stratospheric window region near 200 nm, and decreasing

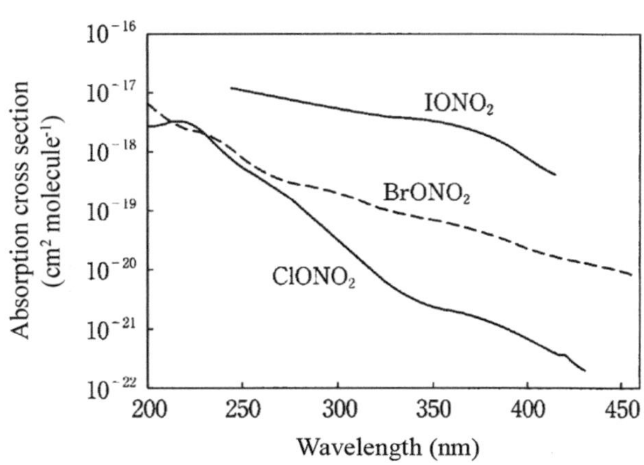

Fig. 4.37 Absorption spectra of $ClONO_2$, $BrONO_2$ and $IONO_2$ (Constructed based on IUPAC Subcommittee Report Vol. III, Atkinson et al. 2007, and NASA/JPL Evaluation No. 17, Sander et al. 2011)

Table 4.30 Absorption cross sections of $ClONO_2$[a], $BrONO_2$[a], $IONO_2$[b] (298 K)

Wavelength	$10^{20}\sigma$ (cm^2 $molecule^{-1}$)			Wavelength	$10^{20}\sigma$ (cm^2 $molecule^{-1}$)		
(nm)	$ClONO_2$	$BrONO_2$	$IONO_2$	(nm)	$ClONO_2$	$BrONO_2$	$IONO_2$
200	282	680	–	330	0.466	9.32	380
205	284	520	–	335	0.367	8.62	374
210	314	361	–	340	0.302	8.06	360
215	342	292	–	345	0.258	7.57	348
220	332	256	–	350	0.229	7.01	334
225	278	230	–	355	0.208	6.52	316
230	208	205	–	360	0.200	5.99	294
235	148	175	–	365	0.180	5.43	270
240	105	140	–	370	0.159	4.89	242
245	76.4	106	1210	375	0.141	4.35	213
250	56.0	79.7	1170	380	0.121	3.85	184
255	43.2	60.0	1060	385	0.106	3.37	153
260	33.8	47.1	946	390	0.091	2.97	130
265	26.5	38.9	880	395	0.076	2.59	103
270	20.5	33.8	797	400	0.064	2.28	78.0
275	15.7	30.5	772	405	0.054	2.01	60.5
280	11.9	27.9	741	410	0.044	1.81	49.6
285	8.80	25.6	691	415	0.036	1.65	41.6
290	6.41	23.2	631	420	0.032	1.50	–
295	4.38	20.8	577	425	0.023	1.38	–
300	3.13	18.6	525	430	0.019	1.29	–
305	2.24	16.5	495	435	–	1.20	–
310	1.60	14.5	462	440	–	1.11	–
315	1.14	12.7	441	445	–	1.03	–
320	0.831	11.3	404	450	–	0.928	–
325	0.613	10.2	396	455	–	0.831	–

Source: [a]IUPAC Subcommittee Report Vol. III, [b]NASA/JPL Panel Evaluation No. 17 (Sander et al. 2011)

monotonically to 380 and 400 nm, respectively. The absorption spectrum of $IONO_2$ has been measured in the wavelength region of 240–415 nm, and is a broad continuum in the region. The absorption cross sections of $ClONO_2$ and $BrONO_2$ are compiled by IUPAC Subcommittee Report Vol. III (Atkinson et al. 2007) taking the average of Burkholder et al. (1994, 1995) and Deters et al. (1998), and those of $IONO_2$ by NASA/JPL Evaluation No. 17 (Sander et al. 2011) based on Mössinger et al. (2002)., Table 4.30 gives the values extracted from them.

The photolytic processes of $ClONO_2$ are thought to be,

$$ClONO_2 + h\nu(\lambda < 1068\ \text{nm}) \rightarrow ClO + NO_2 \quad \Delta H^\circ_{298} = 112\ \text{kJ mol}^{-1} \tag{4.89}$$

$$h\nu(\lambda < 695\ \text{nm}) \rightarrow Cl + NO_3 \quad \Delta H^\circ_{298} = 172\ \text{kJ mol}^{-1} \tag{4.90}$$

$$h\nu(\lambda < 411\ \text{nm}) \rightarrow ClONO + O(^3P) \quad \Delta H^\circ_{298} = 291\ \text{kJ mol}^{-1} \tag{4.91}$$

The quantum yields are recommended as Φ ($Cl + NO_3$) = 0.6 for $\lambda < 308$ nm, $7.143 \times 10^{-3}\ \lambda - 1.60$ for $308 < \lambda < 364$ nm, 1.0 for $\lambda > 364$ nm, and Φ ($ClO+ NO_2$) = 1 − Φ ($Cl + NO_3$) by NASA/JPL Evaluation No. 17 (Sander et al. 2011) based on Goldfarb et al. (1997), Yokelson et al. (1997) and others.

As for $BrONO_2$ the processes,

$$BrONO_2 + h\nu(\lambda < 1049\ \text{nm}) \rightarrow BrO + NO_2 \quad \Delta H^\circ_{298} = 114\ \text{kJ mol}^{-1} \tag{4.92}$$

$$h\nu(\lambda < 836\ \text{nm}) \rightarrow Br + NO_3 \quad \Delta H^\circ_{298} = 143\ \text{kJ mol}^{-1} \tag{4.93}$$

$$h\nu(\lambda < 423\ \text{nm}) \rightarrow BrONO + O(^3P) \quad \Delta H^\circ_{298} = 283\ \text{kJ mol}^{-1} \tag{4.94}$$

can be conceived by analogy to $ClONO_2$ but experiments are sparse (Harwood et al. 1998; Soller et al. 2002). NASA/JPL Evaluation No. 17 (Sander et al. 2011) recommends Φ (total) = 1, Φ ($Br + NO_3$) = 0.85, and Φ ($BrO+ NO_2$) = 0.15 for $\lambda > 300$ nm.

Although the photolytic processes and quantum yields of $IONO_2$ are not studied well, the formation quantum yields of IO and NO_3 are reported as $\Phi(IO) \leq 0.02$, and $\Phi(NO_3) = 0.21 \pm 0.09$ at 248 nm by Joseph et al. (2007). From these results, the main photolytic pathway is presumed to be $I + NO_3$ but the possibility of further decomposition of the formed NO_3 to $NO_2 + O$ is also suggested.

4.4.3 Hydrogen Chloride (HCl), Hydrogen Bromide (HBr), Hydrogen Iodide (HI)

Hydrogen chloride HCl, hydrogen bromide HB, and hydrogen iodide HI are the reservoir molecules formed in the termination reactions of photochemical halogen chain in the troposphere and stratosphere together with $ClONO_2$, $BrONO_2$ and $IONO_2$ mentioned in the previous section. Since their absorption spectra lie in the shorter wavelengths, and cross sections are much smaller than $ClONO_2$, $BrONO_2$ and $IONO_2$, their atmospheric photolysis rates are much smaller and lifetime is much longer as the reservoirs.

As shown in Fig. 4.38, the absorption spectra of HCl and HBr have peaks at 154 and 178 nm in the vacuum ultraviolet region, and decrease monotonically toward longer wavelengths extending to near 230 and 279 nm, respectively. Therefore, the photolyses of HCl and HBr are possible only in the stratosphere. The peak of the absorption band of HI is at 222 nm and the absorption extends to 340 nm so that it can be photolyzed in the troposphere, but the absorption cross sections in the tropospheric actinic flux region is as small as $\leq \sim 3 \times 10^{-20}\text{cm}^2$.

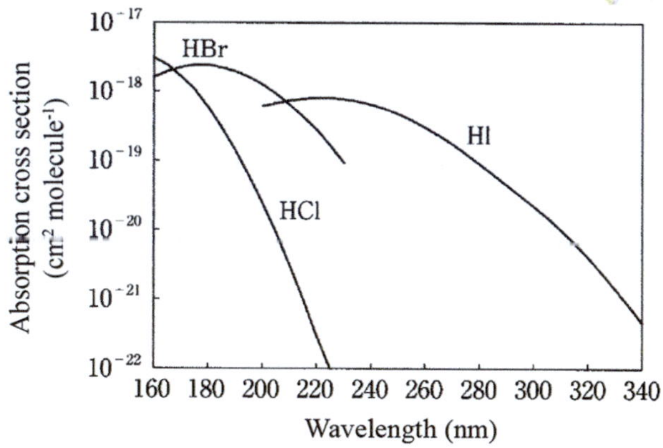

Fig. 4.38 Absorption spectra of HCl, HBr, HI (Constructed based on NASA/JPL Evaluation No. 17, Sander et al. 2011)

Table 4.31 Absorption cross sections of HCl, HBr and HI (298 K)

	10^{20} σ (cm^2 $molecule^{-1}$)				10^{20} σ (cm^2 $molecule^{-1}$)
Wavelength (nm)	HCl	HBr	HI	Wavelength (nm)	HI
150	334	–	–	250	47.0
155	343	131	–	255	38.2
160	306	161	–	260	30.0
165	240	195	–	265	23.0
170	163	225	–	270	17.2
175	106	242	–	275	12.5
180	58.9	242	–	280	8.94
185	29.4	221	–	285	6.37
190	13.8	194	–	290	4.51
195	5.96	161	–	295	3.18
200	2.39	125	61.1	300	2.23
205	0.903	91.8	67.7	305	1.52
210	0.310	64.4	73.8	310	1.01
215	0.101	42.3	78.4	315	0.653
220	0.030	28.0	80.8	320	0.409
225	0.010	16.3	80.4	325	0.247
230	0.0034	9.32	77.4	330	0.145
235	–	–	71.9	335	0.083
240	–	–	64.6	340	0.047
245	–	–	56.1	345	–

Source: NASA/JPL Panel Evaluation No. 17 (Sander et al. 2011)

Table 4.31 cites absorption cross sections of HCl, HBr and HI extracted from NASA/JPL Evaluation No. 17 (Sander et al. 2011). These recommended values are based on Bahou et al. (2001) for HCl, Huebert and Martin (1968) and Nee et al. (1986) for HBr, and Campuzano-Jost and Crowley(1999) for HI.

Total photolytic quantum yields of HCl, HBr and HI in the wavelength region of tropospheric and stratospheric actinic fluxes are thought to be unity. The photolytic processes of each of them,

$$HCl + h\nu(\lambda < 279\ nm) \rightarrow H + Cl(^2P_{3/2}) \quad \Delta H_0^\circ = 428\ kJ\ mole^{-1} \tag{4.95}$$

$$+ h\nu\ (\lambda < 273\ nm) \rightarrow H + Cl^*(^2P_{1/2}) \quad \Delta H_0^\circ = 438\ kJ\ mole^{-1} \tag{4.96}$$

$$HBr + h\nu(\lambda < 330\ nm) \rightarrow H + Br(^2P_{3/2}) \quad \Delta H_0^\circ = 362\ kJ\ mole^{-1} \tag{4.97}$$

$$+ h\nu\ (\lambda < 295\ nm) \rightarrow H + Br^*(^2P_{1/2}) \quad \Delta H_0^\circ = 406\ kJ\ mole^{-1} \tag{4.98}$$

$$HI + h\nu(\lambda < 405\ nm) \rightarrow H + I(^2P_{3/2}) \quad \Delta H_0^\circ = 295\ kJ\ mole^{-1} \tag{4.99}$$

$$+ h\nu\ (\lambda < 311\ nm) \rightarrow H + I^*(^2P_{1/2}) \quad \Delta H_0^\circ = 385\ kJ\ mole^{-1} \tag{4.100}$$

are known to produce both the ground states of Cl, Br and I atoms, $Cl(^2P_{3/2})$, $Br(^2P_{3/2})$, and $I(^2P_{3/2})$, and the spin-orbit excited states of $Cl^*(^2P_{1/2})$, $Br^*(^2P_{1/2})$, and $I^*(^2P_{1/2})$. The formation ratios of the excited and ground state atoms have been studied considerably well. For example, $\Phi[Cl^*(^2P_{1/2})]/\Phi[Cl^*(^2P_{1/2}) + Cl(^2P_{3/2})] =$ 0.42–0.48 at 201–210 nm for HCl (Regan et al. 1999a), $\Phi[Br^*(^2P_{1/2})]/\Phi[Br^*(^2P_{1/2}) + Br(^2P_{3/2})] = 0.15$–$0.23$ at 201–253 nm for HBr (Regan et al. 1999b), and $\Phi[I^*(^2P_{1/2})]/\ \Phi[\ I(^2P_{3/2})] = 0.2$ at 208 nm, 1.7 at 252 nm, and 0.1 at 303 nm for HI (Langford et al. 1998).

4.4.4 *Hypochlorous Acid (HOCl), Hypobromous Acid (HOBr), Hypoiodous Acid (HOI)*

Hypochlorous acid HOCl, hypobromous acid HOBr, and hypoiodous acid HOI are formed in the chain termination reactions between ClO, BrO, IO radicals and HO_2 radicals. Since they have absorption spectra in the ultraviolet to visible region, the photolytic regeneration of the radicals has to be considered in the atmosphere.

As seen in Fig. 4.39(a), absorption spectrum of HOCl consists of a fairly strong continuum band with a peak at around 240 nm and a second continuum which appears as a shoulder at around 300 nm. These transitions have been assigned to $2^1A' \leftarrow 1^1A'$, $1^1A'' \leftarrow 1^1A'$, respectively. Table 4.32 gives absorption cross sections of HOCl extracted from NASA/JPL Evaluation No. 17 (Sander et al. 2011) based on Burkholder (1993) and Barnes et al. (1998).

Considerable disagreement has been noted for the absorption spectrum of HBr due to impurities in samples (Finlayson-Pitts and Pitts 2000). Figure 4.39(b) shows the absorption spectrum by Ingham et al. (1998), and Table 4.32 gives the absorption cross sections recommended by Sander et al. (2011) based on Ingham et al. (1998). In the spectrum of HOBr a peak at 285 nm and a shoulder around 350 nm are seen corresponding to those of HOCl due to the same electronic

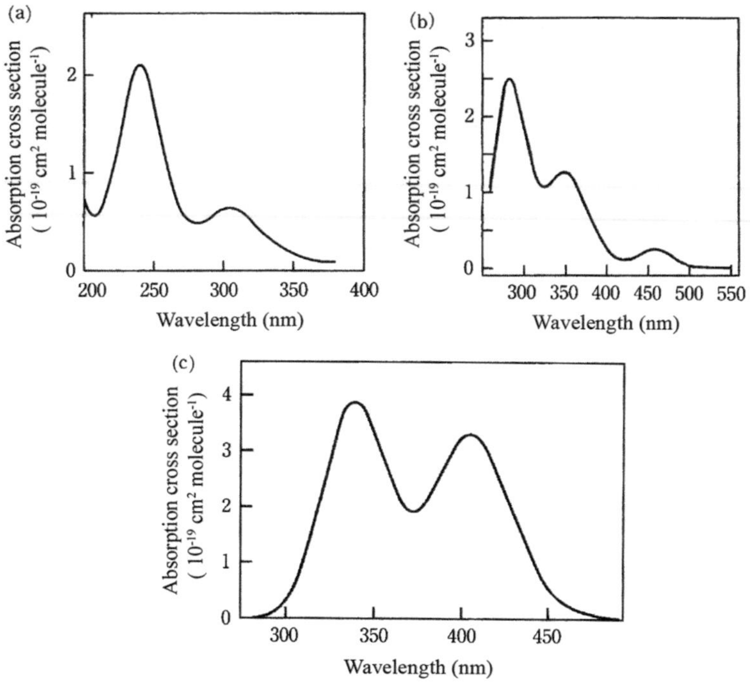

Fig. 4.39 Absorption spectra of (**a**) HOCl, (**b**) HOBr, and (**c**) HOI (Adapted from Burkholder 1993 (HOCl); Ingham et al. 1998 (HOBr); Bauer et al. 1998 (HOI))

transition and shifted to longer wavelength. The absorption range at still longer wavelength is thought due to the forbidden transition to a triplet state (Francisco et al. 1996; Minaev 1999).

The absorption spectrum of HOI has been reported by Bauer et al. (1998) and Rowley et al. (1999). Figure 4.39(c) shows the absorption spectrum by Bauer et al. (1998), and Table 4.32 gives the absorption cross sections recommended by NASA/JPL Evaluation No. 17 (Sander et al. 2011) taking the average of above two studies. As can be seen in Fig. 4.39(c), the absorption spectrum of HOI consists of continuum bands with two peaks at 340 and 408 nm.

The photolyses of HOCl, HOBr and HOI proceeds via the pathways,

$$\mathrm{HOCl} + h\nu(\lambda < 525\ \mathrm{nm}) \rightarrow \mathrm{OH} + \mathrm{Cl} \quad \Delta H_0^\circ = 228\ \mathrm{kJ\ mole^{-1}} \tag{4.101}$$

$$\mathrm{HOBr} + h\nu(\lambda < 589\ \mathrm{nm}) \rightarrow \mathrm{OH} + \mathrm{Br} \quad \Delta H_0^\circ = 203\ \mathrm{kJ\ mole^{-1}} \tag{4.102}$$

$$\mathrm{HOI} + h\nu(\lambda < 572\ \mathrm{nm}) \rightarrow \mathrm{OH} + \mathrm{I} \quad \Delta H_0^\circ = 209\ \mathrm{kJ\ mole^{-1}} \tag{4.103}$$

The quantum yields of unity for each reaction have been confirmed by the experiments of Schindler et al. (1997), Benter et al. (1995) and Bauer et al. (1998) for HOCl, HOBr, and HOI, respectively.

Table 4.32 Absorption cross sections of HOCl, HOBr and HOI (298 K)

	10^{20} σ (cm^2 $molecule^{-1}$)				10^{20} σ (cm^2 $molecule^{-1}$)		
Wavelength (nm)	HOCl[a]	HOBr	HOI[a]	Wavelength (nm)	HOCl	HOBr	HOI
280	4.64	24.3	0.077	390	0.491	4.22	24.8
285	4.74	25.0	0.234	395	0.385	3.23	27.9
290	5.13	24.0	0.608	400	0.288	2.43	30.1
295	5.62	21.9	1.45	405	0.208	1.80	30.9
300	5.99	19.1	3.02	410	0.144	1.36	30.2
305	6.12	16.2	5.77	415	0.097	1.08	28.0
310	5.97	13.6	9.85	420	0.063	0.967	24.7
315	5.56	11.8	15.4	425	–	0.998	20.7
320	4.95	10.8	21.9	430	–	1.15	16.6
325	4.24	10.5	28.6	435	–	1.40	12.7
330	3.50	10.8	34.3	440	–	1.68	9.30
335	2.81	11.3	38.1	445	–	1.96	6.54
340	2.22	11.9	39.2	450	–	2.18	4.40
345	1.77	12.3	37.7	455	–	2.29	2.37
350	1.43	12.4	33.9	460	–	2.28	1.79
355	1.22	12.1	29.1	465	–	2.14	1.09
360	1.06	11.5	24.1	470	–	1.91	0.632
365	0.968	10.5	20.2	475	–	1.62	0.360
370	0.888	9.32	17.8	480	–	1.30	0.196
375	0.804	7.99	17.4	485	–	0.993	–
380	0.708	6.65	18.8	490	–	0.723	–
385	0.602	5.38	21.5	495	–	0.502	–

Source: NASA/JPL Panel Evaluation No. 17 (Sander et al. 2011)

[a]For example, the values at 285 nm, 295 nm are the average at 284 and 286 nm, and 294,296 nm, respectively

4.4.5 Chlorine Monoxide (ClO), Bromine Monoxide (BrO), Iodine Monoxide (IO)

Chlorine monoxide ClO, bromine monoxide BrO, and iodine monoxide IO are the main chain carrier of stratospheric and tropospheric halogen chain reactions. ClO is partially photolyzed only in the stratosphere, and the photolytic rates of BrO and IO are large also in the troposphere. Their photolysis rates have to be considered in the evaluation of ozone depleting chain reactions.

As shown in Fig. 4.40(a), the absorption spectrum of ClO consists of a continuum in the range of 210–265 nm, and vibrational-structured bands in the 265–315 nm (Sander and Friedl 1989; Sander et al. 2011). The absorption cross sections of ClO are dependent on spectral resolution and temperature. Table 4.28 cites the absorption cross sections with an averaged wavelength interval of 1 nm given in NASA/JPL Evaluation No. 17 (Sander et al. 2011) based on the measurement with spectral resolution of 0.3 nm by Sander and Friedl (1989). Although the

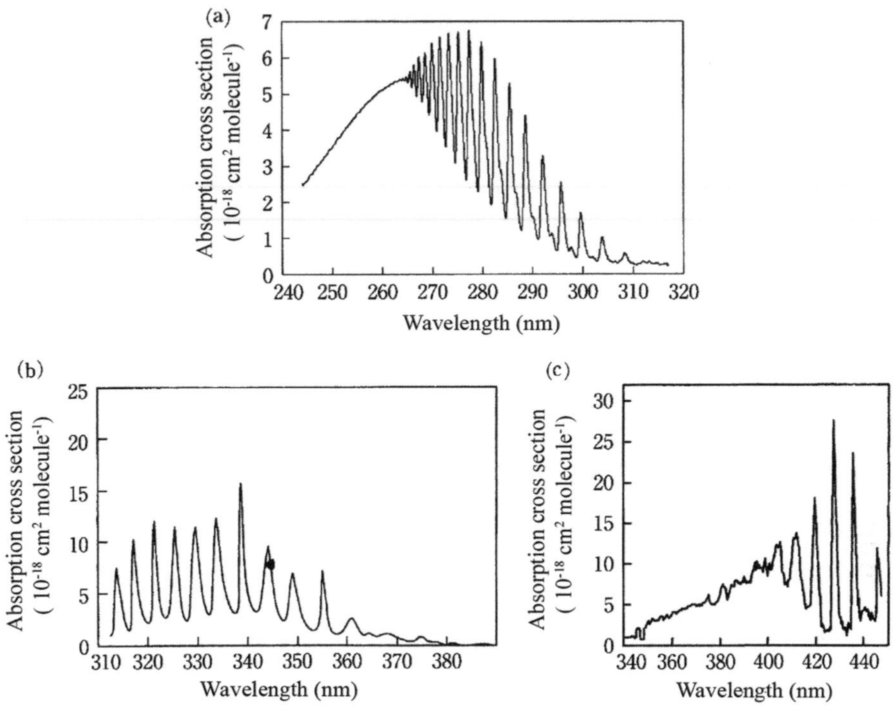

Fig. 4.40 Absorption spectra of (**a**) ClO, (**b**) BrO, and (**c**) IO (Adapted from Sander and Friedl 1989 (ClO); Wahner et al. 1988 (BrO); Lasylo et al. 1995 (IO))

absorption cross sections of ClO at the peak of 265 nm is large, 5.2×10^{-18} cm^2 molecule^{-1}, the region overlaps with the absorption of O_3 in the stratosphere, the photolysis of ClO is mainly brought by the absorption at the long-wavelength tail. Due to this reason, the loss of ClO by the photolysis is thought to be much smaller than the reactions with O atoms and NO (Langhoff et al. 1977).

The quantum yields of the formation of $Cl(2P_{3/2,\ 1/2})$ and $O(^1D)$ in the photolysis of ClO,

$$ClO + h\nu(\lambda < 451\ \text{nm}) \rightarrow Cl + O(^3P) = 265\ \text{kJ mol}^{-1} \quad (4.104)$$

$$+ h\nu\ (\lambda < 263\ \text{nm}) \rightarrow Cl + O(^1D) \quad = 455\ \text{kJ mol}^{-1} \quad (4.105)$$

have been reported to be unity by Schmidt et al. (1998).

The absorption spectrum of BrO has banded structure in the ultraviolet region of 290–380 nm as shown in Fig. 4.40(b), and is assigned to $A^2\Pi_{3/2} \leftarrow X^2\Pi_{3/2}$ transition (Wahner et al. 1988; Sander et al. 2011). The absorption cross sections of BrO depends on spectral resolution and temperature, and Table 4.28 gives the recommended values (Sander et al. 2011) based on Wilmouth et al. (1999). The photolysis of BrO,

$$\mathrm{BrO} + \mathrm{h}\nu(\lambda < 511\ \mathrm{nm}) \rightarrow \mathrm{Br} + \mathrm{O} = 234\ \mathrm{kJ\ mol}^{-1} \quad (4.106)$$

is thought to occur with quantum yields of unity.

As for IO, the absorption bands corresponding to the $A^2\Pi_{3/2} \leftarrow X^2\Pi_{3/2}$ transition similar to BrO can be seen in the ultraviolet and visible region of 338–488 nm. The absorption maximum appears at around 400 nm, and the bands with a vibrational structure overlaps with a continuum at the wavelength side longer than 400 nm. The maximum of absorption cross sections appears for the A-X (4-0) band at 427.2 nm. Figure 4.40(c) depicts the absorption spectrum of IO (Lasylo et al. 1995; Harwood et al. 1997), and Table 4.33 gives the absorption cross sections of 1 nm intervals from NASA/JPL Evaluation No. 17 (Sander et al. 2011) based on Lasylo et al. (1995), Harwood et al. (1997), and Bloss et al. (2001).

Ingham et al. (2000) reported the quantum yields of $O(^3P)$ production in the photolysis of IO as 0.91 at 355 nm. Among the photolytic reactions of

$$\mathrm{IO} + \mathrm{h}\nu(\lambda < 527\ \mathrm{nm}) \rightarrow \mathrm{I} + \mathrm{O}(^3\mathrm{P}) \quad \Delta H_0^\circ = 227\ \mathrm{kJ\ mole}^{-1} \quad (4.107)$$

$$+ \mathrm{h}\nu\ (\lambda < 287\mathrm{nm}) \rightarrow \mathrm{I} + \mathrm{O}(^1\mathrm{D}), \quad \Delta H_0^\circ = 417\ \mathrm{kJ\ mole}^{-1} \quad (4.108)$$

the energetically possible process in the absorption spectrum region of Fig. 4.40(c) is only reaction (4.107), and the formation of I and $O(^3P)$ atoms by this reaction is thought to occur with quantum yields of unity.

4.4.6 *Chlorine Peroxide (ClOOCl)*

Chlorine peroxide ClOOCl is formed by the termolecular recombination reaction of ClO radicals when the concentration of ClO radicals is high. Among the reservoir molecules in the stratosphere over the Antarctica in wintertime, ClOOCl has the highest concentration, and its photolysis in springtime is very important for the formation of the ozone hole.

Figure 4.41 shows the absorption spectrum by Papanastasiou et al. (2009). The absorption spectrum of ClOOCl has a valley at around 218 nm, maximum at 245 nm, and decreases monotonically toward the longer wavelengths side till near 400 nm. For the photochemistry in the lower stratosphere, absorption spectrum in the wavelength region longer than 290 nm is important. However, due to the impurity of Cl_2 in the sample of ClOOCl in the laboratory, the disagreement of the absorption spectrum and cross sections have been much of concern until recently. In order to overcome this problem, Chen et al. (2009) used the molecular beam and measured the photolytic decay by mass spectrometry, and determined the absorption cross sections of ClOOCl. This method is supposed to exclude the effect of

Table 4.33 Absorption cross sections of ClO, BrO and IO (298 K)

Wave-length (nm)	10^{20} σ (cm^2 $molecule^{-1}$) ClO	BrO	Wave-length (nm)	10^{20} σ (cm^2 $molecule^{-1}$) ClO	BrO	Wave-length (nm)	10^{20} σ (cm^2 $molecule^{-1}$) BrO	IO	Wave-length (nm)	10^{20} σ (cm^2 $molecule^{-1}$) IO	Wave-length (nm)	10^{20} σ (cm^2 $molecule^{-1}$) IO
250	352	–	300	133	275	335	652	–	400	671	420	1200
254	425	–	301	56.6	180	336	339	–	391	620	421	681
258	486	–	302	45.2	502	337	222	–	392	617	412	365
262	529	–	303	44.9	217	338	201	–	393	642	423	253
266	549	–	304	87.8	274	339	1296	–	394	684	424	204
270	574	–	305	45.5	466	340	445	118	395	694	425	205
271	489	–	306	33.2	221	341	243	100	396	709	426	302
272	532	–	307	33.1	407	342	235	107	397	701	427	2050
273	515	–	308	47.7	518	343	424	89	398	654	428	1370
274	470	–	309	45.5	227	344	968	96.2	399	671	429	543
275	507	–	310	28.7	396	345	542	86.2	400	671	430	309
276	456	–	311	27.3	659	346	226	126	401	700	431	208
277	418	–	312	33.1	294	347	146	112	402	765	432	173
278	501	–	313	32.5	197	348	258	108	403	859	433	166
279	283	–	314	28.9	901	349	748	142	404	864	434	177
280	538	–	315	27.8	443	350	499	160	400	787	435	653
281	329	–	316	26.8	232	351	272	154	401	667	436	1880
282	311	–	317	–	721	352	182	165	402	606	437	807
283	445	–	318	–	730	353	163	163	403	578	438	381
284	245	–	319	–	345	354	180	181	404	643	439	249
285	292	–	320	–	251	355	789	185	405	787	440	256
286	362	–	321	–	1138	356	276	194	406	667	441	219
287	200	107	322	–	677	357	120	207	407	606	442	168

(continued)

Table 4.33 (continued)

	10^{20} σ (cm^2 $molecule^{-1}$)		Wave-length	10^{20} σ (cm^2 $molecule^{-1}$)		Wave-length	10^{20} σ (cm^2 $molecule^{-1}$)		Wave-length	10^{20} σ (cm^2 $molecule^{-1}$)	Wave-length	10^{20} σ (cm^2 $molecule^{-1}$)
Wave-length (nm)	ClO	BrO	(nm)	ClO	BrO	(nm)	BrO	IO	(nm)	IO	(nm)	IO
288	197	95.0	323	–	301	358	115	223	408	578	443	183
289	337	110	324	–	288	359	144	230	409	643	444	195
290	165	184	325	–	983	360	236	242	410	813	445	957
291	111	134	326	–	838	364	113	268	411	1010	446	805
292	270	157	327	–	312	368	130	326	412	976	447	392
293	161	248	328	–	223	372	39.4	360	413	786	448	214
294	102	140	329	–	789	376	35.4	402	414	589	449	269
295	94.5	294	330	–	1058	380	12.3	504	415	568	450	156
296	206	164	331	–	453	384	3.89	523	416	414	451	96.9
297	83.1	361	332	–	203	388	–	580	417	460	452	102
298	65.1	193	333	–	260	392	–	617	418	734	453	87.3
299	74.8	284	334	–	1294	396	–	709	419	1380	454	100

Source: NASA/JPL Panel Evaluation No. 17 (Sander et al. 2011)

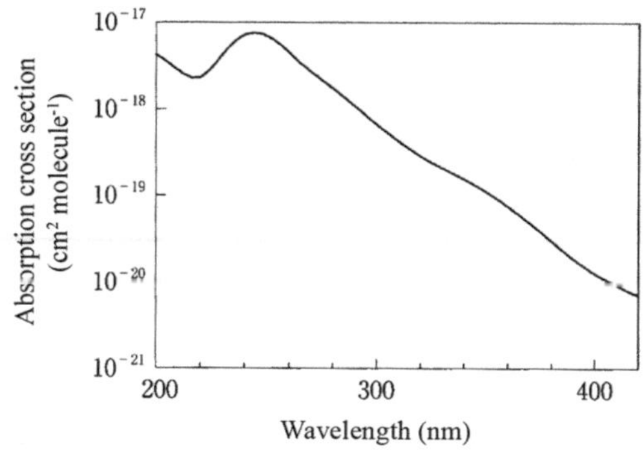

Fig. 4.41 Absorption spectrum of ClOOCl (Adapted from Papanastasiou et al. 2009)

photoabsorption of impurity Cl_2, and the values of 49.0, and 11.2×10^{-20} cm^2 molecule^{-1} are reported at 308 nm, and 351 nm, respectively. These values agree well with those recently reported by Papanastasiou et al. (2009). Based on this data, NASA/JPL Evaluation No. 17 recommended the absorption cross sections averaged over 2 nm and estimated the error as ±35 % (Sander et al. 2011). Table 4.34 cites the recommended absorption cross sections at the temperature range of 190–250 K.

The photolytic pathways of ClOOCl can be conceived as,

$$\mathrm{ClOOCl} + h\nu(\lambda < 1709\ \mathrm{nm}) \rightarrow \mathrm{ClO} + \mathrm{ClO} = 70\ \mathrm{kJ\ mole^{-1}} \quad (4.109)$$

$$+ h\nu\ (\lambda < 357\ \mathrm{nm}) \rightarrow \mathrm{ClO} + \mathrm{Cl} + \mathrm{O} \quad = 335\ \mathrm{kJ\ mole^{-1}} \quad (4.110)$$

$$+ h\nu\ (\lambda < 1375\ \mathrm{nm}) \rightarrow \mathrm{Cl} + \mathrm{ClOO} \quad = 87\ \mathrm{kJ\ mole^{-1}} \quad (4.111)$$

$$+ h\nu\ (\lambda < 1117\ \mathrm{nm}) \rightarrow 2\mathrm{Cl} + \mathrm{O_2}. \quad = 107\ \mathrm{kJ\ mole^{-1}} \quad (4.112)$$

According to the recent study of detecting Cl atoms and ClO radicals directly, the main product is Cl atoms by reaction (4.112) to form $2Cl + O_2$ (Moore et al. 1999; Huang et al. 2011). NASA/JPL Evaluation No. 17 recommended $\Phi(2Cl + O_2) = 0.8 \pm 0.1$, $\Phi(2ClO) + \Phi(ClO + Cl + O) = 0.2 \pm 0.1$ with total quantum yield of unity for the whole spectral range of Fig. 4.44, and $\Phi(ClO + Cl + O) = 0.0 \pm 0.1$ at $\lambda > 300$ nm. Since the difference in the photolytic pathways can cause a big difference in the ozone destruction efficiency, the pathways of this reaction have attracted much interest.

4.4.7 Chlorine Dioxide (OClO)

Chlorine dioxide OClO is formed by the bimolecular cross reactions of ClO by themselves and is a reservoir molecule of chlorine in the stratosphere over wintertime Antarctica the same as ClOOCl. Since OClO has the strong absorption in the visible region, it is photolyzed instantaneously when the solar light starts to irradiate in spring.

Table 4.34 Absorption cross sections of ClOOCl (190–250 K)

Wave-length (nm)	10^{20} σ (cm^2 molecule^{-1})	Wave-length (nm)	10^{20} σ (cm^2 molecule^{-1})	Wave-length (nm)	10^{20} σ (cm^2 molecule^{-1})	Wave-length (nm)	10^{20} σ (cm^2 molecule^{-1})
200	423	260	445	320	28.2	380	2.97
204	362	264	360	324	24.7	384	2.45
208	303	268	294	328	21.9	388	2.04
212	255	272	246	332	19.5	392	1.71
216	228	276	206	336	17.3	396	1.47
220	232	280	173	340	15.4	400	1.26
224	277	284	144	344	13.6	404	1.11
228	366	288	119	348	11.9	408	0.988
232	488	292	98.2	352	10.3	412	0.878
236	618	296	80.5	356	8.82	416	0.778
240	719	300	66.1	360	7.43	420	0.712
244	758	304	54.4	364	6.24		
248	732	308	45.4	368	5.23		
252	651	312	38.2	372	4.35		
256	549	316	32.8	376	3.60		

Source: NASA/JPL Panel Evaluation No. 17 (Sander et al. 2011)

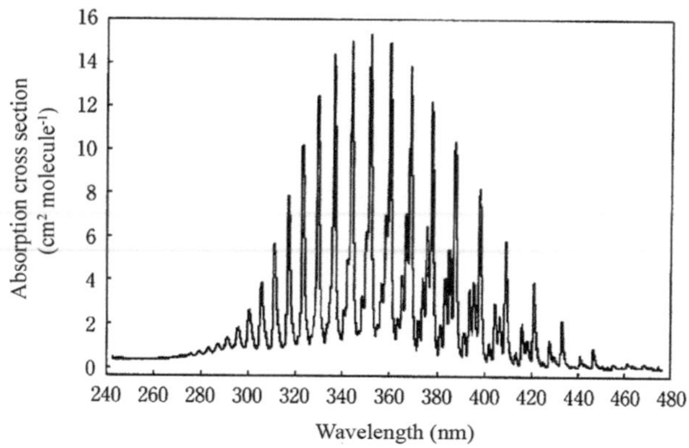

Fig. 4.42 Absorption spectrum of OClO (Adapted from Wahner et al. 1987)

The absorption spectrum consists of well-structured strong bands in the 280–480 nm region (Wahner et al. 1987) as shown in Fig. 4.42. This bands corresponds to the ClO starching vibration ($v' \leftarrow v'' = 0$) of electronic transition A $^2A_2 \leftarrow X^2B_1$. As expected from the banded structure, the shape and cross sections of absorption depends on spectroscopic resolution. The measurements of Wahner et al. (1987) was at 0.25 nm, and those at higher resolution of 0.01–0.02 nm and middle resolution of 0.2–0.4 nm has been made later by Kromminga et al. (2003). NASA/JPL Evaluation No. 17 gives the absorption cross sections averaged over 1 nm interval, and those at the peak of each vibrational band at different temperatures (Sander et al. 2011). Table 4.35 cites the absorption cross sections at the 1 nm interval at 204 K extracted from them.

The photolytic pathways of OClO are known to be

$$\mathrm{OClO} + h\nu(\lambda < 480\ \mathrm{nm}) \rightarrow \mathrm{ClO} + \mathrm{O}(^3\mathrm{P}) \quad \Delta H_0^\circ = 249\ \mathrm{kJ\ mole}^{-1} \tag{4.113}$$

$$+ h\nu\ (\lambda < 1040\ \mathrm{nm}) \rightarrow \mathrm{Cl} + \mathrm{O_2}\ (^1\Delta_g). \quad \Delta H_0^\circ = 115\ \mathrm{kJ\ mole}^{-1} \tag{4.114}$$

The production of $O(^3P)$ is reported to be the main process in the 350–475 nm region including photofragment spectroscopic study (Lawrence et al. 1990; Davis and Lee 1992, 1996; Delmdahl et al. 1998). Although the formation of Cl atoms is also reported at the shorter wavelengths region, it is presumed that the formation of $O(^3P)$ by reaction (4.113) occurs with quantum yields of unity and the formation of Cl atoms is less than 4 % at the wavelength region longer than 350 nm, which is effective in the photolysis of lower stratosphere (Sander et al. 2011).

4.4.8 Nitrosyl Chloride (ClNO), Nitryl Chloride ($ClNO_2$)

Nitrosyl chloride ClNO and nitryl chloride $ClNO_2$ are both formed by the heterogeneous reactions of NO_2 and $N_2O_5$0020on the solid sea salt surface. These are

Table 4.35 Absorption cross sections of OClO (averaged over 1 nm interval, 204 K)

Wave-length (nm)	10^{20} σ (cm^2 $molecule^{-1}$)	Wave-length (nm)	10^{20} σ (cm^2 $molecule^{-1}$)	Wave-length (nm)	10^{20} σ (cm^2 $molecule^{-1}$)	Wave-length (nm)	10^{20} σ (cm^2 $molecule^{-1}$)	Wave-length (nm)	10^{20} σ (cm^2 $molecule^{-1}$)	Wave-length (nm)	10^{20} σ (cm^2 $molecule^{-1}$)
270	44.3	300	226	330	782	360	1210	390	71.4	420	81.4
271	45.7	301	222	331	285	361	477	391	123	421	323
272	49.9	302	143	332	155	362	173	392	109	422	151
273	49.1	303	95.3	333	147	363	179	393	203	423	50.0
274	48.1	304	96.1	334	208	364	207	394	270	424	23.8
275	54.8	305	276	335	335	365	361	395	285	425	23.3
276	58.3	306	328	336	1090	366	403	396	275	426	14.5
277	52.5	307	190	337	782	367	625	397	370	427	43.8
278	54.3	308	116	338	266	368	919	398	53	428	99.5
279	67.4	309	85.4	339	155	369	903	399	225	429	46.9
280	67.2	310	168	340	167	370	268	400	70.1	430	44.3
281	58.3	311	511	341	250	371	107	401	45.6	431	23.3
282	65.4	312	338	342	414	372	180	402	96.9	432	47.0
283	82.4	313	174	343	925	373	170	403	56.3	433	173
284	77.6	314	107	344	1090	374	364	404	196	434	69.6
285	67.2	315	94.2	345	388	375	376	405	194	435	24.6
286	77.7	316	239	346	176	376	554	406	185	436	11.2
287	100	317	686	347	161	377	718	407	160	437	7.68
288	93.7	318	360	348	258	378	881	408	158	438	9.09
289	79.4	319	176	349	320	379	278	409	493	439	5.13
290	90.5	320	114	350	581	380	92.4	410	210	440	12.5
291	127	321	125	351	1100	381	135	411	71.6	441	47.8
292	116	322	279	352	993	382	148	412	34.0	442	23.2
293	90.9	323	873	353	330	383	266	413	46.8	443	14.7

294	94.1	324	443	354	164	384	298	414	44.6	444	7.59
295	147	325	192	355	190	385	440	415	30.0	445	3.96
296	172	326	121	356	276	386	345	416	164	446	46.8
297	122	327	147	357	343	387	762	417	100	447	55.2
298	92.0	328	221	358	597	388	388	418	107	448	18.4
299	106	329	838	359	830	389	173	419	75.1	449	7.17

Source: NASA/JPL Panel Evaluation No. 17 (Sander et al. 2011)

thought to be formed mainly in the marine boundary layer affected by urban polluted plumes, and their photolyses are considered in the halogen chemistry in the troposphere.

As depicted in Fig. 4.43, the ultraviolet and visible spectrum of ClNO has strong absorption maximum at around 200 nm and has very broad continuum extending toward longer wavelength till over 600 nm in the visible region. The evaluation of absorption cross sections by NASA/JPL Evaluation No. 17 (Sander et al. 2011) is based on the data of Tyndall et al. (1987) (190–350 nm) and Roehl et al. (1992) (350–650 nm), and Table 4.36 gives those extracted from it for the 200–500 nm region. The absorption spectrum drawn using these cross sections is shown in Fig. 4.43 for the tropospheric actinic flux region.

The absorption spectrum of $ClNO_2$ also has the maximum at around 215 nm in the ultraviolet region and a broad continuum extends toward visible region as shown in Fig. 4.44. The decay of cross sections are larger than ClNO, and the spectrum terminates at around 400 nm. The absorption cross sections of NASA/JPL Evaluation No. 17 (Sander et al. 2011) takes the average of Illies and Takacs (1976) and Furlan et al. (2000). Table 4.36 cites those extracted from the evaluation together with ClNO. As for the absorption spectrum of $ClNO_2$, new measurement data was reported recently by Ghosh et al. (2011), which gives values somehow different from the above.

The photolyses of ClNO and $ClNO_2$ are confirmed to proceed,

$$\mathrm{ClNO} + h\nu(\lambda < 767\ \mathrm{nm}) \rightarrow \mathrm{Cl} + \mathrm{NO} \quad \Delta H_0^\circ = 156\ \mathrm{kJ\,mole^{-1}} \tag{4.115}$$

$$\mathrm{ClNO_2} + h\nu(\lambda < 867\ \mathrm{nm}) \rightarrow \mathrm{Cl} + \mathrm{NO_2} \quad \Delta H_0^\circ = 138\ \mathrm{kJ\,mole^{-1}} \tag{4.116}$$

by Calvert and Pitts (1966) and Nelson and Johnston (1981). As expected from the absorption spectra, their quantum yields are thought to be unity. As for ClNO, the relative yields of different spin-orbit states of the formed Cl atoms $\Phi[\mathrm{Cl}^*(^2P_{1/2})]/\Phi[\mathrm{Cl}^*(^2P_{1/2}) + \mathrm{Cl}(^2P_{3/2})]$ are also measured. For example, the ratio is reported to be 0.90 ± 0.10 at 351 nm, thus $\mathrm{Cl}^*(^2P_{1/2})$ atoms are mostly formed at this wavelength (Chichinin 1993).

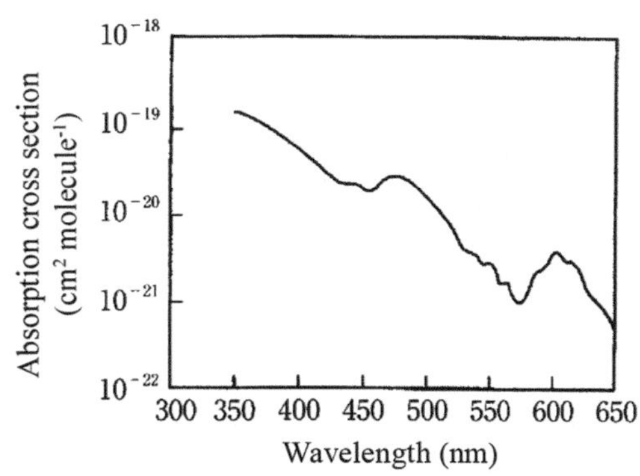

Fig. 4.43 Absorption spectrum of ClNO (Adapted from Roehl et al. 1992)

Table 4.36 Absorption cross sections of ClNO and $ClNO_2$ (298 K)

Wave-length (nm)	10^{20} σ (cm^2 $molecule^{-1}$)		Wave-length (nm)	10^{20} σ (cm^2 $molecule^{-1}$)		Wave-length (nm)	10^{20} σ (cm^2 $molecule^{-1}$)
	ClNO	$ClNO_2$		ClNO	$ClNO_2$		ClNO
200	5860	445	310	11.5	12.1	420	2.89
210	2630	321	320	13.4	9.40	430	2.21
220	896	325	330	14.7	6.79	440	2.20
230	266	221	340	15.2	4.62	450	1.87
240	82.5	132	350	14.2	3.05	460	1.95
250	31.7	90.9	360	12.9	1.86	470	2.50
260	17.5	58.7	370	11.0	1.12	480	2.53
270	12.9	33.7	380	8.86	0.772	490	2.07
280	10.6	20.7	390	6.85	0.475	500	1.50
290	9.64	16.3	400	5.13	0.327		
300	10.0	14.1	410	3.83	–		

Source: NASA/JPL Panel Evaluation No. 17 (Sander et al. 2011)

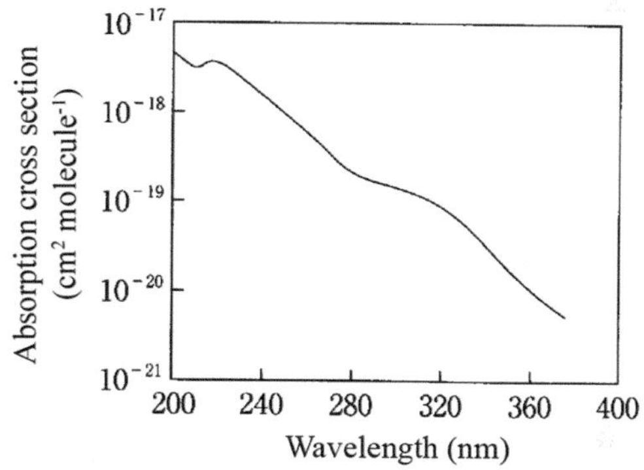

Fig. 4.44 Absorption spectrum of $ClNO_2$ (Adapted from Ganske et al. 1992)

References

Abu-Bajeh, M., Cameron, M., Jung, K.-H., Kappel, C., Laeuter, A., Lee, K.-S., Upadhyaya, H.P., Vasta, R.K., Volpp, H.-R.: Absolute quantum yield measurements for the formation of oxygen atoms after UV laser excitation of SO_2 at 222.4 nm. Proc. Indian Acad. Sci. (Chem. Sci.) **114**, 675–686 (2002)

Adler-Golden, S.M.: Franck-Condon analysis of thermal and vibrational excitation effects on the ozone Hartley continuum. J. Quant. Spectrosc. Radiat. Transf. **30**, 175–185 (1983)

Adler-Golden, S.M., Schweitzer, E.L., Steinfeld, J.I.: Ultraviolet continuum spectroscopy of vibrationally excited ozone. J. Chem. Phys. **76**, 2201–2209 (1982)

Atkinson, R., Lloyd, A.C.: Evaluation of kinetic and mechanistic data for modeling of photochemical smog. J. Phys. Chem. Ref. Data **13**, 315–444 (1984)

Atkinson, R., et al.: Summary of evaluated kinetic and photochemical data for atmospheric chemistry, Data Sheet P7. IUPAC, London (2002)

Atkinson, R., Baulch, D.L., Cox, R.A., Crowley, J.N., Hampson, R.F., Hynes, R.G., Jenkin, M.E., Rossi, M.J., Troe, J.: Evaluated kinetic and photochemical data for atmospheric chemistry:

Volume I – gas phase reactions of Ox, HOx, NOx, and SOx species. Atmos. Chem. Phys. **4**, 1461–1738 (2004)

Atkinson, R., Baulch, D.L., Cox, R.A., Crowley, J.N., Hampson, R.F., Hynes, R.G., Jenkin, M.E., Rossi, M.J., Troe, J.: Evaluated kinetic and photochemical data for atmospheric chemistry: Volume II – gas phase reactions of organic species. Atmos. Chem. Phys. **6**, 3625–4055 (2006)

Atkinson, R., Baulch, D.L., Cox, R.A., Crowley, J.N., Hampson, R.F., Hynes, R.G., Jenkin, M.E., Rossi, M.J., Troe, J.: Evaluated kinetic and photochemical data for atmospheric chemistry: Volume III – gas phase reactions of inorganic halogens. Atmos. Chem. Phys. **7**, 981–1191 (2007)

Atkinson, R., Baulch, D.L., Cox, R.A., Crowley, J.N., Hampson, R.F., Hynes, R.G., Jenkin, M.E., Rossi, M.J., Troe, J., Wallington, T.J.: Evaluated kinetic and photochemical data for atmospheric chemistry: Volume IV – gas phase reactions of organic halogen species. Atmos. Chem. Phys. **8**, 4141–4496 (2008)

Bahou, M., Chung, C.-Y., Lee, Y.-P., Cheng, B.M., Yung, Y.L., Lee, L.C.: Absorption cross sections of HCl and DCl at 135–232 nanometers: implications for photodissociation on Venus. Astrophys. J. **559**, L179–L182 (2001)

Barnes, R.J., Sinha, A., Michelsen, H.A.: Assessing the contribution of the lowest triplet state to the near-UV absorption spectrum of HOCl. J. Phys. Chem. A **102**, 8855–8859 (1998)

Bauer, D., Ingham, T., Carl, S.A., Moortgat, G.K., Crowley, J.N.: Ultraviolet-visible absorption cross sections of gaseous HOI and its photolysis at 355 nm. J. Phys. Chem. **102**, 2857–2864 (1998)

Benter, T., Feldmann, C., Kirchner, U., Schmidt, M., Schmidt, S., Schindler, R.N.: UV/VIS-absorption spectra of HOBr and CH_3OBr; Br(2P3/2) atom yields in the photolysis of H0131. Ber. Bunsenges. Phys. Chem. **99**, 1144–1147 (1995)

Bethke, G.W.: Oscillator strengths in the far ultraviolet. I. Nitric Oxide. J. Chem. Phys. **31**, 662–668 (1959)

Blitz, M.A., Heard, D.E., Pilling, M.J., Arnold, S.R., Chipperfield, M.P.: Pressure and temperature-dependent quantum yields for the photodissociation of acetone between 279 and 327.5 nm. Geophys. Res. Lett. **31**, L06111 (2004). doi:10.1029/2003GL018793

Bloss, W.J., Rowley, D.M., Cox, R.A., Jones, R.L.: Kinetics and products of the IO self-reaction. J. Phys. Chem. **105**, 7840–7854 (2001)

Bogumil, K., Orphal, J., Homann, T., Voigt, S., Spietz, P., Fleischmann, O.C., Vogel, A., Hartmann, M., Bovensmann, H., Frerick, J., Burrows, J.P.: Measurements of molecular absorption spectra with the SCIAMACHY pre-flight model: instrument characterization and reference data for atmospheric remote-sensing in the 230–2380 nm region. J. Photochem. Photobiol. A Chem. **157**, 167–184 (2003)

Bongartz, A., Kames, J., Welter, F., Schurath, U.: Near-UV absorption cross sections and trans/cis equilibrium of nitrous acid. J. Phys. Chem. **95**, 1076–1082 (1991)

Bongartz, A., Kames, J., Schurath, U., George, C., Mirabel, P., Ponche, J.L.: Experimental determination of HONO mass accommodation coefficients using two different techniques. J. Atmos. Chem. **18**, 149–169 (1994)

Brewer, L., Tellinghuisen, J.: Quantum yield for unimolecular dissociation of I2 in visible absorption. J. Chem. Phys. **56**, 3929–3938 (1972)

Brownsword, R.A., Schmiechen, P., Volpp, H.-R., Upadhyaya, H.P., Jung, Y.J., Jung, K.-H.: Chlorine atom formation dynamics in the dissociation of CH_3CF_2Cl (HCFC-142b) after UV laser photoexcitation. J. Chem. Phys. **110**, 11823–11829 (1999)

Brus, L.E., McDonald, J.R.: Time-resolved fluorescence kinetics and $^1B_1(^1\Delta_g)$ vibronic structure in tunable ultraviolet laser excited SO_2 vapor. J. Chem. Phys. **61**, 97–105 (1974)

Burkholder, J.B.: Ultraviolet absorption spectrum of HOCl. J. Geophys. Res. **98**, 2963–2974 (1993)

Burkholder, J.B., Wilson, R.R., Gierczak, T., Talukdar, R., McKeen, S.A., Orlando, J.J., Vaghjiani, G.L., Ravishankara, A.R.: Atmospheric fate of CF_3Br, CF_2Br_2, CF_2ClBr, and CF_2BrCF_2BR. J. Geophys. Res. **96**, 5025–5043 (1991)

Burkholder, J.B., Talukdar, R.K., Ravishankara, A.R., Solomon, S.: Temperature dependence of the HNO_3 UV absorption cross sections. J. Geophys. Res. **98**, 22937–22948 (1993)
Burkholder, J.B., Talukdar, R.K., Ravishankara, A.R.: Temperature dependence of the $ClONO_2$ UV absorption spectrum. Geophys. Res. Lett. **21**, 585–588 (1994)
Burkholder, J.B., Ravishankara, A.R., Solomon, S.: UV/visible and IR absorption cross sections of $BrONO_2$. J. Geophys. Res. **100**, 16793–16800 (1995)
Callear, A.B., Pilling, M.J.: Fluorescence of nitric oxide. Part 7. -Quenching rates of NO C^2 $\Pi(v=0)$, its rate of radiation to NO $A^2\Sigma+$, energy transfer efficiencies, and mechanisms of predissociation. Trans. Faraday Soc. **66**, 1618–1634 (1970)
Calvert, J.G., Pitts, J.N.: Photochemistry, pp. 230–231. Wiley, New York (1966)
Campuzano-Jost, P., Crowley, J.N.: Kinetics of the reaction of OH with HI between 246 and 353 K. J. Phys. Chem. A **103**, 2712–2719 (1999)
Cantrell, C.A., Davidson, J.A., McDaniel, A.H., Shetter, R.E., Calvert, J.G.: Temperature-dependent formaldehyde cross sections in the near-ultraviolet spectral region. J. Phys. Chem. **94**, 3902–3908 (1990)
Carbajo, P.G., Smith, S.C., Holloway, A.-L., Smith, C.A., Pope, F.D., Shallcross, D.E., Orr-Ewing, A.J.: Ultraviolet photolysis of HCHO: absolute HCO quantum yields by direct detection of the HCO radical photoproduct. J. Phys. Chem. A **112**, 12437–12448 (2008)
Carlon, N.R., Papanastasiou, D.K., Fleming, E.L., Jackman, C.H., Newman, P.A., Burkholder, J. B.: UV absorption cross sections of nitrous oxide (N_2O) and carbon tetrachloride (CCl_4) between 210 and 350K and the atmospheric implications. Atmos. Chem. Phys. **10**, 6137–6149 (2010)
Carr, S., Heard, D.E., Blitz, M.A.: Comment on "Atmospheric hydroxyl radical production from electronically excited NO_2 and H_2O". Science **324**, 336 (2009)
Chen, H.-Y., Lien, C.-Y., Lin, W.-Y., Lee, Y.T., Lin, J.J.: UV Absorption cross sections of ClOOCl are consistent with ozone degradation models. Science **324**, 781–784 (2009)
Chichinin, A.I.: Chem. Phys. Lett. **209**, 459–463 (1993)
Clark, R.H., Husain, D.: Quantum yield measurements of $Cl(3^2P_{1/2})$ and $Cl(3^2P_{3/2})$ in the photolysis of C1 chlorofluorocarbons determined by atomic resonance absorption spectroscopy in the vacuum UV. J. Photochem. **24**, 103–115 (1984)
Co, D., Hanisco, T.F., Anderson, J.G., Keutsch, F.N.: Rotationally resolved absorption cross sections of formaldehyde in the 28100–28500 cm^{-1} (351–356 nm) spectral region: implications for in situ LIF measurements. J. Phys. Chem. A **109**, 10675–10682 (2005)
Cooper, I.A., Neill, P.J., Wiesenfeld, J.R.: Relative quantum yield of $O(^1D_2)$ following ozone photolysis between 221 and 243.5 nm. J. Geophys. Res. **98**(12), 795–12,800 (1993)
Cooper, M.J., Wrede, E., Orr-Ewing, A.J., Ashfold, M.N.R.: Ion imaging studies of the $Br(^2P_J)$ atomic products resulting from Br2 photolysis in the wavelength range 260–580 nm. J. Chem. Soc. Faraday Trans. **94**, 2901–2907 (1998)
Cox, R.A., Derwent, R.G.: The ultra-violet absorption spectrum of gaseous nitrous acid. J. Photochem. **6**, 23–34 (1976)
Davis, H.F., Lee, Y.T.: Dynamics and mode specificity in OClO photodissociation. J. Phys. Chem. **96**, 5681–5684 (1992)
Davis, H.F., Lee, Y.T.: Photodissociation dynamics of OClO. J. Chem. Phys. **105**, 8142–8163 (1996)
Davis, H.F., Kim, B., Johnston, H.S., Lee, Y.T.: Dissociation energy and photochemistry of nitrogen trioxide. J. Phys. Chem. **97**, 2172–2180 (1993)
Delmdahl, R.F., Ulrich, S., Gericke, K.-H.: Photofragmentation of $OClO(\tilde{A}^2A_2\ \nu_1\nu_2\nu_3) \rightarrow Cl(^2P_J) + O_2$. J. Phys. Chem. A **102**, 7680–7685 (1998)
DeMore, W.B., Sander, S. P., Golden, D.M., Hampson, R.F., Kurylo, M.J., Howard C.J., Ravishankara, A.R., Kolb, C.E., Molina, M.J.: Chemical kinetics and photochemical data for use in stratospheric modeling, Evaluation Number 12, JPL Publication 97-4, Pasadena, California (1997)

Denzer, W., Hancock, G., Pinot de Moira, J.C., Tyley, P.L.: Spin forbidden dissociation of ozone in the Huggins bands. Chem. Phys. **231**, 109–120 (1998)
Deters, B., Burrows, J.P., Orphal, J.: UV-visible absorption cross sections of bromine nitrate determined by photolysis of $BrONO_2/Br_2$ mixtures. J. Geophys. Res. **103**, 3563–3570 (1998)
Donnelly, V.M., Kaufman, F.: Fluorescence lifetime studies of NO_2. II. Dependence of the perturbed $^2B_2$ state lifetimes on excitation energy. J. Chem. Phys. **69**, 1456–1460 (1978)
Donnelly, V.M., Keil, D.G., Kaufman, F.: Fluorescence lifetime studies of NO_2. III. Mechanism of fluorescence quenching. J. Chem. Phys. **71**, 659–673 (1979)
Douglas, A.E.: Anomalously long radiative lifetimes of molecular excited states. J. Chem. Phys. **45**, 1007–1015 (1966)
Emrich, M., Warneck, P.: Photodissociation of acetone in air: dependence on pressure and wavelength. Behavior of the excited singlet state. J. Phys. Chem. A **104**, 9436–9442 (2000)
Fahr, A., Braun, W., Kurylo, M.J.: Scattered light and accuracy of the cross-section measurements of weak absorptions: gas and liquid phase UV absorption cross sections of CH_3CFCl_2. J. Geophys. Res. **98**, 20467–20472 (1993)
Fahr, A., Nayak, A.K., Kurylo, M.J.: The ultraviolet absorption cross sections of CH_3I temperature dependent gas and liquid phase measurements. Chem. Phys. **197**, 195–203 (1995)
Fang, Q., Han, J., Jiang, J.-L., Chen, X.-B., Fang, W.-H.: The conical intersection dominates the generation of tropospheric hydroxyl radicals from NO_2 and H_2O. J. Phys. Chem. A **114**, 4601–4608 (2010)
Finlayson-Pitts, B.J., Pitts Jr., J.N.: Chemistry of the upper and lower atmosphere. Academic Press, San Diego (2000), 969pp
Francisco, J.S., Hand, M.R., Williams, I.H.: Ab initio study of the electronic spectrum of HOBr. J. Phys. Chem. **100**, 9250–9253 (1996)
Fredlick, J.E., Hudson, R.D.: Predissociayion of nitric oxide in the mesosphere and stratosphere. J. Atmos. Sci. **36**, 737–745 (1979)
Furlan, A., Haeberli, M.A., Huber, R.J.: The 248 nm photodissociation of ClNO2 studied by photofragment translational energy spectroscopy. J. Phys. Chem. **A 104**, 10392–10397 (2000)
Ganske, J.A., Berko, H.N., Finlayson-Pitts, B.J.: Absorption cross sections for gaseous $ClNO_2$ and Cl_2 at 298 K: potential organic oxidant source in the marine troposphere. J. Geophys. Res. **97**, 7651–7656 (1992)
Ghosh, B., Papanastasiou, D.K., Talukdar, R.K., Roberts, J.M., Burkholder, J.B.: Nitryl chloride ($ClNO_2$): UV/Vis absorption spectrum between 210 and 296 K and $O(^3P)$ quantum yield at 193 and 248 nm. J. Phys. Chem. A (2011). doi:10.1021/jp207389y
Gierczak, T., Burkholder, J.B., Bauerle, S., Ravishankara, A.R.: Photochemistry of acetone under tropospheric conditions. Chem. Phys. **231**, 229–244 (1998)
Gillotay, D., Simon, P.C.: Ultraviolet absorption spectrum of trifluoro-bromo-methane, difluoro-dibromo-methane and difluoro-bromo-chloro-methane in the vapor phase. J. Atmos. Chem. **8**, 41–62 (1989)
Gillotay, D., Simon, P.C.: Temperature-dependence of ultraviolet absorption cross-sections of alternative chlorofluoroethanes. J. Atmos. Chem. **12**, 269–285 (1991)
Goldfarb, L., Schmoltner, A.-M., Gilles, M.K., Burkholder, J., Ravishankara, A.R.: Photodissociation of ClONO2: 1. Atomic resonance fluorescence measurements of product quantum yields. J. Phys. Chem. A **101**, 6658–6666 (1997)
Greenblatt, G.D., Ravishankara, A.R.: Laboratory studies on the stratospheric NOx production rate. J. Geophys. Res. **95**, 3539–3547 (1990)
Greenblatt, G.D., Wiesenfeld, J.R.: Time-resolved resonance fluorescence studies of $O(^1D_2)$ yields in the photodissociation of O_3 at 248 and 308 nm. J. Chem. Phys. **78**, 4924–4928 (1983)
Hanf, A., Laüter, A., Volpp, H.-R.: Absolute chlorine atom quantum yield measurements in the UV and VUV gas-phase laser photolysis of CCl_4. Chem. Phys. Lett. **368**, 445–451 (2003)
Harder, J.W., Brault, J.W., Johnston, P.V., Mount, G.H.: Temperature dependent NO_2 cross-sections at high spectral resolution. J. Geophys. Res. D **102**, 3861–3879 (1997)

Harwood, M.H., Jones, R.L., Cox, R.A., Lutman, E., Rattigan, O.V.: Temperature-dependent absorption cross-sections of N_2O_5. J. Photochem. Photobiol. A Chem. **73**, 167–175 (1993)
Harwood, M.H., Burkholder, J.B., Hunter, M., Fox, R.W., Ravishankara, A.R.: Absorption cross sections and self-reaction kinetics of the IO radical. J. Phys. Chem. A **101**, 853–863 (1997)
Harwood, M.H., Burkholder, J.B., Ravishankara, A.R.: Photodissociation of $BrONO_2$ and N_2O_5: quantum yields for NO_3 production at 248, 308, and 352.5 nm. J. Phys. Chem. A **102**, 1309–1317 (1998)
Harwood, M.H., Roberts, J.M., Frost, G.J., Ravishankara, A.R., Burkholder, J.B.: Photochemical studies of CH3C(O)OONO2 (PAN) and CH3CH2C(O)OONO2 (PPN): NO3 quantum yields. J. Phys. Chem. A **107**, 1148–1154 (2003)
Haugen, H.K., Weitz, E., Leone, S.R.: Accurate quantum yields by laser gain vs absorption spectroscopy: Investigation of Br/Br* channels in photofragmentation of Br_2 and IBr. J. Chem. Phys. **83**, 3402–3412 (1985)
Horowitz, A., Calvert, J.G.: Wavelength dependence of the quantum efficiencies of the primary processes in formaldehyde photolysis at 25°C. Int. J. Chem. Kinet. **10**, 805–819 (1978)
Horowitz, A., Calvert, J.G.: Wavelength dependence of the primary processes in acetaldehyde photolysis. J. Phys. Chem. **86**, 3105–3114 (1982)
Huang, W.-T., Chen, A.F., Chen, I.-C., Tsai, C.-H., Lin, J.J.-M.: Photodissociation dynamics of ClOOCl at 248.4 and 308.4 nm. Phys. Chem. Chem. Phys. **13**, 8195–8203 (2011)
Hubrich, C., Stuhl, F.: The ultraviolet absorption of some halogenated methanes and ethanes of atmospheric interest. J. Photochem. **12**, 93–107 (1980)
Huebert, B.J., Martin, R.M.: Gas-phase far-ultraviolet absorption spectrum of hydrogen bromide and hydrogen iodide. J. Phys. Chem. **72**, 3046 (1968)
Hui, M.H., Rice, S.A.: Decay of fluorescence from single vibronic states of SO_2. Chem. Phys. Lett. **17**, 474–478 (1972)
Hynes, A.J., Kenyon, E.A., Pounds, A.J., Wine, P.H.: Temperature dependent absorption cross-sections for acetone and n-butanone – implications for atmospheric lifetimes. Spectrochim. Acta **48A**, 1235–1242 (1992)
Illies, A.J., Takacs, G.A.: Gas phase ultra-violet photoabsorption cross-sections for nitrosyl chloride and nitryl chloride. J. Photochem. **6**, 35–42 (1976)
Imajo, T., Yoshino, K., Esmond, J.R., Parkinson, W.H., Thorne, A.P., Murray, J.E., Learner, R.C. M., Cox, G., Cheung, A.S.-C., Ito, K., Matsui, T.: The application of a VUV Fourier transform spectrometer and synchrotron radiation source to measurements of: II. The δ(1,0) band of NO. J. Chem. Phys. **112–118**, 2251 (2000)
Ingham, T., Bauer, D., Landgraf, J., Crowley, J.N.: Ultraviolet-visible absorption cross sections of gaseous HOBr. J. Phys. Chem. A **102**, 3293–3298 (1998)
Ingham, T., Cameron, M., Crowley, J.N.: Photodissociation of IO (355 nm) and OIO (532 nm): quantum yields for O(3P) and I(2PJ) production. J. Phys. Chem. A **104**, 8001–8010 (2000)
Jimenez, E., Gierczak, T., Stark, H., Burkholder, J.B., Ravishankara, A.R.: Quantum yields of OH, HO_2 and NO_3 in the UV photolysis of HO_2NO_2. Phys. Chem. Chem. Phys. **7**, 342–348 (2005)
Johnston, H.S., Selwyn, G.S.: New cross sections for the absorption of near ultraviolet radiation by nitrous oxide (N_2O). Geophys. Res. Lett. **2**, 549–551 (1975)
Johnston, H.S., Chang, S., Whitten, G.: Photolysis of nitric acid vapor. J. Phys. Chem. **78**, 1–7 (1974)
Johnston, H.S., Davis, H.F., Lee, Y.T.: NO3 photolysis product channels: quantum yields from observed energy thresholds. J. Phys. Chem. **100**, 4713–4723 (1996)
Jones, I.T.N., Bayes, K.D.: Formation of $O_2(a^1\Delta_g)$ by electronic energy transfer in mixtures of NO_2 and O_2. J. Chem. Phys. **59**, 3119–3124 (1973)
Joseph, D.M., Ashworth, S.H., Plane, J.M.C.: On the photochemistry of $IONO_2$: absorption cross section (240–370 nm) and photolysis product yields at 248 nm. Phys. Chem. Chem. Phys. **9**, 5599–5607 (2007)

Kang, W.K., Jung, K.W., Kim, D.-C., Jung, K.-H.: Photodissociation of alkyl iodides and CF_3I at 304 nm: Relative populations of $I(^2P_{1/2})$ and $I(^2P_{3/2})$ and dynamics of curve crossing. J. Chem. Phys. **104**, 5815–5820 (1996)

Katagiri, H., Sako, T., Hishikawa, A., Yazaki, T., Onda, K., Yamanouchi, K., Yoshino, K.: Experimental and theoretical exploration of photodissociation of SO_2 via the C^1B_2 state: identification of the dissociation pathway. J. Mol. Struct. **413–414**, 589–614 (1997)

Katayanagi, H., Mo, Y.X., Suzuki, T.: 223 nm photodissociation of OCS. Two components in $S(^1D_2)$ and $S(^3P_2)$ channels. Chem. Phys. Lett. **247**, 571–576 (1995)

Knight, G.P., Ravishankara, A.R., Burkholder, J.B.: UV absorption cross sections of HO_2NO_2 between 343 and 273 K. Phys. Chem. Chem. Phys. **4**, 1432–1437 (2002)

Kromminga, J., Orphal, J., Spietz, P., Voigt, S., Burrows, J.P.: New measurements of OClO absorption cross-sections in the 325–435 nm region and their temperature dependence between 213 and 293 K. J. Photochem. Photobiol. A Chem. **157**, 149–160 (2003)

Langford, S.R., Regan, P.M., Orr-Ewing, A.J., Ashfold, N.M.R.: On the UV photodissociation dynamics of hydrogen iodide. Chem. Phys. **231**, 245–260 (1998)

Langhoff, S.R., Jaffe, L., Arnold, J.O.: Effective cross sections and rate constants for predissociation of ClO in the earth's atmosphere. J. Quant. Spectrosc. Radiat. Transf. **18**, 227–235 (1977)

Lasylo, B., Kurylo, M.J., Huie, R.E.: Absorption cross sections, kinetics of formation, and self-reaction of the IO radical produced via the laser photolysis of $N_2O/I_2/N_2$ mixtures. J. Phys. Chem. **99**, 11701–11707 (1995)

Lawrence, W.G., Clemitshaw, K.C., Apkarian, V.A.: On the relevance of OClO photodissociation to the destruction of stratospheric ozone. J. Geophys. Res. **95**, 18591–18595 (1990)

Lewis, B.R., Berzins, L., Carver, J.H.: Oscillator strength for the Schumann-Runge bands of O_2. J. Quant. Spectrosc. Radiat. Transf. **36**, 209–232 (1986)

Li, S., Matthew, J., Sunha, A.: Atmospheric hydroxyl radical production from electronically excited NO2 and H2O. Science **319**, 1657–1660 (2008)

Li, A., Suo, B., Wen, Z., Wang, Y.: Potential energy surfaces for low-lying electronic states of SO_2. Sci. China B: Chem. **49**, 289–295 (2006)

Libuda, H.G., Zabel, F.: UV absorption cross section of acetyl peroxynitrate and trifluoroacetyl peroxynitrate at 298 K. Ber. Bunsenges. Phys. Chem. **99**, 1205–1213 (1995)

Lin, C.L., Rohatgi, N.K., DeMore, W.B.: Ultraviolet absorption cross sections of hydrogen peroxide. Geophys. Res. Lett. **5**, 113–115 (1978)

Lindeman, T.G., Wiesenfeld, J.R.: Photodissociation of Br2 in the visible continuum. J. Chem. Phys. **70**, 2882–2888 (1979)

MacLeod, H., Smith, G.P., Golden, D.M.: Photodissociation of pernitric acid (HO_2NO_2) at 248 nm. J. Geophys. Res. **93**, 3813–3823 (1988)

Manatt, S.L., Lane, A.L.: A compilation of the absorption cross-sections of SO_2 from 106 to 403 nm. J. Quant. Spectrosc. Radiat. Transf. **50**, 267–276 (1993)

Maric, D., Burrows, J.P., Meller, R., Moortgat, G.K.: A study of the UV—visible absorption spectrum of molecular chlorine. J. Photochem. Photobiol. A Chem. **70**, 205–214 (1993)

Maric, D., Burrows, J.P., Moortgat, G.K.: A study of the UV—visible absorption spectra of Br_2 and BrCl. J. Photochem. Photobiol. A Chem. **83**, 179–192 (1994)

Martinez, R.D., Buitrago, A.A., Howell, N.W., Hearn, C.H., Joens, J.A.: The near U.V. absorption spectra of several aliphatic aldehydes and ketones at 300 K. Atmos. Environ. **26A**, 785–792 (1992)

Matsumi, Y., Kawasaki, M.: Photolysis of atmospheric ozone in the ultraviolet region. Chem. Rev. **103**, 4767–4781 (2003)

Matsumi, Y., Tonokura, K., Kawasaki, M.: Fine-structure branching ratios and Doppler profiles of $Cl(^2P_j)$ photofragments from photodissociation of the chlorine molecule near and in the ultraviolet region. J. Chem. Phys. **97**, 1065–1071 (1992)

Matsumi, Y., Comes, F.J., Hancock, G., Hofzumahaus, A., Hynes, A.J., Kawasaki, M., Ravishankara, A.R.: Quantum yields for production of $O(^1D)$ in the ultraviolet photolysis of

ozone: recommendation based on evaluation of laboratory data. J. Geophys. Res. **107**(D3), 4024 (2002). doi:10.1029/2001JD000510
Mayor, E., Velasco, A.M., Martin, I.: Photodissociation of the δ(0,0) and δ(1,0) bands of nitric oxide in the stratosphere and the mesosphere: a molecular-adapted quantum defect orbital calculation of photolysis rate constants. J. Geophys. Res. **112**, D13304 (2007). doi:10.1029/2007JD008643
Meller, R., Moortgat, G.K.: Temperature dependence of the absorption cross sections of formaldehyde between 223 and 323 K in the wavelength range 225–375 nm. J. Geophys. Res. D **105**, 7089–7101 (2000)
Mérienne, M.F., Coquart, B., Jenouvier, A.: Temperature effect on the ultraviolet absorption of CFCl2, CF2Cl2 and N2O. Planet. Space Sci. **38**, 617–625 (1990)
Meyrahn, H., Moortgat, G.K., Warneck, P.: Photolysis of CH_3CHO in the range 250–330 nm. J. Photochem. **17**, 138 (1981)
Meyrahn, H., Pauly, J., Schneider, W., Warneck, P.: Quantum yields for the photodissociation of acetone in air and an estimate for the lifetime of acetone in the lower troposphere. J. Atmos. Chem. **4**, 277–291 (1986)
Minaev, B.F.: Physical properties and spectra of IO, IO^- and HOI studied by ab initio methods. J. Phys. Chem. A **103**, 7294–7309 (1999)
Minschwaner, K., Siskind, D.E.: A New calculation of nitric oxide photolysis in the stratosphere, mesosphere, and lower thermosphere. J. Geophys. Res. **98**(D11), 20,401–20,412 (1993)
Molina, L.T., Molina, M.J.: UV absorption cross sections of HO_2NO_2 vapor. J. Photochem. **15**, 97–108 (1981)
Molina, L.T., Molina, M.J.: Absorption cross sections of ozone in the 185–350 nm wavelength range. J. Geophys. Res. **91**, 14501–14508 (1986)
Molina, L.T., Lamb, J.J., Molina, M.J.: Temperature dependent UV absorption cross sections for carbonyl sulfide. Geophys. Res. Lett. **8**, 1008–1011 (1981)
Molina, L.T., Molina, M.J., Rowland, F.S.: Ultraviolet absorption cross sections of several brominated methanes and ethanes of atmospheric interest. J. Phys. Chem. **86**, 2672–2676 (1982)
Moore, T.A., Okumura, M., Seale, J.W., Minton, T.K.: UV photolysis of ClOOCl. J. Phys. Chem. A **103**, 1692–1695 (1999)
Moortgat, G.K., Warneck, P.: CO and H_2 quantum yields in the photodecomposition of formaldehyde in air. J. Chem. Phys. **70**, 3639–3651 (1979)
Moortgat, G.K., Seiler, W., Warneck, P.: Photodissociation of HCHO in air: CO and H_2 quantum yields at 220 and 300 K. J. Chem. Phys. **78**, 1185–1190 (1983)
Mössinger, J.C., Rowley, D.M., Cox, R.A.: The UV-visible absorption cross-sections of $IONO_2$. Atmos. Chem. Phys. **2**, 227–234 (2002)
Nee, J.B., Suto, M., Lee, L.C.: Quantitative spectroscopy study of HBr in the 105–235 nm region. J. Chem. Phys. **85**, 4919–4924 (1986)
Nelson, H.H., Johnston, H.S.: J. Phys. Chem. **85**, 3891–3896 (1981)
Nicolet, M., Cieslik, S.: The photodissociation of nitric oxide in the mesosphere and stratosphere. Planet. Space Sci. **28**, 105–115 (1980)
Nicovich, J.M., Wine, P.H.: Temperature-dependent absorption cross sections for hydrogen peroxide vapor. J. Geophys. Res. **93**, 2417 (1988)
Nishida, S., Takahashi, K., Matsumi, Y., Taniguchi, N., Hayashida, S.: Formation of $O(^3P)$ atoms in the photolysis of N_2O at 193 nm and $O(^3P) + N_2O$ product channel in the reaction of $O(^1D) + N_2O$. J. Phys. Chem. A **108**, 2451–2456 (2004)
Okabe, H.: Fluorescence and predissociation of sulfur dioxide. J. Am. Chem. Soc. **93**, 7095–7096 (1971)
Okabe, H.: Photochemistry of small molecules. Wiley, New York (1978)
Orkin, V.L., Kasimovskaya, E.E.: Ultraviolet absorption spectra of some Br-containing haloalkanes. J. Atmos. Chem. **21**, 1–11 (1995)

Orlando, J.J., Tyndall, G.S., Moortgat, G.K., Calvert, J.G.: Quantum yields for NO_3 photolysis between 570–635 nm. J. Phys. Chem. **97**, 10996–11000 (1993)

Orphal, J.: A critical review of the absorption cross-sections of O_3 and NO_2 in the ultraviolet and visible. J. Photochem. Photobiol. A: Chem. **157**, 185–209 (2003)

Orphal, J., Fellows, C.E., Flaud, J.-M.: The visible absorption spectrum of NO_3 measured by high-resolution Fourier transform spectroscopy. J. Geophys. Res. **108**, 4077 (2003). doi:10.1029/2002JD002489

Osborne, B.A., Marston, G., Kaminski, L., Jones, N.C., Gingell, J.M., Mason, N.J., Walker, I.C., Delwiche, J., Hubin-Franskin, M.-J.: Vacuum ultraviolet spectrum of dinitrogen pentoxide. J. Quant. Spectrosc. Radiat. Transf. **64**, 67–74 (2000)

Osthoff, H.D., Pilling, M.J., Ravishankara, A.R., Brown, S.S.: Temperature dependence of the NO3 absorption cross-section above 298 K and determination of the equilibrium constant for $NO_3 + NO_2 \leftrightarrow N_2O_5$ at atmospherically relevant conditions. Phys. Chem. Chem. Phys. **9**, 5785–5793 (2007)

Pagsberg, P., Bjergbakke, E., Sillesen, A., Ratajczak, E.: Kinetics of the gas phase reaction $OH + NO(+M) \rightarrow HONO(+M)$ and the determination of the UV absorption cross sections of HONO. Chem. Phys. Lett. **272**, 383–390 (1997)

Papanastasiou, D.K., Papadimitriou, V.C., Fahey, D.W., Burkholder, J.B.: UV absorption spectrum of the ClO dimer (Cl_2O_2) between 200 and 420 nm. J. Phys. Chem. A **113**, 13711–13726 (2009)

Paraskevopoulos, G., Cvetanovic, R.J.: Competitive reactions of the excited oxygen atoms, $O(^1D)$. J. Am. Chem. Soc. **91**, 7572–7577 (1969)

Park, J., Lee, Y., Flynn, G.W.: Tunable diode laser probe of chlorine atoms produced from the photodissociation of a number of molecular precursors. Chem. Phys. Lett. **186**, 441–449 (1991)

Park, M.-S., Jung, Y.-J., Lee, S.-H., Kim, D.-C., Jung, K.-H.: The role of $3\Pi_0+$ in the photodissociation of BrCl at 235 nm. Chem. Phys. Lett. **322**, 429–438 (2000)

Peterson, A.B., Smith, I.W.M.: Yields of Br* ($4^2P_{1/2}$) as a function of wavelength in the photodissociation of Br_2 and IBr. Chem. Phys. **30**, 407–413 (1978)

Pope, F.D., Smith, C.A., Ashfold, M.N.R., Orr-Ewing, A.J.: High-resolution absorption cross sections of formaldehyde at wavelengths from 313 to 320 nm. Phys. Chem. Chem. Phys. **7**, 79–84 (2005a)

Pope, F.D., Smith, C.A., Davis, P.R., Shallcross, D.E., Ashfold, M.N.R., Orr-Ewing, A.J.: Photochemistry of formaldehyde under tropospheric conditions. J. Chem. Soc. Faraday Disc **130**, 59–73 (2005b)

Prahlad, V., Kumar, V.: Temperature dependence of photoabsorption cross-sections of sulfur dioxide at 188–220 nm. J. Quant. Spectrosc. Radiat. Transf. **57**, 719–723 (1997)

Preston, K.F., Barr, R.F.: Primary processes in the photolysis of nitrous oxide. J. Chem. Phys. **54**, 3347–3348 (1971)

Rattigan, O., Lutman, E., Jones, R.L., Cox, R.A., Clemitshaw, K., Williams, J.: Temperature-dependent absorption cross-sections of gaseous nitric acid and methyl nitrate. J. Photochem. Photobiol. A Chem. **69**, 125–126 (1992)

Ravishankara, A.R., Hancock, G., Kawasaki, M., Matsumi, Y.: Photochemistry of ozone: surprises and recent lessons. Science **280**, 60–61 (1998). doi:10.1126/science.280.5360.60

Regan, P.M., Langford, S.R., Ascenzi, D., Cook, P.A., Orr-Ewing, A.J., Ashfold, M.N.R.: Spin–orbit branching in Cl(2P) atoms produced by ultraviolet photodissociation of HCl. Phys. Chem. Chem. Phys. **1**, 3247–3251 (1999a)

Regan, P.M., Langford, S.R., Orr-Ewing, A.J., Ashfold, M.N.R.: The ultraviolet photodissociation dynamics of hydrogen bromide. J. Chem. Phys. **110**, 281–288 (1999b)

Robbins, D.E.: Photodissociation of methyl chloride and methyl bromide in the atmosphere. Geophys. Res. Lett. **3**, 213–216 (1976). (Erratum, GRL, 3, 757, 1976.)

Roberts, J.M., Fajer, R.W.: UV absorption cross sections of organic nitrates of potential atmospheric importance and estimation of atmospheric lifetimes. Environ. Sci. Technol. **23**, 945–951 (1989)

Roehl, C.M., Orlando, J.J., Calvert, J.G.: The temperature dependence of the UV-visible absorption cross-sections of NOCl. J. Photochem. Photobiol. A Chem. **69**, 1–5 (1992)

Roehl, C.M., Orlando, J.J., Tyndall, G.S., Shetter, R.E., Vazquez, G.J., Cantrell, C.A., Calvert, J. G.: Temperature dependence of the quantum yields for the photolysis of NO_2 near the dissociation limit. J. Phys. Chem. **98**, 7837–7843 (1994)

Roehl, C.M., Burkholder, J.B., Moortgat, G.K., Ravishankara, A.R., Crutzen, P.J.: Temperature dependence of UV absorption cross sections and atmospheric implications of several alkyl iodides. J. Geophys. Res. **102**, 12819–12829 (1997)

Roehl, C.M., Mazely, T.L., Friedl, R.R., Li, Y.M., Francisco, J.S., Sander, S.P.: NO_2 quantum yield from the 248 nm photodissociation of peroxynitric acid (HO_2NO_2). J. Phys. Chem. A **105**, 1592–1598 (2001)

Rogers, J.D.: Ultraviolet absorption cross sections and atmospheric photodissociation rate constants of formaldehyde. J. Phys. Chem. **94**, 4011–4015 (1990)

Rowley, D.M., Mössinger, J.C., Cox, R.A., Jones, R.L.: The UV-visible absorption cross-sections and atmospheric photolysis rate of HOI. J. Atmos. Chem. **34**, 137–151 (1999)

Rudolph, R.N., Inn, E.C.Y.: OCS photolysis and absorption in the 200- to 300-nm region. J. Geophys. Res. **86**, 9891–9894 (1981)

Rufus, J., Stark, G., Smith, P.L., Pickering, J.C., Thorne, A.P.: High-resolution photoabsorption cross section measurements of SO_2, 2: 220 to 325 nm at 295 K. J. Geophys. Res. **108**, 5011 (2003). doi:10.1029/2002JE001931

Saiz-Lopez, A., Saunders, R.W., Joseph, D.M., Ashworth, S.H., Plane, J.M.C.: Absolute absorption cross-section and photolysis rate of I_2. Atmos. Chem. Phys. **4**, 1443–1450 (2004)

Samartzis, P.C., Sakellariou, I., Gougousi, T.: Photofragmentation study of Cl2 using ion imaging. J. Chem. Phys. **107**, 43–48 (1997)

Sander, S.P.: Temperature dependence of the nitrogen trioxide absorption spectrum. J. Phys. Chem. **90**, 4135–4142 (1986)

Sander, S.P., Friedl, R.R.: Kinetics and product studies of the reaction chlorine monoxide + bromine monoxide using flash photolysis-ultraviolet absorption. J. Phys. Chem. **93**, 4764–4771 (1989)

Sander, S.P., Golden, D.M., Kurylo, M.J., Moortgat, G.K., Ravishankara, A.R., Kolb, C.E., Molina, M.J., Finlayson-Pitts, B.J.: Chemical kinetics and photochemical data for use in atmospheric studies, Evaluation Number 14, JPL Publication 02-25, Pasadena, California, 2003. Website: http://jpldataeval.jpl.nasa.gov/pdf/JPL_02-25_rev02

Sander, S.P., Baker, R., Golden, D.M., Kurylo, M.J., Wine, P.H., Abatt, J.P.D., Burkholder, J.B. Kolb, C.E, Moortgat, G.K., Huie, R.E., Orkin, V.L.: Chemical kinetics and photochemical data for use in atmospheric studies, Evaluation Number 17, JPL Publication 10-6, Pasadena, California, 2011. Website: http://jpldataeval.jpl.nasa.gov/

Schindler, R., Liesner, M., Schmidt, S., Kirschner, U., Benter, T.: Identification of nascent products formed in the laser photolysis of CH3OCl and HOCl at 308 nm and around 235 nm. Total Cl-atom quantum yields and the state and velocity distributions of $Cl(^2P_j)$. J. Photochem. Photobiol. A Chem. **107**, 9–19 (1997)

Schmidt, S., Benter, T., Schindler, R.N.: Photodissociation dynamcis of ClO radicals in the range $(237 \leq \lambda \leq 270)$ nm and at 205 nm and the velocity distribution of $O(^1D)$ atoms. Chem. Phys. Lett. **282**, 292–298 (1998)

Selwyn, G., Podolske, J., Johnston, H.S.: Nitrous oxide ultraviolet absorption spectrum at stratospheric temperatures. Geophys. Res. Lett. **4**, 427–430 (1977)

Simon, P.C., Gillotay, D., Vanlaethem-Meuree, N., Wisemberg, J.: Ultraviolet absorption cross-sections of chloro and chlorofluoro-methanes at stratospheric temperatures. J. Atmos. Chem. **7**, 107–135 (1988)

Singer, R.J., Crowley, J.N., Burrows, J.P., Schneider, W., Moortgat, G.K.: Measurement of the absorption cross-section of peroxynitric acid between 210 and 330 nm in the range 253–298 K. J. Photochem. Photobiol. **48**, 17–32 (1989)

Slanger, T.G., Cosby, P.C.: O_2 spectroscopy below 5.1 eV. J. Phys. Chem. **92**, 267–282 (1988)

Smith, G.D., Molina, L.T., Molina, M.J.: Measurement of radical quantum yields from formaldehyde photolysis between 269 and 339 nm. J. Phys. Chem. A **106**, 1233–1240 (2002)

Smith, C.A., Pope, F.D., Cronin, B., Parkes, C.B., Orr-Ewing, A.J.: Absorption cross sections of formaldehyde at wavelengths from 300 to 340 nm at 294 and 245 K. J. Phys. Chem. A **110**, 11645–11653 (2006)

Soller, R., Nicovich, J.M., Wine, P.H.: Bromine nitrate photochemistry: quantum yields for O, Br, and BrO over the wavelength range 248–355 nm. J. Phys. Chem. A **106**, 8378–8385 (2002)

Stevens, C.G., Swagel, M.W., Wallace, R., Zare, R.N.: Analysis of polyatomic spectra using tunable laser-induced fluorescence: applications to the NO_2 visible band system. Chem. Phys. Lett. **18**, 465–469 (1973)

Stockwell, W.R., Calvert, J.G.: The near ultraviolet absorption spectrum of gaseous HONO and N_2O_3. J. Photochem. **8**, 193–203 (1978)

Stutz, J., Kim, E.S., Platt, U., Bruno, P., Perrino, C., Febo, A.: UV-visible absorption cross section of nitrous acid. J. Geophys. Res. **105**(D11), 14,585–14,592 (2000)

Su, F., Bottenheim, J.W., Thorsell, D.L., Calvert, J.G., Damon, E.K.: The efficiency of the phosphorescence decay of the isolated SO_2 ($^3B_1$) molecule. Chem. Phys. Lett. **49**, 305–311 (1977)

Takacs, G., Willard, J.: Primary products and secondary reactions in photodecomposition of methyl halides. J. Phys. Chem. **81**, 1343–1349 (1977)

Takahashi, K., Matsumi, Y., Kawasaki, M.: Photodissociation processes of ozone in the Huggins band at 308–326 nm: direct observation of O($^1D_2$) and O(^3P_j)products. J. Phys. Chem. **100**, 4084–4089 (1996)

Takahashi, K., Kishigami, M., Taniguchi, N., Matsumi, Y., Kawasaki, M.: Photofragment excitation spectrum for O(1D) from the photodissociation of jet-cooled ozone in the wavelength range 305–329 nm. J. Chem. Phys. **106**, 6390–6397 (1997)

Takahashi, K., Taniguchi, N., Matsumi, Y., Kawasaki, M., Ashfold, M.N.R.: Wavelength and temperature dependence of the absolute O(1D) yield from the 305–329 nm photodissociation of ozone. J. Chem. Phys. **108**, 7161–7172 (1998)

Takahashi, K., Hayashi, S., Matsumi, Y., Taniguchi, N., Hayashida, S.: Quantum yields of O(^{1}D) formation in the photolysis of ozone between 230 and 308 nm. J. Geophys. Res. **107**(D20), ACH-11 (2002). doi:10.1029/2001JD002048

Taketani, F., Takahashi, K., Matsumi, Y.: Quantum yields for Cl(^{2}Pj) atom formation from the photolysis of chlorofluorocarbons and chlorinated hydrocarbons at 193.3 nm. J. Phys. Chem. A **109**, 2855–2860 (2005)

Talukdar, R.K., Vashjiani, G.L., Ravishankara, A.R.: Photodissociation of bromocarbons at 193, 222, and 248 nm: quantum yields of Br atom at 298 K. J. Chem. Phys. **96**, 8194–8201 (1992)

Talukdar, R.K., Burkholder, J.B., Schmoltner, A.-M., Roberts, J.M., Wilson, R., Ravishankara, A. R.: Investigation of the loss processes for peroxyacetyl nitrate in the atmosphere: UV photolysis and reaction with OH. J. Geophys. Res. **100**, 14163–14173 (1995)

Talukdar, R.K., Hunter, M., Warren, R.F., Burkholder, J.B., Ravishankara, A.R.: UV laser photodissociation of CF_2ClBr and CF_2Br_2 at 298 K: quantum yields of Cl, Br, and CF_2. Chem. Phys. Lett. **262**, 669–674 (1996)

Talukdar, R.K., Burkholder, J.B., Hunter, M., Gilles, M.K., Roberts, J.M., Ravishankara, A.R.: Atmospheric fate of several alkyl nitrates Part 2: UV absorption cross-sections and photodissociation quantum yields. J. Chem. Soc. Faraday Trans. **93**, 2797–2805 (1997a)

Talukdar, R.K., Gilles, M.K., Battin-Leclerc, F., Ravishankara, A.R.: Photolysis of ozone at 308 and 248 nm: quantum yield of O(1D) as a function of temperature. Geophys. Res. Lett. **24**, 1091–1094 (1997b)

Talukdar, R.K., Longfellow, C.A., Gilles, M.K., Ravishankara, A.R.: Quantum yields of O(^{1}D) in the photolysis of ozone between 289 and 329 nm as a function of temperature. Geophys. Res. Lett. **25**, 143–146 (1998)

Taniguchi, N., Takahashi, K., Matsumi, Y., Dylewski, S.M., Geiser, J.D., Houston, P.L.: Determination of the heat of formation of O_3 using vacuum ultraviolet laser-induced fluorescence spectroscopy and two-dimensional product imaging techniques. J. Chem. Phys. **111**, 6350–6355 (1999)

Taniguchi, N., Takahashi, K., Matsumi, Y.: Photodissociation of O_3 around 309 nm. J. Phys. Chem. **104**, 8936–8944 (2000)

Taylor, W.D., Allston, T.D., Moscato, M.J., Fazekas, G.B., Koslowski, R., Takacs, G.A.: Atmospheric photodissociation lifetimes for nitromethane, methyl nitrite, and methyl nitrate. Int. J. Chem. Kinet. **12**, 231–240 (1980)

Troe, J.: Are primary quantum yields of NO2 photolysis at 398 nm smaller than unity. Z. Phys. Chem. **214**, 573–581 (2000)

Troe, J.: Analysis of quantum yields for the photolysis of formaldehyde at $\lambda > 310$ nm. J. Phys. Chem. A **111**, 3868–3874 (2007)

Turnipseed, A.A., Vaghjiani, G.L., Gierczak, T., Thompson, J.E., Ravishankara, A.R.: The photochemistry of ozone at 193 and 222 nm. J. Chem. Phys. **95**, 3244–3251 (1991)

Tyndall, G.S., Stedman, K.M., Schneider, W., Burrows, J.P., Moortgat, G.K.: The absorption spectrum of ClNO between 190 and 350 nm. J. Photochem. **36**, 133–139 (1987)

Vaghjiani, G.L., Ravishankara, A.R.: Absorption cross sections of CH_3OOH H_2O_2, and D_2O_2 vapors between 210 and 365 nm at 297 K. J. Geophys. Res. **94**, 3487–3492 (1989)

Vandaele, A.C., Simon, P.C., Guilmot, G.M., Carleer, M., Colin, R.: SO_2 absorption cross section measurement in the UV using a Fourier transform spectrometer. J. Geophys. Res. **99**, 25599–25605 (1994)

Vandaele, A.C., Hermans, C., Simon, P.C., Carleer, M., Colin, R., Fally, S., Mérienne, M.F., Jenouvrier, A., Coquart, B.: Measurements of the NO_2 absorption cross-section from 42 000 cm^{-1} to 10 000 cm^{-1} (238–1000 nm) at 220 K and 294 K. J. Quant. Spectr. Rad. Trans. **59** (3–5), 171–184 (1998)

Vandaele, A.C., Hermans, C., Fally, S., Carleer, M., Colin, R., Meerienne, M.-F., Jenouvrier, A., Coquart, B.: High-resolution Fourier transform measurement of the NO_2 visible and near-infrared absorption cross sections: temperature and pressure effects. J. Geophys. Res. **107** (D18), 4348 (2002). doi:10.1029/2001JD000971

Vasudev, R.: Absorption spectrum and solar photodissociation of gaseous nitrous acid in actinic wavelength region. Geophys. Res. Lett. **17**, 2153–2155 (1990)

Vasudev, R., Zare, R.N., Dixon, R.N.: State-selected photodissociation dynamics: complete characterization of the OH fragment ejected by the HONO state. J. Chem. Phys. **80**, 4863–4878 (1984)

Wahner, A., Tyndall, G.S., Ravishankara, A.R.: Absorption cross sections for symmetric chlorine dioxide as a function of temperature in the wavelength range 240–480nm. J. Phys. Chem. **91**, 2734–2738 (1987)

Wahner, A., Ravishankara, A.R., Sander, S.P., Friedl, R.R.: Absorption cross section of BrO between 312 and 385 nm at 298 and 223 K. Chem. Phys. Lett. **152**, 507–512 (1988)

Wallace, L., Hunten, D.N.: Dayglow of the oxygen A band. J. Geophys. Res. **73**, 4813–4834 (1968)

Wayne, R.P., Barnes, I., Burrows, J.P., Canosa-Mas, C.E., Hjorth, J., Le Bras, G., Moortgat, G.K., Perner, D., Poulet, G., Restelli, G., Sidebottom, H.: The nitrate radical: physics, chemistry and atmosphere. Atmos. Environ. **25**, 1–203 (1991)

Wilmouth, D.M., Hanisco, T.F., Donahue, N.M., Anderson, J.G.: Fourier transform ultraviolet spectroscopy of the A $^2\Pi_{3/2} \leftarrow$ X $^2\Pi_{3/2}$ transition of BrO. J. Phys. Chem. A **103**, 8935–8945 (1999)

Wollenhaupt, M., Carl, S.A., Horowitz, A., Crowley, J.N.: Rate coefficients for reaction of OH with acetone between 202 and 395 K. J. Phys. Chem. A **104**, 2695–2705 (2000)

Wu, C.Y.R., Chen, F.Z., Judge, D.L.: Temperature-dependent photoabsorption cross section of OCS in the 2000–2600 Å region. J. Quant. Spectrosc. Radiat. Transf. **61**, 265–271 (1999)

Yao, F., Wilson, I., Johnston, H.: Temperature-dependent ultraviolet absorption spectrum for dinitrogen pentoxide. J. Phys. Chem. **86**, 3611–3615 (1982)

Yokelson, R.J., Burkholder, J.B., Fox, R.W., Talukdar, R.K., Ravishankara, A.R.: Temperature dependence of the NO_3 absorption spectrum. J. Phys. Chem. **98**, 13144–13150 (1994)

Yokelson, R.J., Burkholder, J., Fox, R.W., Ravishankara, A.R.: Photodissociation of ClONO2: 2. Time-resolved absorption studies of product quantum yields. J. Phys. Chem. A **101**, 6667–6678 (1997)

Yoshino, K., Freeman, D.F., Esmond, J.R., Parkinson, W.H.: High resolution absorption cross section measurements and band oscillator strengths of the (1, 0) – (12, 0) Schumann-Runge bands of O_2. Planet. Space Sci. **31**, 339–353 (1983)

Zhao, Z., Stickel, R.E., Wine, P.H.: Quantum yield for carbon monoxide production in the 248 nm photodissociation of carbonyl sulfide (OCS). Geophys. Res. Lett. **22**, 615–618 (1995)

Chapter 5
Homogeneous Elementary Reactions in the Atmosphere and Rate Constants

Atmospheric chemical systems consist of numerous chemical reactions. Fundamental reactions that can be considered with interests in microscopic processes focusing on the changes of chemical bonds among atoms, molecules and free radicals, are often called elementary reactions. Homogeneous elementary reactions are the subjects of theoretical analysis, for example by the transition state theory described in Chap. 2, and also the subjects of research on molecular dynamics. In this chapter, among the gas phase homogeneous reactions related to atmospheric chemistry, basic elementary reactions are picked up, and their reaction products and rate constants are presented including brief information on quantum chemical knowledge on reaction pathways and reaction rates.

Sequential oxidation reaction mechanisms of many organic molecules emitted into the atmosphere and the overall reaction system with HO_x chain reaction mechanism in the troposphere are treated in Chap. 7, and the HO_x, NO_x, ClO_x chain reaction systems in the stratosphere are described in Chap. 8.

Regarding the reaction rate constants, evaluation by the NASA/JPL panel has been conducted from 1977, and the recommended rate constants have been renewed every few years. The most recent evaluation was obtained in Evaluation 17 (Sander et al. 2011). Another evaluation has been reported by the IUPAC Subcommittee on Gas Kinetic Data Evaluation for Atmospheric Chemistry as Vol. I Inorganic Species, Vol. II Organic Species, Vol. III Inorganic Halogen Species, Vol. IV Organic Halogen Species in 2004, 2006, 2007 and 2008, separately (Atkinson et al. 2004, 2006, 2007, 2008). More recent data was updated on a web page (Wallington et al. 2012). These recommend values are extremely useful for atmospheric chemistry research, and those values are fully quoted in this chapter. In principle, the temperature ranges to be applied for temperature dependent rate constant formulas are taken from the IUPAC subcommittee report, and the values of enthalpy of reactions ΔH° are given at 298 K calculated from the heat of formation given in Table 2.5 at the end of this book.

H. Akimoto, *Atmospheric Reaction Chemistry*, Springer Atmospheric Sciences,
DOI 10.1007/978-4-431-55870-5_5

5.1 Reactions of $O(^3P)$ and $O(^1D)$ Atoms

Reactions of oxygen atoms formed in the photolysis of atmospheric O_2, O_3, NO_2 are initial triggers to drive stratospheric and tropospheric chemistry system. Oxygen atoms targeted in the stratospheric and tropospheric chemistry are $O(^3P)$ in the ground state, and $O(^1D)$ in the lowest excited state at 190 kJ mol^{-1} higher than the ground state.

$O(^3P)$ can react with many organic and inorganic molecules but the reactivity is much lower than $O(^1D)$. Thus, although hydrocarbons existing in the clean and polluted troposphere can all react with $O(^3P)$, the reactions can in general be neglected as compared to those with OH radicals when atmospheric concentrations of $O(^3P)$ and reaction rates are considered, and only the reaction with the atmospheric main species O_2 is necessary to be considered as the $O(^3P)$ reaction in the troposphere. In the stratosphere, reactions with O_2 and O_3 are predominant, but those with OH, HO_2, NO_2, ClO are also important in the chain reaction system.

On the other hand, $O(^1D)$ is much more reactive than $O(^3P)$, and reacts with many atmospheric trace molecules that do not react with $O(^3P)$. Among them, the most important is the reaction with water vapor (H_2O) to form OH radicals. As for the $O(^1D)$ reactions, it is necessary to consider only this reaction in the troposphere, other reactions with N_2O, CH_4, CFC are also important in the stratosphere.

Table 5.1 gives reaction rate constants of $O(^3P)$ and $O(^1D)$ with atmospheric species, that are excerpted from the NASA/JPL panel evaluation No. 17 (Sander et al. 2011) and the IUPAC subcommittee report (Atkinson et al. 2004). In this section, reactions of $O(^3P)$ with O_2, O_3, OH, HO_2, NO_2, ClO, and those of $O(^1D)$ with H_2O, N_2O, CH_4, CFC that are particularly important in atmospheric chemistry, are described.

5.1.1 $O\,(^3P) + O_2 + M$

The reaction of $O(^3P)$ and O_2 is a typical termolecular reaction (see 2.3.1) forming ozone O_3,

$$O(^3P) + O_2 + M \rightarrow O_3 + M. \quad \Delta H^{\circ}_{298} = -107 \text{ kJ mol}^{-1} \tag{5.1}$$

The slightly different low pressure limit rate constant k_0 of reaction (5.1) has been obtained in laboratory for O_2 and N_2 acting as M (Lin and Leu 1982; Hippler et al. 1990). The recommended values of the IUPAC subcommittee (Atkinson et al. 2004) are,

Table 5.1 Rate constants at 298 K and Arrhenius parameters for the reactions of $O(^3P)$ and $O(^1D)$ atoms

Reactions	k (298 K) ($cm^3\ molecule^{-1}\ s^{-1}$)	A Factor ($cm^3\ molecule^{-1}\ s^{-1}$)	E_a/R (K)	Ref.
$O(^3P) + O_2 + M \rightarrow O_3 + M$	6.0×10^{-34} $[O_2]$ (k_0)	6.0×10^{-34} $(T/300)^{-2.6}$ $[O_2]$ (k_0)		(a1)
	5.6×10^{-34} $[N_2]$ (k_0)	5.6×10^{-34} $(T/300)^{-2.6}$ $[N_2]$ (k_0)		
$O(^3P) + O_3 \rightarrow 2\ O_2$	8.0×10^{-15}	8.0×10^{-12}	2060	(a1)
$O(^3P) + OH \rightarrow H + O_2$	3.5×10^{-11}	2.4×10^{-11}	−110	(a1)
$O(^3P) + HO_2 \rightarrow OH + O_2$	5.8×10^{-11}	2.7×10^{-11}	−220	(a1)
$O(^3P) + NO_2 \rightarrow NO + O_2$	1.0×10^{-11}	5.5×10^{-12}	−190	(a1)
$O(^3P) + NO_2 + M \rightarrow NO_3 + M$	1.3×10^{-31} $[N_2]$ (k_0)	1.3×10^{-31} $(T/300)^{-1.5}$ $[N_2]$ (k_0)		(a1)
	2.3×10^{-11} (k_∞)	2.3×10^{-11} $(T/300)^{-0.24}$ (k_∞)		
$O(^3P) + ClO \rightarrow Cl + O_2$	3.7×10^{-11}	2.5×10^{-11}	−110	(a2)
$O(^3P) + BrO \rightarrow Br + O_2$	4.1×10^{-11}	1.9×10^{-11}	−230	(a2)
$O(^1D) + N_2 + M \rightarrow N_2O + M$	2.8×10^{-36} $[N_2]$ (k_0)	2.8×10^{-36} $(T/300)^{-0.9}$ $[N_2]$ (k_0)		(a1, b)
$O(^1D) + N_2 \rightarrow O(^3P) + N_2$	3.1×10^{-11}	2.2×10^{-11}	−110	(b)
$+ O_2 \rightarrow O(^3P) + O_2$	4.0×10^{-11}	3.3×10^{-11}	−60	(b)
$O(^1D) + H_2 \rightarrow OH + H$	1.2×10^{-10}	1.2×10^{-10}	0	(b)
$O(^1D) + H_2O \rightarrow 2OH$	2.0×10^{-10}	1.6×10^{-10}	−60	(b)
$O(^1D) + N_2O \rightarrow$ overall	1.3×10^{-10}	1.2×10^{-10}	−20	(b)
$\rightarrow N_2 + O_2$	5.0×10^{-11}	4.6×10^{-11}	−20	(b)
$\rightarrow 2\ NO$	7.8×10^{-11}	7.3×10^{-11}	−20	(b)
$O(^1D) + CH_4 \rightarrow$ overall	1.8×10^{-10}	1.8×10^{-10}	0	(b)
$\rightarrow CH_3 + OH$	1.3×10^{-10}	1.3×10^{-10}	0	(b)
$\rightarrow CH_2OH + H$	0.35×10^{-10}	0.35×10^{-10}	0	(b)
$\rightarrow HCHO + H_2$	0.09×10^{-10}	0.09×10^{-10}	0	(b)
$O(^1D) + CCl_3F \rightarrow$ overall	2.3×10^{-10}	2.3×10^{-10}	0	(b)
$O(^1D) + CCl_2F_2 \rightarrow$ overall	1.4×10^{-10}	1.4×10^{-10}	0	(b)

(continued)

Table 5.1 (continued)

Reactions	k (298 K) (cm^3 $molecule^{-1}$ s^{-1})	A Factor (cm^3 $molecule^{-1}$ s^{-1})	E_a/R (K)	Ref.
$\rightarrow ClO + CClF_2$	1.2×10^{-10}	1.2×10^{-10}	0	(a3)
$\rightarrow O(^3P) + CCl_2F_2$	2.4×10^{-11}	2.4×10^{-11}	0	(a3)
$O(^1D) + CClF_3 \rightarrow$ overall	8.7×10^{-11}	8.7×10^{-11}	0	(b)

(a1, a2, a3) IUPAC subcommittee report Vol. I, III, IV (Atkinson et al. 2004, 2007, 2008), respectively
(b) NASA/JPL panel evaluation No.17 (Sander et al. 2011)

$$k_{0,5.1}(T, N_2) = 5.6 \times 10^{-34}[N_2]\left(\frac{T}{300}\right)^{-2.6} \quad \text{cm}^3\text{molecule}^{-1}\text{s}^{-1} \quad (100 - 300\text{K})$$
$$k_{0,5.1}(T, O_2) = 6.0 \times 10^{-34}[O_2]\left(\frac{T}{300}\right)^{-2.6} \quad \text{cm}^3\text{molecule}^{-1}\text{s}^{-1} \quad (100 - 300\text{K})$$

Under atmospheric conditions, reaction (5.1) is in the low-pressure limit, where the rate constants are proportional to pressure, and the apparent bimolecular rate constant at 298 K and 1 atm ($M = 2.69 \times 10^{19}$ molecules cm^{-3}) is 1.6×10^{-14} cm^3 $molecule^{-1}$ s^{-1}.

5.1.2 *O (^{3}P) + O$_3$*

The reaction of (^{3}P) and O_3 are typical bimolecular reaction with large exothermicity.

$$O(^3P) + O_3 \rightarrow 2\ O_2. \quad \Delta H^\circ_{298} = -320 \text{ kJ mol}^{-1} \tag{5.2}$$

Wine et al. (1983) determined the rate constants of reaction (5.2) at 237–477 K by direct measurement of time decay of O(^{3}P) using a resonance fluorescence method. These values agreed well with previous experiments, and the IUPAC subcommittee (Atkinson et al. 2004) recommends the rate constant at 298 K as $k(298\ K) = 8.0 \times 10^{-15}$ cm^3 $molecule^{-1}$ s^{-1}, and the Arrhenius's formula for temperature dependence as,

$$k_{5.2}(T) = 8.0 \times 10^{-12}\exp\left(-\frac{2060}{T}\right) \quad \text{cm}^3\text{molecule}^{-1}\text{s}^{-1}(200 - 400\ \text{K}).$$

The activation energy of reaction (5.2) is considerably large, 17.1 kJ $mole^{-1}$, so that it is necessary to use proper values considering the temperature conditions in the stratosphere.

Quantum chemical calculation of transition state theory for reaction (5.2) has been reported by Balakrishnan and Billing (1996), and the obtained theoretical value agreed with the above experimental value reasonably well. According to the theoretical calculation, excess energy released by the large exothermicity is retained in one of the produced O_2 exclusively, which is formed as vibrationally highly excited molecule with up to $v = 27$.

5.1.3 *O (^{3}P) + OH, HO$_2$, NO$_2$, ClO*

Reactions of O(^{3}P) with OH, HO_2, NO_2, and ClO are important in the upper troposphere where the concentration of O(^{3}P) is high enough to constitute ozone dissipation cycles (Sect. 8.2).

These reactions have common characteristics that the reaction partners are all radicals with unpaired electron. Each reaction is an oxygen atom transfer reaction expressed as,

$$O(^3P) + OH \rightarrow H + O_2 \quad \Delta H^\circ_{298} = -68 \text{ kJ mol}^{-1} \tag{5.3}$$

$$+ HO_2 \rightarrow OH + O_2 \quad \Delta H^\circ_{298} = -226 \text{ kJ mol}^{-1} \tag{5.4}$$

$$+ NO_2 \rightarrow NO + O_2 \quad \Delta H^\circ_{298} = -192 \text{ kJ mol}^{-1} \tag{5.5}$$

$$+ ClO \rightarrow Cl + O_2 \quad \Delta H^\circ_{298} = -226 \text{ kJ mol}^{-1}. \tag{5.6}$$

Since the reactions form atoms and radicals as products, they play a role as propagation reaction of ozone destruction chain reactions in the stratosphere (see Sect. 8.2).

The atom-radical reactions such as those of O (^{3}P) with OH, HO_2, NO_2, and ClO, have characteristics of very large rate constants with nearly zero activation energy. The rate constants of reactions (5.3), (5.4), (5.5), and (5.6) at 298 K are known to be 3.5, 5.8, 1.0, 3.7×10^{-11} cm^3 molecule^{-1} s^{-1}, respectively, all having similar magnitude (Atkinson et al. 2004, 2007). The recommended values of the IUPAC subcommittee including temperature dependences are,

$$k_{5.3}(T) = 2.4 \times 10^{-11} \exp\left(\frac{110}{T}\right) \quad \text{cm}^3\text{molecule}^{-1}\text{s}^{-1} \quad (150-500\ \text{K})$$
$$k_{5.4}(T) = 2.7 \times 10^{-11} \exp\left(\frac{224}{T}\right) \quad \text{cm}^3\text{molecule}^{-1}\text{s}^{-1} \quad (220-400\ \text{K})$$
$$k_{5.5}(T) = 5.5 \times 10^{-12} \exp\left(\frac{188}{T}\right) \quad \text{cm}^3\text{molecule}^{-1}\text{s}^{-1} \quad (220-420\ \text{K})$$
$$k_{5.6}(T) = 2.5 \times 10^{-11} \exp\left(\frac{110}{T}\right) \quad \text{cm}^3\text{molecule}^{-1}\text{s}^{-1} \quad (150-500\ \text{K}),$$

and all have small negative temperature dependence. Recommended temperature dependence of rate constants are based on O(^{3}P) + OH: Lewis and Watson (1980), Howard and Smith (1981),O(^{3}P) + HO_2: Keyser (1982), Nicovich and Wine (1987), O(^{3}P) + NO_2: Ongstad and Birks (1986), Geers-Müller and Stuhl (1987), Gierczak et al. (1999),O(^{3}P) + ClO: Ongstad and Birks (1986), Nicovich et al. (1988), Goldfarb et al. (2001), and previous experiments.

As for the O(^{3}P) + NO_2 reaction, termolecular process,

$$O(^3P) + NO_2 + M \rightarrow NO_3 + M \quad \Delta H^\circ_{298} = -209 \text{ kJ mol}^{-1} \tag{5.7}$$

is known to occur in addition to above reaction (5.5). Since this reaction is in fall-off region between the low- and high-pressure limit under the atmospheric conditions, the following formula described in Chap. 2 (Eq. (2.54)),

$$k_{ter}([M],T) = \left[\frac{k_0(T)[M]}{1+\frac{k_0(T)[M]}{k_\infty(T)}}\right] F_c^{\left\{1+\left[\log_{10}\left(\frac{k_0(T)[M]}{k_\infty(T)}\right)^2\right]\right\}^{-1}} \tag{5.8}$$

has to be applied to the temperature and pressure dependence. The recommended parameters by IUPAC subcommittee (Atkinson et al. 2004) are by taking $F_c = 0.6$,

$$k_{0,5.5b}(T) = 1.3 \times 10^{-31}\left(\frac{T}{300}\right)^{-1.5} \quad \text{cm}^6\text{molecule}^{-2}\text{s}^{-1}(200-400\ \text{K})$$
$$k_{\infty,5.5b}(T) = 2.3 \times 10^{-11}\left(\frac{T}{300}\right)^{-0.24} \quad \text{cm}^3\text{molecule}^{-1}\text{s}^{-1}(200-400\ \text{K}),$$

which are based on Burkholder and Ravishankara (2000), Hahn et al. (2000) and previous data.

5.1.4 *O (^{1}D) + H$_2$O*

As for the reaction pathways of the reaction of O (^{1}D) and H_2O, other than the formation of OH, the formation of $H_2 + O_2$ and the deactivation (also called quenching) can also be conceived.

$$O(^1D) + H_2O \rightarrow 2\ OH \quad \Delta H^\circ_{298} = -121\ \text{kJ mol}^{-1} \tag{5.9}$$

$$\rightarrow H_2 + O_2 \quad \Delta H^\circ_{298} = -197\ \text{kJ mol}^{-1} \tag{5.10}$$

$$\rightarrow O(^3P) + H_2O \quad \Delta H^\circ_{298} = -190\ \text{kJ mol}^{-1} \tag{5.11}$$

However, the reaction ratio of (5.10) and (5.11) has been reported to be 0.6 % (Glinski and Birks 1985), and 0.3 % (Carl 2005), respectively, and almost negligible against reaction (5.9).

The recent measurement of rate constant for the reaction O(^{1}D) + H_2O at 298 K by Dunlea and Ravishankara (2004b) agreed well with previous values, and the NASA/JPL panel evaluation No. 17 (Sander et al. 2011) recommends $k_{5.9}$ (298 K) = 2.0×10^{-10} cm^3 molecule^{-1} s^{-1} based on these values. The IUPAC subcommittee report Vol. I (Atkinson et al. 2004) recommends temperature independent value of $k_{5.9} = 2.2 \times 10^{-10}$ cm^3 molecule^{-1} s^{-1} at 200–350 K, which agreed with the above value within 10 %. More recent measurements by Vranckx et al. (2010) for wider temperature range also agreed well with previous results.

Thus, reaction (5.9) is the very fast reaction that is close to collision frequency with nearly zero activation energy. For this reason, this reaction can produce OH radicals under the condition with enough concentration of H_2O in the troposphere competing well with the deactivation reactions (Sander et al. 2011),

$$O(^1D) + N_2 \rightarrow O(^3P) + O_2 \quad k_{5.12}(298\ K) = 3.1 \times 10^{-11} cm^3 molecule^{-1} s^{-1} \tag{5.12}$$

$$O(^1D) + O_2 \rightarrow O(^3P) + O_2 \quad k_{5.13}(298\ K) = 4.0 \times 10^{-11} cm^3 molecule^{-1} s^{-1}. \tag{5.13}$$

The quantum chemical calculation of potential surface of the $O(^1D) + H_2O$ reaction has been conducted by Sayós et al. (2001). The lowest energy pathway for the main reaction channel to form OH + OH have no energy barrier agreeing with the experimental results.

5.1.5 $O\ (^1D) + N_2O$

For the reaction of $O(^1D)$ and N_2O, three pathways,

$$O(^1D) + N_2O \rightarrow N_2 + O_2 \quad \Delta H^\circ_{298} = -521\ kJ\ mol^{-1} \tag{5.14}$$

$$\rightarrow 2\ NO \quad \Delta H^\circ_{298} = -340\ kJ\ mol^{-1} \tag{5.15}$$

$$\rightarrow O(^3P) + N_2O \quad \Delta H^\circ_{298} = -190\ kJ\ mol^{-1} \tag{5.16}$$

can be conceived. Although the formation ratio of $O(^3P)$ by reaction (5.16) is <0.01 and can be neglected in general (Vranckx et al. 2008a), the reaction pathways, (5.14) and (5.15), are known to be both important. In atmospheric chemistry, reaction (5.15) is important as the formation of reactive nitrogen (also called odd nitrogen) in the stratosphere, and the overall reaction rate constant combining reaction (5.14) and (5.15) is important for the estimation of atmospheric lifetime of N_2O.

Many measurements of rate constants of the reaction of $O(^1D)$ and N_2O have been made mostly before 2000 (Blitz et al. 2004; Dunlea and Ravishankara 2004a; Carl 2005; Takahashi et al. 2005; Vranckx et al. 2008a), and the NASA/JPL panel evaluation No, 17 recommends $k_{5.10}(298\ K) = 1.3 \times 10^{-10}$ cm^3 molecule^{-1} s^{-1} based on these values (Sander et al. 2011). The recommended rate constant by the IUPAC subcommittee agrees well with this value. Similar to the reaction of $O(^1D)$ and H_2O, this reaction also has nearly zero activation energy and there is no temperature dependence in the range of 200–350 K.

As for the ratio of reaction (5.15), it is recommended the value of $k_{5.15}/k_{5.10} = 0.61$ based on Cantrell et al. (1994), and reaction rate constant for the NO formation reaction is $k_{5.15} = 7.8 \times 10^{-11}$ cm^3 molecule^{-1} s^{-1} (Sander et al. 2011). Reaction (5.15) has a large exothermicity and it is known experimentally that the formed NO is in vibrationally and rotationally excited. Akagi et al. (1999) found from the experiments of $^{18}O(^1D) + N_2{}^{16}O \rightarrow N^{18}O + N^{16}O$ using the oxygen isotopes that the $N^{16}O$ in the original $N^{16}O$ molecule is formed mostly in v = 0,

1 without vibrational excitation and the newly formed $N^{18}O$ is produced in highly vibrationally excited levels in v = 4–15. Further, Tokel et al. (2010) recently reported in the molecular beam experiment that the NO vibrational levels (v = 0 ~ 9) showed inverted population, which suggests that there are two reaction pathways, one gives the statistical distribution and another gives the inverted distribution.

Quantum chemical theoretical calculations for this reaction have been conducted by several groups. Takayanagi and Wada (2001) showed that the exothermic energy is mainly distributed in the vibrational energy of the newly formed NO and not in the NO in the original N_2O, but the latter NO is not a complete spectator but a part of excess energy is also distributed in the old NO. This result of theoretical calculation agrees well with the above experimental results using the oxygen isotopes. According to the trajectory calculation on the potential energy surface by Takayanagi and Akagi (2002), the ratio of reactions (5.15) forming 2NO and (5.14) forming $N_2 + O_2$ is affected by the collision energy of O(^{1}D) and approaching angle of O(^{1}D) toward N – N – O.

5.1.6 *O (^{1}D) + CH$_4$*

The pathways of the reaction of O(^{1}D) and CH_4 has been thought to be,

$$O(^1D) + CH_4 \rightarrow CH_3 + OH \quad \Delta H^\circ_{298} = -181 \text{ kJ mol}^{-1} \tag{5.17}$$

$$\rightarrow CH_2OH + H \quad \Delta H^\circ_{298} = -163 \text{ kJ mol}^{-1} \tag{5.18}$$

$$\rightarrow HCHO + H_2 \quad \Delta H^\circ_{298} = -473 \text{ kJ mol}^{-1} \tag{5.19}$$

$$\rightarrow O(^3P) + CH_4 \quad \Delta H^\circ_{298} = -190 \text{ kJ mol}^{-1}. \tag{5.20}$$

Overall reaction rate constant is recommended as $k_{5.17-5.20}(298\ K) = 1.8 \times 10^{-10}$ cm^3 molecule^{-1} s^{-1} by the NASA/JPL panel evaluation No. 17 (Sander et al. 2011), and as $k_{5.11-5.20}(298\ K) = 1.5 \times 10^{-10}$ cm^3 molecule^{-1} s^{-1} with no temperature dependence for 200–350 K by the IUPAC subcommittee report Vol. II(Atkinson et al. 2006) based on experiments by Davidson et al. (1977), Blitz et al. (2004), Dillon et al. (2007), Vranckx et al. (2008b) et al. Similar to the reaction of O(^{1}D) with H_2O and N_2O, this reaction is also a very fast reaction with nearly collisional frequency and near zero activation energy.

As for the ratio of reaction pathways of (5.17) (5.18) and (5.19), spectroscopic studies have been conducted using molecular beams (Casavecchia et al. 1980; Lin et al. 1999; Matsumi et al. 1993; Chen et al. 2005). The branching ratios and rate constants of each reaction pathway at 298 K are,

$$\frac{k_{5.17}}{k_{5.17-5.20}} = 0.75 \pm 0.15, \quad k_{5.17} = 1.31 \times 10^{-10} \text{cm}^3 \text{molecule}^{-1} \text{s}^{-1}$$
$$\frac{k_{5.18}}{k_{5.17-5.20}} = 0.20 \pm 0.10, \quad k_{5.18} = 1.35 \times 10^{-10} \text{cm}^3 \text{molecule}^{-1} \text{s}^{-1}$$
$$\frac{k_{5.19}}{k_{5.17-5.20}} = 0.05 \pm 0.05, \quad k_{5.19} = 0.09 \times 10^{-10} \text{cm}^3 \text{molecule}^{-1} \text{s}.$$

For reaction (5.18) it has been shown from a crossed molecular beam experiment that the main process is $CH_2OH + H$, and the process of $CH_3O + H$ is not important (Lin et al. 1998). Also for the physical deactivation pathway (5.20), Wine and Ravishankara (1982) and Takahashi et al. (1996) reported that the ratio is less than a few%, while the recent high precision experiment of Vranckx et al. (2008b) by using the chemiluminescence method showed that it is 0.2 ± 0.3 % and that it is negligible as an atmospheric reaction.

Regarding the reaction of oxygen atoms with CH_4 and other alkanes, in contrast to that, the reactions of $O(^3P)$ that is an H atom abstraction, the reactions of $O(^1D)$ have been discussed to be mainly insertion to C-H bonds. Lin et al. (1999) showed from an angular distribution of products in the crossed molecular beam experiment, that H atom forming pathway (5.18) and H_2 molecule forming pathway (5.19) both go through reaction intermediate $CH_3OH^{\ddagger}$ with a rather long lifetime. Also, the relative distribution of vibrational levels of the OH radical formed in reaction (5.17) has been shown by LIF experiment to be 0.18, 0.29, 0.37, 0.15 and 0.01 for $v = 0$, 1, 2, 3 and 4, respectively, and an inversion from the statistical distribution has been known for $v = 0$, 1 and 2 (Cheskis et al. 1989). Furthermore, from the experiment of changing the collision energy of $O(^1D)$ by González et al. (2000), it has been shown that the inversion of the vibrational levels of OH are observed reflecting the insertion reaction when the collision energy of less than 57.8 kJ mol^{-1} (0.6 eV), but the inversion is not seen for the higher collision energy. From the trajectory calculation on the potential surface it has also been shown that insertion is the main reaction when the collision energy is small and abstraction reaction increases as the collision energy increases, which agrees well with experimental evidence (Sayós et al. 2002). Theoretical calculation by Yu and Muckerman (2004) considering configuration interaction, give the ratios of pathways of (5.17) (5.18) and (5.19) to form OH, H and H_2, respectively, as 0.725, 0.186 and 0.025, respectively, reproducing well the experimental results, and also suggested that the reaction pathway, $CH_2 + H_2O$ exists in the ratio of 0.064.

5.1.7 *O (^{1}D) + CFCs*

The reactions of $O(^1D)$ with CFCs (chlorofluorocarbons) and halons are known to have large rate constants in the order of 10^{-10} cm^3 molecule^{-1} s^{-1}, and could possibly affect the stratospheric lifetime of these ozone layer depleting substances. However, decomposition rates of CFCs and halons in the stratosphere are in general determined mainly by photolysis rate, and the reaction with $O(^1D)$ is almost negligible for $CFCl_3$ (CFC – 11) and halons whose photolysis rates are high. However, as the photolysis rate decreases for such as CF_2Cl_2 (CFC – 12), CF_3Cl (CFC – 13), the importance of $O(^1D)$ reaction increases.

The pathways of the reactions of $O(^1D)$ and CFCs can be conceived as ClO radical formation and physical deactivation. For example, for CCl_2F_2 the following reaction pathways are known to be,

$$O(^1D) + CF_2Cl_2 \rightarrow ClO + CF_2Cl \quad \Delta H^\circ_{298} = -123 \text{ kJ mol}^{-1} \tag{5.21}$$

$$\rightarrow O(^3P) + CF_2Cl_2. \quad \Delta H^\circ_{298} = -190 \text{ kJ mol}^{-1} \tag{5.22}$$

The measurements of rate constants for the reaction of $O(^1D)$ with $CFCl_3$ and CF_2Cl_2 are made by Davidson et al. (1978) and Force and Wiesenfeld (1981), and with CF_3Cl by Ravishankara et al. (1993), which all showed no temperature dependence. Table 5.1 shows the recommended values by the NASA/JPL panel and the IUPAC subcommittee based on these experiments. The ratio for the production of ClO against quenching are obtained to be 88 ± 18, 87 ± 18 and 85 ± 18 % for $CFCl_3$, CF_2Cl_2 and CF_3Cl, respectively, by Takahashi et al. (1996).

Although $O(^1D)$ also reacts with halons to form BrO, FO is not formed from the reaction with CFC and PFCs (perfluoro carbons).

5.2 Reactions of OH Radicals

OH radicals react with almost all trace molecules in the atmospheric excluding CO_2, N_2O, CFC, etc., and drives atmospheric photochemical reaction system while the atmospheric lifetime of most of chemical species are determined by the reaction rate with OH, so that OH is the most important reactive species in the troposphere. In this sense, the rate constants of atmospheric molecules and OH are of unequivocal importance. In the stratosphere, inorganic reactions of OH in the HO_x cycle and in the cross chain reactions with NO_x and ClO_xcycles are also important. The OH reaction rate constants and their temperature dependence with almost all molecules of atmospheric interest have been measured in laboratory.

In this section, among the numerous atmospheric OH reactions, those with interest as elementary reactions are selected and explained. Table 5.2 gives the rate constants and their temperature dependence of particular importance in the atmosphere are excerpted from the recommendation by the IUPAC subcommittee (Atkinson et al. 2004, 2006) and the NASA/JPL panel (Sander et al. 2011). The equilibrium constants of typical reversible reactions in the atmosphere are shown in Table 5.3.

Table 5.2 Rate constants at 298 K and Arrhenius parameters for the reactions of OH radicals

Reactions	k(298 K) (cm^3 $molecule^{-1}$ s^{-1})	A Factor (cm^3 $molecule^{-1}$ s^{-1})	E_a/R (K)	Ref.
$OH + O_3 \rightarrow HO_2 + O_2$	7.3×10^{-14}	1.7×10^{-12}	940	(a1)
$OH + HO_2 \rightarrow H_2O + O_2$	1.1×10^{-10}	4.8×10^{-11}	−250	(a1)
$OH + H_2O_2 \rightarrow H_2O + HO_2$	1.7×10^{-12}	2.9×10^{-12}	160	(a1)
$OH + CO \rightarrow$ overall	2.3×10^{-13} (1 atm)	1.4×10^{-13} $(1 + [N_2]/4.2 \times 10^{19})$		(a2)
$\rightarrow H + CO_2$	1.5×10^{-13}	1.5×10^{-13} $(T/300)^{0.6}$ (k_0)		(b)
$+ M \rightarrow HOCO + M$		5.9×10^{-33} $(T/300)^{-1.4}$ $[N_2]$ (k_0)		(b)
		1.1×10^{-12} $(T/300)^{1.3}$ (k_∞)		
$OH + NO + M \rightarrow HONO + M$	$7.4 \times 10^{-31}[N_2]$ (k_0)	7.4×10^{-31} $(T/300)^{-2.4}$ $[N_2]$ (k_0)		(a1)
	3.3×10^{-11} (k_∞)	3.3×10^{-11} $(T/300)^{0.3}$ (k_∞)		
$OH + NO_2 + M \rightarrow HOOH + M$	1.2×10^{-11} (1 atm)	1.8×10^{-30} $(T/300)^{-3.0}$ $[M]$ (k_0)		(b)
		2.8×10^{-11} $(T/300)^{0}$ (k_∞)		
$\rightarrow HOONO + M$	1.2×10^{-12} (1 atm)	9.1×10^{-32} $(T/300)^{-3.9}$ $[M]$ (k_0)		(b)
		4.2×10^{-11} $(T/300)^{-0.5}$ (k_∞)		
$OH + HONO \rightarrow H_2O + NO_2$	6.0×10^{-12}	2.5×10^{-12}	−260	(a1)
$OH + HONO_2 \rightarrow H_2O + NO_3$	1.5×10^{-13} (1 atm)	2.4×10^{-14}	−460	(a1) (b)
$OH + HO_2NO_2 \rightarrow$ products	4.6×10^{-12}	1.3×10^{-12}	−380	(b)
$OH + NH_3 \rightarrow H_2O + NH_2$	1.6×10^{-13}	1.7×10^{-12}	710	(b)
$OH + SO_2 + M \rightarrow HOSO_2 + M$	1.1×10^{-12} (1 atm)	4.5×10^{-31} $(T/300)^{-3.9}$ $[N_2]$ (k_0)		(a1)
		1.3×10^{-12} $(T/300)^{0.7}$ (k_∞)		
$OH + OCS \rightarrow$ products	2.0×10^{-15}	1.1×10^{-13}	1200	(a1)
$OH + CS_2 + M \rightarrow HOCS_2 + M$	1.2×10^{-12} (1 atm)	4.9×10^{-31} $(T/300)^{-3.5}$ $[N_2]$ (k_0)		(b)
		1.4×10^{-11} $(T/300)^{-1}$ (k_∞)		
$OH + H_2S \rightarrow SH + H_2O$	4.7×10^{-12}	6.1×10^{-12}	80	(b)

$OH + CH_3SH \rightarrow CH_3S + H_2O$	3.3×10^{-11}	9.9×10^{-12}	−360	(b)
$OH + CH_3SCH_3 \rightarrow CH_3SCH_2 + H_2O$	3.1×10^{-11}	1.2×10^{-11}	280	(b)
$OH + CH_3SCH_3 + M \rightarrow (CH_3)_2SOH + M$	5.7×10^{-12} (1 atm)	2.9×10^{-31} $(T/300)^{-6.2}$ [M] (k_0)		(a1) (b)
$OH + CH_3SSCH_3 \rightarrow$ products	2.3×10^{-10}	6.0×10^{-11}	−400	(b)
$OH + CH_4 \rightarrow CH_3 + H_2O$	6.4×10^{-15}	1.9×10^{-12}	1690	(a2)
$OH + C_2H_6 \rightarrow C_2H_5 + H_2O$	2.4×10^{-13}	6.9×10^{-12}	1000	(a2)
$OH + C_3H_8 \rightarrow C_3H_7 + H_2O$	1.1×10^{-12}	7.6×10^{-12}	590	(a2)
$OH + C_2H_4 + M \rightarrow HOCH_2CH_2 + M$	7.9×10^{-12} (1 atm)	8.6×10^{-29} $(T/300)^{-3.1}$ $[N_2]$ (k_0)		(a2)
		9.0×10^{-12} $(T/300)^{-0.85}$ (k_∞)		
$OH + C_3H_6 + M \rightarrow HOC_3H_6 + M$	2.9×10^{-11} (1 atm)	8×10^{-27} $(T/300)^{-3.5}$ $[N_2]$ (k_0)		(a2)
		3.0×10^{-11} $(T/300)^{-1}$ (k_∞)		
$OH + C_5H_8$ (isoprene) $\rightarrow$ products	1.0×10^{-10}	3.1×10^{-11}	−350	(b)
$OH + C_{10}H_{16}$ (α-pinene) $\rightarrow$ products	5.3×10^{-11}	1.2×10^{-11}	−440	(a2)
$OH + C_2H_2 + M \rightarrow HOCHCH + M$	7.8×10^{-13} (1 atm)	5.5×10^{-30} $(T/300)^{0}$ $[N_2]$ (k_0)		(a2) (b)
		$8.3 \times 10^{-13}$$(T/300)^2$ (k_∞)		
$OH + HCHO \rightarrow H_2O + HCO$	8.5×10^{-12}	5.4×10^{-12}	−140	(a2)
$OH + CH_3CHO \rightarrow H_2O + CH_3CO$	1.5×10^{-11}	4.4×10^{-12}	−370	(a2)
$OH + (CHO)_2 \rightarrow H_2O + CH(O)CO$	1.1×10^{-11}	–	–	(a2)
$OH + CH_3C(O)CH_3 \rightarrow H_2O + CH_2C(O)CH_3$	1.8×10^{-13}	8.8×10^{-12}	1320	(a2)
$OH + CH_3OH \rightarrow$ products	9.0×10^{-13}	2.9×10^{-12}	350	(a2)
$OH + CH_3OOH \rightarrow$ products	5.5×10^{-12}	2.9×10^{-12}	−190	(a2)
$OH + HC(O)OH \rightarrow$ products	4.5×10^{-13}	4.5×10^{-13}	0	(a2)
$OH + CH_3ONO_2 \rightarrow$ products	2.3×10^{-14}	4.0×10^{-13}	850	(a2)
$OH + HCl \rightarrow H_2O + Cl$	7.8×10^{-13}	1.8×10^{-12}	250	(b)
$OH + CH_3Cl \rightarrow H_2O + CH_2Cl$	3.6×10^{-14}	2.4×10^{-12}	1250	(b)
$OH + CH_3CCl_3 \rightarrow H_2O + CH_2CCl_3$	1.0×10^{-14}	1.6×10^{-12}	1520	(b)

(continued)

Table 5.2 (continued)

Reactions	k(298 K) (cm^3 $molecule^{-1}$ s^{-1})	A Factor (cm^3 $molecule^{-1}$ s^{-1})	E_a/R (K)	Ref.
$OH + CHF_2Cl \rightarrow H_2O + CF_2Cl$	4.8×10^{-15}	1.1×10^{-12}	1600	(b)
$OH + HBr \rightarrow H_2O + Br$	1.1×10^{-11}	5.5×10^{-12}	−200	(b)
$OH + CH_3Br \rightarrow H_2O + CH_2Br$	3.0×10^{-14}	2.4×10^{-12}	1300	(b)
$OH + HI \rightarrow H_2O + I$	7.0×10^{-11}	1.6×10^{-11}	−440	(a3)
$OH + CH_3I \rightarrow H_2O + CH_2I$	7.2×10^{-14}	2.9×10^{-12}	1100	(b)

(a1, a2, a3) IUPAC subcommittee report Vol. I, II, III (Atkinson et al. 2004, 2006, 2007), respectively
(b) NASA/JPL panel evaluation No. 17 (Sander et al. 2011)

Table 5.3 Equilibrium constants at 298 K and temperature dependence parameters for reversible reactions

Reactions	K_{eq} (298 K) (cm^3 $molecule^{-1}$)	A (cm^3 $molecule^{-1}$)	B (°K)
$HO + NO_2 \rightleftarrows HOONO$	2.2×10^{-12}	3.5×10^{-27}	10140
$HO_2 + NO_2 \rightleftarrows HO_2NO_2$	1.6×10^{-11}	2.1×10^{-27}	10900
$HO_2 + H_2O \rightleftarrows HO_2{\bullet}H_2O$	5.2×10^{-19}	2.4×10^{-25}	4350
$NO_2 + NO_3 \rightleftarrows N_2O_5$	2.9×10^{-11}	2.7×10^{-27}	11000
$CH_3O_2 + NO_2 \rightleftarrows CH_3O_2NO_2$	2.2×10^{-12}	9.5×10^{-29}	11230
$CH_3C(O)O_2 + NO_2 \rightleftarrows CH_3C(O)O_2NO_2$	2.3×10^{-8}	9.0×10^{-29}	14000
$OH + CS_2 \rightleftarrows CS_2OH$	1.4×10^{-17}	4.5×10^{-25}	5140
$CH_3S + O_2 \rightleftarrows CH_3SOO$	2.2×10^{-19}	1.8×10^{-25}	5550
$Cl + O_2 \rightleftarrows ClOO$	2.9×10^{-21}	6.6×10^{-25}	2500
$ClO + ClO \rightleftarrows Cl_2O_2$	6.9×10^{-15}	1.7×10^{-27}	8650

K_{eq} (T) (cm^3 $molecule^{-1}$) = A exp (B/T) (200 < T < 300 K)
Source: NASA/JPL panel evaluation No.17 (Sander et al. 2011)

5.2.1 *OH + O₃*

The reaction of OH radicals and O_3 is an important reaction that converts OH to HO_2 in the HO_x cycle of the ozone dissipating chain process in the lower stratosphere (Sect. 8.2.1). In the troposphere, HO_2 is reproduced via the OH + CO reaction even in the clean atmosphere, the contribution of OH + O_3 reaction for the reproduction of HO_2 is small.

The reaction of OH with O_3 proceeds,

$$OH + O_3 \rightarrow HO_2 + O_2 \quad \Delta H^\circ_{298} = -167 \text{ kJ mol}^{-1} \tag{5.23}$$

to form a HO_2 radical. The IUPAC subcommittee (Atkinson et al. 2004) recommends the rate constants of this reaction at 298 K as $k_{5.23}$ (298 K) = 7.3×10^{-14} cm^3 $molecule^{-1}$ s^{-1}, and the temperature dependence as,

$$k_{5.23}(T) = 1.7 \times 10^{-12}\exp\left(-\frac{940}{T}\right)\text{cm}^3\text{molecule}^{-1}\text{s}^{-1}$$

based on Ravishankara et al. (1979), Nickolaisen et al. (2000), etc. Thus, the reaction of OH and O_3 has the activation energy, 7.8 kJ mol^{-1}, and is not very fast at near room temperature.

Meanwhile, according to quantum chemical calculations, this reaction has one or two transition states depending on the level of calculation, and the calculated activation barrier is larger than experimental values (Peiró-García and Nebot-Gil, 2003a) so that further theoretical study is needed.

5.2.2 $OH + HO_2$

The reaction of HO and HO_2 forms stable molecules from two radicals acting as a termination reaction of the HO_x cycle in the middle to upper stratosphere, which is important as affecting the efficiency of the ozone destruction process (Sect. 8.2.1).

The reaction of OH and O_3 proceeds as,

$$OH + HO_2 \rightarrow H_2O + O_2 \quad \Delta H^{\circ}_{298} = -295 \text{ kJ mol}^{-1} \tag{5.24}$$

to form H_2O and O_2. The measurement of this reaction rate had been difficult experimentally, and previously measured values had large uncertainty. Keyser (1988) identified that the cause of the uncertainty is due to side reactions by H and O atoms formed secondarily, and reported more accurate values by the using chemical model analysis. The recommended value at 298 K by the IUPAC sub-committee is $k_{5.24}$ (298 K) $= 1.1 \times 10^{-10}$ cm^3 molecule^{-1} s^{-1} (Atkinson et al. 2004). Thus, it is a very fast radical-radical reaction. The temperature dependence is recommended in the Arrhenius equation,

$$k_{5.14}(T) = 4.8 \times 10^{-11}\exp\left(\frac{250}{T}\right) \quad \text{cm}^3\text{molecule}^{-1}\text{s}^{-1}$$

with a small negative activation energy (Atkinson et al. 2004). It has also been confirmed that there is no pressure dependence in the range of 1–1000 torr (Keyser 1988).

This reaction has been concerned from the perspective of combustion chemistry and rate constants at high temperature have been measured. Anomalous temperature dependence, decrease of rate constant at 1100 K, has been reported by Hippler et al. (1995), and the possibility of existence of intermediate complex has been suggested.

5.2.3 $OH + CO$

The reaction of OH and CO is important for OH radical as the main reaction in the clean troposphere (Sect. 7.1), and for CO to determine its atmospheric lifetime. There are many studies on the rate constants of this reaction, and it has been revealed that the reaction proceeds via a process,

$$\begin{array}{c} OH + CO \leftrightarrows HOCO^{\dagger} \rightarrow H + CO_2 \\ \downarrow +M \\ HOCO. \end{array} \tag{5.25}$$

The reaction scheme involving the intermediate HOCO has been proposed from the experimental evidence that the rate constants of OH + CO reaction increase with pressure (Smith 1977). Later, HOCO has been detected directly by infrared absorption (Petty et al. 1993), and photoionization spectrometer (Miyoshi et al. 1994), and confirmed theoretically as described below.

According to the above scheme, the rate constants of OH + CO reaction can be described by the following two processes,

$$\mathrm{OH + CO \rightarrow H + CO_2} \quad \Delta H^\circ_{298} = -102 \ \mathrm{kJ\ mol^{-1}} \tag{5.26}$$

$$\mathrm{OH + CO + M \rightarrow HOCO + M.} \quad \Delta H^\circ_{298} = -115 \ \mathrm{kJ\ mol^{-1}} \tag{5.27}$$

The HOCO formed in reaction (5.27) reacts with O_2,

$$\mathrm{HOCO + O_2 \rightarrow HO_2 + CO_2} \tag{5.28}$$

with a rate constant, ~1.5×10^{-12} cm^3 molecule^{-1} s^{-1} (Miyoshi et al. 1994). The H atom formed in the reaction (5.26) reacts exclusively with O_2 under the atmospheric conditions and also gives HO_2. Thus, the products of OH + CO can be thought as $HO_2 + CO_2$ in the atmosphere regardless either of pathway, (5.26) or (5.27) is taken.

Recent measurements of the reaction rate constants following the above scheme have been conducted by Golden et al. (1998), McCabe et al. (2001), etc., and the data including these studies have been compiled by the NASA/JPL evaluation No. 17 (Sander et al. 2011). At the low-pressure limit, the reaction proceeds via the path (5.26), and the bimolecular reaction rate constant is recommended as,

$$k_{5.26}(T) = 1.5 \times 10^{-13} \left(\frac{T}{300}\right)^{0.6} \mathrm{cm^3 molecule^{-1} s^{-1}},$$

and the temperature dependence is small. On the other hand, reaction (5.27) is a termolecular reaction and Eq. (5.7) is applied for the temperature and pressure dependences of the rate constant. The recommended low-pressure and high-pressure limit values with $F_c = 0.6$ (Sander et al. 2011) are,

$$k_{0,5.27}(T) = 5.9 \times 10^{-33} \left(\tfrac{T}{300}\right)^{-1.4} \mathrm{cm^6 molecule^{-2} s^{-1}}$$
$$k_{\infty,5.27}(T) = 1.1 \times 10^{-12} \left(\tfrac{T}{300}\right)^{-1.3} \mathrm{cm^3 molecule^{-1} s^{-1}}.$$

Under the atmospheric conditions, reaction (5.27) is in the intermediate region between the low-pressure and high-pressure limit, and the pressure dependence has to be calculated by using Eq. (5.7). The IUPAC subcommittee report Vol. II proposed the approximate equation for the overall rate constants combining reaction (5.26) and (5.27),

$$k_{5.26-5.27} = 1.44 \times 10^{-33} \left(1 + \frac{[N_2]}{4.2 \times 10^{19}}\right) \mathrm{cm^3 molecule^{-1} s^{-1}}$$

for the pressure range of 0–1 atm (N_2), and temperature range 200–300 K where the temperature dependence is small (Atkinson et al. 2006). From these results, the rate constant at 298 K and 1 atm is obtained as $k_{5.26\text{–}5.27}$ (298 K, 1 atm) $\approx 2.4 \times 10^{-13}$ cm^3 molecule^{-1} s^{-1}.

This reaction has attracted interest from a theoretical point of view as a four-centered reaction model involving three heavy atoms, and many studies have been conducted. The calculations of the potential energy surface of the association reaction of OH + CO have been conducted by several groups (Yu et al. 2001; Zhu et al. 2001; Valero et al. 2004), and the reaction of OH + CO is thought to initially form trans-HOCO, and it decomposes into H + CO_2 after trans- to cis- isomerization. Theoretical calculations of the reaction rate constants have been performed using such potential surface (Valero et al. 2004; Medvedev et al. 2004), and unimolecular decomposition theory (see Sect. 2.3.2) applying to HOCO† (Troe 1998; Zhu et al. 2001; Senosiain et al. 2003; Chen and Marcus 2005; Joshi and Wang 2006). Other than atmospheric chemistry, the reaction of OH + CO is also important in combustion chemistry, so that studies have been conducted in wide temperature range in 80–2800 K. The temperature dependence of the obtained bimolecular rate constant is known to show large deviation from the Arrhenius plot between, below and above 500 K. Although the temperature dependence is very small below 300 K as noted above, the temperature dependence, pressure dependence, and isotope effects cannot be reproduced well with recent theoretical studies, and the discussion is still going on the energy height of the transition states (TS_1, TS_2) of decomposition pathways of HOCO and the existence of the tunneling effect.

5.2.4 *OH + NO_2 + M*

The recombination reaction of OH radical and NO_2 is the most important as HO_x chain termination reaction in the troposphere (Sect. 7.3.2). The reaction of OH + NO_2used to be described as a termolecular reaction forming nitric acid ($HONO_2$) (although nitric acid is usually written as HNO_3, it is written as $HONO_2$ in this chapter for helping to understand the reaction path more easily). However, in the measurements of reaction rate constant of this reaction using pulse photolysis, the time profile of HO shows double exponential decay, and the pathway to form HOONO (peroxynitrous acid) in addition to $HONO_2$ has been suggested (Burkholder et al. 1987; Hippler et al. 2002), and Golden and Smith (2000) proposed the importance of this reaction pathway.

$$\mathrm{OH} + \mathrm{NO_2} + \mathrm{M} \rightarrow \mathrm{HONO_2} + \mathrm{M} \quad \Delta H^\circ_{298} = -208\ \mathrm{kJ\ mol^{-1}} \tag{5.29}$$

$$\rightarrow \mathrm{HOONO} + \mathrm{M} \quad \Delta H^\circ_{298} = -93\ \mathrm{kJ\ mol^{-1}} \tag{5.30}$$

Figure 5.1 shows an example of time profile of OH signal intensity in the reaction of OH + NO_2 by Hippler et al. (2002). As shown in Fig. 5.1, the OH decay is single exponential at 300 K, but shows apparent double exponential at 340 K suggesting

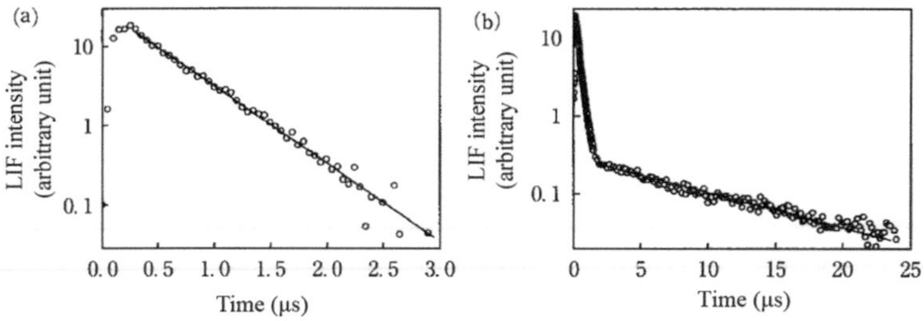

Fig. 5.1 Decay curves of OH signals in the experiments of $OH + NO_2$ reaction (**a**) 300 K, (**b**) 430 K (Adapted from Hippler et al. 2002)

the existence of the equilibrium reaction, $OH + NO_2 \rightleftarrows HOONO$ experimentally. The existence of reaction (5.30) confirmed by such experimental evidence and theoretical consideration implies that the rate constant used for $OH + NO_2$ for a long time was supposed to be overestimation as a rate of reaction (5.29), which may affect model calculation of ozone formation, etc. (Golden and Smith 2000). The spectroscopic detection of HOONO has been reported by using infrared spectroscopy (Nizkorodov and Wennberg 2002; Pollack et al. 2003), and cavity ring-down spectroscopy (Bean et al. 2003).

The NASA/JPL panel evaluation No. 17 (Sander et al. 2011) recommends the low-pressure and high-pressure limit rate constants of reaction (5.29) and (5.30) as,

$$k_{0,5.29}(T) = 1.8 \times 10^{-30}\left(\frac{T}{300}\right)^{-3.0}\text{cm}^6\text{molecule}^{-2}\text{s}^{-1}$$
$$k_{\infty,5.29}(T) = 2.8 \times 10^{-11}\left(\frac{T}{300}\right)^{0}\text{cm}^3\text{molecule}^{-1}\text{s}^{-1}$$
$$k_{0,5.30}(T) = 9.1 \times 10^{-32}\left(\frac{T}{300}\right)^{-3.9}\text{cm}^6\text{molecule}^{-2}\text{s}^{-1}$$
$$k_{\infty,5.30}(T) = 4.2 \times 10^{-11}\left(\frac{T}{300}\right)^{-0.5}\text{cm}^3\text{molecule}^{-1}\text{s}^{-1},$$

based on studies by Brown et al. (1999), D'Ottone et al. (2001), Hippler et al. (2002) and others. The formation rate constant of $HONO_2$ at 298 K, and 1 atm using the recommended value is $k_{5.29} = 1.1 \times 10^{-11}$ cm^6 molecule^{-2} s^{-1}, and the formation ratio of HOONO is 5 – 15 %.

Recently, Mollner et al. (2010) reported more accurate values of overall rate constant of $k_{5.29} + k_{5.30}$ at 20–900 torr by high sensitivity LIF, and branching ratio of reaction (5.29) and (5.30) by detecting $HONO_2$ and HOONO individually using cavity ring-down spectroscopy. From these results, there is some difference in the efficiency of N_2 and O_2 as a third body M, and the collisional efficiency of air is 94 % of N_2. The rate constants of reaction (5.29) and (5.30) at 298 K in air are reported as $k_{0,\ 5.29} = 1.51 \times 10^{-30}$ cm^6 molecule^{-2} s^{-1}, $k_{\infty,\ 5.29} = 1.84 \times 10^{-11}$ cm^3 molecule^{-1} s^{-1}, $k_{0,\ 5.30} = 6.2 \times 10^{-32}$ cm^6 molecule^{-2} s^{-1}, and $k_{\infty,\ 5.30} = 8.1 \times 10^{-11}$ cm^3 molecule^{-1} s^{-1}. According to these results, formation rate constant of $HONO_2$ under atmospheric pressure is $k_{5.29} = 9.2\ (\pm 0.4) \times 10^{-12}$ cm^3 molecule^{-1} s^{-1}, and the branching ratio of the formation of HOONO is $k_{\infty,\ 5.30}/k_{\infty,\ 5.29} = 0.142\ (\pm 0.012)$. The $HONO_2$ formation rate constant obtained here is 14 %

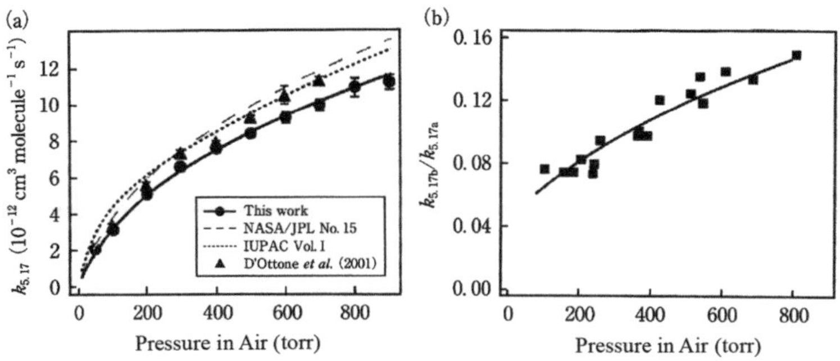

Fig. 5.2 Pressure dependence of the rate constants for the reaction of $OH + NO_2$: (**a**) k_{17}($OH + NO_2$); (**b**) k_{17a}/k_{17b} (Adapted from Mollner et al. 2010)

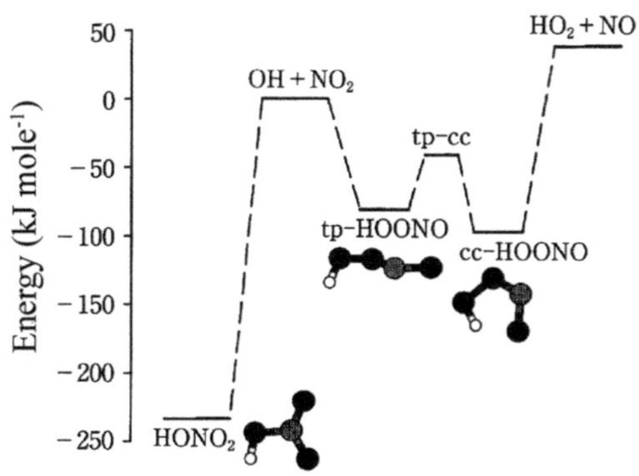

Fig. 5.3 Schematic diagram of potential energy surface for the reaction of $OH + NO_2$ (Adapted from Pollack et al. 2003)

lower than the recommended value of the NASA/JPL evaluation No. 17 (Sander et al. 2011). Figure 5.2 shows the pressure dependence of the rate constants, k($OH + NO_2$), $k_{5.29}$, and the branching ratio $k_{5.30}/k_{5.29}$.

Figure 5.3 depicts potential energy surface of $OH + NO_2$ reaction obtained by quantum chemical calculation (Pollack et al. 2003). Reaction rate constants calculated by RRKM calculation using the electronic structure of the transition state has been compared with the observed values (Sumathi and Peyerimhoff 1997; Chakraborty et al. 1998), and Golden et al. (2003) reported that recently calculated rate constants reproduced well the temperature and pressure dependence obtained by experiments. It has not been elucidated yet, however, if the reaction intermediate HOONO isomerizes to $HONO_2$ or it regenerates more reactive chemical species by photolysis or reaction with other reactive species in the atmosphere, which would affect the ozone formation efficiency in the troposphere.

5.2.5 *OH + HONO₂ (HNO₃)*

The reaction of OH radical with nitric acid ($HONO_2$, HNO_3) in the stratosphere is important as it reproduces active nitrogen from the reservoir molecule $HONO_2$ in the NO_x cycle. Although in the troposphere, water-soluble nitric acid is mainly removed by wet deposition into cloud and fog, and dry deposition on earth's surface, the OH reaction as well as photolysis are also important as removal processes and as active nitrogen regenerating process in the upper troposphere where clouds are not abundant.

It has been known experimentally that the rate constants of the OH and $HONO_2$ reaction show pressure dependence and negative temperature dependence deviating largely from Arrhenius plot (Margitan and Watson 1982; Smith et al. 1984; Devolder et al. 1984; Stachnik et al. 1986; Brown et al. 1999). From this experimental evidence, it has been suggested that the reaction proceeds through two paths, one through a reaction intermediate $OH - HONO_2$, and another by a direct reaction (Smith et al. 1984; Brown et al. 1999).

$$OH + HONO_2 \rightarrow H_2O + NO_3 \quad \Delta H^\circ_{298} = -70 \ \text{kJ mol}^{-1} \tag{5.31}$$

$$\rightarrow [OH\text{–}HONO_2] \rightarrow H_2O + NO_3 \quad \Delta H^\circ_{298} = -70 \ \text{kJ mol}^{-1} \tag{5.32}$$

The existence of $OH - HONO_2$ has been confirmed theoretically as a six-membered ring molecule (Xia and Lin 2001), and experimentally by direct detection with infrared spectroscopy (O'Donnell et al. 2008). The pathways of reaction (5.32) can be expressed as,

$$\begin{array}{ccc} OH + HONO_2 & \rightleftarrows & OH\text{–}HONO_2^{\dagger} \rightarrow H_2O + NO_3 \\ & (5.35) & \downarrow + M \quad (5.34) \\ & & OH\text{–}HONO_2 \rightarrow H_2O + NO_3 \end{array} \tag{5.33}$$

Figure 5.4 shows the energy scheme of the reaction (Brown et al. 1999).

From these considerations, the overall reaction rate constants of the OH + $HONO_2$ can be expressed by the sum of pressure-independent $k_{5.31}$ and pressure-dependent $k_{5.32}$ described by the Lindeman mechanism formula,

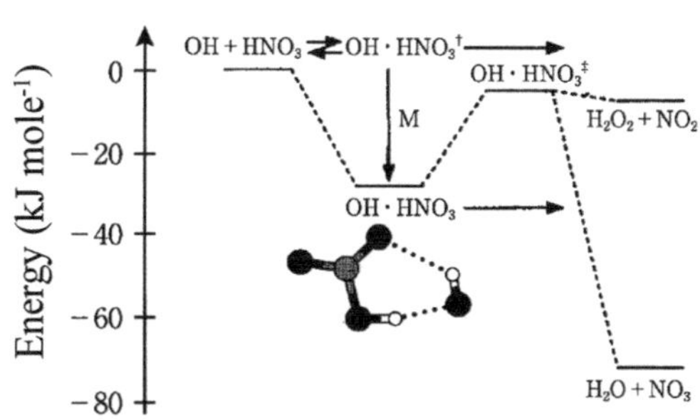

Fig. 5.4 Schematic diagram of potential energy surface for the reaction of OH + $HONO_2$ (Adapted from Brown et al. 1999)

$$k_{5.33} = k_{5.31}(T) + k_{5.32}([M], T) \tag{5.36}$$

$$k_{5.32}(M, T) = \frac{k_{5.35}[M]}{1 + \frac{k_{5.35}[M]}{k_{5.34}}}. \tag{5.37}$$

Recommended values by the NASA/JPL panel evaluation No. 17 (Sander et al. 2011) for rate parameters are,

$$k_{5.31}(T) = 2.4 \times 10^{-14} exp\left(\frac{460}{T}\right) \mathrm{cm^3 molecule^{-1} s^{-1}}$$
$$k_{5.34}(T) = 2.7 \times 10^{-17} exp\left(\frac{2199}{T}\right) \mathrm{cm^3 molecule^{-1} s^{-1}}$$
$$k_{5.35}(T) = 6.5 \times 10^{-34} exp\left(\frac{1335}{T}\right) \mathrm{cm^3 molecule^{-1} s^{-1}},$$

and the same values are adopted by the IUPAC subcommittee report (Atkinson et al. 2004). Since the same products are given by either of the reaction path (5.31) and (5.32), the reaction yield of NO_3 is unity. From the above reaction rate parameters, it can be seen that the pathway via $OH - HONO_2$ gets more important as the temperature goes down.

5.2.6 *OH + SO₂ + M*

The main atmospheric reactions of SO_2 are homogenous gas phase oxidation reaction by OH radicals and liquid phase oxidation reaction by H_2O_2 and O_3 in water droplets of cloud and fog. In this paragraph, the reaction with OH is described an important homogeneous gas phase reaction.

The reaction of OH and SO_2 is the termolecular reaction expressed by,

$$\mathrm{OH + SO_2 + M \rightarrow HOSO_2 + M.} \quad \Delta H^\circ_{298} = -125 \ \mathrm{kJ\ mol^{-1}} \tag{5.38}$$

As for the rate constant of reaction (5.38), the low-pressure and high-pressure limit equations given by Wine et al. (1984) adopting $F_c = 0.525$,

$$k_{0,5.22}(T) = 4.5 \times 10^{-31} [N_2] \left(\frac{T}{300}\right)^{-3.9} \mathrm{cm^3 molecule^{-1} s^{-1}}$$
$$k_{\infty,5.22}(T) = 1.3 \times 10^{-12} \left(\frac{T}{300}\right)^{-0.7} \mathrm{cm^3 molecule^{-1} s^{-1}}$$

are recommended by the IUPAC subcommittee (Atkinson et al. 2004). The value at 298 K and 1 atm is ~1×10^{-12} $\mathrm{cm^3\ molecule^{-1}\ s^{-1}}$, which is nearly one order of magnitude smaller than the rate constant for the reaction of NO_2 by OH.

Quantum chemical calculations for the pathways of reaction (5.38) and the product $HOSO_2$ have been conducted by Li and McKee (1997), Somnitz (2004) and others. The exact thermochemical values for the $HOSO_2$ radical was not obtained experimentally, Klopper et al. (2008) recently reported the heat of formation as $\Delta H^\circ_{f,0}(HOSO_2) = -366.6 \pm 2.5$ and $\Delta H^\circ_{f,298}(HOSO_2) = -374.1 \pm 3.0$ kJ mol^{-1}.

The $HOSO_2$ radicals formed in the reaction of OH and SO_2 are thought to react with O_2 in the atmosphere,

$$HOSO_2 + O_2 \rightarrow HO_2 + SO_3 \quad \Delta H^\circ_{298} = -9.5 \text{ kJ mol}^{-1} \tag{5.39}$$

to form HO_2 and SO_3. The enthalpy of the reaction was obtained from the above value of heat of formation as $\Delta H^\circ_{f,0} = -8.5 \pm 3.0, \Delta H^\circ_{f,298} = -9.5 \pm 3.0$ kJ mol^{-1}. Experimentally, $HOSO_2$ has been detected in the gas phase (Egsgaard and Carlsen 1988) and in the low temperature matrix (Hashimoto et al. 1984; Kuo et al. 1991).

5.2.7 *OH + CH_4, C_2H_6, C_3H_8*

Although the OH radicals react with all alkanes, methane (CH_4), ethane (C_2H_6), and propane (C_3H_8) are picked up here as typical alkanes. Since the atmospheric lifetimes of alkanes are determined by the reaction rates with OH, for example the accurate rate constant of OH + CH_4 is important for the evaluation of global warming.

The reactions of OH with alkanes are H-atom abstraction such as,

$$OH + CH_4 \rightarrow CH_3 + H_2O \quad \Delta H^\circ_{298} = -58 \text{ kJ mol}^{-1} \tag{5.40}$$

$$+ C_2H_6 \rightarrow C_2H_5 + H_2O \quad \Delta H^\circ_{298} = -74 \text{ kJ mol}^{-1} \tag{5.41}$$

$$+ C_3H_8 \rightarrow CH_3CH_2CH_2 + H_2O \quad \Delta H^\circ_{298} = -74 \text{ kJ mol}^{-1} \tag{5.42}$$

$$\rightarrow CH_3CHCH_3 + H_2O \quad \Delta H^\circ_{298} = -88 \text{ kJ mol}^{-1} \tag{5.43}$$

Many measurements OH and CH_4 reaction have been reported within the temperature range of 200–420 K, and the IUPAC subcommittee report Vol. II (Atkinson et al. 2006) cited the value at 298 K, $k_{5.40}$ (298 K) = 6.4×10^{-15} cm^3 molecule^{-1} s^{-1}, and the temperature dependence as,

$$k_{5.40}(T) = 1.85 \times 10^{-12} \exp\left(-\frac{1690}{T}\right) \text{cm}^3 \text{molecule}^{-1} \text{s}^{-1}.$$

The NASA/JPL panel evaluation No. 17 (Sander et al. 2011) recommends more detailed formula with three parameters,

$$k_{5.40}(T) = 2.80 \times 10^{-14} T^{0.667} \exp\left(-\frac{1575}{T}\right) \mathrm{cm^3 molecule^{-1} s^{-1}}$$

for applying to model calculation in the lower and upper troposphere. Recent values of the rate constant of the OH and CH_4 reaction reported by Gierczak et al. (1997) and Bonard et al. (2002) agree with previous values.

Similarly, many studies on measurements for rate constants of the reaction of OH with C_2H_6, C_3H_8 has been reported, and the recommended values by the IUPAC subcommittee report Vol. II (Atkinson et al. 2006) based on these are $k_{5.41} = 2.4 \times 10^{-13}$, $k_{5.42+5.53} = 1.1 \times 10^{-12}$ cm^3 molecule^{-1} s^{-1}at 298 K and the temperature dependences are,

$$k_{5.41}(T) = 6.9 \times 10^{-12} \exp\left(-\frac{1000}{T}\right) \mathrm{cm^3 molecule^{-1} s^{-1}}$$

$$k_{5.42+5.43}(T) = 7.6 \times 10^{-12} \exp\left(-\frac{585}{T}\right) \mathrm{cm^3 molecule^{-1} s^{-1}}.$$

Comparing at 298 K, the rate constants for C_2H_6 and C_3H_8 are by one and two orders of magnitude larger than for CH_4, respectively. The difference is found to be due to the difference in the activation energy of each reaction.

For the reaction of alkanes of $>C_3$, hydrogen atom abstraction by OH can occur either from the primary (one adjacent carbon atom), secondary (two adjacent carbon atoms), and tertiary (three adjacent carbon atoms) carbon atom. In case of C_3H_8, the temperature dependence for the two reaction paths (5.42) and (5.43) are given by Droege and Tully (1986) as,

$$k_{5.42}(T) = 6.3 \times 10^{-12} \exp\left(-\frac{1050}{T}\right) \mathrm{cm^3 molecule^{-1} s^{-1}}$$

$$k_{5.43}(T) = 6.3 \times 10^{-12} \exp\left(-\frac{580}{T}\right) \mathrm{cm^3 molecule^{-1} s^{-1}},$$

and the ratio of hydrogen abstraction from the primary and secondary carbon atoms are 0.17:0.83 at 298 K. It has been known that the hydrogen atoms on tertiary, secondary, and primary are easier to be abstracted in this order, and the reason is due to the difference in the activation energy as seen in the above equation.

The difference in reactivity of OH for different alkanes has long been discussed in terms of physical parameters of molecules, and it has been known that the ionization potentials (IP) have good correlation with the rate constants (Grosjean 1990) as in the reaction of OH, O_3 and NO_3 with olefins described below. Furthermore, comparison of rate constants using the calculated $\Delta S^{\ddagger}$ and $\Delta H^{\ddagger}$ by conventional transition state theory with experimental values has also been made (Cohen 1982).

5.2.8 $OH + C_2H_4 + M$

Alkenes such as ethylene (ehene, C_2H_4) have large reaction rate constants with OH among hydrocarbons, and have large contribution to photochemical ozone formation in urban air considering their mixing ratios.

The reactions of OH with alkenes are in general addition, and the reaction of OH and C_2H_4 is expressed by termolecular reaction such as,

$$OH + C_2H_4 + M \rightarrow HOCH_2CH_2 + M. \quad \Delta H^\circ_{298} = -23 \text{ kJ mol}^{-1} \tag{5.44}$$

Many experiments on the temperature and pressure dependence of reaction (5.44) have been reported. The IUPAC subcommittee report Vol. II (Atkinson et al. 2006) recommends the low-pressure and high-pressure limit rate constants in 200–300 K as,

$$k_{0,5.44}(T) = 8.6 \times 10^{-29}\left(\frac{T}{300}\right)^{-3.1} \text{cm}^6\text{molecule}^{-2}\text{s}^{-1}$$
$$k_{\infty,5.44}(T) = 9.0 \times 10^{-12}\left(\frac{T}{300}\right)^{-0.85} \text{cm}^3\text{molecule}^{-1}\text{s}^{-1}$$

based on Zellner and Lorenz (1984), Klein et al. (1984), Kuo and Lee (1991), Fulle et al. (1997), Vakhtin et al. (2003), etc. As seen in the above formula, this reaction has large temperature dependence at low-pressure limit while there is almost no temperature dependence at the high-pressure limit. Quantum chemical calculation has been made for the $OH + C_2H_4$ reaction, and used for the analysis of experimental values (Cleary et al. 2006; Taylor et al. 2008).

The $OH + C_2H_4$ reaction is nearly in the high-pressure limit at the atmospheric condition of 1 atm, but it is in the intermediate region between the low- and high-pressure limit in the upper atmosphere, and the rate constants have to be calculated by Eq. (5.8). The $HOCH_2CH_2$ radicals formed in reaction (5.44) are stabilized in the atmospheric conditions, and react with O_2 to form peroxy-radicals (see Chap. 7).

The reaction of OH with higher alkenes with more carbon atoms are almost in the high-pressure limit, and the IUPAC subcommittee report Vol. II recommends the rate constants at 298 K and 1 atm as 7.9×10^{-12}, 2.9×10^{-11} and 1.0×10^{-10} cm^3 molecule^{-1} s^{-1} for C_2H_4, C_3H_6 and C_5H_8 (isoprene), respectively (Atkinson et al. 2006). Thus, the reaction rate constants of OH and alkenes increases significantly with carbon number in general. The correlation of rate constants of the OH-alkene reactions with parameters of electronic structure of molecules has been studied, and it has been shown that the ionization potential (Grosjean 1990) and the highest occupied orbit (HOMO) energy (King et al. 1999) have a very good correlation.

5.2.9 $OH + C_2H_2 + M$

Alkynes represented by acetylene (C_2H_2) are chain hydrocarbons containing a triple bond in a molecule. Although the reaction rate of alkynes with OH is smaller than alkenes, C_2H_2 in particular has relatively high mixing ratio in the polluted atmosphere and its reaction cannot be ignored for the photochemical ozone production in general.

The reactions of OH with alkynes are addition reactions near at the room temperature and atmospheric pressure similar to alkenes, and the reaction of OH and C_2H_2 can be written as the termolecular reaction,

$$OH + C_2H_2 + M \rightarrow HOCHCH + M. \quad \Delta H^\circ_{298} = -145 \text{ kJ mol}^{-1} \tag{5.45}$$

The rate constants of reaction (5.45) have been measured experimentally, and the IUPAC subcommittee report Vol. II recommends based on Bohn et al. (1996), Fulle et al. (1997) and other previous studies,

$$k_{0,5.45}(T, N_2) = 5 \times 10^{-30}[N_2]\left(\frac{T}{300}\right)^{-1.5} \text{cm}^3\text{molecule}^{-1}\text{s}^{-1}$$
$$k_{\infty,5.45}(298\ K) = 1.0 \times 10^{-12}\ \text{cm}^3\text{molecule}^{-1}\text{s}^{-1}$$

for the low-pressure limit at 300–800 K and high-pressure limit at 298 K, respectively (Atkinson et al. 2006). This reaction does not reach to the high-pressure limit under the atmospheric pressure, and the recommended value of the rate constant at room temperature and 1 atm. is

$$k_{5.45}(298\ \text{K}, 1\ \text{atm}) = 7.8 \times 10^{-13}\text{cm}^3\text{molecule}^{-1}\text{s}^{-1}.$$

Quantum chemical calculation for the OH and C_2H_2 reaction has recently been made by Senosiain et al. (2005) and the theoretical rate constants has been obtained. According to the calculation, it has been concluded that the reaction pathway to form ketene,

$$OH + C_2H_2 \rightarrow CH_2CO + H \tag{5.46}$$

is predominant rather than the hydrogen atom abstraction at high temperature, and collisional stabilization of HOCHCH formed in the reaction (5.45) is the main path at low temperature.

5.2.10 $OH + C_6H_6$, $C_6H_5CH_3$

The reaction of OH with a typical aromatic hydrocarbon, benzene (C_6H_6), is known to be addition to the benzene ring (Atkinson and Arey 2003),

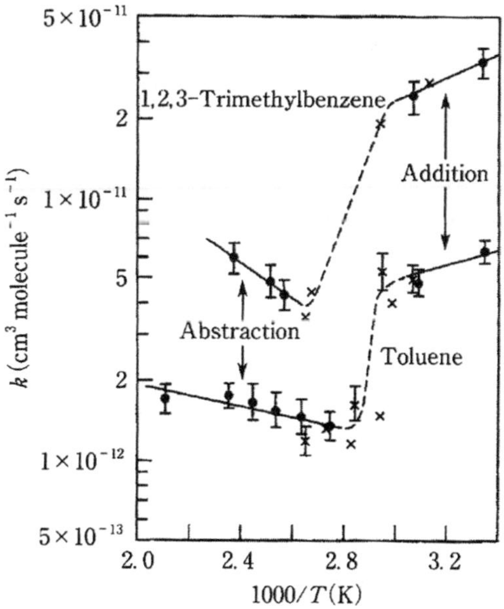

Fig. 5.5 Arrhenius plots of rate constants for the reactions of OH + toluene and 1,2,3 trimethylbenzene (Adapted from Perry et al. 1977)

$$OH + C_6H_6 + M \rightleftarrows HO\ C_6H_6 + M. \quad (5.47)$$

For aromatic hydrocarbons with side chain alkyl group such as toluene ($C_6H_5CH_3$), the addition reaction to the benzene ring and hydrogen atom abstraction reaction from the alkyl group can occur simultaneously.

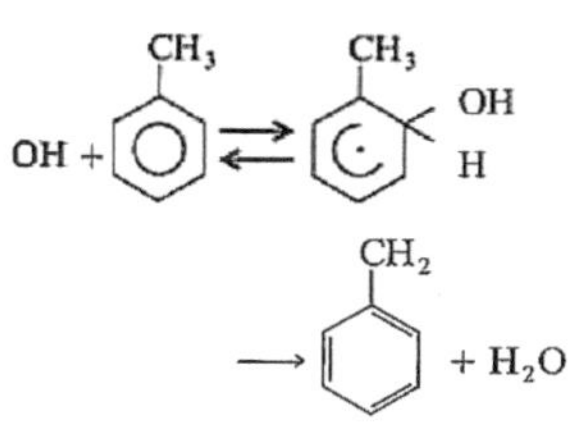

(5.48 and 5.49)

Figure 5.5 shows the Arrhenius plots for the reaction of OH and toluene and 1,2,3-trimethybenzene (Perry et al. 1977). From this figure, it can be seen that at the high temperature regime of >~350 K for toluene and >380 K for trimethylbenzene, their Arrhenius plots show normal negative slope (positive activation energy), but at the low temperature regime of <~330 K for both compounds they give positive slope (negative activation energy) and there a big gap in rate constants between both temperature regimes. Furthermore, it is reported that the logarithmic plot of the decay rate of OH is linear with time in the high- and low-temperature range, where the Arrhenius plot is linear, non-exponential decay is seen in the gap region (Perry et al. 1977). This implies that the initial reactions between OH and aromatic hydrocarbons are hydrogen abstraction and addition reactions in the high- and low-temperature range, respectively, and both processes are important in the intermediate temperature range, where OH radicals are regenerated at certain time delay causing nonexponential decay of OH. Also, the reason that the hydrogen atom abstraction can be seen exclusively from the side chain rather than the benzene ring reflects the fact that the C-H bond energy of the alkyl group (360 kJ mol^{-1}) is much smaller than the benzene ring (460 kJ mol^{-1}) (Uc et al. 2006).

The adduct radical formed by the OH addition to the benzene ring in reactions (5.47) and (5.48) are called cyclohexadienyl radicals, and their existence has been confirmed experimentally by their UV absorption spectrum (Grebenkin and Krasnoperov 2004; Johnson et al. 2005).

The addition of OH to toluene can be thought to occur at the position of ipso (carbon atom bonded to the methyl group), ortho (carbon atom adjacent to the methyl group), meta (one carbon apart from the methyl group), and para (opposite carbon to the methyl group). Actually, the OH addition reaction occurs predominantly to the ortho position followed by to the para position (ortho-para orientation), which has been confirmed by the yields of cresol isomers (Smith et al. 1998; Klotz et al. 1998), and also by supported by theoretical calculations (Bartolotti and Edney 1995; Suh et al. 2002). The reaction of cyclohexadienyl radicals with O_2 in the atmosphere will be described in Sect. (7.2.8).

The rate constants of the reaction of OH and benzene has been measured by Perry et al. (1977), Tully et al. (1981), Goumri et al. (1991), Bohn and Zetzsch (1999), etc. They showed that the reaction is in the high-pressure limit where the OH adduct radicals are stabilized by collision at the pressure of >100 torr, and in the fall-off region at the lower pressure than this. The IUPAC subcommittee (Wallington et al. 2012) recommends the temperature dependence of the rate constant at atmospheric pressure (high-pressure limit) at 230–350 K,

$$k_{5.47}(T) = 2.3 \times 10^{-12} \exp\left(-\frac{190}{T}\right) \mathrm{cm^3 molecule^{-1} s^{-1}},$$

and the value at 298 K,

$$k_{5.47}(298\ K) = 1.2 \times 10^{-12}\ \mathrm{cm^3 molecule^{-1} s^{-1}}.$$

The rate constants for reaction of OH and toluene has also been measured by Perry et al. (1977), Tully et al. (1981), Bohn (2001), and others. Based on these studies, the IUPAC subcommittee (Wallington et al. 2012) recommends the overall rate constant for the reactions (5.48 and 5.49) at 210–350 K and at 298 K,

$$k_{5.48+5.49}(T) = 1.8 \times 10^{-12} \exp\left(-\frac{340}{T}\right) \mathrm{cm^3 molecule^{-1} s^{-1}}$$
$$k_{5.48+5.49}(298\ K) = 5.6 \times 10^{-12}\ \mathrm{cm^3 molecule^{-1} s^{-1}}.$$

The ratio of abstraction reaction (5.49) is given as $k_{5.49}/\ (k_{5.48+5.49}) = 0.063$ at 298 K, reflecting the results of product analysis obtained by the experiments (Smith et al. 1998; Klotz et al. 1998).

5.2.11 *OH + HCHO, CH₃CHO*

The reaction of OH and aldehydes are thought to be H atom abstraction from the aldehyde group in general. The reactions of HCHO and CH_3CHO are described here as one of typical atmospheric reaction of the OH with organic compounds, in parallel with the H atom abstraction reactions from alkyl groups of alkanes and addition reactions to C-C double bonds of alkenes taken up so far.

The reaction of OH and HCHO has been confirmed to be H atom abstraction from aldehyde group,

$$OH + HCHO \rightarrow H_2O + HCO \quad \Delta H^\circ_{298} = -127 \text{ kJ mol}^{-1} \tag{5.50}$$

base on experimental evidence (Niki et al. 1984; Butkovskaya and Setser 1998; Sivakumaran et al. 2003). For this reaction, theoretical consideration for the possibility of addition reaction to form HCOOH (formic acid),

$$OH + HCHO \rightarrow H_2C(O)OH \rightarrow HCOOH + H \tag{5.51}$$

has been made (D'Anna et al. 2003). Figure 5.6 depicts the energy diagram for the OH and HCHO obtained by quantum chemical calculation of D'Anna et al. (2003). According to this calculation, there is no energy barrier for the transition state of the H atom abstraction reaction of OH and HCHO, while the transition state of the addition reaction has a positive barrier of 30 kJ mol^{-1}, agreeing with the experimental evidence that HCOOH is not seen in the product.

In the case of the reaction of OH and CH_3CHO, two possibilities of H atom abstraction,

$$OH + CH_3CHO \rightarrow H_2O + CH_3CO \quad \Delta H^\circ_{298} = -123 \text{ kJ mol}^{-1} \tag{5.52}$$

$$\rightarrow H_2O + CH_2CHO \quad \Delta H^\circ_{298} = -133 \text{ kJ mol}^{-1} \tag{5.53}$$

can be conceived, but it has been known that a major process is the H atom abstraction from aldehyde group given in (5.52). Cameron et al. (2002) reported

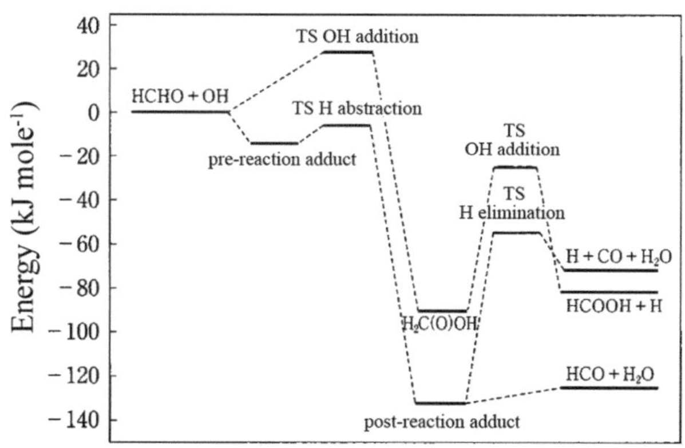

Fig. 5.6 Energy diagram for the reaction system of OH + HCHO (Adapted from D'Anna et al. 2003)

the yield of this reaction as (93 ± 18)% based on the direct measurement of CH_3CO radicals. Furthermore, the ratio of H atom abstraction from CH_3 group expressed by the reaction of (5.53) has been reported as ca. 5.1 % based on the direct measurement of CH_2CHO radicals by Butkovskaya et al. (2004).

The HCO, CH_3CO, and CH_2CHO radicals formed in the reactions (5.50), (5.52), and (5.53) react with O_2 in the atmosphere to form peroxy radicals,

$$HCO + O_2 \rightarrow HO_2 + CO \quad \Delta H^\circ_{298} = -139 \text{ kJ mol}^{-1} \tag{5.54}$$

$$CH_3CO + O_2 + M \rightarrow CH_3C(O)O_2 + M \quad \Delta H^\circ_{298} = -162 \text{ kJ mol}^{-1} \tag{5.55}$$

$$CH_2CHO + O_2 \rightarrow HCHO + HO_2 \quad \Delta H^\circ_{298} = -76 \text{ kJ mol}^{-1} \tag{5.56}$$

Here, acetylperoxy radicals formed in the reaction (5.55) reacts with NO_2 in polluted atmosphere to give peculiar compound called peroxyacetyl nitrate (PAN) $CH_3C(O)O_2NO_2$ (see Sect. 7.2.9).

The temperature dependence of the reaction rate constant of OH and HCHO has been measured by Atkinson and Pitts (1978), Stief et al. (1980), Sivakumaran et al. (2003), etc. and the IUPAC subcommittee (Atkinson et al. 2006) recommends the Arrhenius formula based on these results,

$$k_{5.50}(T) = 5.4 \times 10^{-12} \exp\left(\frac{135}{T}\right) \text{cm}^3 \text{molecule}^{-1} \text{s}^{-1}. \quad (200 - 300 \text{ K})$$

As seen in the above formula, small negative activation energy is obtained agreeing with the result of theoretical calculation for the abstraction reaction (D'Anna et al. 2003). However, it should be noted that the rate constants deviate from the above Arrhenius formula toward the positive activation energy at higher temperature than 330 K, and the above formula should be applied only at atmospheric conditions (Atkinson et al. 2006).

The similar Arrhenius formula for the reaction of CH_3CHO is recommended as,

$$k_{5.52+5.53}(T) = 4.6 \times 10^{-12} \exp\left(\frac{350}{T}\right) \text{cm}^3 \text{molecule}^{-1} \text{s}^{-1}. \quad (200 - 300 \text{ K})$$

by the IUPAC subcommittee (Atkinson et al. 2006) based on Sivakumaran and Crowley (2003), etc. This value agrees well with the recommendation by NASA/JPL panel evaluation No. 17 which includes the recent result of Zhu et al. (2008).

There are no large differences in the pre-exponential factor and small negative activation energy in the reaction rate constants of OH with C_1 and C_2 aldehydes, which is very much different from the case of OH reaction with alkanes and alkenes. The rate constants for HCHO and CH_3CHO at 298 K are given as $k_{5.50}(298 \text{ K}) = 8.5 \times 10^{-12}$, $k_{5.52+5.53}(298 \text{ K}) = 1.5 \times 10^{-11}$ cm^3 molecule^{-1} s^{-1}, respectively, both by the IUPAC subcommittee report Vol. II (Atkinson et al. 2004) and NASA/JPL panel evaluation No. 17 (Sander et al. 2011). Therefore, the reaction of OH with aldehydes has large rate constants at room temperature even for HCHO with the least carbon number.

5.3 Reactions of HO_2, CH_3O_2 Radicals

HO_2 radicals are one of the major chemical species constituting chain reaction cycles, together with OH radicals in the troposphere and stratosphere. In the troposphere, other than HO_2, organic peroxy radicals RO_2 are also important chain propagation intermediates. In this section, among the reactions of peroxy radicals, the reaction of HO_2 and O_3, which affects formation and dissipation of O_3 in the stratosphere and troposphere, HO_2 and CH_3O_2 with NO, which is directly related to the tropospheric ozone formation, and radical-radical reaction of HO_2 and CH_3O_2, are taken up. Table 5.4 cites rate constants and their temperature dependence for the reactions HO_2, CH_3O_2 and other organic radicals extracted from the IUPAC subcommittee report Vol. II (Atkinson et al. 2004), and the NASA/JPL panel evaluation No. 17 (Sander et al. 2011).

5.3.1 *$HO_2 + O_3$*

The ,reaction of HO_2 and O_3 is important in the stratosphere as a reaction to convert HO_2 to OH in the HO_x cycle. In the troposphere, it is a main reaction to convert HO_2 to OH in the marine boundary layer and free troposphere where the concentration of NO is low.

The pathway of the $HO_2 + O_3$ reaction is,

$$HO_2 + O_3 \rightarrow OH + 2\ O_2. \quad \Delta H^\circ_{298} = -118\ \text{kJ mol}^{-1} \tag{5.57}$$

As for the reaction rate constants, the NASA/JPL panel evaluation No. 17 (Sander et al. 2011) recommends the Arrhenius formula,

$$k_{5.38}(T) = 1.0 \times 10^{-14} \exp\left(-\frac{490}{T}\right) \text{cm}^3\text{molecule}^{-1}\text{s}^{-1},$$

and the IUPAC subcommittee (Atkinson et al. 2004) recommends the temperature dependent rate formula,

$$k_{5.38}(T) = 2.0 \times 10^{-16} \left(\tfrac{T}{300}\right)^{4.57} \exp\left(\frac{693}{T}\right) \text{cm}^3\text{molecule}^{-1}\text{s}^{-1} \quad (250 - 340\ \text{K})$$

based on the measurements by Zahniser and Howard (1980), Sinha et al. (1987),

Table 5.4 Rate constants at 298 K and Arrhenius parameters for the reactions of HO_2 and organic radicals

Reactions	k(298 K) (cm^3 molecule^{-1} s^{-1})	A Factor (cm^3 molecule^{-1} s^{-1})	E_a/R (K)	Ref.
$HO_2 + O_3 \rightarrow OH + 2O_2$	1.9×10^{-15}	1.0×10^{-14}	490	(b)
$HO_2 + NO \rightarrow OH + NO_2$	8.0×10^{-12}	3.3×10^{-12}	−270	(b)
$CH_3O_2 + NO \rightarrow CH_3O + NO_2$	7.7×10^{-12}	2.3×10^{-12}	360	(a)
$HO_2 + NO_2 + M \rightarrow HO_2NO_2 + M$	2.0×10^{-31} [N$_2$] (k_0)	2.0×10^{-31} $(T/300)^{-3.4}$ [N$_2$] (k_0)		(b)
	2.9×10^{-12} (k_∞)	2.9×10^{-12} $(T/300)^{-1.1}$ (k_∞)		
$HO_2 + HO_2 \rightarrow H_2O_2 + O_2 + M \rightarrow HO_2NO_2 + M$	1.4×10^{-12}	3.0×10^{-13}	−460	(b)
	4.6×10^{-32} [M]	2.1×10^{-33} [M]	−920	
$HO_2 + CH_3O_2 \rightarrow CH_3OOH + O_2$	5.2×10^{-12}	4.1×10^{-13}	−750	(b)
$CH_3O_2 + CH_3O_2 \rightarrow$ products	3.5×10^{-13}	1.0×10^{-13}	−370	(a)
$HCO + O_2 \rightarrow HO_2 + CO$	5.1×10^{-12}	5.1×10^{-12}	0	(a)
$CH_3CO + O_2 + M \rightarrow CH_3C(O)O_2 + M$	5.1×10^{-12} (k_∞)	5.1×10^{-12}	0	(a)
$CH_2OH + O_2 \rightarrow HCHO + HO_2$	9.7×10^{-12}	–	–	(a)
$CH_3O + O_2 \rightarrow HCHO + HO2$	1.9×10^{-15}	7.2×10^{-14}	1080	(a)
$CH_3O + NO + M \rightarrow CH_3ONO + M$	2.6×10^{-29} [N$_2$] (k_0)	2.6×10^{-29} $(T/300)^{-2.8}$ [N$_2$] (k_0)		(a)
	3.3×10^{-11} (k_∞)	3.3×10^{-11} $(T/300)^{-0.6}$ (k_∞)		
$CH_3O + NO_2 + M \rightarrow CH_3ONO + M$	8.1×10^{-29} [N$_2$] (k_0)	8.1×10^{-29} $(T/300)^{-4.5}$ [N$_2$] (k_0)		(a)
	2.1×10^{-11} (k_∞)	2.1×10^{-11} (k_∞)		

(a) IUPAC subcommittee report Vol. II (Atkinson et al. 2006)
(b) NASA/JPL panel evaluation No. 17 (Sander et al. 2011)

Wang et al. (1988), Herndon et al. (2001), etc. The rate constants at 298 K is $k_{5.57}$ (298 K) = 2.0×10^{-15} cm^3 molecule^{-1} s^{-1}, which is relatively slow as compared to the reaction of OH + O_3 (see 5.2.1) by an order of magnitude. Furthermore, ,the Arrhenius plot shows a curved feature as shown in the above formula, which is more pronounced under 250 K. This implies the activation energy decreases at low temperature.

As for the reaction mechanism this reaction, Sinha et al. (1987) detected OH by LIF in the experiments using isotope labeled $H^{18}O_2$ and $^{16}O_3$, and found that most of OH (75 ± 10 %) is ^{16}OH. This means that the hydrogen abstraction by O_3 is the main reaction. Nelson and Zahniser (1994) determined the formation ratio of ^{16}OH and ^{18}OH by LIF in a similar experiment, which showed the H-atom abstraction by O_3 occurs 94 ± 5 and 85 ± 5 % at 226 K and 355 K, respectively, indicating that the temperature dependence is small.

Quantum chemical calculation for the reaction of HO_2 and O_3 has been performed to show that the energy barrier for the pathway of H-atom abstraction by forming HO_3 via $O_3 - HO_2$ complex is lower than that of O-atom abstraction via $O_3 - O_2H$, which agrees well with the branching, ratio obtained by the experiments (Xu and Lin 2007; Varandas' and Viegas 2011).

5.3.2 HO_2 + NO

The reaction of HO_2 and NO to form OH and NO_2 is the important reaction to complete the OH – HO_2 chain reaction in the troposphere and stratosphere. Particularly in the troposphere, it is the fundamental reaction to produce photochemical ozone together with the reaction of RO_2 and NO that will be described later (see Sect. 7.3.1 and 7.3.2).

The rate constants of the reaction of HO_2 + NO,

$$HO_2 + NO \rightarrow HO + NO_2 \quad \Delta H^\circ_{298} = -35 \text{ kJ mol}^{-1} \tag{5.58}$$

have been measured in many studies since the first measurement of absolute rate using LMR (Laser Magnetic Resonance) by Howard and Evenson (1977). The studies after 1990 are Jemi-Alade and Thrush (1990), Seeley et al. (1996a), Bohn and Zetzsch (1997), Bardwell et al. (2003) and others. The NASA/JPL panel evaluation No. 17 (Sander et al. 2011) recommends the value at 298 K as $k_{5.58}$ (298 K) = 8.0×10^{-12} cm^3 molecule^{-1} s^{-1}, and the temperature dependence as,

$$k_{5.58}(T) = 3.3 \times 10^{-12} \exp\left(\frac{270}{T}\right) \text{cm}^3\text{molecule}^{-1}\text{s}^{-1}$$

The IUPAC subcommittee recommends a 10 % larger value at 200–400 K (Atkinson et al. 2004). It has been known that the reaction has negative activation energy and does not have pressure dependence.

Studies on potential energy ,surface of the reaction of HO_2 and NO by quantum chemical calculation has been made by Sumathi and Peyerimhoff (1997), Chakraborty et al. (1998), Zhang and Donahue (2006), etc. According to these studies, the reaction of HO_2 and NO proceeds the HOONO intermediate, and the possibilities of reaction pathways other than reaction (5.58) to form $HONO_2$, and HNO,

$$HO_2 + NO \rightarrow HOONO^{\dagger} \rightarrow HO + NO_2 \tag{5.58}$$

$$\rightarrow HONO_2 \tag{5.59}$$

$$\rightarrow HNO + O_2 \tag{5.60}$$

are also suggested. Particularly concerned is that the pernitric acid (HOONO) is a common intermediate with the reaction of OH and NO_2 described in Sect. (5.2.4), and it is interesting to see if the reaction of $HO_2 + NO$ partially leads to formation of nitric acid ($HONO_2$).

In this regard, Butkovskaya et al. (2005, 2007, 2009) reported the formation of $HONO_2$ in the $HO_2 + NO$ reaction experimentally. Butkovskaya et al. (2009) detected $HONO_2$ by using chemical ionization mass spectrometer, and showed that the reaction rate to form $HONO_2$ increases with the decrease of temperature and the increase of pressure in the range of 223–323 K, and 72–600 Torr, respectively, in the absence of water vapor. The ratio of reaction (5.58) in the absence of water vapor, $\alpha\ (T, P) = k_{5.59}/\ k_{5.58}$, is given, by,

$$\alpha(T, P) = \frac{k_{5.59}}{k_{5.58}} = \frac{530}{T} + 6.4 \times 10^{-4} P(torr) - 1.731. \tag{5.61}$$

In the presence of water vapor, the increase ratio f of $HONO_2$formation is represented as,

$$f = \left(1 + 2 \times 10^{-17} [H_2O]\right) \tag{5.62}$$

where the unit of $[H_2O]$ is molecule cm^{-3}. From this equation the yield of $HONO_2$ under the relative humidity of 50 % at 298 K ($[H_2O] = 4 \times 10^{17}$ molecule cm^{-3}) increases by a factor of 8 as compared to the case of $[H_2O] = 0$. If the increase of the reaction by water vapor is considered to be due to the reaction involving $HO_2 \cdot H_2O$ complex,

$$HO_2 \cdot H_2O + NO \rightarrow HONO_2 + H_2O, \tag{5.63}$$

the rate constant of reaction (5.63) is deduced to be $k_{5.63} = 6 \times 10^{-13}$ cm^3 molecule^{-1} s^{-1} at 298 K and 1 atm., which is larger than the rate constant of , CH_3O_2 Radicalsreaction (5.59) in the absence of water vapor by a factor of 40. Butkovskaya et al. (2009) discussed the possibility of contribution of heterogeneous reaction is low, and suggested that the formation of $HONO_2$ could be important in the reaction of $HO_2 \cdot H_2O$ complex with NO in the lower troposphere.

Quantum ,chemical calculation for the $HO_2 + NO$ reaction has also been reported by Zhu and Lin (2003a) and the major reaction pathway is the $OH + NO_2$ by the direct decomposition of HOONO, and the isomerization of HOONO to $HONO_2$ is disadvantageous energetically by 21.7 kJ mol^{-1}. The theoretical calculation shows that there is no pressure dependence for the $OH + NO_2$ reaction under 10 atm, agreeing with experimental results.

5.3.3 $CH_3O_2 + NO$

The reactions of CH_3O_2 and other organic peroxy radicals with NO are also important in tropospheric chemistry as HO_x chain propagation reaction together with HO_2 and NO mentioned in the previous paragraph. Particularly in urban and forest air where anthropogenic and biogenic hydrocarbon concentration is high, the contribution of the reaction of $RO_2 + NO$ is large for local photochemical ozone formation. Here, CH_3O_2 is picked up as a representative organic peroxy radical, and other alkyl peroxy radicals will be mentioned where appropriate.

The products, of the reaction of CH_3O_2 and NO is known to be CH_3O and NO_2 (Ravishankara et al. 1981; Zellner et al. 1986; Bacak et al. 2004), so that the main reaction path is,

$$CH_3O_2 + NO \rightarrow CH_3O + NO_2 \quad \Delta H^\circ_{298} = -49 \ \text{kJ mol}^{-1}. \tag{5.64}$$

Many measurements of the, rate constants of this reaction has been reported, and the IUPAC subcommittee (Atkinson et al. 2006) recommends $k_{5.64}$ (298 K) = 7.7×10^{-12} cm^3 molecule^{-1} s^{-1} at 298 K and temperature dependent formula,

$$k_{5.64}(T) = 2.3 \times 10^{-12} \exp\left(\frac{360}{T}\right) \text{cm}^3 \text{molecule}^{-1} \text{s}^{-1}$$

based on Scholtens et al. (1999), Bacak et al. (2004) and previous values. As shown in the above formula, this reaction has negative activation energy and the rate constants are known to increase with pressure a low temperature. From these evidence, the pathways of this reaction is thought to be similar to the $HO_2 + NO$ reaction (Scholtens et al. 1999),

$$CH_3O_2 + NO \leftrightarrow [CH_3OONO]^\dagger \tag{5.65}$$

$$[CH_3OONO]^\dagger \rightarrow CH_3O + NO_2 \tag{5.66}$$

$$+ M \rightarrow CH_3ONO_2. \tag{5.67}$$

The scheme, of isomerization reaction of CH_3OONO to CH_3ONO_2 is also expected from the quantum chemical calculation by Zhang et al. (2004). However, the formation of CH_3ONO_3 is not seen experimentally, and the upper limit is given as 0.3 % by Scholtens et al. (1999).

The yields of nitric ester (alkyl nitrate) in the above scheme is known to become larger as the number of carbon atoms in the alkyl radical increases. For example, the reaction of ethyl peroxy radical, $C_2H_5O_2$ and NO,

$$C_2H_5O_2 + NO \rightarrow C_2H_5O + NO_2 \quad \Delta H^\circ_{298} = -45.2 \text{ kJ mol}^{-1} \tag{5.68}$$

$$\rightarrow C_2H_5ONO_2 \quad \Delta H^\circ_{298} = -217.0 \text{ kJ mol}^{-1} \tag{5.69}$$

gives the yield of $C_2H_5ONO_2$ as $k_{5.69}/k_{5.68+5.69} \leq 0.014$ at 298 K (Ranschaert et al. 2000), and the yields of $n-C_3H_7ONO_2$ and $i-C_3H_7ONO_2$ for $n-C_3H_7O_2$ and $i-C_3H_7O_2$ as 0.020 and 0.042, respectively (Carter and Atkinson 1989) as recommended by the IUPAC subcommittee report Vol. II (Atkinson et al. 2006). The production yield of $RONO_2$ for each alkane has been tabulated in Lightfoot et al. (1992), Tyndall et al. (2001) and Finlayson-Pitts and Pitts (2000). The production ratios increase for higher than C_4 alkanes, and reach over 0.3 for C_7 and C_8.

In general, the reaction of RO_2 and NO,

$$RO_2 + NO \rightarrow RO + NO_2 \tag{5.70}$$

$$\rightarrow RONO_2 \tag{5.71}$$

forms RO radicals and $RONO_2$. From the RO radicals HO_2 is regenerated by

$$RO + O_2 \rightarrow R'CHO + HO_2 \tag{5.72}$$

to propagate the chain reaction, while the formation of $RONO_2$ act as the termination of the chain reaction. Therefore, the production yield of $RONO_2$ in the reaction of RO_2 and NO is a very important parameter affecting the photochemical production efficiency of each hydrocarbon.

Quantum chemical ,calculation for the reaction of RO_2 ($R = CH_3$, C_2H_5, $n-C_3H_7$, $i-C_3H_7$, $2-C_5H_{11}$) + NO has been conducted by Lohr et al. (2003) and Barker et al. (2003). According to these theoretical calculations, the main products of the reaction are $RO + NO_2$, and the energy barrier for the isomerization from ROONO to $RONO_2$ is very high and could not explain the experimentally obtained yield of $RONO_2$.

5.3.4 $HO_2 + NO_2 + M$

The production of peroxy nitric acid (HO_2NO_2) by the reaction of HO_2 and NO_2 has a possibility to act as the termination reaction of $OH - HO_2$ radical chain reaction, but since HO_2NO_2 is thermally unstable and decomposes back to HO_2 and NO_2 at around the room temperature, it generally does not affect the formation of ozone near the surface. However, when HO_2NO_2 is formed in the troposphere at low temperature, it acts as a reservoir species to transport NO_x to long distance so that it is important to include this reaction in chemical transport models.

The reaction of HO_2 and NO_2 is in the thermal equilibrium,

$$HO_2 + NO_2 + M \rightarrow HO_2NO_2 + M \quad \Delta H^\circ_{298} = -101 \ \text{kJ mol}^{-1} \tag{5.73}$$

$$HO_2NO_2 + M \rightarrow HO_2 + NO_2 + M \quad \Delta H^\circ_{298} = 101 \ \text{kJ mol}^{-1} \tag{5.74}$$

The reaction (5.73) is the termolecular reaction, and the low- and high-pressure limit rate constants have been described by Eq. (5.6) taking e.g. $F_c = 0.6$. The NASA/JPL evaluation No. 17 (Sander et al. 2011) recommends,

$$k_{0,5.73}(T) = 2.0 \times 10^{-31} \left(\tfrac{T}{300}\right)^{-3.4} \text{cm}^6\text{molecule}^{-2}\text{s}^{-1}$$
$$k_{\infty,5.73}(T) = 2.9 \times 10^{-12} \left(\tfrac{T}{300}\right)^{-1.1} \text{cm}^3\text{molecule}^{-1}\text{s}^{-1}$$

based on the data of Christensen et al. (2004). Meanwhile for the reverse reaction the IUPAC subcommittee (Atkinson et al. 2004) recommends,

$$k_{0,5.74}(T, N_2) = 4.1 \times 10^{-5}[N_2]\left(-\frac{10650}{T}\right) \text{s}^{-1}$$
$$k_{\infty,5.74}(T) = 4.8 \times 10^{15} \exp\left(-\frac{11170}{T}\right) \text{s}^{-1}$$

for the ,temperature range in 260–300 K based on the data of Graham et al. (1977) and Zabel (1995). The values at 298 K are $k_{0,\ 5.74} = 1.3 \times 10^{-20}[N_2]\ \text{s}^{-1}$ and $k_{\infty,\ 5.74} = 0.25\ \text{s}^{-1}$, and from these values the atmospheric lifetime of HO_2NO_2 can be estimated to be a few seconds. From the rate constants of these reverse reactions and reaction (5.38), the NASA/JPL panel (Sander et al. 2011) recommends the equilibration constant of reaction (5.73) and (5.74) as,

$$k_{5.73/5.74}(T) = 2.1 \times 10^{-27} \exp\left(-\frac{10900}{T}\right) \text{cm}^3\text{molecule}^{-1}$$

and the equilibrium constant $K_{5.73/5.74}$ (298 K) $= 1.6 \times 10^{-11}$ cm^3 molecule^{-1} at 298 K.

Although quantum chemical calculation by Bai et al. (2005) suggested the formation pathway of HONO, the production of HONO in this reaction has not been confirmed experimentally (Dransfield et al. 2001).

5.3.5 $HO_2 + HO_2$ (+ M)

The ,self-reaction of HO_2 is important in the radical chain reaction in the troposphere and stratosphere since the concentration of HO_2 is in general the highest among atmospheric radicals. H_2O_2 formed in this reaction act as an important oxidizing agent in the liquid phase oxidation reaction of SO_2 in cloud and fog (see Sect. 7.6.2).

The self-reaction of HO_2 is known to proceed both via bimolecular and termolecular reaction in parallel (Kircher and Sander 1984; Kurylo et al. 1986; Takacs and Howard 1986; Lightfoot et al. 1988),

$$HO_2 + HO_2 \rightarrow H_2O_2 + O_2 \quad \Delta H^\circ_{298} = -166 \text{ kJ mol}^{-1} \tag{5.75}$$

$$+ M \rightarrow H_2O_2 + O_2 + M \quad \Delta H^\circ_{298} = -166 \text{ kJ mol}^{-1} \tag{5.76}$$

The recommended values of rate constants at 298 K by the IUPAC subcommittee (Atkinson et al. 2004) is $k_{5.75}$ (298 K) = 1.6×10^{-12} cm^3 molecule^{-1} s^{-1} for the bimolecular reaction and $k_{5.76}$ (298 K) = 5.2×10^{-32} [N_2], 4.6×10^{-32} [O_2] cm^3 molecule^{-1} s^{-1} for the termolecular reaction. The NASA/JPL panel evaluation (Sander et al. 2011) recommends $k_{5.75}$ (298 K) = 1.4×10^{-12} cm^3 molecule^{-1} s^{-1} for the bimolecular reaction, and $k_{5.76}$ (298 K) = 4.6×10^{-32} [M] cm^3 molecule^{-1} s^{-1} for the termolecular reaction in common for N_2, O_2 as the third body. They also recommend the temperature dependent rate equation for the bimolecular and termolecular reaction as,

$$k_{5.75}(T) = 3.0 \times 10^{-13} \exp\left(\frac{460}{T}\right) \text{cm}^3\text{molecule}^{-1}\text{s}^{-1}$$

$$k_{5.76}(T) = 2.1 \times 10^{-33} [M] \exp\left(\frac{920}{T}\right) \text{cm}^3\text{molecule}^{-1}\text{s}^{-1},$$

respectively. Thus, this reaction has negative activation energy and is not very fast as a radical-radical reaction.

The rate constants of this reaction is known to increase in the presence of water vapor. The effect of water vapor for this reaction was found by Hamilton (1975) for the first time, and reaction rate equation including the temperature dependence has been given by Lii et al. (1981) and Kircher and Sander (1984). The increasing factor $f_{5.50}$ in the presence of water vapor given by Kircher and Sander (1984) is,

$$f_{5.75+5.76}(T) = 1 + 1.4 \times 10^{-21} [H_2O] \exp\left(\frac{2200}{T}\right) \text{cm}^3\text{molecule}^{-1}\text{s}^{-1},$$

which is also ,recommended by the IUPAC subcommittee (Atkinson et al. 2004).

The effect of water vapor for the $HO_2 + HO_2$ reaction has been studied in detail both theoretically and experimentally, and it has been revealed that HO_2 forms a complex $HO_2 - H_2O$ with H_2O as,

$$HO_2 + H_2O \rightarrow HO_2{\cdot}H_2O \tag{5.77}$$

and the reaction rate constant of this complex with HO_2,

$$HO_2 + HO_2{\cdot}H_2O \rightarrow H_2O_2 + O_2 + H_2O \tag{5.78}$$

is faster than the rate constant of $HO_2 + HO_2$ (Aloisio and Francisco 1998; Aloisio et al. 2000; Zhu and Lin 2002). Kanno et al. (2005, 2006) determined the equilibrium constant for HO_2 and $HO_2 - H_2O$ from the decrease of infrared absorption of HO_2 radicals in the presence of H_2O and obtained the temperature dependent rate constant of reaction (5.78),

$$k_{5.52}(T) = 5.4 \times 10^{-11} \exp\left(-\frac{410}{T}\right) \text{cm}^3 \text{molecule}^{-1} \text{s}^{-1}$$

from the experimental decay rate of HO_2 by subtraction the contribution due to reaction (5.50). The value of $k_{5.78}$ at 298 K is 1.4×10^{-11} cm^3 molecule^{-1} s^{-1}, which is one order of magnitude larger than the rate constant of reaction (5.50) without H_2O. The NASA/JPL panel (Sander et al. 2011) recommends this value for the rate constant of reaction (5.78).

The enthalpy ΔH°_{298} and entropy ΔS°_{298} of the reaction (5.78), calculated from the equilibrium constant obtained by Kanno et al. (2006) are -31 ± 4 kJ mol^{-1} and -83 ± 14 mol^{-1} K^{-1}, respectively, agreeing well with previous experimental (Aloisio et al. 2000) and theoretical (Hamilton and Naleway 1976; Aloisio and Francisco 1998) values. From the obtained equilibrium constant, the concentration ratio of $[HO_2 - H_2O]/[HO_2]$ is estimated as 0.19 ± 0.11 at a relative humidity of 50 % at 297 K (Kanno et al. 2005).

The quantum ,chemical calculation for the self-reaction of HO_2 has been reported by Zhu and Lin (2002). They showed the pathways though both of singlet and triplet potential surface are possible and a H_2O molecule decreases the potential barrier in both paths.

5.3.6 $HO_2 + CH_3O_2$

In the free troposphere with low NO_x concentration, the chain termination reaction by the cross radical reaction between HO_2 and CH_3O_2 formed in the oxidation of CH_4 is important in addition to the self–reaction of HO_2. In the polluted atmosphere where concentrations of organic peroxy radicals (RO_2) are high, their cross radical reactions with HO_2 also need to be considered in the model calculation of photochemical ozone formation. Here, as a representative radical-radical reaction of RO_2, the reaction of HO_2 and CH_3O_2 is described.

The two reaction pathways of HO_2 and CH_3O_2 are known to be,

$$HO_2 + CH_3O_2 \rightarrow CH_3OOH + O_2 \quad \Delta H^\circ_{298} = -155 \text{ kJ mol}^{-1} \tag{5.79}$$

$$\rightarrow HCHO + H_2O + O_2 \quad \Delta H^\circ_{298} = -374 \text{ kJ mol}^{-1}. \tag{5.80}$$

The scatters of previous experimental values of the rate constants of this reaction were mainly due to the scatters of UV absorption cross-section of the reactant radicals. The IUPAC subcommittee (Atkinson et al. 2006) and NASA/JPL panel (Sander et al. 2011) recommends the rate constant at 298 K and temperature dependence equation based on Cox and Tyndall (1980), Dagaut et al. (1988), Lightfoot et al. (1991), Boyd et al. (2003), etc., and a review of Tyndall et al. (2001). The recommended value by the both evaluation for 298 K is $k_{5.79+5.80}$ (298 K) = 5.2×10^{-12} cm^3 molecule^{-1} s^{-1}. The reaction rate constant at 298 K is shown to be pressure independent in the range of 13–1013 hPa by Lightfoot et al. (1990), etc.

The recommended ,Arrhenius formula by the NASA/JPL panel (Sander et al. 2011) is

$$k_{5.79+5.80}(T) = 4.1 \times 10^{-13} \exp\left(\frac{750}{T}\right) \text{cm}^3 \text{molecule}^{-1} \text{s}^{-1},$$

which agrees with the recommended parameters by the IUPAC subcommittee (Atkinson et al. 2006) within 10 %.

The branching ratio of the two reaction paths has been obtained by Wallington and Japar (1990) and Elrod et al. (2001). The yield of reaction (5.80) is small at 298 K, $k_{5.80}/k_{5.53}$ (298 K) = 0.1. However, the branching ratio has been found to be temperature dependent (Elrod et al. 2001),and the IUPAC subcommittee (Atkinson et al. 2006) recommends the temperature dependent ratio as,

$$\frac{k_{5.80}}{k_{5.79+5.80}} = \frac{1}{1 + 498\exp(-1160/T)}$$

Thus, the branching ratio of reaction (5.80) increases with decreasing temperature and is 0.31 at 218 K.

According to the quantum chemical calculation for this reaction, there are singlet and triplet paths involving a complex CH_3OOOOH, which are related to the formation of singlet and triplet O_2 in reaction (5.79) (Zhou et al. 2006).

5.4 Reactions of O_3

Although ozone (O_3) is a reactive species among stable molecules in the atmosphere, partner molecules of homogeneous gas phase reaction are not so many. Important atmospheric reactions are reactions with halogen atoms in the

Table 5.5 Rate constants at 298 K and Arrhenius parameters for the reactions of ozone

Reactions	k(298 K) (cm^3 molecule^{-1} s^{-1})	A Factor (cm^3 molecule^{-1} s^{-1})	E_a/R (K)	Ref.
$O_3 + NO \rightarrow NO_2 + O_2$	1.8×10^{-14}	1.0×10^{-12}	1310	(a1)
$O_3 + NO_2 \rightarrow NO_3 + O_2$	3.5×10^{-17}	1.4×10^{-13}	2470	(a1)
$O_3 + C_2H_2 \rightarrow$ products	1.0×10^{-20}	1.0×10^{-14}	4100	(b)
$O_3 + C_2H_4 \rightarrow$ products	1.7×10^{-18}	1.2×10^{-14}	2630	(b)
$O_3 + C_3H_6 \rightarrow$ products	1.1×10^{-17}	6.5×10^{-15}	1900	(b)
$O_3 + i\text{-}C_4H_8 \rightarrow$ products	1.1×10^{-17}	2.7×10^{-15}	1630	(a2)
$O_3 + 1\text{-}C_4H_8 \rightarrow$ products	9.6×10^{-18}	3.6×10^{-15}	1750	(a2)
$O_3 + cis\text{-}2\text{-}C_4H_8 \rightarrow$ products	1.3×10^{-16}	3.2×10^{-15}	970	(a2)
$O_3 + trans\text{-}2\text{-}C_4H_8 \rightarrow$ products	1.9×10^{-16}	6.6×10^{-15}	1060	(a2)
$O_3 + C_5H_8$ (isoprene) $\rightarrow$ products	1.3×10^{-17}	1.0×10^{-14}	1970	(b)
$O_3 + C_{10}H_{16}$ (α-pinene) $\rightarrow$ products	9.0×10^{-17}	6.3×10^{-16}	580	(a2)

(a1, a2) IUPAC subcommittee report Vol. I, V (Atkinson et al. 2004; Wallington et al. 2012), respectively
(b) NASA/JPL panel evaluation No. 17 (Sander et al. 2011)

stratosphere, NO, NO_2, OH and HO_2 in both troposphere and stratosphere, and alkenes (olefins), dienes (diolefines), and biogenic cyclic hydrocarbons, etc. in the troposphere. Among these, the reactions with NO, NO_2, and C_2H_4 as a typical alkene are described in detail in this section, and the reaction with Cl atom, and biogenic hydrocarbons will be described in Sects. (5.6.1) and (7.2.6), respectively. The reactions with OH and HO_2 have already given in Sects. (5.2.1 and 5.3.1). Rate constants of fundamental O_3 reactions are summarized in Table 5.5.

5.4.1 $O_3 + NO$

The, reaction of O_3 and NO in the troposphere dissipate O_3 temporarily, and is known as a "titration reaction," which is important at near NO_xsources and in urban air. In the stratosphere, it is important as a reaction to constitute NO_x cycle to bring about the net destruction of O_3 (see Sect. 8.2.2).

The reaction of O_3 and NO can be written as,

$$O_3 + NO \rightarrow NO_2 + O_2. \quad \Delta H^\circ_{298} = -200 \text{ kJ mol}^{-1} \tag{5.81}$$

The rate constant of this reaction has been measured by many studies including Lippmann et al. (1980), Ray and Watson (1981), Borders and Birks (1982), Moonen et al. (1998). Based on these, the IUPAC subcommittee report Vol. I (Atkinson et al. 2004) and the NASA/JPL panel evaluation No. 17 (Sander et al. 2011) gives the rate constants and their temperature dependence. The IUPAC subcommittee recommends the overall reaction rate constants adding reaction (5.81) and (5.82) described below as $k_{5.81+5.82}$ (298 K) = 1.8×10^{-14} cm^3 molecule^{-1} s^{-1} at 298 K, and Arrhenius formula as,

$$k_{5.81+5.82}(T) = 1.4 \times 10^{-12} \exp\left(-\frac{1310}{T}\right) \quad \mathrm{cm^3 molecule^{-1} s^{-1}}$$

in the temperature range of 195–308 K (Atkinson et al. 2004).

Meanwhile, this reaction has been known to be a chemiluminescent reaction (Clyne et al. 1964; Clough and Thrush 1967),

$$O_3 + NO \rightarrow NO_2{}^* + O_2, \tag{5.82}$$

and the activation energy of reaction (5.82) has been shown to be larger than reaction (5.81) (Michael et al. 1981; Schurath et al. 1981). The quantum yield of $NO_2{}^*$ is obtained as 0.20 at 290 K by Schurath et al. (1981). As for the atmospheric reaction, only the overall reaction rate combining reaction (5.81) and (5.82) is important, since almost all of $NO_2{}^*$ is deactivated to the ground sate NO_2. The chemiluminescence by reaction (5.82) has been utilized for an instrument measuring O_3 in the atmosphere (Fontijn et al. 1970).

Quantum chemical calculation for the O_3 + NO reaction has been made by classical trajectory method on potential surfaces by Viswanathan and Raff (1983), and reaction cross section and internal energy distribution among products was obtained, although no recent study has been reported.

5.4.2 $O_3 + NO_2$

The reaction of O_3 and NO_2 is important as an NO_3radical forming reaction in tropospheric chemistry. The reactions of NO_3 with other atmospheric species are described in Sect. (5.5).

The reaction pathway of $O_3 + NO_2$ is,

$$O_3 + NO_2 \rightarrow NO_3 + O_2 \quad \Delta H^\circ_{298} = -102 \ \mathrm{kJ\ mol^{-1}} \tag{5.83}$$

to form NO_3. The rate constants of this reaction at 298 K and an Arrhenius formula have been recommended by the IUPAC subcommittee (Atkinson et al. 2004) as $k_{5.54}$ (298 K) = 3.5×10^{-17} cm^3 molecule^{-1}, and

$$k_{5.83}(T) = 1.4 \times 10^{-13} \exp\left(-\frac{2470}{T}\right) \mathrm{cm^3 molecule^{-1} s^{-1}}, \quad (230 - 360\ \mathrm{K})$$

respectively, based on the measurements of Davis et al. (1974), Graham and Johnston (1974), Huie and Herron (1974) and Cox and Coker (1983). This reaction has an activation energy of 20.5 kJ mole^{-1} and is rather slow at room temperature.

Quantum chemical calculation has been reported by Peiró-García and Nebot-Gil (2003b), and activation energy, reaction enthalpy, and reaction rate constants have been obtained, agreeing well with experiments.

5.4.3 $O_3 + C_2H_4$

Important reactions of O_3 in the polluted atmosphere are those with organic compounds with double bonds such as alkenes, dienes, terpenes, etc. Here, the most fundamental elemental reactions with ethylene (C_2H_4) are described. Reactions with other alkenes and biogenic hydrocarbons are treated in Sects. (7.2.4) and (7.2.6).

The rate constants of the reaction of O_3 and C_2H_4,

$$O_3 + C_2H_4 \rightarrow \text{products} \qquad (5.84)$$

have been measured in wide range of temperature. The IUPAC subcommittee (Atkinson et al. 2006) recommends the Arrhenius formula,

$$k_{5.56}(T) = 9.1 \times 10^{-15} \exp\left(-\frac{2580}{T}\right) \text{cm}^3\text{molecule}^{-1}\text{s}^{-1} \quad (180 - 360\text{K})$$

and rate constant at 298 K as $k_{5.84}$ (298 K) = 1.6×10^{-18} cm^3 molecule^{-1} based on measurements of Bahta et al. (1984) and Treacy et al. (1992). Thus, although the reaction of O_a and C_2H_4 is a relatively slow reaction with activation energy of 21.5 kJ mol^{-1}, it is important as a dissipation process of C_2H_4 and also as a radical forming reaction as described below in the polluted atmosphere where the concentrations of O_3 and C_2H_4 are high.

The activation energy of O_3 and the alkene reaction decreases with the increase of the number of carbon atoms, and reaction rate constants increase rapidly. For example, the activation energy for the reaction with propylene (C_3H_6) and α-pinene ($C_{10}H_{16}$) decreases to 15.6 and 4.8 kJ mol^{-1}, and the rate constants are 1.0×10^{-17} and 9.0×10^{-17} cm^3 molecule^{-1}, respectively, which are 1–2 orders of magnitude larger than compared to with ethylene (Atkinson et al. 2006).

The initial reaction of O_3 and alkenes are known to be the formation of carbonyl compounds and carbonyl oxide via primary ozonide formed by cyclic addition of O_3 to double bond. In the case of ethylene, the reaction formula can be represented as,

$O_3 + C_2H_4 \rightarrow$

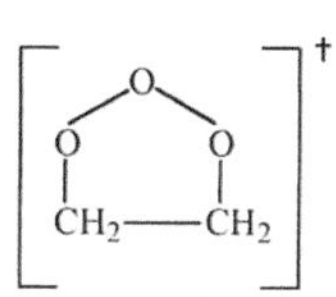

(A)

$\rightarrow HCHO + [CH_2OO]^{\dagger}$. (5.85)

Historically, this type of reaction was first proposed to interpret the formation of secondary ozonide in solvent cage in the liquid phase (Criegee 1975),

$$\mathrm{HCHO} + \mathrm{CH_2OO} \rightarrow \qquad (5.86)$$

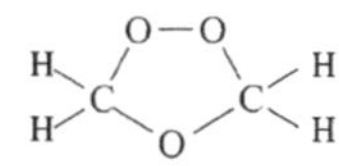

and now recognized well also in the gas phase ozone-alkene reactions (Finlayson-Pitts and Pitts 2000). The formation of primary ozonide (A) in the gas phase reaction of ethylene has been confirmed by micro-wave spectroscopy and its molecular structure has been determined (Gillies et al. 1988, 1989). Also, quantum chemical calculation for the primary ozonide has been made by McKee and Rohlfing (1989) and obtained geometrical molecular structure agrees very well with experiments. Form these experimental and theoretical studies, the structure of the primary ozonide has confirmed that two methylene planes of C_2H_4 and O-O-O plane of O_3 have parallel conformation. Taking the theoretical value for the heat of formation of primary ozonide, as -51 kJ mole^{-1} (Olzmann et al. 1997), reaction (5.85) is exothermic by 246 kJ mol^{-1} and it is expected that the formed primary ozonide has excess energy to decompose to HCHO and CH_2OO as shown in reaction (5.85). Experimentally, the production yield of HCHO is known to be unity (Grosjean and Grosjean 1996).

The RR'COO type species (CH_2OO in case of C_2H_4) formed by the decomposition of primary ozonide in the reaction of ozone-olefin reactions are in general called carbonyl oxide or Criegee intermediate after the name of Criegee who first proposed the mechanism. As for the Criegee intermediate, although the existence of the species has been well recognized, including in theoretical studies, direct measurement has not been made in the gas phase for a long time. Very recently, Taatjes et al. (2008) and Welz et al. (2012) detected CH_2OO (formaldehyde oxide) by photoionization mass spectrometer in use of synchrotron radiation directly, and its UV and IR absorption spectrum has been reported (Bearnes et al. 2012; Su et al. 2013).

In liquid phase, CH_2OO is recognized as in zwitterion structure $^{+}CH_2OO^{-}$ from its reactivity. Meanwhile, there has been a long discussion about whether CH_2OO in gas phase is in either biradical $\cdot CH_2OO\cdot$ or zwitterion structure (Wald and Goddard III 1975; Johnson and Marston 2008), and recent quantum chemical calculation showed it is a biradical with iconicity (Sander 1990; Cremer et al. 1993). Heat of formation of the ground state CH_2OO ($1A'$) has been reported as, $\Delta H^{\circ}_{f,\,298} = 1.2$ and $\Delta H^{\circ}_{f,\,0} = 12.5$ kJ mol^{-1} from a theoretical study (Nguyen et al. 2007).

The CH_2OO produced in reaction (5.85) is known to be vibrationally excited, and partially undergoes unimolecular decomposition and partially participates bimolecular reaction with other molecules under the atmospheric conditions (Atkinson et al. 2006).

$$[CH_2OO]^\dagger + M \rightarrow CH_2OO + M \quad (5.87)$$

$\rightarrow$ 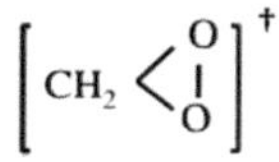 (5.88)

$$\rightarrow HCO + OH \quad (5.89)$$

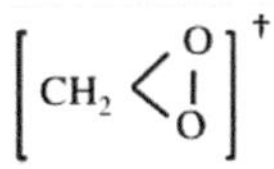 $\rightarrow [HCOOH]^\dagger \rightarrow CO_2 + H_2$ (5.90)

$$\rightarrow CO + H_2O \quad (5.91)$$

$$\rightarrow CO_2 + 2H \quad (5.92)$$

$$\rightarrow HCO + OH \quad (5.93)$$

The yield of CH_2OO stabilized under the atmospheric conditions has been previously obtained as Φ $(CH_2OO) = 0.35–0.39$ from experiments by capturing it by HCHO, SO_2 and other molecules at 1 atm and room temperature (Niki et al. 1981; Kan et al. 1981; Hatakeyama et al. 1984, 1986; Hasson et al. 2001), while recent experiments give larger values, Φ $(CH_2OO) = 0.47$-0.50 (Horie and Moortgat 1991: Neeb et al. 1996, 1998; Horie et al. 1999: Alam et al. 2011). The yield of stabilized CH_2OO is thought to be pressure dependent. Hatakeyama et al. 1986 reported Φ $(CH_2OO) = 0.20 \pm 0.03$ even at extrapolated zero pressure, which implies that a part of CH_2OO is formed without excess energy in reaction (5.85).

Decomposition products of vibrationally excited CH_2OO are known to include OH radicals and H atoms in addition to CO_2, H_2, CO, H_2O (Atkinson et al. 2006; Finlayson-Pitts and Pitts 2000). Many theoretical studies on the decomposition path of CH_2OO presume that the decomposition occurs from vibrationally excited formic acid, HCOOH, formed via dioxirane, a cyclic isomer as shown in reactions (5.90), (5.91), (5.92), and (5.93) (Anglada et al. 1996; Gutbrod et al. 1996; Olzmann et al. 1997; Qi et al. 1998).

Among these decomposition processes, formation of OH radicals and H atoms are important in atmospheric chemistry. The yield of OH has been obtained by experiments with tracers or direct measurements with LIF method (Paulson et al. 1999; Rickard et al. 1999; Kroll et al. 2001), and the IUPAC subcommittee recommends Φ $(OH) = 0.16$ (Atkinson et al. 2006). A recent value by the LIF detection using the EUPHORE chamber (see column in page 278) is Φ $(OH) = 0.17 \pm 0.09$ agreeing with the above recommendation (Alam et al. 2011). As for the reaction pathway of HCO + OH, existence of a direct decomposition path, reaction (5.89) has also been suggested in addition to the path via dioxirane mentioned above (Alam et al. 2011). Information on the OH yields in the reactions of OH with alkene other than C_2H_4 will be described in Sect. (7.2.4).

As for the yields of H atoms, they have been obtained as the yields of HO_2 since H atoms formed in reaction (5.92) react with O_2 to be transformed into HO_2 under atmospheric conditions. However, the HCO radicals formed in reaction (5.93) also produce HO_2 via reaction with O_3 by reaction (5.35), measured yields of HO_2 corresponds to the sum, $2\Phi_{5.92} + \Phi_{5.93}$. Also under certain experimental conditions, formed OH also gives HO_2 by its reaction with O_3 via reaction (5.13), which could give error to the initial yields of HO_2. The yields of HO_2 have been reported to be 0.39 ± 0.03 (Mihelcic et al. 1999), 0.38 ± 0.02 (Qi et al. 2006), 0.27 ± 0.07 (Alam et al. 2011) by the methods of low temperature matrix, chemical amplification method (PERCA), etc.

As for the stabilized CH_2OO formed in reaction (5.87), bimolecular reactions with other atmospheric molecules can occur. The bimolecular reactions so far conceived are,

$$CH_2OO + H_2O \rightarrow HCOOH + H_2O \tag{5.94}$$

$$+ NO \rightarrow HCHO + NO_2 \tag{5.95}$$

$$+ NO_2 \rightarrow HCHO + NO_3 \tag{5.96}$$

$$+ SO_2 \rightarrow HCHO + SO_3. \tag{5.97}$$

The reported rate constants of these reactions, however, used to have large uncertainties since they were indirect values assuming the above reaction mechanism for CH_2OO. Recently, Welz et al. (2012) and Stone et al. (2014) reported the following values based on the direct measurement of CH_2OO as: SO_2: 3.9×10^{-11}, 3.4×10^{-11}; NO_2: 7×10^{-12}, 1.5×10^{-12}; NO: $< 6 \times 10^{-14}$, $< 2 \times 10^{-13}$; H_2O: $< 4 \times 10^{-15}$, $< 9 \times 10^{-17}$ cm^3 molecule^{-1} s^{-1}, respectively. The values for SO_2 and NO_2 are large enough for these reactions to affect in the polluted atmosphere.

The reaction of $O_3$with other alkenes and derived carbonyl oxides are described in Sect. (7.2.4).

5.5 Reactions of NO_3 Radicals

Nitrate radicals (NO_3) are formed by the reaction of O_3 and NO_2 (Sect. 5.4.2) and play an important role in atmospheric chemistry at nighttime in polluted air. NO_3 has an absorption spectrum in the visible region as seen in Sect. (4.2.4) so that daytime concentration is very low since it is easily photodecomposed by sun light. Simultaneously, since the reaction rate constant of NO_3 with NO is large, it returns easily to NO_2 by NO so that its concentration near NO sources is also very low. NO_3 reacts with alkenes and aldehydes to form dinitrates and OH/HO_2 radicals at nighttime. Rate constants of fundamental reactions of atmospheric NO_3 and related N_2O_5 are cited in Table 5.6.

Table 5.6 Rate constants at 298 K and Arrhenius parameters for the reactions of NO_3 and N_2O_5

Reactions	k(298 K) (cm3 molecule-1 s-1)	A Factor (cm3 molecule-1 s-1)	Ea/R (K)	Ref.
$NO_3 + NO \rightarrow 2\ NO_2$	2.6×10^{-11}	1.8×10^{-11}	−110	(a1)
$NO_3 + NO_2 + M \rightarrow N_2O_5 + M$	3.6×10^{-30} $[N_2]$ (k_0)	3.6×10^{-30} $(T/300)^{-4.1}$ $[N_2]$ (k_0)		(a1)
	1.9×10^{-12} (k_∞)	1.9×10^{-12} $(T/300)^{0.2}$ (k_∞)		
$N_2O_5 + M \rightarrow NO_3 + NO_2$	$1.2 \times 10^{-19} [N_2]$ (k_0/s^{-1})	1.3×10^{-3} $(T/300)^{-3.5} \times \exp(-11000/T)$ $[N_2]$ (k_0/s^{-1})		(a1)
	6.9×10^{-2} (k_∞/s^{-1})	9.7×10^{14} $(T/300)^{0.1} \times \exp(-11080/T)$ (k_∞/s^{-1})		
$N_2O_5 + H_2O \rightarrow 2\ HONO_2$	2.5×10^{-22}	–	–	(a1)
$N_2O_5 + 2\ H_2O \rightarrow 2\ HONO_2 + H_2O$	1.8×10^{-39} (b)	–	–	(a1)
$NO_3 + C_2H_4 \rightarrow$ products	2.1×10^{-16}	3.3×10^{-12}	2880	(a2)
$NO_3 + C_3H_6 \rightarrow$ products	9.5×10^{-15}	4.6×10^{-13}	1160	(a2)
$NO_3 +$ i-$C_4H_8 \rightarrow$ products	3.4×10^{-13}	–	–	(a3)
$NO_3 +$ 1-$C_4H_8 \rightarrow$ products	1.3×10^{-14}	3.2×10^{-13}	950	(a3)
$NO_3 +$ cis-2-$C_4H_8 \rightarrow$ products	3.5×10^{-13}	–	–	(a3)
$NO_3 +$ trans-2-$C_4H_8 \rightarrow$ products	3.9×10^{-13}	–	–	(a3)
$NO_3 + C_5H_8$ (isoprene) $\rightarrow$ products	7.0×10^{-13}	3.2×10^{-12}	450	(a2)
$NO_3 + C_{10}H_{16}$ (α-pinene) $\rightarrow$ products	6.2×10^{-12}	1.2×10^{-12}	−490	(a2)
$NO_3 +$ n-$C_4H_{10} \rightarrow$ products	4.6×10^{-17}	2.8×10^{-12}	3280	(a3)
$NO_3 +$ i-$C_4H_{10} \rightarrow$ products	1.1×10^{-16}	3.0×10^{-12}	3050	(a3)
$NO_3 + HCHO \rightarrow HONO_2 + HCO$	5.6×10^{-16}	–	–	(a2)
$NO_3 + CH_3CHO \rightarrow HONO_2 + CH_3CO$	2.7×10^{-15}	1.4×10^{-12}	1860	(a2)
$NO_3 + C_2H_5CHO \rightarrow HONO_2 + C_2H_5CO$	6.4×10^{-15}	–	–	(a2)
$NO_3 +$ pinonaldehyde $\rightarrow$ products	2.0×10^{-14}	–	–	(a3)

(a1, a2, a3) IUPAC subcommittee report Vol. I, II (Atkinson et al. 2004, 2006), and Wallington et al. 2012), respectively
(b) Unit:cm6 molecule − 2 s − 1

5.5.1 $NO_3 + NO$

The reaction of NO_3 and NO is a simple oxygen atom transfer reaction,

$$NO_3 + NO \rightarrow 2\ NO_2. \quad \Delta H^\circ_{298} = -98\ \text{kJ mol}^{-1} \tag{5.98}$$

Measured values for reaction (5.98) by Hammer et al. (1986), Sander and Kircher (1986), Tyndall et al. (1991), Brown et al. (2000) agree very well, and based on these values the IUPAC subcommittee (Atkinson et al. 2004) recommends $k_{5.98}$ (298 K) = 2.6×10^{-11} cm^3 molecule^{-1}, and an Arrhenius formula,

$$k_{5.98}(T) = 1.8 \times 10^{-11} \exp\left(\frac{110}{T}\right) \text{cm}^3\text{molecule}^{-1}\text{s}^{-1} \quad (220 - 420\ \text{K})$$

in the above temperature range. Thus, the reaction of NO_3 and NO is a fast reaction with negative activation energy and frequency factor of one tenth of collisional frequency.

5.5.2 $NO_3 + NO_2 + M$

The reaction of NO_3 radicals and NO_2 in the nighttime polluted atmosphere removes NO_3 to form N_2O_5, which is transformed into nitric acid, $HONO_2$, by reacting with H_2O. Therefore, this reaction is important as a process removing NO_x from the chain reaction system and forms a $HONO_2$reservoir together with the daytime reaction of OH + NO_2 + M (Sect. 5.2.4).

The reaction of NO_3 and NO_2 is an equilibrium reaction,

$$NO_3 + NO_2 + M \rightarrow N_2O_5 + M \quad \Delta H^\circ_{298} = -96\ \text{kJ mol}^{-1} \tag{5.99}$$

$$N_2O_5 + M \rightarrow NO_3 + NO_2 + M, \quad \Delta H^\circ_{298} = 96\ \text{kJ mol}^{-1} \tag{5.100}$$

and the IUPAC subcommittee (Atkinson et al. 2004) recommends low- and high-pressure limit formulas of the $NO_3 + NO_2 + M$ reaction as,

$$k_{0,5.99}(T, N_2) = 3.6 \times 10^{-30}[N_2]\left(-\tfrac{T}{300}\right)^{-4.1} \text{cm}^3\text{molecule}^{-1}\text{s}^{-1} \quad (200 - 300\ \text{K})$$
$$k_{\infty,5.99}(T) = 1.9 \times 10^{-12}\left(\tfrac{T}{300}\right)^{0.2} \text{cm}^3\text{molecule}^{-1}\text{s}^{-1}, \quad (200 - 400\ \text{K})$$

respectively, by taking $F_c = 0.35$ based on Orlando et al. (1991), Hahn et al. (2000), and other previous data. Also, the recommended low- and high-pressure limit formulas of the $N_2O_5 + M$ reaction (5.100) are,

$$k_{0,5.100}(T,N_2) = 1.3 \times 10^{-3}[N_2]\left(-\frac{T}{300}\right)^{-3.5}\exp\left(-\frac{11000}{T}\right)\text{s}^{-1} \quad (200 - 400\ \text{K})$$

$$k_{\infty,5.100}(T) = 9.7 \times 10^{14}\left(\frac{T}{300}\right)^{0.1}\exp\left(\frac{-11000}{T}\right)\text{s}^{-1}, \quad (200 - 400\ \text{K})$$

respectively, based on Cantrell et al. (1993). The NASA/JPL pane evaluation No. 17 (Sander et al. 2011) recommends the equilibrium constants of reaction (5.99) and (5.100),

$$K_{5.99/5.100}(T) = 2.7 \times 10^{-27}\exp\left(-\frac{11000}{T}\right)\text{cm}^3\text{molecule}^{-1}$$

$$K_{5.99/5.100}(298\ K) = 2.9 \times 10^{-11}\ \text{cm}^3\text{molecule}^{-1}$$

as cited in Table 5.3.

The atmospheric lifetimes of N_2O_5 at room temperature and atmospheric pressure is calculated from the above formula as about 10 s, and during this lifetime N_2O_5 is converted to $HONO_2$ either by homogeneous reaction with H_2O molecule,

$$N_2O_5 + H_2O \rightarrow 2\ HONO_2 \quad \Delta H^\circ_{298} = -39\ \text{kJ mol}^{-1} \tag{5.101}$$

$$N_2O_5 + 2H_2O \rightarrow 2\ HONO_2 + H_2O \quad \Delta H^\circ_{298} = -39\ \text{kJ mol}^{-1} \tag{5.102}$$

or by heterogeneous reaction on aerosols. The homogeneous rate constants of reactions (5.101) and (5.102) are recommended as $k_{5.101}$ (290 K) = 2.5×10^{-22} cm^3 molecule^{-1} s^{-1} and $k_{5.102}$ *(290 K)* = 1.8×10^{-39} cm^6 molecule^{-2} s^{-1} by IPAC subcommittee based on the measurement of Wahner et al. (1998). In real polluted atmosphere, heterogeneous reaction is in general thought to be more important for the formation of $HONO_2$. The conversion of N_2O_5 to $HONO_2$ by heterogeneous reaction is treated in Chap. 6.

Quantum chemical theoretical calculations for the equilibrium reaction between NO_3, NO_2 and N_2O_5 have been made by Jitariu and Hirst (2000), Glendening, and Halpern (2007). Jitariu and Hirst (2000) obtained the molecular structure of N_2O_5 and transition state of the unimolecular decomposition, and suggested that the NO_3-NO_2 reaction could lead to decomposition into $NO_2 + NO + O_2$ via peroxy type complex, ONO---ONOO. Experimentally, however, the importance of the process to form $NO_2 + NO + O_2$ has not been confirmed (Sander et al. 2011).Glendening and Halpern (2007) obtained ΔH°, ΔG°, ΔS° for the reactions (5.99), (5.100) and other equilibrium reactions of nitrogen oxides by theoretical calculations, and compared with the values in NIST/JANAF (1988). As for the hydrolysis reaction (5.101) and (5.102), Hanway and Tao (1998) made theoretical calculations, and showed that there are two low-energy reaction pathways, one involves one H_2O molecule and another with two H_2O molecules. According to this calculation. it has been shown that the activation energy for the reaction with one H_2O molecule is 84 kJ mol^{-1}, while it reduces to nearly a half value for the reaction involving two H_2O molecules, and the heterogeneous hydrolysis process proceeds more efficiently than the homogeneous process.

5.5.3 $NO_3 + C_2H_4$

NO_3 radicals are known to react with alkenes and aldehydes among organic compounds. Here the reaction with C_2H_4 is described as a representative of the reactions with alkenes.

The reactions of NO_3 with C_2H_4 and other alkenes are known to be initiated by the addition, and proceed as,

$NO_3 + C_2H_4 \rightarrow$ 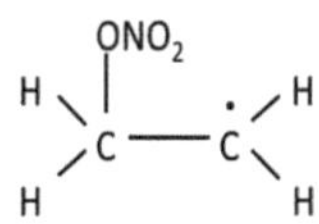 (5.103)

 (5.104)

Thus, epoxides (epoxyethane in the case of C_2H_4) are formed from the NO_3-alkene adducts (Benter et al. 1994; Skov et al. 1994). The lifetime of the adduct is long enough to allow molecular rotation around C-C bond, and cis- and trans-epoxides are formed in the same ratio irrespective of starting with either cis- or trans-asymmetrical alkenes such as cis- and trans-2-butenes (Benter et al. 1994). The yields of epoxides are reported to be 0.50, 0.95, 0.20 for cis-, tran-2-butene. The yields of epoxides are 0.50, 0.95 and 0.20 for 2,3-dimethyl-2-butene, and isoprene, respectively (Skov et al. 1994).

Meanwhile, the NO_3-alkene adduct reacts with O_2 in the atmosphere to form peroxy radical, from which dinitrate is formed,

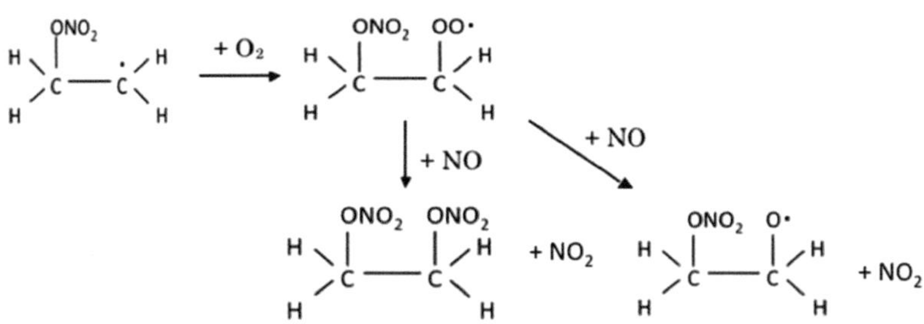 (5.105)

As for the rate constants of the reaction of NO_3 and C_2H_4, the IUPAC subcommittee report Vol. II (Atkinson et al. 2006) recommends the Arrhenius formula,

$$k_{5.103}(T) = 3.3 \times 10^{-12} \exp\left(-\frac{2880}{T}\right) \quad \text{cm}^3\text{molecule}^{-1}\text{s}^{-1} \quad (270 - 340\ \text{K})$$

and the rate constant at 298 K, $k_{5.103}(298\ K) = 2.1 \times 10^{-16}$ cm^3 molecule^{-1} s^{-1} based on Canosa-Mas et al. (1988a, b) and others. Although the reaction of NO_3 and

C_2H_4 is addition reaction as described above, pressure dependence is not seen in this reaction implying the reaction is in the high-pressure limit under the atmospheric conditions. Although the reaction rate constants of NO_3 and C_2H_4 reaction is rather small with relatively high activation energy of 23.9 kJ mol^{-1}, the activation energy decreases with the increase of carbon number and the rate constants at 298 K become large as 3.5×10^{-13}, 5.7×10^{-11} and 7.0×10^{-13} cm^3 $molecule^{-1}$ s^{-1}, for cis-2-butene, 2,3-dimethyl-2-butene and isoprene, respectively (Atkinson et al. 2006).

Quantum chemical calculation for the reaction of NO_3 and C_2H_4 was made by Nguyen et al. (2011). They confirmed the picture that the initial pathway is the electrophilic addition of O atom of NO_3 to C-C double bond to form an open-chain adduct, as shown in reaction (5.103). Form this calculation, 80-90 % of the formed adduct is stabilized in this form, and the remaining 10-20 % forms epoxyethane. The calculated rate constants agreed well with the experimental values.

5.5.4 *NO_3 + HCHO*

NO_3 reacts with aldehydes other than alkenes among organic compounds. Since HO_2 radicals are formed from acyl radicals (RCO) generated in the reaction, the NO_3-aldehydes reactions are important as HO and HO_2 radical source at nighttime. In this paragraph, the reactions of NO_3 with HCHO and CH_3CHO are described as representative example of aldehydes.

The reactions of NO_3 and aldehydes are H atom abstraction from aldehyde group. In the case of HCHO and CH_3CHO, the reactions proceed as,

$$NO_3 + HCHO \rightarrow HONO_2 + HCO \quad \Delta H^\circ_{298} = -57 \text{ kJ mol}^{-1} \tag{5.106}$$

$$NO_3 + CH_3CHO \rightarrow HONO_2 + CH_3CO \quad \Delta H^\circ_{298} = -53 \text{ kJ mol}^{-1} \tag{5.107}$$

to produce formyl and acetyl radicals, HCO and CH_3CO, respectively, from which peroxy radicals, HO_2 and $CH_3C(O)OO$ are generated with the reaction with O_2 as seen in reaction (5.54), (5.55), and (5.56) (Sect. 5.2.11).

Measurements of rate constants of the reaction of NO_3 with HCHO and CH_3CHO are scarce. The IUPAC subcommittee (Atkinson et al. 2006) recommends the rate constants at 298 K for the reactions of HCHO and CH_3CHO as $k_{5.106}$ (298 K) $= 5.6 \times 10^{-16}$, and $k_{5.107}$ (298 K) $= 2.7 \times 10^{-15}$ cm^3 $molecule^{-1}$ s^{-1}, respectively, based on Cantrell et al. (1985) using DOAS for the direct measurement of NO_3. The measurement of temperature dependence has been made only for reaction (5.107) with CH_3CHO, and the IUPAC subcommittee (Atkinson et al. 2006) recommends

$$k_{5.107}(T) = 1.4 \times 10^{-12} \exp\left(-\frac{1860}{T}\right) \quad \text{cm}^3\text{molecule}^{-1}\text{s}^{-1}$$

based on Dlugokencky and Howard (1989). The activation energy given here is 15.6 kJ mol^{-1}. Measurement of temperature dependence of rate constants for the reaction with HCHO has not been reported, and if the similar value is assumed for the pre-exponential factor as the reaction of CH_3CHO, the Arrhenius formula for HCHO is assumed to be $k_{5.106}(T) = \sim 2 \times 10^{-12} \exp(-2440/T)$ cm^3 molecule^{-1} s^{-1} with larger activation energy of 20.3 kJ mol^{-1} (Atkinson et al. 2006).

Quantum chemical calculations for the reactions of NO_3 with HCHO and CH_3CHO has been made by Mora-Diez and Boyd (2002), and the transition states and their energy levels for the H-atom abstraction reaction (5.106) and (5.107) has been obtained. The activation energy for the reaction of CH_3CHO obtained theoretically is 18.5 kJ mol^{-1}, which agrees reasonably well with the experimental value of 15.6 kJ mol^{-1}.

5.6 Reactions of Cl Atoms and ClO Radicals

The reactions of halogen atoms and radicals are of fundamental importance in stratospheric chemistry (see Sects. 4.4 and 8.2, 8.3, and 8.4), and the halogen cycle is also of interest in the marine boundary layer in the troposphere (see Sect. 7.5). In this section, among the atmospheric reactions of halogen atoms and radicals, fundamental homogeneous reactions of Cl atoms and ClO radicals are described, and the reactions of bromine and iodine atoms and radicals are discussed in the more phenomenological discussions in Chaps. 7 and 8.

Table 5.7 cites the rate constants and their temperature dependence for the reactions of Cl, ClO and other halogen atoms and radicals excerpted from the NASA/JPL panel evaluation No. 17 (Sander et al. 2011), and the IUPAC subcommittee reports Vol. III (Atkinson et al. 2007).

5.6.1 $Cl + O_3$

The reaction of Cl atoms and O_3 is the direct dissipation reaction of O_3 molecules in the ClO_x cycle (Sect. 8.2.3), and is important as a rate-determining step to reduce the equilibrium concentration of ozone by CH_3Cl in the natural stratosphere.

The reaction of Cl and O_3,

$$Cl + O_3 \rightarrow ClO + O_2 \quad \Delta H^\circ_{298} = -162 \text{ kJ mol}^{-1} \tag{5.108}$$

is known to produce chlorine oxide (ClO) radical with the yield of nearly unity

Table 5.7 Rate constants and Arrhenius parameters for the reactions of halogen atoms and radicals

Reactions	k(298 K) (cm^3 $molecule^{-1}$ s^{-1})	A Factor (cm^3 $molecule^{-1}$ s^{-1})	E_a/R (K)	References
$F + O_2 + M \rightarrow FOO + M$	5.8×10^{-33} $[N_2]$ (k_0)	5.8×10^{-33} $(T/300)^{-3.9}$ $[N_2]$ (k_0)		(a)
	1.2×10^{-10} (k_∞)	1.2×10^{-10} (k_∞)		
$Cl + O_2 + M \rightarrow ClOO + M$	1.4×10^{-33} $[N_2]$ (k_0)	1.4×10^{-33} $(T/300)^{-3.9}$ $[N_2]$ (k_0)		(a)
	$1.6 \times 10^{-33}$$[O_2]$ (k_0)	1.6×10^{-33} $(T/300)^{-2.9}$ $[O_2]$ (k_0)		
$F + O_3 \rightarrow FO + O_2$	1.0×10^{-11}	2.2×10^{-11}	230	(b)
$Cl + O_3 \rightarrow ClO + O_2$	1.2×10^{-11}	2.3×10^{-11}	200	(b)
$Br + O_3 \rightarrow BrO + O_2$	1.2×10^{-12}	1.6×10^{-11}	780	(b)
$I + O_3 \rightarrow BrO + O_2$	1.2×10^{-12}	2.3×10^{-11}	870	(b)
$F + H_2 \rightarrow HF + H$	2.6×10^{-11}	1.4×10^{-10}	500	(b)
$F + CH_4 \rightarrow HF + CH_3$	6.7×10^{-11}	1.6×10^{-10}	260	(b)
$F + H_2O \rightarrow HF + OH$	1.4×10^{-11}	1.4×10^{-11}	0	(b)
$Cl + CH_4 \rightarrow HCl + CH_3$	1.0×10^{-13}	7.3×10^{-12}	1280	(b)
$FO + O \rightarrow F + O_2$	2.7×10^{-11}	–	–	(a)
$ClO + O \rightarrow Cl + O_2$	3.7×10^{-11}	2.8×10^{-11}	−90	(b)
$BrO + O \rightarrow Br + O_2$	4.1×10^{-11}	1.9×10^{-11}	−230	(b)
$IO + O \rightarrow I + O_2$	1.2×10^{-10}	–	–	(b)
$ClO + OH \rightarrow HO_2 + Cl$	1.8×10^{-11}	7.4×10^{-12}	−270	(b)
$\rightarrow HCl + O_2$	1.3×10^{-12}	6.0×10^{-13}	−230	
$BrO + OH \rightarrow$ products	3.9×10^{-11}	1.7×10^{-12}	−250	(b)
$ClO + HO_2 \rightarrow HOCl + O_2$	6.9×10^{-12}	2.6×10^{-12}	−290	(b)
$BrO + HO_2 \rightarrow HOBr + O_2$	2.1×10^{-11}	4.5×10^{-12}	−460	(b)
$IO + HO_2 \rightarrow HOI + O_2$	8.4×10^{-11}	–	–	
$FO + NO \rightarrow NO_2 + F$	2.2×10^{-11}	8.2×10^{-12}	−300	(b)
$ClO + NO \rightarrow NO_2 + Cl$	1.7×10^{-11}	6.4×10^{-12}	−290	(b)

(continued)

Table 5.7 (continued)

Reactions	k(298 K) (cm^3 $molecule^{-1}$ s^{-1})	A Factor (cm^3 $molecule^{-1}$ s^{-1})	E_a/R (K)	References
$BrO + NO \rightarrow NO_2 + Br$	2.1×10^{-11}	8.8×10^{-12}	−260	(b)
$IO + NO \rightarrow NO_2 + I$	2.0×10^{-11}	9.1×10^{-12}	−240	(b)
$FO_2 + NO \rightarrow FNO + O_2$	7.5×10^{-13}	7.5×10^{-12}	690	(b)
$ClO + NO_2 + M \rightarrow ClONO_2 + M$	1.8×10^{-31} [M] (k_0)	1.8×10^{-31} $(T/300)^{-3.4}$ [M] (k_0)		(b)
	1.5×10^{-11} (k_∞)	1.5×10^{-11} $(T/300)^{-1.9}$ (k_∞)		
$BrO + NO_2 + M \rightarrow BrONO_2 + M$	5.2×10^{-31} [M] (k_0)	5.2×10^{-31} $(T/300)^{-32}$ [M] (k_0)		(b)
	6.9×10^{-12} (k_∞)	6.9×10^{-12} $(T/300)^{-2.9}$ (k_∞)		
$ClO + ClO + M \rightarrow Cl_2O_2 + M$	1.6×10^{-32} [M] (k_0)	1.6×10^{-32} $(T/300)^{-4.5}$ [M] (k_0)		(b)
	3.0×10^{-12} (k_∞)	3.0×10^{-12} $(T/300)^{-2.0}$ (k_∞)		
$ClO + ClO \rightarrow Cl_2 + O_2$	4.8×10^{-15}	1.0×10^{-12}	1590	(a)
$\rightarrow Cl + ClOO$	8.0×10^{-15}	3.0×10^{-11}	2450	
$\rightarrow Cl + OClO$	3.5×10^{-15}	3.5×10^{-13}	1370	
$BrO + ClO \rightarrow Br + OClO$	6.0×10^{-12}	9.5×10^{-13}	−550	(b)
$\rightarrow Br + ClOO$	5.5×10^{-12}	2.3×10^{-12}	−260	
$\rightarrow BrCl + O_2$	1.1×10^{-12}	4.1×10^{-13}	−290	
$BrO + BrO \rightarrow$ products	3.2×10^{-12}	1.5×10^{-12}	−230	(b)
$BrO + IO \rightarrow$ products	6.9×10^{-11}	–	–	(b)
$IO + IO \rightarrow$ products	8.0×10^{-11}	1.5×10^{-11}	−500	(b)

(a) IUPAC subcommittee report Vol. III (Atkinson et al. 2007)
(b) NASA/JPL panel evaluationNo.17 (Sander et al. 2011)

(Atkinson et al. 2007; Sander et al. 2011). Numerous measurements of the rate constants of reaction (5.108) have been made due to its importance related to ozone layer destruction by CFCs. The NASA/JPL panel evaluation No. 17 (Sander et al. 2011) recommends the Arrhenius formula,

$$k_{5.108}(T) = 2.3 \times 10^{-11} \exp\left(\frac{200}{T}\right) \text{cm}^3 \text{molecule}^{-1} \text{s}^{-1} \quad (180 - 300 \text{ K})$$

and the value at 298 K, $k_{5.108}$ $(298\,K) = 1.2 \times 10^{-11}$ cm^3 molecule^{-1} s^{-1} based on relatively recent measurements by Nicovich et al. (1990), Seeley et al. (1996b), Beach et al. (2002), and other previous data. The recommended value of $k_{5.108}$ $(298\,K)$ by the IUPAC subcommittee (Atkinson et al. 2008) is the same as above and the Arrhenius parameters are consistent with the above formula within an error range.

Reaction (5.108) is an exothermic reaction with large excess energy, and Matsumi et al. (1996) showed the vibrationally excited ClO is formed with inverse distribution of 0.8 : 1.0 : 1.3 : 2.4 : 2.9 : 2.7 for $v'' = 1$, 2, 3, 4, 5. Castillo et al. (2011) determined potential energy surface of this reaction by quantum chemical calculation, and showed that the energy level of the transition state is lower than the reactants, which agrees well with the experimental evidence that the reaction has small negative activation energy with no energy barrier. The calculated reaction rate constants and strongly inverted distribution of vibrationally excited levels obtained by the theoretical calculation agrees reasonably well with the experiments.

5.6.2 *Cl + CH₄*

Although the CH_4 dissipation reaction by Cl atoms has an important role not only in the stratosphere but also possibly in the troposphere, the ratio of its importance relative to OH reaction has not been estimated quantitatively since the global emissions of inorganic chlorine from sea salt has not been evaluated yet.

The reaction of Cl and CH_4 is a simple H-atom abstraction,

$$\text{Cl} + \text{CH}_4 \rightarrow \text{HCl} + \text{CH}_3. \quad \Delta H^\circ_{298} = 7.6 \text{ kJ mol}^{-1} \tag{5.109}$$

The measurements of rate constant of this reaction have been made extensively including relatively recent measurements for temperature dependence (Seeley et al. 1996b; Pilgrim et al. 1997; Wang and Keyser 1999; Bryukov et al. 2002). Based on these and other measurements at 298 K, the NASA/JPL panel evaluation No. 17 (Sander et al. 2011) recommends,

$$k_{5.109}(T) = 7.3 \times 10^{-12} \exp\left(-\frac{1280}{T}\right) \mathrm{cm^3 molecule^{-1} s^{-1}} \quad (200 - 300\ \mathrm{K})$$

and the value at 298 K, $k_{5.109}$ (298 K) = 1.0×10^{-13} cm^3 molecule^{-1} s^{-1}. Recommended value of $k_{5.109}$ (298 K) by the IUPAC subcommittee (Atkinson et al. 2006) is the same as above and the Arrhenius parameters are also consistent within error range. The activation energy of this reaction is reported to be different at below and above 300 K, and the above formula is recommended for conditions at lower than room temperature.

Although the reaction of Cl + CH_4 is slightly endothermic, its rate constant at 298 K is larger than that of the OH + CH_4 reaction (5.40) described in Sect. (5.2.7) by a factor of 15 since the activation energy is smaller by 3.4 kJ and the pre-exponential is larger by a factor of 4 than the latter. Thus, this reaction may not be negligible even though the concentration of Cl is lower than OH.

The Cl + CH_4 reaction has been attracted much interest theoretically, and the effect of excitation of inner-vibrational mode of CH_4 and difference of spin-orbit state of Cl atoms ($^2P_{3/2}$, $^2P_{1/2}$) on reactivity, and the distribution of internal energy of formed HCl have been studied by molecular dynamic experiments using crossed molecular beams (Yoon et al. 2002; Bechtel et al. 2004; Zhou et al. 2004; Bass et al. 2005). Many studies on quantum chemical calculations for the potential energy surface, rate constants, molecular dynamic parameters have been made, and comparisons with experiments have been discussed (Corchado et al. 2000; Troya et al. 2002; Yang et al. 2008).

5.6.3 ClO + OH

The reaction of ClO and OH is a cross reaction between the ClO_x and HO_x cycles together with the reaction of Cl and HO_2 described in the next paragraph.

For the ClO and OH reaction, two reaction pathways,

$$\mathrm{OH + ClO \rightarrow HO_2 + Cl} \quad \Delta H^\circ_{298} = -4.9\ \mathrm{kJ\ mol^{-1}} \tag{5.110}$$

$$\mathrm{\rightarrow HCl + O_2} \quad \Delta H^\circ_{298} = -233\ \mathrm{kJ\ mol^{-1}} \tag{5.111}$$

have been confirmed experimentally. Reaction (5.110) is a chain propagating reaction by regenerating Cl atoms while reaction (5.111) is a chain termination reaction by forming metastable HCl molecules so that their function in the chain reaction is quite different. The branching ratio of these reactions has been obtained based on the direct measurements of products by Bedjanian et al. (2001), Wang and Keyser (2001a) and Lipson et al. (1999), and the temperature independent ratio for reaction (5.111) has been reported to be 0.035 ± 0.010, 0.07 ± 0.03 and 0.090 ± 0.04, respectively. The IUPAC subcommittee recommends the ratio of the formation of $HCl + O_2$ as 0.06 by taking the average of these values (Atkinson et al. 2007).

Many measurements of the rate constant of the ClO + OH reaction have been reported, and studies after the 1990s are by Lipson et al. (1999), Kegley-Owen et al. (1999), Bedjanian et al. (2001), Wang and Keyser (2001b). Based on these and previous studies, the NASA/JPL panel evaluation No. 17 recommends,

$$k_{5.110}(T) = 7.4 \times 10^{-12} \exp\left(\frac{270}{T}\right) \text{cm}^3 \text{molecule}^{-1} \text{s}^{-1} \quad (200 - 380 \text{ K})$$

$$k_{5.111}(T) = 6.0 \times 10^{-13} \exp\left(\frac{230}{T}\right) \text{cm}^3 \text{molecule}^{-1} \text{s}^{-1} \quad (200 - 380 \text{ K})$$

and $k_{5.110}$ (298 K) = 1.8×10^{-11}, $k_{5.111}$ (298 K) = 1.3×10^{-12} cm^3 molecule^{-1} s^{-1} (Sander et al. 2011). The recommendation by the IUPAC subcommittee matches well these values (Atkinson et al. 2007).

Quantum chemical calculations for the reaction of ClO + OH has been reported by Zhu et al. (2002). According to the study, this reaction proceeds mainly on the singlet potential energy surface and forms $HO_2 + Cl$ and $HCl + O_2$ ($^1\Delta$) according to reaction path (5.110) and (5.111), and the formation ratio of $HCl + O_2$ is 0.073, agreeing well with the experimental values.

5.6.4 *ClO + HO₂*

The reaction of ClO and HO_2 is also a cross reaction of ClO_x and HO_x cycles in the stratosphere as well as the reaction with OH in the previous paragraph.

As for the pathway of the ClO and HO_2 reaction,

$$HO_2 + ClO \rightarrow HOCl + O_2 \quad \Delta H^\circ_{298} = -195 \text{ kJ mol}^{-1} \tag{5.112}$$

is the sole path substantially. Other than the above pathway,

$$HO_2 + ClO \rightarrow HCl + O_3 \quad \Delta H^\circ_{298} = -65.8 \text{ kJ mol}^{-1} \tag{5.113}$$

has been considered, but several experiments have reported that the formation of HCl by reaction (5.113) has not been observed including recent study of Knight et al. (2000). Theoretical study by Nickolaisen et al. (2000) also concluded that the contribution of this reaction is zero.

The measurements of the rate constant of reaction (5.78) have been reported rather recently by Nickolaisen et al. (2000), Knight et al. (2000) and Hickson et al. (2007). The recommended Arrhenius formula and value at 298 K by the NASA/JPL panel evaluation No. 17 (Sander et al. 2011) based on these and previous studies are,

$$k_{5.112}(T) = 2.6 \times 10^{-12} \exp\left(\frac{290}{T}\right) \quad \mathrm{cm^3 molecule^{-1} s^{-1}} \quad (230 - 300\ \mathrm{K})$$

and $k_{5.78}$ (298 K) = 6.9×10^{-12} cm^3 molecule^{-1} s^{-1}. The recommendations by the IUPAC subcommittee agree mostly with the above (Atkinson et al. 2007). Thus, this reaction has nearly the same small negative activation energy and pre-exponential factor, and the rate constant at 298 K is about 50 % as compared to the ClO + OH reaction.

According to the quantum chemical calculations (Nickolaisen et al. 2000; Kaltsoyannis and Rowley 2002; Xu et al. 2003), the ClO + HO_2 reaction proceeds on a singlet potential energy surface as,

$$\mathrm{ClO + HO_2 + M \rightarrow HOOOCl + M.} \quad (5.114)$$

This reaction forms an adduct which is more stable than the reactants by 64 kJ mol^{-1}, and since there are large potential barrier to decompose into either HOCl + O_2 ($^1\Delta$) or HCl + O_3, it has been suggested that HOOOCl can have relatively long lifetime in the atmosphere. On the other hand, a reaction along the triplet energy surface gives HOCl + $O_2(^3\Sigma)$ as shown in reaction (5.112) and the small negative activation energy of 10 kJ mol^{-1} is given. These theoretical results agree well with the experimental evidence that thus reaction has small negative activation energy and the reaction path is solely (5.112) and the path (5.113) for HCl + O_3 is not seen.

5.6.5 *ClO + NO$_2$*

The formation of $ClONO_2$ by the reaction of ClO and NO_2 is the termination reaction of the ClO_x radical chain. The reaction is also a cross reaction of ClO_x and NO_x cycles in the stratospheric ozone dissipation reaction (see Sect. 8.2.3).

The reaction of ClO + NO_2 is a termolecular recombination reaction similar to OH + NO_2,

$$\mathrm{ClO + NO_2 + M \rightarrow ClONO_2 + M,} \quad \Delta H^\circ_{298} = -111.9\ \mathrm{kJ\ mol^{-1}} \quad (5.115)$$

and the rate constant is in the fall-off region under the atmospheric pressure. The measurements of the rate constants of this reaction were made mostly prior to the 1990s, and the NASA/JPL panel evaluation (Sander et al. 2011) recommends the low- and high-pressure limit formulas as,

$$k_{0,5.115}(T) = 1.8 \times 10^{-31}\left(\frac{T}{300}\right)^{-3.4}\text{cm}^3\text{molecule}^{-1}\text{s}^{-1},$$
$$k_{\infty,5.115}(T) = 1.5 \times 10^{-11}\left(\frac{T}{300}\right)^{-1.9}\text{cm}^6\text{molecule}^{-2}\text{s}^{-1},$$

respectively, based on Handwerk and Zellner (1984), Wallington and Cox (1986), Percival et al. (1997) and more previous studies. However, the high-pressure limit equation has not been obtained directly from experiments under high pressure, but from the curvilinear regression value based on the measurement under 1 atm using Eq. (5.7) with $F_c = 0.6$. The IUPAC subcommittee (Atkinson et al. 2007) recommends the temperature independent high pressure limit value, $k_{\infty,\,5.115} = 7 \times 10^{-11}$ cm^3 molecule^{-1} s^{-1}, using more plausible value of $F_c = 0.4$ based on Cobos and Troe (2003).

Quantum chemical calculations for the reaction of $ClO + NO_2$ have been performed by Kovacic et al. (2005), Zhu and Lin (2005). According to these theoretical calculations, this reaction forms $ClONO_2$ by the association of the O atom of ClO and N atom of NO_2 without energy barrier. Also, although the formation of cis-ClOONO and trans-ClOONO which correspond to HOONO in the reaction of $OH + NO_2$ is thermo-neutral or slightly exothermic, there is a large energy barrier for the isomerization from ClOONO to $ClONO_2$, and it was concluded that $ClONO_2$ may not be formed by this process.

5.6.6 ClO + ClO

The self-reaction of ClO becomes important under the condition of very high concentration of ClO such as in the stratospheric ozone hole.

The most important pathway of the ClO + ClO reaction is,

$$\text{ClO} + \text{ClO} + \text{M} \rightarrow \text{ClOOCl} + \text{M} \quad \Delta H^\circ_{298} = -75 \text{ kJ mol}^{-1} \tag{5.116}$$

to form dichlorine peroxide, ClOOCl, a dimer of ClO by the termolecular association reaction (Birk et al. 1989; Trolier et al. 1990). This termolecular reaction is in the fall-off region under the stratospheric conditions, and the low- and high-pressure expressions has been obtained by the curvilinear regression using equation (5.7). The NASA/JPL panel evaluation (Sander et al. 2011) recommends the rate equations by taking $F_c = 0.6$,

$$k_{0,5.116}(T) = 1.6 \times 10^{-32}\left(\frac{T}{300}\right)^{-4.5}\text{cm}^6\text{molecule}^{-2}\text{s}^{-1}$$
$$k_{\infty,5.116}(T) = 3.0 \times 10^{-12}\left(\frac{T}{300}\right)^{-2.0}\text{cm}^3\text{molecule}^{-1}\text{s}^{-1}$$

based on Trolier et al. (1990), Nickolaisen et al. (1994), Bloss et al. (2001), Boakes et al. (2005), and others. Meanwhile, the IUPAC subcommittee report (Atkinson et al. 2007) recommends,

$$k_{0,5.116}(T) = 2.0 \times 10^{-32}\left(\frac{T}{300}\right)^{-4}\text{cm}^6\text{molecule}^{-2}\text{s}^{-1} \quad (190-300\ \text{K})$$
$$k_{\infty,5.116} = 1.0 \times 10^{-11}\text{cm}^3\text{molecule}^{-1}\text{s}^{-1} \quad (190-300\ \text{K})$$

based on Trolier et al. (1990) and Bloss et al. (2001). Particularly, they presented the pressure independent value by taking $F_c = 0.45$ as above.

As for the ClO + ClO reaction, other than the above pathway, bimolecular processes,

$$\text{ClO} + \text{ClO} \rightarrow \text{Cl}_2 + \text{O}_2 \quad \Delta H^\circ_{298} = -203\ \text{kJ mol}^{-1} \tag{5.117}$$
$$\rightarrow \text{Cl} + \text{ClOO} \quad \Delta H^\circ_{298} = 16\ \text{kJ mol}^{-1} \tag{5.118}$$
$$\rightarrow \text{Cl} + \text{OClO} \quad \Delta H^\circ_{298} = 13\ \text{kJ mol}^{-1} \tag{5.119}$$

are known, and the NASA/JPL panel and the IUPAC subcommittee recommends the measured values by Nickolaisen et al. (1994) (Sander et al. 2011; Atkinson et al. 2007) as,

$$k_{5.117}(T) = 1.0 \times 10^{-12}\exp\left(-\frac{1590}{T}\right)\text{cm}^3\text{molecule}^{-1}\text{s}^{-1} \quad (260-390\ \text{K})$$
$$k_{5.118}(T) = 3.0 \times 10^{-11}\exp\left(-\frac{2450}{T}\right)\text{cm}^3\text{molecule}^{-1}\text{s}^{-1} \quad (260-390\ \text{K})$$
$$k_{5.119}(T) = 3.5 \times 10^{-13}\exp\left(-\frac{1370}{T}\right)\text{cm}^3\text{molecule}^{-1}\text{s}^{-1} \quad (260-390\ \text{K})$$

The rate constants at 298 K are 4.8, 8.0, and 3.5 × 10^{-15} cm^3 molecule^{-1} s^{-1} for $k_{5.117}$, $k_{5.118}$ and $k_{5.119}$, respectively, and the branching ratios of reaction (5.117), (5.118) and (5.119) are 0.39, 0.41 and 0.20, respectively. Since these bimolecular reactions have large activation energy, the rate constants are very small as a radical-radical reaction, and the termolecular reaction (5.116) is predominant by far under the stratospheric conditions. However, the ClOOCl molecule formed by the termolecular reaction is thermally unstable, and is in equilibrium with ClO,

$$\text{ClO} + \text{ClO} + \text{M} \rightarrow \text{ClOOCl} + \text{M} \tag{5.116}$$
$$\text{ClOOCl} + \text{M} \rightarrow \text{ClO} + \text{ClO} + \text{M} \tag{5.120}$$

with the equilibrium constant (Sander et al. 2011)

$$K_{5.116/5.120}(T) = 1.72 \times 10^{-27}\exp(-8650/T)\text{cm}^3\text{molecule}^{-1}.$$

Thus, the termolecular reaction (5.116) becomes less important above 285 K (Atkinson et al. 2007).

According to the quantum chemical calculation, this reaction process mainly along the singlet energy surface. As for the structure of the dimer, ClOOCl, ClOClO and ClClOO are conceived, but ClOOCl is the most stable energetically, agreeing with the experimental results (Lee et al. 1992; Zhu and Lin 2003b; Liu and Barker 2007).

References

Akagi, H., Fujimura, Y., Kajimoto, O.: Energy partitioning in two kinds of NO molecules generated from the reaction of $O(^1D)$ with N_2O: Vibrational state distributions of "new" and "old" NO's. J. Chem. Phys. **111**, 115–122 (1999)

Alam, M.S., Camredon, M., Rickard, A.R., Carr, T., Wyche, K.P., Hornsby, K.E., Monks, P.S., Bloss, W.J.: Total radical yields from tropospheric ethene ozonolysis. Phys. Chem. Chem. Phys. **13**, 11002–11015 (2011)

Aloisio, S., Francisco, J.S.: Existence of a hydroperoxy and water ($HO_2 \cdot H_2O$) radical complex. J. Phys. Chem. A **102**, 1899–1902 (1998)

Aloisio, S., Francisco, J.S., Friedl, R.R.: Experimental evidence for the existence of the $HO_2 - H_2O$ complex. J. Phys. Chem. A **104**, 6597–6601 (2000)

Anglada, J.M., Bofill, J.M., Olivella, S., Solé, A.: Unimolecular isomerizations and oxygen atom loss in formaldehyde and acetaldehyde carbonyl oxides. A theoretical investigation. J. Am. Chem. Soc. **118**, 4636–4647 (1996)

Atkinson, R., Arey, J.: Atmospheric degradation of volatile organic compounds. Chem. Rev. **103**, 4605–4638 (2003)

Atkinson, R., Pitts, J.: J. N., Kinetics of the reactions of the OH radical with HCHO and CH_3CHO over the temperature range 299–426 K. J. Chem. Phys. **68**, 3581–3590 (1978)

Atkinson, R., Baulch, D.L., Cox, R.A., Crowley, J.N., Hampson, R.F., Hynes, R.G., Jenkin, M.E., Rossi, M.J., Troe, J.: Evaluated kinetic and photochemical data for atmospheric chemistry: Volume I – gas phase reactions of Ox, HOx, NOx, and SOx species. Atmos. Chem. Phys. **4**, 1461–1738 (2004)

Atkinson, R., Baulch, D.L., Cox, R.A., Crowley, J.N., Hampson, R.F., Hynes, R.G., Jenkin, M.E., Rossi, M.J., Troe, J.: Evaluated kinetic and photochemical data for atmospheric chemistry: Volume II – gas phase reactions of organic species. Atmos. Chem. Phys. **6**, 3625–4055 (2006)

Atkinson, R., Baulch, D.L., Cox, R.A., Crowley, J.N., Hampson, R.F., Hynes, R.G., Jenkin, M.E., Rossi, M.J., Troe, J.: Evaluated kinetic and photochemical data for atmospheric chemistry: Volume III – gas phase reactions of inorganic halogens. Atmos. Chem. Phys. **7**, 981–1191 (2007)

Atkinson, R., Baulch, D.L., Cox, R.A., Crowley, J.N., Hampson, R.F., Hynes, R.G., Jenkin, M.E., Rossi, M.J., Troe, J., Wallington, T.J.: Evaluated kinetic and photochemical data for atmospheric chemistry: Volume IV – gas phase reactions of organic halogen species. Atmos. Chem. Phys. **8**, 4141–4496 (2008)

Bacak, A., Bardwell, M.W., Raventos, M.T., Percival, C.J., Sanchez-Reyna, G., Shallcross, D.E.: Kinetics of the reaction of $CH_3O_2 + NO$: A temperature and pressure dependence study with chemical ionization mass spectrometry. J. Phys. Chem. A **108**, 10681–10687 (2004)

Bahta, A., Simonaitis, R., Heicklen, J.: Reactions of ozone with olefins: Ethylene, allene, 1,3-butadiene, and *trans*-1,3-pentadiene. Int. J. Chem. Kinet. **16**, 1227–1246 (1984)

Bai, H.-T., Huang, X.-R., Wei, Z.-G., Li, J.-L., Sun, J.-Z.: Theoretical study on the reaction of HO_2 radical with NO_2 by density functional theory method. Huaxue. Xuebao. **63**, 196–202 (2005)

Balakrishnan, N., Billing, G.D.: Quantum-classical reaction path study of the reaction $O(^3P) + O_3$ $(^1A_1) \rightarrow 2O_2$ $(X^3\Sigma_g^-)$. J. Chem. Phys. **104**, 9482–9494 (1996)

Bardwell, M.W., Bacak, A., Raventos, M.T., Percival, C.J., Sanchez-Reyna, G., Shallcross, D.E.: Kinetics of the $HO_2 + NO$ reaction: A temperature and pressure dependence study using chemical ionization mass spectrometry. Phys. Chem. Chem. Phys. **5**, 2381–2385 (2003)

Barker, J.R., Lohr, L.L., Shroll, R.M., Reading, S.: Modeling the organic nitrate yields in the reaction of alkyl peroxy radicals with nitric oxide. 2. Reaction simulations. Phys. Chem. A **107**, 7434–7444 (2003)

Bartolotti, L.J., Edney, E.O.: Density functional theory derived intermediates from the OH initiated atmospheric oxidation of toluene. Chem. Phys. Lett. **245**, 119–122 (1995)

Bass, M.J., Brouard, M., Cireasa, R., Clark, A.P., Vallance, C.: Imaging photon-initiated reactions: A study of the $Cl(^2P_{3/2}) + CH_4 \rightarrow HCl + CH_3$ reaction. J. Chem. Phys. **123**, 094301 (2005). 12 pages

Beach, S.D., Smith, I.W.M., Tuckett, R.P.: Rate constants for the reaction of Cl atoms with O_3 at temperatures from 298 to 184 K. Int. J. Chem. Kinet. **34**, 104–109 (2002)

Bean, B.D., Mollner, A.K., Nizkorodov, S.A., Nair, G., Okumura, M., Sander, S.P., Peterson, K. A., Francisco, J.S.: Cavity ringdown spectroscopy of *cis* – *cis* HOONO and the HOONO/ $HONO_2$ branching ratio in the reaction $OH + NO_2 + M$. J. Phys. Chem. A **107**, 6974–6985 (2003)

Bearnes, J.M., Liu, F., Lu, L., Lester, M.I.: Ultraviolet spectrum and photochemistry of the simplest Criegee intermediate CH_2OO. J. Am. Chem. Soc. **134**, 20045–20048 (2012)

Bechtel, H.A., Camden, J.P., Brown, D.J.A., Zare, R.N.: Comparing the dynamical effects of symmetric and antisymmetric stretch excitation of methane in the $Cl + CH_4$ reaction. J. Chem. Phys. **120**, 5096–5103 (2004)

Bedjanian, Y., Riffault, V., Le Bras, G.: Kinetics and mechanism of the reaction of OH with ClO. Int. J. Chem. Kinet. **33**, 587–599 (2001)

Benter, T., Liesner, M., Schindler, R.N., Skov, H., Hjorth, J., Restelli, G.: REMPI-MS and FTIR study of NO_2 and oxirane formation in the reactions of unsaturated hydrocarbons with NO_3 radicals. J. Phys. Chem. **98**, 10492–10496 (1994)

Birk, M., Friedl, R.R., Cohen, E.A., Pickett, H.M., Sander, S.P.: The rotational spectrum and structure of chlorine peroxide. J. Chem. Phys. **91**, 6588–6597 (1989)

Blitz, M.A., Dillon, T.J., Heard, D.E., Pilling, M.J., Trought, I.D.: Laser induced fluorescence studies of the reactions of $O(^1D_2)$ with N_2, O_2, N_2O, CH_4, H_2, CO_2, Ar, Kr and $n-C_4H_{10}$. Phys. Chem. Chem. Phys. **6**, 2162–2171 (2004)

Bloss, W.J., Nickolaisen, S.L., Salawitch, R.J., Friedl, R.R., Sander, S.P.: Kinetics of the ClO self-reaction and 210 nm absorption cross section of the ClO dimer. J. Phys. Chem. A **105**, 11226–11239 (2001)

Boakes, G., Mok, W.H.H., Rowley, D.M.: Kinetic studies of the ClO + ClO association reaction as a function of temperature and pressure. Phys. Chem. Chem. Phys. **7**, 4102–4113 (2005)

Bohn, B.: Formation of peroxy radicals from OH – toluene adducts and O_2. J. Phys. Chem. A **105**, 6092–6101 (2001)

Bohn, B., Zetzsch, C.: Rate constants of $HO_2 + NO$ covering atmospheric conditions. 1. HO_2 formed by $OH + H_2O_2$. J. Phys. Chem. A **101**, 1488–1493 (1997)

Bohn, B., Zetzsch, C.: Kinetics of the reaction of hydroxyl radical with benzene and toluene. Phys. Chem. Chem. Phys. **1**, 5097–5107 (1999)

Bohn, B., Siese, M., Zetzsch, C.: Kinetics of the $OH + C_2H_2$ reaction in the presence of O_2. J. Chem. Soc. Faraday Trans. **92**, 1459–1466 (1996)

Bonard, A., Daele, V., Delfau, J.-L., Vovelle, C.: Kinetics of OH radical reactions with methane in the temperature range 295–660 K and with dimethyl ether and methyl – *tert* – butyl ether in the temperature range 295–618 K. J. Phys. Chem. A **106**, 4384–4389 (2002)

Borders, R.A., Birks, J.W.: High-precision measurements of activation energies over small temperature intervals: curvature in the Arrhenius plot for the reaction $NO + O_3 \rightarrow NO_2 + O_2$. J. Phys. Chem. **86**, 3295–3302 (1982)

Boyd, A.A., Flaud, P.-M., Daugey, M., Lesclaux, R.: Rate constants for $RO_2 + HO_2$ reactions measured under a large excess of HO_2. J. Phys. Chem. A **107**, 818–821 (2003)

Brown, S.S., Talukdar, R.K., Ravishankara, A.R.: Reconsideration of the rate constant for the reaction of hydroxyl radicals with nitric acid. J. Phys. Chem. A **103**, 3031–3037 (1999)

Brown, S.S., Ravishankara, A.R., Stark, H.: Simultaneous kinetics and ring-down: Rate coefficients from single cavity loss temporal profiles. J. Phys. Chem. A **104**, 7044–7052 (2000)

Bryukov, M.G., Slagle, I.R., Knyazev, V.D.: Kinetics of reactions of Cl atoms with methane and chlorinated methanes. J. Phys. Chem. A **106**, 10532–10542 (2002)

Burkholder, J.B., Ravishankara, A.R.: Rate Coefficient for the Reaction: O $+ NO_2 + M \rightarrow NO_3 + M$. J. Phys. Chem. A **104**, 6752–6757 (2000)

Burkholder, J.B., Hammer, P.D., Howard, C.J.: Product analysis of the $OH + NO_2 + M$ reaction. J. Phys. Chem. **91**, 2136–2144 (1987)

Butkovskaya, N.I., Setser, D.W.: Infrared chemiluminescence study of the reactions of hydroxyl radicals with formaldehyde and formyl radicals with H, OH, NO, and NO_2. J. Phys. Chem. A **102**, 9715–9728 (1998)

Butkovskaya, N.I., Kukui, A., Le Bras, G.: Branching fractions for H_2O forming channels of the reaction of OH radicals with acetaldehyde. J. Phys. Chem. A **108**, 1160–1168 (2004)

Butkovskaya, N.I., Kukui, A., Pouvesle, N., Le Bras, G.: Formation of nitric acid in the gas-phase $HO_2 + NO$ reaction: Effects of temperature and water vapor. J. Phys. Chem. A **109**, 6509–6520 (2005)

Butkovskaya, N., Kukui, A., Le Bras, G.: HNO_3 Forming channel of the $HO_2 + NO$ reaction as a function of pressure and temperature in the ranges of 72–600 Torr and 223–323 K. J. Phys. Chem. A **111**, 9047–9053 (2007)

Butkovskaya, N., Rayez, M.-T., Rayez, J.-C., Kukui, A., Le Bras, G.: Water vapor effect on the HNO_3 yield in the $HO_2 + NO$ reaction: experimental and theoretical evidence. J. Phys. Chem. A **113**, 11327–11342 (2009)

Cameron, M., Sivakumaran, V., Dillon, T.J., Crowley, J.: Reaction between OH and CH_3CHO Part 1. Primary product yields of CH_3 (296 K), CH_3CO (296 K), and H (237–296 K). Phys. Chem. Chem. Phys. **4**, 3628–3638 (2002)

Canosa-Mas, C., Smith, S.J., Toby, S., Wayne, R.P.: Reactivity of the nitrate radical towards alkynes and some other molecules, J. Chem. Soc. Faraday Trans. **2**(84), 247–262 (1988a)

Canosa-Mas, C., Smith, S.J., Toby, S., Wayne, R.P.: Temperature dependences of the reactions of the nitrate radical with some alkynes and with ethylene, J. Chem. Soc. Faraday Trans. **2**(84), 263–272 (1988b)

Cantrell, C.A., Stockwell, W.R., Anderson, L.G., Busarow, K.L., Perner, D., Schmeltekopf, A., Calvert, J.G., Johnston, H.S.: Kinetic study of the nitrate free radical (NO_3)-formaldehyde reaction and its possible role in nighttime tropospheric chemistry. J. Phys. Chem. **89**, 139–146 (1985)

Cantrell, C.A., Shetter, R.E., Calvert, J.G., Tyndall, G.S., Orlando, J.J.: Measurement of rate coefficients for the unimolecular decomposition of dinitrogen pentoxide. J. Phys. Chem. **97**, 9141–9148 (1993)

Cantrell, C.A., Shetter, R.E., Calvert, J.G.: Branching ratios for the $O(^1D) + N_2O$ reaction. J. Geophys. Res. **99**, 3739–3743 (1994)

Carl, S.A.: A highly sensitive method for time-resolved detection of $O(^1D)$ applied to precise determination of absolute $O(^1D)$ reaction rate constants and $O(^3P)$ yields. Phys. Chem. Chem. Phys. **7**, 4051–4053 (2005)

Carter, W.P.L., Atkinson, R.: Alkyl nitrate formation from the atmospheric photoxidation of alkanes; a revised estimation method. J. Atmos. Chem. **8**, 165–173 (1989)

Casavecchia, P., Buss, R.J., Sibener, S.J., Lee, Y.T.: A crossed molecular beam study of the O $(^1D_2) + CH_4$ reaction. J. Chem. Phys. **73**, 6351–6352 (1980)

Castillo, J.F., Aoiz, F.J., Martínez-Haya, B.: Theoretical study of the dynamics of $Cl + O_3$ reaction I. Ab initio potential energy surface and quasiclassical trajectory results. Phys. Chem. Chem. Phys. **13**, 8537–8548 (2011)

Chakraborty, D., Park, J., Lin, M.C.: Theoretical study of the $OH + NO_2$ reaction: formation of nitric acid and the hydroperoxyl radical. Chem. Phys. **231**, 39–49 (1998)

Chen, W.-C., Marcus, R.A.: On the theory of the $CO + OH$ reaction, including H and C kinetic isotope effects. J. Chem. Phys. **123**(094307), 1–16 (2005)

Chen, H.-B., Thweatt, W.D., Wang, J., Glass, G.P., Curl, R.F.: IR kinetic spectroscopy, investigation of the $CH_4 + O(^1D)$ reaction. J. Phys. Chem. A **109**, 2207–2216 (2005)
Cheskis, S.G., Iogansen, A.A., Kulakov, P.V., Razuvaev, I.Y., Sarkisov, O.M., Titov, A.A.: OH vibrational distribution in the reaction $O(^1D) + CH_4$. Chem. Phys. Lett. **155**, 37–42 (1989)
Christensen, L.E., Okumura, M., Sander, S.P., Friedl, R.R., Miller, C.E., Sloan, J.J.: Measurements of the rate constant of $HO_2 + NO_2 + N_2 \rightarrow HO_2NO_2 + N_2$ using near-infrared wavelength-modulation spectroscopy and UV – Visible absorption spectroscopy. J. Phys. Chem. A **108**, 80–91 (2004)
Cleary, P.A., Romero, M.T.B., Blitz, M.A., Heard, D.E., Pilling, M.J., Seakins, P.W., Wang, L.: Determination of the temperature and pressure dependence of the reaction $OH + C_2H_4$ from 200–400 K using experimental and master equation analyses. Phys. Chem. Chem. Phys. **8**, 5633–5642 (2006)
Clough, P.N., Thrush, B.A.: Mechanism of chemiluminescent reaction between nitric oxide and ozone. Trans. Faraday Soc. **63**, 915–925 (1967)
Clyne, M.A.A., Thrush, B.A., Wayne, R.P.: Kinetics of the chemiluminescent reaction between nitric oxide and ozone. Trans. Faraday Soc. **60**, 359–370 (1964)
Cobos, C.J., Troe, J.: Prediction of reduced falloff curves for recombination reactions at low temperatures. Z. Phys. Chem. **217**, 1031–1044 (2003)
Cohen, N.: The use of transition-state theory to extrapolate rate coefficients for reactions of OH with alkanes. Int. J. Chem. Kinet. **14**, 1339–1362 (1982)
Corchado, J.C., Truhlar, D.G., Espinosa-García, J.: Potential energy surface, thermal, and state-selected rate coefficients, and kinetic isotope effects for $Cl + CH_4 \rightarrow HCl + CH_3$. J. Chem. Phys. **112**, 9375–9389 (2000)
Cox, R.A., Coker, G.B.: Kinetics of the reaction of nitrogen dioxide with ozone. J. Atmos. Chem. **1**, 53–63 (1983)
Cox, R.A., Tyndall, G.S.: Rate constants for the reactions of CH_3O_2 with HO_2, NO and NO_2 using molecular modulation spectrometry. J. Chem. Soc., Faraday Trans. **2**(76), 153–163 (1980)
Cremer, D., Gauss, J., Kraka, E., Stanton, J.F., Bartlett, R.J.: A CCSD (T) investigation of carbonyl oxide and dioxirane. Equilibrium geometries, dipole moments, infrared spectra, heats of formation and isomerization energies. Chem. Phys. Lett. **209**, 547–556 (1993)
Criegee, R.: Mechanism of Ozonolysis. Angew. Chem. Int. Ed. Engl. **14**, 745–752 (1975)
D'Anna, B., Bakke, V., Beuke, J.A., Nielsen, C.J., Brudnik, K., Jodkowski, J.T.: Experimental and theoretical studies of gas phase NO_3 and OH radical reactions with formaldehyde, acetaldehyde and their isotopomers. Phys. Chem. Chem. Phys. **5**, 1790–1805 (2003)
Dagaut, P., Wallington, T.J., Kurylo, M.J.: The temperature dependence of the rate constant for the $HO_2 + CH_3O_2$ gas-phase reaction. J. Phys. Chem. **92**, 3833–3836 (1988)
Davidson, J.A., Schiff, H.I., Streit, G.E., McAfee, J.R., Schmeltekopf, A.L., Howard, C.J.: Temperature dependence of $O(^1D)$ rate constants for reactions with N_2O, H_2, CH_4, HCl, and NH_3. J. Chem. Phys. **67**, 5021–5025 (1977)
Davidson, J.A., Schiff, H.I., Brown, T.J., Howard, C.J.: Temperature dependence of the rate constants for reactions of $O(^1D)$ atoms with a number of halocarbons. J. Chem. Phys. **69**, 4277–4279 (1978)
Davis, D.D., Prusazcyk, J., Dwyer, M., Kim, P.: Stop-flow time-of-flight mass spectrometry kinetics study. Reaction of ozone with nitrogen dioxide and sulfur dioxide. J. Phys. Chem. **78**, 1775–1779 (1974)
Devolder, P., Carlier, M., Pauwels, J.F., Sochet, L.R.: Rate constant for the reaction of OH with nitric acid: A new investigation by discharge flow resonance fluorescence. Chem. Phys. Lett. **111**, 94–99 (1984)
Dillon, T.J., Horowitz, A., Crowley, J.N.: Absolute rate coefficients for the reactions of $O(^1D)$ with a series of n – alkanes. Chem. Phys. Lett. **443**, 12–16 (2007)
Dlugokencky, E.J., Howard, C.J.: Studies of nitrate radical reactions with some atmospheric organic compounds at low pressures. J. Phys. Chem. **93**, 1091–1096 (1989)
D'Ottone, L., Campuzano-Jost, P., Bauer, D., Hynes, A.J.: A pulsed laser photolysis – pulsed laser induced fluorescence study of the kinetics of the gas-phase reaction of OH with NO_2. J. Phys. Chem. A **105**, 10538–10543 (2001)

Dransfield, T.J., Donahue, N.M., Anderson, J.G.: High-pressure flow reactor product study of the reactions of $HO_2 + NO_2$: The role of vibrationally excited intermediates. J. Phys. Chem. A **105**, 1507–1514 (2001)

Droege, A.T., Tully, F.P.: Hydrogen-atom abstraction from alkanes by hydroxyl. 3. Propane. J. Phys. Chem. **90**, 1949–1954 (1986)

Dunlea, E.J., Ravishankara, A.R.: Kinetic studies of the reactions of $O(^1D)$ with several atmospheric molecules. Phys. Chem. Chem. Phys. **6**, 2152–2161 (2004a)

Dunlea, E.J., Ravishankara, A.R.: Measurement of the rate coefficient for the reaction of $O(^1D)$ with H_2 O and re-evaluation of the atmospheric OH production rate. Phys. Chem. Chem. Phys. **6**, 3333–3340 (2004b)

Egsgaard, H., Carlsen, L.: Experimental evidence for the gaseous HSO_3• radical. The key intermediate in the oxidation of SO_2 in the atmosphere. Chem. Phys. Lett. **148**, 537–540 (1988)

Elrod, M.J., Ranschaert, D.L., Schneider, N.J.: Direct kinetics study of the temperature dependence of the CH_2O branching channel for the $CH_3O_2 + HO_2$ reaction. Int. J. Chem. Kinet. **33**, 363–376 (2001)

Finlayson-Pitts, B.J., Pitts Jr., J.N.: Chemistry of the Upper and Lower Atmosphere, p. 969. Academic, San Diego (2000)

Fontijn, A., Sabadell, A.J., Ronco, R.J.: Homogeneous chemiluminescent measurement of nitric oxide with ozone. Implications for continuous selective monitoring of gaseous air pollutants. Anal. Chem. **42**, 575–579 (1970)

Force, A.P., Wiesenfeld, J.R.: Collisional deactivation of oxygen($^1D_2$) by the halomethanes. Direct determination of reaction efficiency. J. Phys. Chem. **85**, 782–785 (1981)

Fulle, D., Hamann, H.F., Hippler, H., J¨ansch, C.P.: The high pressure range of the addition of OH to C_2H_2 and C_2H_4. Ber. Bunsenges. Phys. Chem. **101**, 1433–1442 (1997)

Geers-Müller, R., Stuhl, F.: On the kinetics of the reactions of oxygen atoms with NO_2, N_2O_4, and N_2O_3 at low temperatures. Chem. Phys. Lett. **135**, 263–268 (1987)

Gierczak, T., Talukdar, R.K., Herndon, S.C., Vaghjiani, G.L., Ravishankara, A.R.: Rate coefficients for the reactions of hydroxyl radicals with methane and deuterated methanes. J. Phys. Chem. A **101**, 3125–3134 (1997)

Gierczak, T., Burkholder, J.B., Ravishankara, A.R.: Temperature dependent rate coefficient for the reaction $O(^3P) + NO_2 \rightarrow NO + O_2$. J. Phys. Chem. A **103**, 877–883 (1999)

Gillies, J.Z., Gillies, C.W., Suenram, R.D., Lovas, F.J.: The ozonolysis of ethylene. Microwave spectrum, molecular structure, and dipole moment of ethylene primary ozonide (1,2,3-trioxolane). J. Am. Chem. Soc. **110**, 7991–7999 (1988)

Gillies, J.Z., Gillies, C.W., Suenram, R.D., Lovas, F.J., Stahl, W.: The microwave spectrum and molecular structure of the ethylene-ozone van der Waals complex. J. Am. Chem. Soc. **111**, 3073–3074 (1989)

Glendening, E.D., Halpern, A.M.: Ab initio calculations of nitrogen oxide reactions: Formation of N_2O_2, N_2O_3, N_2O_4, N_2O_5, and N_4O_2 from NO, NO_2, NO_3, and N_2O. J. Chem. Phys. **127**, 164307–164317 (2007)

Glinski, R.J., Birks, J.W.: Yields of molecular hydrogen in the elementary reactions hydroperoxo (HO_2) + HO_2 and atomic oxygen ($^1D_2$) + water. J. Phys. Chem. **89**, 3449–3453 (1985)

Golden, D.M., Smith, G.P.: Reaction of $OH + NO_2 + M$: A new view. J. Phys. Chem. A **104**, 3991–3997 (2000)

Golden, D.M., Smith, G.P., McEwen, A.B., Yu, C.-L., Eiteneer, B., Frenklach, M., Vaghjiani, G. L., Ravishankara, A.R., Tully, F.P.: OH(OD) + CO: Measurements and an optimized RRKM Fit. J. Phys. Chem. A **102**, 8598–8606 (1998)

Golden, D.M., Barker, J.R., Lohr, L.L.: Master Equation Models for the Pressure- and Temperature-Dependent Reactions $HO + NO_2 \rightarrow HONO_2$ and $HO + NO_2 \rightarrow HOONO$. J. Phys. Chem. A **107**, 11057–11071 (2003)

Goldfarb, L., Burkholder, J.B., Ravishankara, A.R.: Kinetics of the O + ClO Reaction. J. Phys. Chem. A **105**, 5402–5409 (2001)

González, M., Puyuelo, M.P., Hernando, J., Sayós, R., Enríquez, P.A., Guallar, J., Baños, I.: Influence of the collision energy on the $O(^1D) + RH \rightarrow OH(X^2\Pi) + R$ ($RH = CH_4$, C_2H_6, C_3H_8) reaction dynamics: A laser-induced fluorescence and quasiclassical trajectory Study. J. Phys. Chem. A **104**, 521–529 (2000)
Goumri, A., Pauwels, J.F., Devolder, P.: Rate of the $OH + C_6H_6 + He$ reaction in the fall-off range by discharge flow and OH resonance fluorescence. Can. J. Chem. **69**, 1057–1064 (1991)
Graham, R.A., Johnston, H.S.: Kinetics of the gas-phase reaction between ozone and nitrogen dioxide. J. Chem. Phys. **60**, 4628–4629 (1974)
Graham, R.A., Winer, A.M., Pitts Jr., J.N.: Temperature dependence of the unimolecular decomposition of pernitric acid and its atmospheric implications. Chem. Phys. Lett. **51**, 215–220 (1977)
Grebenkin, S.Y., Krasnoperov, L.N.: Kinetics and thermochemistry of the hydroxycyclohexadienyl radical reaction with O_2 : $C_6H_6OH + O_2 \rightleftharpoons C_6H_6$ (OH) OO. J. Phys. Chem. **108**, 1953–1963 (2004)
Grosjean, D.: Atmospheric chemistry of toxic contaminants 1. Reaction rates and atmospheric persistence. J. Air Waste Manag. Assoc. **40**, 1397–1402 (1990)
Grosjean, E., Grosjean, D.: Carbonyl products of the gas phase reaction of ozone with symmetrical alkenes. Environ. Sci. Technol. **30**, 2036–2044 (1996)
Gutbrod, R., Schindler, R.N., Kraka, E., Cremer, D.: Formation of OH radicals in the gas phase ozonolysis of alkenes: the unexpected role of carbonyl oxides. Chem. Phys. Lett. **252**, 221–229 (1996)
Hahn, J., Luther, K., Troe, J.: Experimental and theoretical study of the temperature and pressure dependences of the recombination reactions $O + NO_2$ (+M) $\rightarrow NO_3$ (+M) and NO_2+ NO_3 (+M) $\rightarrow N_2O_5$ (+M). Phys. Chem. Chem. Phys. **2**, 5098–5104 (2000)
Hamilton, E.J.: Water vapor dependence of the kinetics of the self-reaction of HO_2 in the gas phase. J. Chem. Phys. **63**, 3682–3683 (1975)
Hamilton Jr., E.J., Naleway, C.A.: Theoretical calculation of strong complex formation by the HO_2 radical: $HO_2{\cdot}H_2O$ and $HO_2{\cdot}NH_3$. J. Phys. Chem. **80**, 2037–2040 (1976)
Hammer, P.D., Dlugokencky, E.J., Howard, C.J.: Kinetics of the nitric oxide-nitrate radical gas-phase reaction NO + NO3 → 2NO2. J. Phys. Chem. **90**, 2491–2496 (1986)
Handwerk, V., Zellner, R.: Pressure and temperature dependence of the reaction $ClO + NO_2$ (+ N_2) $\rightarrow ClONO_2$ (+ N_2). Ber. Bunsenges. Phys. Chem. **88**, 405–409 (1984)
Hanway, D., Tao, F.-M.: A density functional theory and ab initio study of the hydrolysis of dinitrogen pentoxide. Chem. Phys. Lett. **285**, 459–466 (1998)
Hashimoto, S., Inoue, G., Akimoto, H.: Infrared spectroscopic detection of the $HOSO_2$ radical in argon matrix at 11 K. Chem. Phys. Lett. **107**, 198–202 (1984)
Hasson, A.S., Orzechowska, G., Paulson, S.E.: Production of stabilized Criegee intermediates and peroxides in the gas phase ozonolysis of alkenes 1. Ethene, *trans*-2-butene, and 2,3-dimethyl-2-butene. J. Geophys. Res. **106**, 34131–34142 (2001)
Hatakeyama, S., Kobayashi, H.: Z.-Y., H. Takagi and H. Akimoto, Mechanism for the reaction of peroxymethylene with sulfur dioxide. J. Phys. Chem. **90**, 4131–4135 (1986)
Hatakeyama, S., Kobayashi, H., Akimoto, H.: Gas-phase oxidation of sulfur dioxide in the ozone-olefin reactions. J. Phys. Chem. **88**, 4736–4739 (1984)
Herndon, S.C., Malta, P.W.V., Nelson, D.D., Jayne, J.T., Zahniser, M.S.: Rate constant measurements for the reaction of HO_2 with O_3 from 200 to 300 K using a turbulent flow reactor. J. Phys. Chem. A **105**, 1583–1591 (2001)
Hickson, K.M., Keyser, L.F., Sander, S.P.: Temperature dependence of the $HO_2 + ClO$ reaction. 2. Reaction kinetics using the discharge-flow resonance-fluorescence technique, J. Phys. Chem. A **111**, 8126–8138 (2007)
Hippler, H., Rahn, R., Troe, J.: Temperature and pressure dependence of ozone formation rates in the range 1–1000 bar and 90–370 K. J. Chem. Phys. **93**, 6560 (1990)
Hippler, H., Neunaber, N., Troe, J.: Shock wave studies of the reactions $HO + H_2O_2 \rightarrow H_2O + HO_2$ and $HO + HO_2 \rightarrow H_2O + O_2$ between 930 and 1680 K. J. Chem. Phys. **103**, 3510–3516 (1995)

Hippler, H., Nasterlack, S., Striebel, F.: Reaction of $OH + NO_2 + M$: Kinetic evidence of isomer formation. Phys. Chem. Chem. Phys. **4**, 2959–2964 (2002)

Horie, O., Moortgat, G.K.: Decomposition pathways of the excited Criegee intermediates in the ozonolysis of simple alkenes. Atmos. Environ. **25A**, 1881–1896 (1991)

Horie, O., Schafer, C., Moortgat, G.K.: High reactivity of hexafluoro acetone toward Criegee intermediates in the gas-phase ozonolysis of simple alkenes. Int. J. Chem. Kinet. **31**, 261–269 (1999)

Howard, C.J., Evenson, K.M.: Kinetics of the reaction of HO_2 with NO. Geophys. Res. Lett. **4**, 437–440 (1977)

Howard, M.J., Smith, I.W.M.: Direct rate measurements on the reactions $N + OH \rightarrow NO + H$ and $O + OH \rightarrow O_2 + H$ from 250 to 515 K. J. Chem. Soc. Faraday Trans. **2**(77), 997–1008 (1981)

Huie, R.E., Herron, J.T.: The rate constant for the reaction $O_3 + NO_2 \rightarrow O_2 + NO_3$ over the temperature range 259–362 K. Chem. Phys. Lett. **27**, 411–414 (1974)

Jemi-Alade, A.A., Thrush, B.A.: Reactions of HO_2 with NO and NO_2 studied by mid-infrared laser magnetic resonance. J. Chem. Soc. Faraday Trans. **2**(86), 3355–3363 (1990)

Jitariu, L.C., Hirst, D.M.: Theoretical investigation of the $N_2O_5 \rightleftarrows NO_2 + NO_3$ equilibrium by density functional theory and ab initio calculations. Phys. Chem. Chem. Phys. **2**, 847–852 (2000)

Johnson, D., Marston, G.: The gas-phase ozonolysis of unsaturated volatile organic compounds in the troposphere. Chem. Soc. Rev. **37**, 699–716 (2008)

Johnson, D., Raoult, S., Lesclaux, R., Krasnoperov, L.N.: UV absorption spectra of methyl-substituted hydroxy-cyclohexadienyl radicals in the gas phase. J. Photochem. Photobiol. A **176**, 98–106 (2005)

Joshi, V.A., Wang, H.: Master equation modeling of wide range temperature and pressure dependence of $CO + OH \rightarrow$ products. Int. J. Chem. Kinet. **38**, 57–73 (2006)

Kaltsoyannis, N., Rowley, D.M.: Ab initio investigations of the potential energy surfaces of the $XO + HO_2$ reaction (X = chlorine or bromine). Phys. Chem. Chem. Phys. **4**, 419–427 (2002)

Kan, C.S., Su, F., Calvert, J.G., Shaw, J.H.: Mechanism of the ozone-ethene reaction in dilute N_2/O_2 mixtures near 1-atm pressure. J. Phys. Chem. **85**, 2359–2363 (1981)

Kanno, N., Tonokura, K., Tezaki, A., Koshi, M.: Water dependence of the HO_2 self reaction: Kinetics of the $HO_2 - H_2O$ complex. J. Phys. Chem. A **109**, 3153–3158 (2005)

Kanno, N., Tonokura, K., Koshi, M.: Equilibrium constant of the $HO_2 - H_2O$ complex formation and kinetics of $HO_2 + HO_2 - H_2O$: Implications for tropospheric chemistry. J. Geophys. Res. **111**(D20312), 1–7 (2006)

Kegley-Owen, C.S., Gilles, M.K., Burkholder, J.B., Ravishankara, A.R.: Rate coefficient measurements for the reaction $OH + ClO \rightarrow$ Products. J. Phys. Chem. A **103**, 5040–5048 (1999)

Keyser, L.F.: Kinetics of the reaction $O + HO_2 \rightarrow OH + O_2$ from 229 to 372 K. J. Phys. Chem. **86**, 3439–3446 (1982)

Keyser, L.F.: Kinetics of the reaction $OH + HO_2 \rightarrow H_2O + O$ from 254 to 382 K. J. Phys. Chem. **92**, 1193–1200 (1988)

King, M.D., Canosa-Mas, C.E., Wayne, R.P.: Frontier molecular orbital correlations for predicting rate constants between alkenes and the tropospheric oxidants NO_3, OH and O_3. Phys. Chem. Chem. Phys. **1**, 2231–2238 (1999)

Kircher, C.C., Sander, S.P.: Kinetics and mechanism of HO_2 and DO_2 disproportionations. J. Phys. Chem. **88**, 2082–2091 (1984)

Klein, T., Barnes, I., Becker, K.H., Fink, E.H., Zabel, F.: Pressure dependence of the rate constants for the reactions of ethene and propene with hydroxyl radicals at 295 K. J. Phys. Chem. **88**, 5020–5025 (1984)

Klopper, W., Tew, D.P., González-García, N., Olzmann, M.: Heat of formation of the $HOSO_2$ radical from accurate quantum chemical calculations. J. Chem. Phys. **129**, 114308 (2008)

Klotz, B., Sørensen, S., Barnes, I., Becker, K.H., Etzkorn, T., Volkamer, R., Platt, U., Wirtz, K., Martín-Reviejo, M.: Atmospheric oxidation of toluene in a large-volume outdoor photoreactor:

In situ determination of ring-retaining product yields. J. Phys. Chem. A **102**, 10289–10299 (1998)
Knight, G.P., Beiderhase, T., Helleis, F., Moortgat, G.K., Crowley, J.N.: Reaction of HO_2 with ClO: flow tube studies of kinetics and product formation between 215 and 298 K. J. Phys. Chem. A **104**, 1674–1685 (2000)
Kovacic, S., Lesar, K., Hodoscek, S.M.: Quantum mechanical study of the potential energy surface of the $ClO+NO_2$ reaction. J. Chem. Inf. Model. **45**, 58–64 (2005)
Kroll, J.H., Hanisco, T.F., Donahue, N.M., Demerjian, K.L., Anderson, J.G.: Accurate, direct measurements of OH yields from gas-phase ozone-alkene reactions using an LIF Instrument. Geophys. Res. Lett. **28**, 3863–3866 (2001)
Kuo, C.H., Lee, Y.P.: Kinetics of the reaction hydroxyl + ethene in helium, nitrogen, and oxygen at low pressure. J. Phys. Chem. **95**, 1253–1257 (1991)
Kuo, Y.P., Cheng, B.M., Lee, Y.P.: Production and trapping of $HOSO_2$ from the gaseous reaction $OH+SO_2$: the infrared absorption of $HOSO_2$ in solid argon. Chem. Phys. Lett. **177**, 195–199 (1991)
Kurylo, M.J., Ouellette, P.A., Laufer, A.H.: Measurements of the pressure dependence of the hydroperoxy (HO_2) radical self-disproportionation reaction at 298 K. J. Phys. Chem. **90**, 437–440 (1986)
Lee, T.J., Rohlfing, C.M., Rice, J.E.: An extensive ab initio study of the structures, vibrational spectra, quadratic force fields, and relative energetics of three isomers of Cl_2O_2. J. Chem. Phys. **97**, 6593–6605 (1992)
Lewis, R.S., Watson, R.T.: Temperature dependence of the reaction $O(^3P)+OH(^2II)\rightarrow O_2+H$. J. Phys. Chem. **84**, 3495–3503 (1980)
Li, W.-K., McKee, M.L.: Theoretical study of OH and H_2O addition to SO_2. J. Phys. Chem. A **101**, 9778–9782 (1997)
Lightfoot, P.D., Veyret, B., Lesclaux, R.: The rate constant for the HO_2+HO_2 reaction at elevated temperatures. Chem. Phys. Lett. **150**, 120–126 (1988)
Lightfoot, P.D., Veyret, B., Lesclaux, R.: Flash photolysis study of the $CH_3O_2+HO_2$ reaction between 248 and 573 K. J. Phys. Chem. **94**, 708–714 (1990)
Lightfoot, P.D., Roussel, P., Caralp, F., Lesclaux, R.: Flash photolysis study of the CH_3O_2+ CH_3O_2 and CH_3O_2+ HO_2 reactions between 600 and 719 K: unimolecular decomposition of methylhydroperoxide. J. Chem. Soc. Faraday Trans. **87**, 3213–3220 (1991)
Lightfoot, P.D., Cox, R.A., Crowly, J.N., Destriau, M., Hayman, G.D., Jenkin, M.E., Moortgat, G. K., Zabel, F.: Organic peroxy radicals: kinetics, spectroscopy and tropospheric chemistry. Atmos. Environ. **26A**, 1805–1961 (1992)
Lii, R.-R., Sauer Jr., M.C., Gordon, S.: Temperature dependence of the gas-phase self- reaction of HO_2 in the presence of H_2O. J. Phys. Chem. **85**, 2833–2834 (1981)
Lin, C.L., Leu, M.T.: Temperature and third-body dependence of the rate constant for the reaction $O+O_2+M\rightarrow O_3+M$. Int. J. Chem. Kinet. **14**, 417 (1982)
Lin, J.J., Lee, Y.T., Yang, X.: Crossed molecular beam studies of the $O(^1D)+CH_4$ reaction: Evidences for the CH_2OH+H channel. J. Chem. Phys. **109**, 2975–2978 (1998)
Lin, J.J., Harich, S., Lee, Y.T., Yang, X.: Dynamics of the $O(^1D)+CH_4$ reaction: Atomic hydrogen channel vs molecular hydrogen channel. J. Chem. Phys. **110**, 10821–10829 (1999)
Lippmann, H.H., Jesser, B., Schurath, U.: The rate constant of $NO+O_3\rightarrow NO_2+O_2$ in the temperature range of 283–443 K. Int. J. Chem. Kinet. **12**, 547–554 (1980)
Lipson, J.B., Beiderhase, T.W., Molina, L.T., Molina, M.J., Olzmann, M.: Production of HCl in the OH + ClO reaction: Laboratory measurements and statistical rate theory calculations. J. Phys. Chem. A **103**, 6540–6551 (1999)
Liu, J.Y., Barker, J.R.: On the Chaperon mechanism: Application to ClO + ClO $(+N_2)\rightarrow$ ClOOCl $(+N_2)$. J. Phys. Chem. A **111**, 8689–8698 (2007)
Lohr, L.L., Barker, J.R., Shroll, R.M.: Modeling the organic nitrate yields in the reaction of alkyl peroxy radicals with nitric oxide. 1. Electronic structure calculations and thermochemistry. J. Phys. Chem. A **107**, 7429–7433 (2003)

Margitan, J.J., Watson, R.T.: Kinetics of the reaction of hydroxyl radicals with nitric acid. J. Phys. Chem. **86**, 3819–3824 (1982)

Matsumi, Y., Tonokura, K., Inagaki, Y., Kawasaki, M.: Isotopic branching ratios and translational energy release of hydrogen and deuterium atoms in reaction of oxygen (^{1}D) atoms with alkanes and alkyl chlorides. J. Phys. Chem. **97**, 6816–6821 (1993)

Matsumi, Y., Nomura, S., Kawasaki, M., Imamura, T.: Vibrational Distribution of ClO Radicals Produced in the Reaction $Cl + O_3 \rightarrow ClO + O_2$. J. Phys. Chem. **100**, 176–179 (1996)

McCabe, D.C., Gierczak, T., Talukdara, R., Ravishankara, A.R.: Kinetics of the reaction OH + CO under atmospheric conditions. Geophys. Res. Lett. **28**, 3135–3138 (2001)

McKee, M.J., Rohlfing, C.M.: An ab initio study of complexes between ethylene and ozone. J. Am. Chem. Soc. **111**, 2497–2500 (1989)

Medvedev, D., Gray, S.K., Goldfield, E.M., Lakin, M.J., Troya, D., Schatz, G.C.: Quantum wave packet and quasiclassical trajectory studies of OH + CO: Influence of the reactant channel well on thermal rate constants. J. Chem. Phys. **120**, 1231–1238 (2004)

Michael, J.V., Allen Jr., J.E., Brobst, W.D.: Temperature dependence of the nitric oxide + ozone reaction rate from 195 to 369 K. J. Phys. Chem. **85**, 4109–4117 (1981)

Mihelcic, D., Heitlinger, M., Kley, D., Musgen, P., Volz-Thomas, A.: Formation of hydroxyl and hydroperoxy radicals in the gas-phase ozonolysis of ethene. Chem. Phys. Lett. **301**, 559–564 (1999)

Miyoshi, A., Matsui, H., Washida, N.: Detection and reactions of the HOCO radical in gas phase. J. Chem. Phys. **100**, 3532–3539 (1994)

Mollner, A.K., Valluvadasan, S., Feng, L., Sprague, M.K., Okumura, M., Milligan, D.B., Bloss, W.J., Sander, S.P., Martien, P.T., Harley, R.A., McCoy, A.B., Carter, W.P.L.: Rate of gas phase association of hydroxyl radical and nitrogen dioxide. Science **330**, 646–649 (2010)

Moonen, P.C., Cape, J.N., Storeton-West, R.L., McColm, R.: Measurement of the $NO + O_3$ reaction rate at atmospheric pressure using realistic mixing ratios. J. Atmos. Chem. **29**, 299–314 (1998)

Mora-Diez, N., Boyd, R.J.: A computational study of the kinetics of the NO_3 Hydrogen -abstraction reaction from a series of aldehydes (XCHO: X) F, Cl, H, CH_3. J. Phys. Chem. A **106**, 384–394 (2002)

Neeb, P., Horie, O., Moortgat, G.K.: Gas-phase ozonolysis of ethene in the presence of hydroxylic compounds. Int. J. Chem. Kinet. **28**, 721–730 (1996)

Neeb, P., Horie, O., Moortgat, G.K.: The Ethene – ozone reaction in the gas phase. J. Phys. Chem. A **102**, 6778–6785 (1998)

Nelson Jr., D.D., Zahniser, M.S.: A mechanistic study of the reaction of HO_2 radical with ozone. J. Phys. Chem. **98**, 2101–2104 (1994)

Nguyen, M.T., Nguyen, T.L., Ngan, V.T., Nguyen, H.M.T.: Heats of formation of the Criegee formaldehyde oxide and dioxirane. Chem. Phys. Lett. **448**, 183–188 (2007)

Nguyen, T.L., Park, J.H., Lee, K.J., Song, K.Y., Barker, J.R.: Mechanism and kinetics of the reaction $NO_3 + C_2H_4$. J. Phys. Chem. A **115**, 4894–4901 (2011)

Nickolaisen, S.L., Friedl, R.R., Sander, S.P.: Kinetics and mechanism of the chlorine oxide ClO + ClO reaction: pressure and temperature dependences of the bimolecular and termolecular channels and thermal decomposition of chlorine peroxide. J. Phys. Chem. **98**, 155–169 (1994)

Nickolaisen, S.L., Roehl, C.M., Blakeley, L.K., Friedl, R.R., Francisco, J.S., Liu, R.F., Sander, S. P.: Temperature dependence of the $HO_2 + ClO$ reaction. 1. Reaction kinetics by pulsed photolysis-ultraviolet absorption and ab initio studies of the potential surface. J. Phys. Chem. A **104**, 308–319 (2000)

Nicovich, J.M., Wine, P.H.: Temperature dependence of the $O + HO_2$ rate coefficient. J. Phys. Chem. **91**, 5118–5123 (1987)

Nicovich, J.M., Wine, P.H., Ravishankara, A.R.: Pulsed laser photolysis kinetics study of the O (^3P) + ClO reaction. J. Chem. Phys. **89**, 5670–5679 (1988)

Nicovich, J.M., Kreutter, K.D., Wine, P.H.: Kinetics of the reactions of Cl(^2PJ) and Br($^2P_{3/2}$) with O_3. Int. J. Chem. Kinet. **22**, 399–414 (1990)

Niki, H., Maker, P.D., Savage, C.M., Breitenbach, L.P.: A FT IR study of a transitory product in the gas-phase ozone-ethylene reaction. J. Phys. Chem. **85**, 1024–1027 (1981)

Niki, H., Maker, P.D., Savage, C.M., Breitenbach, L.P.: An Fourier transform infrared study of the kinetics and mechanism for the reaction of hydroxyl radical with formaldehyde. J. Phys. Chem. **88**, 5342–5344 (1984)

NIST-JANAF: Thermochemical tables, Edited by M. W. Chase, Jr., American Chemical Society and American Institute of Physics, Woodbury, 4th ed., J. Phys. Chem. Ref. Data Monograph No. 9, (1988)

Nizkorodov, S.A., Wennberg, P.O.: First spectroscopic observation of gas-phase HOONO. J. Phys. Chem. A **106**, 855–859 (2002)

O'Donnell, B.A., Li, E.X.J., Lester, M.I., Francisco, J.S.: Spectroscopic identification and stability of the intermediate in the $OH + HONO_2$ reaction, Proc. Natl. Acad. Sci. USA **105**, 12647–12648 (2008)

Olzmann, M., Kraka, E., Cremer, D., Gutbrod, R., Andersson, S.: Energetics, kinetics, and product distributions of the reactions of ozone with ethene and 2,3-dimethyl-2-butene. J. Phys. Chem. A **101**, 9421–9429 (1997)

Ongstad, A.P., Birks, J.W.: Studies of reactions of importance in the stratosphere. VI. Temperature dependence of the reactions $O + NO_2 \rightarrow NO + O_2$ and $O + ClO \rightarrow Cl + O_2$. J. Chem. Phys. **85**, 3359–3368 (1986)

Orlando, J.J., Tyndall, G.S., Cantrell, C.A., Calvert, J.G.: Temperature and pressure dependence of the rate coefficient for the reaction $NO_3 + NO_2 + N_2 \rightarrow N_2O_5 + N_2$. J. Chem. Soc. Faraday Trans. **87**, 2345–2349 (1991)

Paulson, S.E., Fenske, J.D., Sen, A.D., Callahan, T.W.: A novel small-ratio relative-rate technique for measuring OH formation yields from the reactions of O_3 with alkenes in the gas phase, and Its application to the reactions of ethene and propene. J. Phys. Chem. A **103**, 2050–2059 (1999)

Peiró-García, J., Nebot-Gil, I.: Ab initio study on the mechanism of the atmospheric reaction OH $+ O_3 \rightarrow HO_2 + O_2$. Chem. Phys. Chem. **4**, 843–847 (2003a)

Peiró-García, J., Nebot-Gil, I.: Ab initio study of the mechanism of the atmospheric reaction: $NO_2 + O_3 \rightarrow NO_3 + O_2$. J. Comput. Chem. **24**, 1657–1663 (2003b)

Percival, C.J., Smith, G.D., Molina, L.T., Molina, M.J.: Temperature and pressure dependence of the rate constant for the $ClO + NO_2$ reaction. J. Phys. Chem. A **101**, 8830–8833 (1997)

Perry, R.A., Atkinson, R., Pitts Jr., J.N.: Kinetics and mechanism of the gas phase reaction of hydroxyl radicals with aromatic hydrocarbons over the temperature range 296–473 K. J. Phys. Chem. **81**, 296–304 (1977)

Petty, J.T., Harrison, J.A., Moore, C.B.: Reactions of trans – HOCO studied by infrared spectroscopy. J. Phys. Chem. **97**, 11194–11198 (1993)

Pilgrim, J.S., McIlroy, A., Taatjes, C.A.: Kinetics of Cl Atom reactions with methane, ethane, and propane from 292 to 800 K. J. Phys. Chem. A **101**, 1873–1880 (1997)

Pollack, I.B., Konen, I.M., Li, E.X.J., Lester, M.I.: Spectroscopic characterization of HOONO and its binding energy via infrared action spectroscopy. J. Chem. Phys. **119**, 9981–9984 (2003)

Qi, B., Su, K.-H., Wang, Y.-B., Wen, Z.-Y., Tang, X.-Y.: Gaussian-2 Calculations of the thermochemistry of Criegee intermediates in gas phase reactions. Acta Phys. Chim. Sin. **14**, 1033–1039 (1998)

Qi, B., Sato, K., Imamura, T., Takami, A., Hatakeyama, S., Ma, Y.: Production of the radicals in the ozonolysis of ethene: A chamber study by FT-IR and PERCA. Chem. Phys. Lett. **427**, 461–465 (2006)

Ranschaert, D.L., Schneider, N.J., Elrod, M.J.: Kinetics of the $C_2H_5O_2 + NO_x$ reactions: Temperature dependence of the overall rate constant and the $C_2H_5ONO_2$ branching channel of $C_2H_5O_2 + NO$. J. Phys. Chem. A **104**, 5758–5765 (2000)

Ravishankara, A.R., Wine, P.H., Langford, A.O.: Absolute rate constant for the reaction OH $(v = 0) + O_3 \rightarrow HO_2 + O_2$ over the temperature range 238–357 K. J. Chem. Phys. **70**, 984–989 (1979)

Ravishankara, A.R., Eisele, F.L., Kreutter, N.M., Wine, P.H.: Kinetics of the reaction of CH_3O_2 with NO. J. Chem. Phys. **74**, 2267–2274 (1981)

Ravishankara, A.R., Solomon, S., Turnipseed, A.A., Warren, R.F.: Atmospheric lifetimes of long-lived halogenated species. Science **259**, 194–199 (1993)

Ray, G.W., Watson, R.T.: Kinetics of the reaction $NO + O_3 \rightarrow NO_2 + O_2$ from 212 to 422 K. J. Phys. Chem. **85**, 1673–1676 (1981)

Rickard, A.R., Johnson, D., McGill, C.D., Marston, G.: OH yields in the gas-phase reactions of ozone with alkenes. J. Phys. Chem. A **103**, 7656–7664 (1999)

Sander, W.: Carbonyl oxides: Zwitterions or diradicals? Angew. Chem. Int. Ed. Engl. **29**, 344–354 (1990)

Sander, S.P., Kircher, C.C.: Temperature dependence of the reaction $NO + NO_3 \rightarrow 2NO_2$. Chem. Phys. Lett. **126**, 149–152 (1986)

Sander, S.P., Baker R., Golden, D.M., Kurylo, M.J., Wine, P.H., Abatt, J.P.D., Burkholder, J.B., Kolb, C.E., Moortgat, G.K., Huie, R.E., Orkin, V.L.: Chemical Kinetics and Photochemical Data for Use in Atmospheric Studies, Evaluation Number 17, JPL Publication 10–6, Pasadena, California. http://jpldataeval.jpl.nasa.gov/ (2011)

Sayós, R., Olive, C., González, M.: Ab initio CASPT2/CASSCF study of the $O(^1D) + H_2O(X^1A_1)$ reaction. J. Chem. Phys. **115**, 8826–8835 (2001)

Sayós, R., Hernando, J., Puyuelo, M.P., Enríquez, P.A., González, M.: Influence of collision energy on the dynamics of the reaction $O(^1D) + CH_4(X^1A_1) \rightarrow OH(X\ ^2\Pi) + CH_3(X\ ^2A_2)$. Phys. Chem. Chem. Phys. **4**, 288–294 (2002)

Scholtens, K.W., Messer, B.M., Cappa, C.D., Elrod, M.J.: Kinetics of the $CH_3O_2 + NO$ reaction: Temperature dependence of the overall rate constant and an improved upper limit for the CH_3ONO_2 branching channel. J. Phys. Chem. A **103**, 4378–4384 (1999)

Schurath, U., Lippmann, H.H., Jesser, B.: Temperature dependence of the chemiluminescent reaction (1), $NO + O_3 \rightarrow NO_3\ (^2A_1;\ ^2B_{1,\ 2}) + O_3$, and quenching of the excited product. Ber. Bunsenges. Phys. Chem. **85**, 807–813 (1981)

Seeley, J.V., Meads, R.F., Elrod, M.J., Molina, M.J.: Temperature and pressure dependence of the rate constant for the $HO_2 + NO$ reaction. J. Phys. Chem. **100**, 4026–4031 (1996a)

Seeley, J.V., Jayne, J.T., Molina, M.J.: Kinetic studies of chlorine atom reactions using the turbulent flow tube technique. J. Phys. Chem. **100**, 4019–4025 (1996b)

Senosiain, J.P., Musgrave, C.B., Golden, D.M.: Temperature and pressure dependence of the reaction of OH and CO: Master equation modeling on a high-level potential energy surface. Int. J. Chem. Kinet. **35**, 464–474 (2003)

Senosiain, J.P., Klippenstein, S.J., Miller, J.A.: The reaction of acetylene with hydroxyl radicals. J. Phys. Chem. A **109**, 6045–6055 (2005)

Sinha, A., Lovejoy, E.R., Howard, C.J.: Kinetic study of the reaction of HO_2 with ozone. J. Chem. Phys. **87**, 2122–2128 (1987)

Sivakumaran, V., Crowley, J.N.: Reaction between OH and CH_3CHO Part 2. Temperature dependent rate coefficients (201–348 K). Phys. Chem. Chem. Phys. **5**, 106–111 (2003)

Sivakumaran, V., Holscher, D., Dillon, T.J., Crowley, J.: Reaction between OH and HCHO: temperature dependent rate coefficients (202–399 K) and product pathways (298 K). Phys. Chem. Chem. Phys. **5**, 4821–4827 (2003)

Skov, H., Benter, T., Schindler, R.N., Hjorth, J., Restelli, G.: Epoxide formation in the reactions of the nitrate radical with 2,3-dimethyl-2-butene, cis- and trans-2-butene and isoprene. Atmos. Environ. **28**, 1583–1592 (1994)

Smith, I.W.M.: The mechanism of the $OH + CO$ reaction and the stability of the HOCO radical. Chem. Phys. Lett. **49**, 112–115 (1977)

Smith, C.A., Molina, L.T., Lamb, J.J., Molina, M.J.: Kinetics of the reaction of OH with pernitric and nitric acids. Int. J. Chem. Kinet. **16**, 41–55 (1984)

Smith, D.F., McIver, C.D., Kleindienst, T.E.: Primary product distribution from the reaction of hydroxyl radicals with toluene at ppb NOx mixing ratios. J. Atmos. Chem. **30**, 209–228 (1998)

Somnitz, H.: Quantum chemical and dynamical characterization of the reaction OH $+ SO_2 \rightarrow HOSO_2$ over an extended range of temperature and pressure. Phys. Chem. Chem. Phys. **6**, 3844–3851 (2004)

Stachnik, R.A., Molina, M.J., Molina, L.T.: Pressure and temperature dependences of the reaction of hydroxyl radical with nitric acid. J. Phys. Chem. **90**, 2777–2780 (1986)

Stief, L.J., Nava, D.F., Payne, W.A., Michael, J.V.: Rate constant for the reaction of hydroxyl radical with formaldehyde over the temperature range 228–362 K. J. Chem. Phys. **73**, 2254–2258 (1980)

Stone, D., Blitz, M., Doubney, L., Howes, N.U.M., Seakins, P.: Kinetics of CH_2OO reactions with SO_2, NO_2, NO, H_2O and CH_3CHO as a function of pressure. Phys. Chem. Chem. Phys. **16**, 1139–1149 (2014)

Su, Y.-T., Huang, Y.-H., Witek, H.A., Lee, Y.-P.: Infrared absorption spectrum of the simplest Criegee intermediate CH_2OO. Science **340**, 174–176 (2013)

Suh, I., Zhang, D., Zhang, R.: L. T Molina, M. J Molina, Theoretical study of OH addition reaction to toluene. Chem. Phys. Lett. **364**, 454–462 (2002)

Sumathi, R., Peyerimhoff, S.D.: An *ab initio* molecular orbital study of the potential energy surface of the $HO_2 + NO$ reaction. J. Chem. Phys. **107**, 1872–1880 (1997)

Taatjes, C.A., Meloni, G., Selby, T.M., Trevitt, A.J., Osborn, D.L., Percival, C.J., Shallcross, D.E.: Direct observation of the gas phase Criegee intermediate (CH_2OO). J. Am. Chem. Soc. **103**, 11883–11885 (2008)

Takacs, G.A., Howard, C.J.: Temperature dependence of the reaction $HO_2 + HO_2$ at low pressures. J. Phys. Chem. **90**, 687–690 (1986)

Takahashi, K., Wada, R., Matsumi, Y., Kawasaki, M.: Product branching ratios for $O(^3P)$ atom and ClO radical formation in the reactions of $O(^1D)$ with chlorinated compounds. J. Phys. Chem. **100**, 10145–10149 (1996)

Takahashi, K., Takeuchi, Y., Matsumi, Y.: Rate constants of the $O(^1D)$ reactions with N_2, O_2, N_2O, and H_2O at 295 K. Chem. Phys. Lett. **410**, 196–200 (2005)

Takayanagi, T., Akagi, H.: Translational energy dependence of $NO + NO/N_2 + O_2$ product branching in the $O(^1D) + N_2O$ reaction: a classical trajectory study on a new global potential energy surface for the lowest $1A'$ state. Chem. Phys. Lett. **363**, 298–306 (2002)

Takayanagi, T., Wada, A.: Reduced dimensionality quantum reactive scattering calculations on the ab initio potential energy surface for the $O(^1D) + N_2O \rightarrow NO + NO$ reaction. Chem. Phys. **269**, 37–47 (2001)

Taylor, S.E., Goddard, A., Blitz, M.A., Cleary, P.A., Heard, D.E.: Pulsed Laval nozzle study of the kinetics of OH with unsaturated hydrocarbons at very low temperatures. Phys. Chem. Chem. Phys. **10**, 422–437 (2008)

Tokel, O., Chen, J., Ulrich, C.K., Houston, P.L.: $O(^1D) + N_2O$ reaction: NO vibrational and rotational distributions. J. Phys. Chem. A **114**, 11292–11297 (2010)

Treacy, J., El Hag, M., O'Farrell, D., Sidebottom, H.: Reactions of ozone with unsaturated organic compounds. Ber. Bunsenges. Phys. Chem. **96**, 422–427 (1992)

Troe, J.: Modeling the temperature and pressure dependence of the reaction $HO + CO \rightleftarrows HOCO \rightleftarrows H + CO_2$. Proc. Combust. Inst. **27**, 167–175 (1998)

Trolier, M., Mauldin III, R.L., Ravishankara, A.R.: Rate coefficient for the termolecular channel of the self-reaction of chlorine monoxide. J. Phys. Chem. **94**, 4896–4907 (1990)

Troya, D., Millán, J., Baños, I., González, M.: Ab initio, kinetics, and dynamics study of Cl $+ CH_4 \rightarrow HCl + CH_3$. J. Chem. Phys. **117**, 5730–5741 (2002)

Tully, F.P., Ravishankara, A.R., Thompson, R.L., Nicolvich, J.M., Shah, R.C., Kreulter, N.M.: Kinetics of the reaction of hydroxyl radical with benzene and toluene. J. Phys. Chem. **85**, 2262–2269 (1981)

Tyndall, G.S., Orlando, J.J., Cantrell, C.A., Shetter, R.E., Calvert, J.G.: Rate coefficient for the reaction $NO + NO_3 \rightarrow 2NO_2$ between 223 and 400 K. J. Phys. Chem. **95**, 4381–4386 (1991)

Tyndall, G.S., Cox, R.A., Granier, C., Lesclaux, R., Moortgat, G.K., Pilling, M.J., Ravishankara, A.R., Wallington, T.J.: Atmospheric chemistry of small organic peroxy radicals. J. Geophys. Res. **106**, 12157–12182 (2001)

Uc, V.H., Alvarez-Idaboy, J.R., Galano, A., García-Cruz, I., Vivier-Bunge, A.: Theoretical determination of the rate constant for OH hydrogen abstraction from toluene. J. Phys. Chem. A **110**, 10155–10162 (2006)

Vakhtin, A.B., Murphy, J.E., Leone, S.R.: Low-temperature kinetics of reactions of OH radical with ethene, propene, and 1 − butene. J. Phys. Chem. A **107**, 10055–10062 (2003)

Valero, R., van Hemert, M.C., Kroes, G.-J.: Classical trajectory study of the HOCO system using a new interpolated ab initio potential energy surface. Chem. Phys. Lett. **393**, 236–244 (2004)

Varandas, A.J.C., Viegas, L.P.: The $HO_2 + O_3$ reaction: Current status and prospective work. Comp. Theoret. Chem. **965**, 291–297 (2011)

Viswanathan, R., Raff, L.M.: Theoretical investigations of the reaction dynamics of polyatomic gas-phase systems: the ozone + nitric oxide reaction. J. Phys. Chem. **87**, 3251–3266 (1983)

Vranckx, S., Peeters, J., Carl, S.A.: Absolute rate constant and $O(^3P)$ yield for the $O(^1D) + N_2O$ reaction in the temperature range 227 K to 719 K. Atmos. Chem. Phys. **8**, 6261–6272 (2008a)

Vranckx, S., Peeters, J., Carl, S.: A temperature dependence kinetic study of $O(^1D) + CH_4$: overall rate coefficient and product yields. Phys. Chem. Chem. Phys. **10**, 5714–5722 (2008b)

Vranckx, S., Peeters, J., Carl, S.: Kinetics of $O(^1D) + H_2O$ and $O(^1D) + H_2$: absolute rate coefficients and $O(^3P)$ yields between 227 and 453 K. Phys. Chem. Chem. Phys. **28**, 9213–9221 (2010)

Wahner, A., Mentel, T.F., Sohn, M.: Gas-phase reaction of N_2O_5 with water vapor: Importance of heterogeneous hydrolysis of N_2O_5 and surface desorption of HNO_3 in a large Teflon chamber. Geophys. Res. Lett. **25**, 2169–2172 (1998)

Wald, W.R., Goddard III, W.A.: The electronic structure of the Criegee intermediate. Ramifications for the mechanism of ozonolysis. J. Am. Chem. Soc. **97**, 3004–3021 (1975)

Wallington, T.J., Cox, R.A.: Kinetics and product of the gas-phase reaction of ClO with NO_2. J. Chem. Soc. Faraday Trans. **2**(82), 275–289 (1986)

Wallington, T.J., Japar, S.M.: Reaction of $CH_3O_2 + HO_2$ in air at 295 K: A product study. Chem. Phys. Lett. **167**, 513–518 (1990)

Wallington, T., Ammann M., Atkinson, R., Cox, R.A., Crowley J.N., Hynes R., Jenkin, M.E., Mellouki, W., Rossi, M.J., Troe, J.: IUPAC Subcommittee for Gas Kinetic Data Evaluation for Atmospheric Chemistry, Evaluated Kinetic Data, Gas-phase Reactions. http://www.iupac-kinetic.ch.cam.ac.uk/ (2012)

Wang, J.J., Keyser, L.F.: Kinetics of the $Cl(^2P_j) + CH_4$ reaction: effects of secondary chemistry below 300 K. J. Phys. Chem. A **103**, 7460–7469 (1999)

Wang, J.J., Keyser, L.F.: HCl yield from the OH + ClO reaction at temperatures between 218 and 298 K. J. Phys. Chem. A **105**, 6479–6489 (2001a)

Wang, J.J., Keyser, L.F.: Absolute rate constant of the OH + ClO reaction at temperatures between 218 and 298 K. J. Phys. Chem. A **105**, 10544–10552 (2001b)

Wang, X., Suto, M., Lee, L.C.: Reaction rate constants of $HO_2 + O_3$ in the temperature range 233–400 K. J. Chem. Phys. **88**, 896–899 (1988)

Welz, O., Savee, J.D., Osborn, D.L., Vasu, S.S., Percival, C.J., Shallcross, D.E., Taatjes, C.A.: Direct kinetic measurements of Criegee intermediate (CH_2OO) formed by reaction of CH_2I with O_2. Science **335**, 204–207 (2012)

Wine, P.H., Ravishankara, A.R.: O_3 photolysis at 248 nm and $O(^1D_2)$ quenching by H_2O, CH_4, H_2, and N_2O: $O(^3P_J)$ yields. Chem. Phys. **69**, 365–373 (1982)

Wine, P.H., Nicovich, J.M., Thompson, R.J., Ravishankara, A.R.: Kinetics of atomic oxygen (^3P_J) reactions with hydrogen peroxide and ozone. J. Phys. Chem. **87**, 3948–3954 (1983)

Wine, P.H., Thompson, R.J., Ravishankara, A.R., Semmes, D.H., Gump, C.A., Torabi, A., Nicovich, J.M.: Kinetics of the reaction $OH + SO_2 + M \rightarrow HOSO_2 + M$. Temperature and pressure dependence in the fall-off region. J. Phys. Chem. **88**, 2095–2104 (1984)

Xia, W.S., Lin, M.C.: A multifacet mechanism for the $OH + HNO_3$ reaction: An ab initio molecular orbital/statistical theory study. J. Chem. Phys. **114**, 4522–4532 (2001)
Xu, Z.F., Lin, M.C.: Ab initio study on the kinetics and mechanisms for O_3 reactions with HO_2 and HNO. Chem. Phys. Lett. **440**, 12–18 (2007)
Xu, Z.F., Zhu, R.S., Lin, M.C.: Ab initio studies of ClO_x reactions: VI. Theoretical prediction of total rate constant and product branching probabilities for the $HO_2 + ClO$ reaction. J. Phys. Chem. A **107**, 3841–3850 (2003)
Yang, M.-U., Yang, C.-L., Chen, J.-Z., Zhang, Q.-G.: Modified potential energy surface and time-dependent wave packet dynamics study for $Cl + CH_4 \rightarrow HCl + CH_3$ reaction. Chem. Phys. **354**, 180–185 (2008)
Yoon, S., Henton, S., Zivkovic, A.N., Crim, F.F.: The relative reactivity of the stretch–bend combination vibrations of CH_4 in the Cl $(^2P_{3/2}) + CH_4$ reaction. J. Chem. Phys. **116**, 10744–19752 (2002)
Yu, H.G., Muckerman, J.T.: MRCI calculations of the lowest potential energy surface for CH_3OH and direct ab initio dynamics simulations of the $O(^1D) + CH_4$ reaction. J. Phys. Chem. A **108**, 8615–8623 (2004)
Yu, H.-G., Muckerman, J.T., Sears, T.J.: A theoretical study of the potential energy surface for the reaction $OH + CO \rightarrow H + CO_2$. Chem. Phys. Lett. **349**, 547–554 (2001)
Zabel, F.: Unimolecular decomposition of peroxynitrates. Z. Physik. Chem. **188**, 119–142 (1995)
Zahniser, M.S., Howard, C.J.: Kinetics of the reaction of HO_2 with ozone. J. Chem. Phys. **73**, 1620–1626 (1980)
Zellner, R., Lorenz, K.: Laser photolysis/resonance fluorescence study of the rate constants for the reactions of hydroxyl radicals with ethene and propene. J. Phys. Chem. **88**, 984–989 (1984)
Zellner, R., Fritz, B., Lorenz, K.: Methoxy formation in the reaction of CH_3O_2 radicals with NO. J. Atmos. Chem. **4**, 241–251 (1986)
Zhang, J., Donahue, N.M.: Constraining the mechanism and kinetics of $OH + NO_2$ and $HO_2 + NO$ using the multiple-well master equation. J. Phys. Chem. A **110**, 6898–6911 (2006)
Zhang, J., Dransfield, T., Donahue, N.M.: On the mechanism for nitrate formation via the peroxy radical + NO reaction. J. Phys. Chem. A **108**, 9082–9095 (2004)
Zhou, J., Lin, J.J., Zhang, B., Liu, K.: On the Cl^* $(^2P_{1/2})$ reactivity and the effect of bond excitation in the $Cl + CH_4/CD_4$ reaction. J. Phys. Chem. **108**, 7832–7836 (2004)
Zhou, X.-M., Zhou, Z.-Y., Wu, Q.-Y., Jalbout, A.F., Zhang, N.: Reaction of CH3O2 and HO2: Ab initio characterization of dimer structure and vibrational mode analysis for reaction mechanisms. Int. J. Quant. Chem. **106**, 514–525 (2006)
Zhu, R.S., Lin, M.C.: Ab initio studies of ClOx reactions: prediction of the rate constants of ClO $+ NO_2$ for the forward and reverse processes. Chem. Phys. Chem. **12**, 1514–1521 (2005)
Zhu, R.S., Lin, M.C.: Ab initio study of the catalytic effect of H_2O on the self-reaction of HO_2. Chem. Phys. Lett. **354**, 217–226 (2002)
Zhu, R.S., Lin, M.C.: Ab initio study of the $HO_2 + NO$ reaction: Prediction of the total rate constant and product branching ratios for the forward and reverse processes. J. Chem. Phys. **119**, 10667–10677 (2003a)
Zhu, R.S., Lin, M.C.: Ab initio studies of ClOx reactions. IV. Kinetics and mechanism for the self-reaction of ClO radicals. J. Chem. Phys. **118**, 4094–4106 (2003b)
Zhu, R.S., Diau, E.G.W., Lin, M.C., Mebel, A.M.: A computational study of the OH(OD) + CO reactions: Effects of pressure, temperature, and quantum-mechanical tunneling on product formation. J. Phys. Chem. A **105**, 11249–11259 (2001)
Zhu, R.S., Xu, Z.F., Lin, M.C.: Ab initio studies of Cl*Ox* reactions. I. Kinetics and mechanism for the OH + ClO reaction. J. Chem. Phys. **116**, 7452–7460 (2002)
Zhu, L., Talukdar, R.K., Burkholder, J.B., Ravishankara, A.R.: Rate coefficients for the OH + acetaldehyde (CH_3CHO) reaction between 204 and 373 K. Int. J. Chem. Kinet. **40**, 635–646 (2008)

Chapter 6
Heterogeneous Reactions in the Atmosphere and Uptake Coefficients

Although most of chemical processes constituting a chemical system in the atmosphere consist of the photolysis of gaseous molecules, and gas phase homogeneous reactions that are described in Chaps. 4 and 5, there are some phenomena in which multiphase processes, such as uptake of atmospheric molecules on the solid and liquid surface, heterogeneous reactions on the surface, reactions in the bulk liquids, etc. are important. In the stratosphere, chemical reactions on the polar stratospheric clouds (PSC) are the most prominent example, and are of essential importance in the formation of the "ozone hole". In the troposphere, multiphase reactions in clouds and fog have long been investigated as being related to acid rain historically. Recently, the surface reactions on sea salt in the marine boundary layer and on snow and ice related to tropospheric halogen chemistry, and the uptake and heterogeneous reactions of HO_2 radicals and nitrogeneous compounds on aerosols related to tropospheric ozone chemistry, and heterogeneous oxidation reactions of organic aerosols related to their aging process are paid attention. In this chapter, uptake coefficients of atmospheric species on water droplet, sea salt, mineral particle and soot as important surfaces for tropospheric heterogeneous processes, and reactive uptake on PSC important in the stratosphere are described. The roles of multiphase reactions in the troposphere and stratosphere are described in Chaps. 7 and 8, respectively.

Since most homogeneous reaction processes have already been well established in principle, much attention has recently been focused on chemical processes at the gas-solid and gas-liquid interfaces in the atmosphere and studies are now going on intensively. However, such heterogeneous processes have a problem that the uptake coefficients and reaction probabilities are not unequivocally determined as a constant, being different from homogeneous processes, due to the diversity of chemical and morphological structure of surfaces. The constant to describe loss rate of gas phase molecules on liquid and solid particle surfaces can be expressed as,

H. Akimoto, *Atmospheric Reaction Chemistry*, Springer Atmospheric Sciences,
DOI 10.1007/978-4-431-55870-5_6

$$J_{het} = \frac{1}{4}\gamma u_{av} N_g \tag{6.1}$$

as described by Eq. (2.84) in Sect. 2.4. Here, J (molecules cm^{-2} s^{-1}) is the dissipation flux, γ is the uptake coefficient (the ratio of the loss number of gaseous molecule to the number of colliding molecule to the particles (see Eq. (2.81)), N_g (molecules cm^{-1} s^{-1}) is a molecular density (molecules cm^{-1} s^{-1}) and u_{av} (cm s^{-1}) is the average kinetic velocity of gaseous molecules. In general, the loss process of gaseous molecules on the liquid and solid surfaces is the combined process of gas phase diffusion, accommodation coefficient, Henry's law constant, interaction at the interface, reaction in the bulk liquid, and so on, and the uptake coefficient γ contains their simultaneous equations. Since the general solution of such equations cannot be obtained, approximate equation based on a resistant model is used for expressing γ (see Sect. 2.4). The γ for the case of uptake accompanying heterogeneous reactions which are treated in this chapter can be expressed as shown by Eq. (2.88) before,

$$\frac{1}{\gamma} = \frac{1}{\Gamma_g} + \frac{1}{\alpha} + \frac{1}{\Gamma_{sol} + \Gamma_{rxn}}. \tag{6.2}$$

Here, α is the mass accommodation coefficient expressing the probability of a gaseous molecule to stay on the surface when it collides with the particle surface, and Γ_g, Γ_{sol}, and Γ_{rxn} are conductance corresponding to the gas phase diffusion, liquid phase diffusion and chemical reaction, respectively. When we neglect Γ_g, and $\Gamma_{sol} + \Gamma_{rxn}$ are expressed as Γ_{rs}, Eq. (6.2) can be simplified as,

$$\frac{1}{\gamma} = \frac{1}{\alpha} + \frac{1}{\Gamma_{rs}} \tag{6.3}$$

In this chapter, the experimental values of γ and α defied by the above formula are treated.

However, in some literature, the values under the conditions where Γ_g cannot be neglected, have been reported as they are, as γ and α. Discussion will be made for some heterogeneous processes in such a case. Meanwhile, for example, for the reactions of PSCs, it is thought that $\alpha = 1$ and γ is almost determined by the Γ_{rs}. Furthermore, γ is in general dependent on the density of coexisting molecules which have been taken to the surface in the past and are therefore dependent on the reaction time, initial and stationary state uptake coefficients are expressed as γ_0 and γ_{ss}, respectively, and γ is simply used when they are not distinguished. However, the time for reaching the stationary state depends on the surface process of each reaction and molecular density, it is not necessarily clear whether γ_0 or γ_{ss} should be used in the model for the actual atmosphere. Moreover, in the case of porous surface many orders of magnitude difference appears whether geometric or BET (Brunauer-Emmett-Teller) surface area is used for the calculation of γ. Further, reactive uptake coefficient γ_r is used when the uptake of a molecule accompanies reactions on the surface or in bulk liquid and formation rate of reaction products are concerned.

As for the heterogeneous processes, the NASA/JPL panel evaluation No. 17 (Sander et al. 2011), and Report V (Crowley et al. 2010) and Data Sheet (Wallington et al. 2012) of the IUPAC Subcommittee on Gas Kinetics Data Evaluation for Atmospheric Chemistry provide useful reviews. Due to the above reasons, however, there are many cases in which the recommended values of uptake coefficients are not given since the they are still in process of research. Table 6.1 gives the (reactive) uptake coefficients taken from the above evaluations or recent literature on the heterogeneous processes treated in this chapter. In Table 6.1, the values of α, γ, γ_0, or $\gamma_{0,\ BET}$ are given on a case by case basis, according to each literature. There are high possibilities that they will be renewed in the future, and this chapter will have to be revised. The heterogeneous processes of organic molecules that are now under the progress of research are not included in this book, being related to aging of organic aerosols.

6.1 Uptake on Water Droplet

The uptake coefficients of gaseous molecules into water $H_2O(l)$ are an important constant that determines their removal process on clouds and fogs, and removal rates by deposition over the ocean. The measurements of uptake coefficients for water surface have been made for O_3, H_2O, H_2O_2, NO_2, HONO, HNO_3, and many water soluble organic molecules among atmospheric species. Table 6.1 cites accommodation coefficients or uptake coefficients of typical inorganic atmospheric molecules for water surface. Simulation of transport through the gas-liquid interface of water has been interested theoretically, and a recent review is available in Garrett et al. (2006).

6.1.1 H_2O

Since the uptake and evaporation of gaseous H_2O molecules (water vapor) are very important as a microscopic process of cloud physics, many measurements and theoretical studies have been made from this viewpoint. From the perspective of atmospheric chemistry, it is worthy of study as a most fundamental process of uptake to the water surface among other many atmospheric molecules.

Recent measurements of accommodation coefficients of $H_2O(g)$ to $H_2O(l)$ have been made mainly using a liquid droplet train flow reactor or liquid droplet growth method. Although the obtained values converge within $0.1 < \alpha < 1$, it has been pointed out that there is some discrepancy between the values obtained by both methods. For example, the values of α obtained by the liquid droplet train method of Li et al. (2001) are 0.17 ± 0.03 and 0.32 ± 0.04 at 280 and 258 K, while the value by

Table 6.1 Uptake coefficients on water droplet, sea salt, soil and mineral dust, and soot

Gas phase molecules	Surface	Products	Uptake (accommodation) coefficients	Temperature (K)	References
H_2O	Water	–	$\alpha > 0.3$	250–290	(c)
OH	Water	–	$\alpha > 0.1$	275–310	(a)
HO_2	Water	–	$\alpha > 0.5$	270–300	(a)
	Water	–	$\gamma = 0.1$	290–300	(a)
O_3	Water	–	$\alpha \geq 0.04$	195–262	(b)
	NaCl (s)	–	$\gamma < 1 \times 10^{-2}$	223–300	(b)
	Natural Salt	–	$\gamma = 10^{-3} - 10^{-2}$	~298	(d)
	Al_2O_3	–	$\gamma_{O,\ BET} = (1.2 \pm 0.4) \times 10^{-4}$	296	(e)
	Soil	–	$\gamma_{O,\ BET} = (3\text{–}6) \times 10^{-5}$	296	(e)
	Soot	–	$\gamma_O = 10^{-4}\text{–}10^{-3}$	298	(f)
NO_2	Soot	HONO, NO	$\gamma_{O,\ BET} = (3\text{–}5) \times 10^{-5}$	240–350	(g)
N_2O_5	Water	HNO_3	$\gamma = 0.01\text{–}0.06$	260–295	(b)
	NaCl (s)	$ClNO_2$, HNO_3	$\gamma_O = 0.005$	~ 298	(h)
	NaCl (aq)	–	$\gamma_O = 0.02$	260–300	(a)
	Synthetic Salt	$ClNO_2$	$\gamma_O = 0.02\text{–}0.03$	~ 298	(h)
	Soil	–	$\gamma_{ss} = 0.01\text{–}0.04$	296	(i)
	Soot	HNO_3	$\gamma = (4 \pm 2) \times 10^{-4}$	294	(j)
	Soot	$NO + NO_2$	$\gamma_{ss} = 5.0 \times 10^{-3}$	298	(k)
HNO_3	Water	–	$\alpha \geq 0.05$	250–300	(b)
	NaCl (s)	$NaNO_3$	$\gamma_O = 0.002$	295–298	(b)
	Synthetic Salt	$NaNO_3$	$\gamma_O = 0.07\text{–}0.75$	298	(l)
	$\alpha\text{-}Al_2O_3$	–	$\gamma_O < 0.2$	295–300	(b)
	Soil	–	$\gamma_O \approx 0.1$	298	(m)
	Soot	NO, NO_2	$\gamma_O = (2.0 \pm 0.1) \times 10^{-2}$	298	(n)
$ClONO_2$	NaCl (s)	–	$\gamma_O = 0.23 \pm 0.06$	298	(o)
	Synthetic Salt	–	$\gamma_O = 0.42$	298	(o)

(continued)

Table 6.1 (continued)

Gas phase molecules	Surface	Products	Uptake (accommodation) coefficients	Temperature (K)	References
SO_2	Water	–	$\alpha \geq 0.12$	260–298	(b)
	γ-Al_2O_3	–	$\gamma_0 > 5 \times 10^{-3}$	298	(b)
	Soil	–	$\gamma_0 = (7.6 \pm 0.5) \times 10^{-2}$	298	(b)
	Soot	–	$\gamma \approx 2 \times 10^{-3}$	298	(q)

(a) IUPAC Subcommittee Report Sheet (Wallington et al. 2012)
(b) NASA/JPL Panel Evaluation No.17 (Sander et al. 2011)
(c) Voiglander et al. (2007)
(d) Mochida et al. (2000)
(e) Michel et al. (2003)
(f) Fendel et al. (1995) and Rogaski et al. (1997)
(g) Lelievre et al. (2004)
(h) Thornton and Abbatt (2005)
(i) Wagner et al. (2008, 2009)
(j) Saathoff et al. (2001)
(k) Kargulian and Rossi (2007)
(l) De Haan and Finlayson-Pitts (1997)
(m) Seisel et al. (2004)
(n) Salgado-Muñoz and Rossi (2002)
(o) Gebel and Finlayson-Pitts (2001)
(p) Seisel et al. (2006)
(q) Koehler et al. (1999)

liquid droplet evaporation method of Smith et al. (2006) is 0.62 ± 0.09 at 255–295 K. Meanwhile, Laaksonen et al. (2005), Winkler et al. (2006) reported the values, $0.4 < \alpha < 1$, obtained by the liquid droplet growth rate in the expansion chamber and recommended $\alpha \approx 1$. The discussion on the discrepancy between the values obtained by those methods has been given by Davidovits et al. (2004). In turn, a recent study, combining the liquid droplet growth experiment, by the fluid mechanics model calculation of Voigländer et al. (2007) α is given as close to unity, and the lower limit is reported as 0.3, taking into consideration experimental error. The recommended value by the NASA/JPL panel evaluation No. 17 is $\alpha > 0.1$ (Sander et al. 2011).

The uptake of an H_2O molecule to the liquid H_2O surface has been attracted interest from the molecular dynamics point of view and theoretical analysis has been made (Morita et al. 2003, 2004b; Vieceli et al. 2004). From these theoretical analysis, $\alpha \approx 1$ has been obtained and the reason that experiments by the droplet train method gives the smaller values has been ascribed to the diffusion effect in the gas phase.

In the case that water droplet is coated by organic compounds, α is known to be decreased (Chakraborty and Zachariah 2011; Takahama and Russell 2011; Sakaguchi et al. 2012), which is paid attention to from the point of indirect effect of climate sensitivity of aerosols.

6.1.2 OH

Uptake of OH and HO_2 on $H_2O(l)$ has attracted interest and been investigated from a theoretical point of view since they are the main free radicals carrying the most important HO_x chain reaction in the troposphere as well as being hydrophilic. Since the uptake rate of OH and HO_2 are thought to be dependent on how quickly they are removed from the interface by self-reactions on the $H_2O(l)$ surface, uptake coefficients are conceived to be dependent on contact time of the radicals to the interface, pH of water, co-existing substances, etc. Therefore, values obtained directly by experiments are those under the specific experimental conditions, and it is necessary to convert them by simulation taking into account these interface reactions in order to obtain the accommodation coefficient α.

As for the uptake of OH on $H_2O(l)$, Hanson et al. (1992a) gave $\gamma > 3.5 \times 10^{-3}$ for pure water at 275 K by the measurement using wetted wall flow tube reactor. Takami et al. (1998) obtained $\gamma = (4.2 \pm 2.8) \times 10^{-3}$ for pure water at 293 K by the impinging flow method, and reported that the value decreases with gas-liquid contact time and increases by a factor of 2–3 for the acidic and alkaline water with pH = 1 and 11. They estimated that the accommodation coefficient α is close to unity by the simulation using the rate constant of OH in the aqueous phase and Henry's law constants, which agrees with the result of $\alpha = 0.83$ (300 K) by molecular dynamics calculation by Roeselová et al. (2004). The recommended values of the IUPAC subcommittee and the NASA/JPL panel are $\alpha > 0.1$ and 0.02, respectively (Wallington et al. 2012; Sander et al. 2011).

6.1.3 HO_2

As for the uptake of HO_2 into pure water, Hanson et al. (1992) reported an accommodation coefficient $\alpha > 0.01$ at 275 K by the measurement with the wetted wall flow reactor. Meanwhile, Mozurkewich et al. (1987) reported $\alpha > 0.2$ by taking the obtained value by a fine particle flow system using concentrated NH_4HSO_4 solution doped with Cu^{2+} to facilitate the uptake as an accommodation coefficient. Morita et al. (2004a) obtained $\alpha \approx 1$ by molecular dynamics calculation and deduced that the upper limit of uptake coefficient γ is close to unity. The recommended values by the IUPAC subcommittee and the NASA/JPL panel are $\alpha > 0.5$ and $\alpha > 0.02$, respectively (Wallington et al. 2012; Sander et al. 2011).

6.1.4 O_3

The uptake rate of O_3 at the surface of pure water is very small and cannot be directly measured by experiments, so that all the measurements of the accommodation coefficient have so far been made on aqueous solution containing ions e.g. I^- as O_3 scavenger. From these measurements, it has been known that the accommodation coefficient of O_3 to H_2O (l) has negative temperature dependence, and α becomes larger at the lower temperature between 275 and 300 K. Measured values by Magi et al. (1997) are in the range of 0.92×10^{-2} to 2.08×10^{-2} between 281 and 261 K. Measurements by Müller and Heal (2002) using the wetted wall reactor gave the value of $\alpha = 4 \times 10^{-3}$ at 293 K. They proposed an estimate of $\alpha \geq 0.01$ or the possibility of even larger value considering the experimental errors, while Schutze and Herrmann (2002) reported a value of $\alpha \geq 0.02$ at 298 K by using the single drops flow tube reactor.

From the theoretical calculation of molecular dynamics, $\alpha \approx 0.1$, and 0.047 are given by Roeselová et al. (2003), and Vieseli et al. (2005), respectively. In these studies, they made a comparison between the collisions of OH and O_3 on H_2O (l) surface and showed that OH stays on the surface for 100 ps while O_3 stays less than 50 ps when they collide with thermal energy at room temperature, reflecting the difference in affinity to H_2O (l). The value recommended by the NASA/JPL panel is $\alpha \geq 0.04$ (Sander et al. 2011).

6.1.5 N_2O_5

The uptake of N_2O_5 to water droplets and aerosols forms nitric acid $HONO_2$ (HNO_3), which removes NO_x form the gas phase. Since the process could have large effects on the production rate of O_3 and OH in the troposphere, many studies have been conducted recently.

The reactive uptake processes of N_2O_5 on pure water and salt solutions have been summarized by Bertram and Thornton (2009) as follows,

$$N_2O_5(g) \rightleftarrows N_2O_5(aq) \tag{6.4}$$

$$N_2O_5(aq) + H_2O(l) \rightarrow H_2ONO_2^+(aq) + NO_3^-(aq) \tag{6.5}$$

$$H_2ONO_2^+(aq) + NO_3^-(aq) \rightarrow N_2O_5(aq) + H_2O(l) \tag{6.6}$$

$$H_2ONO_2^+(aq) + H_2O(l) \rightarrow H_3O^+(aq) + HONO_2(aq). \tag{6.7}$$

Succeeding to the absorptive dissolution process (6.4) of N_2O_5 on H_2O (l), the formation of protonated nitric acid $H_2ONO_2^+$ (aq) by the reaction of N_2O_5 (aq) and H_2O (l) is assumed (Eq. 6.5), and $H_2ONO_2^+$ (aq) forms $HONO_2$ (aq) by the reaction with H_2O (l). Also, it is thought that the presence of NO_3^- (aq) increases N_2O_5 (aq) and decreases the uptake rate of N_2O_5 (g).

Van Doren et al. (1990) found that the uptake coefficients of N_2O_5 onto the surface of pure water have negative temperature dependence such that

$\gamma = 0.057 \pm 0.003$ at 271 K and 0.036 ± 0.004 at 282 K by using the droplet train flow reactor method. George et al. (1994) gave similar values of 0.030 ± 0.002 (262 K), 0.013 ± 0.008 (277 K) by using the same experimental methodology. Schütze and Herrmann (2002) obtained $\alpha = 0.011$ (293 K) after the correction of gas phase diffusion by using the single drops flow tube reactor.

On the other hand, many measurements have been made for the uptake of N_2O_5 on aqueous solutions of inorganic salts. The NASA/JPL panel evaluation No. 17 mentions that the reported values are in good agreement within the range of $\gamma = 0.02$–0.04 at room temperature considering the temperature and humidity effects including pioneering study of Mozurkewich and Calvert (1988) for NH_3/H_2SO_4/H_2O and more recent studies of Hallquist et al. (2003) ($(NH_4)_2SO_4$, $NaNO_3$) and Griffiths et al. (2009) (organic acid + $(NH_4)_2SO_4$). Bertram and Thornton (2009) proposed a new parameterization for the uptake of N_2O_5 on a mixed droplet according to the reaction scheme (6.4)–(6.7) based on the experiment using the chemical ionization mass spectrometer for the mixed solution droplets of H_2O, NO_3^-, Cl^-, and organic acid showing that the inhibition effect by NO_3^- is negated by the co-existence of Cl^-.

It has been known that the uptake coefficient of N_2O_5 decreases when the water droplet surface is coated by organics, as in the case of the uptake of H_2O on H_2O (l) (Folkers et al. 2003; Anttila et al. 2006; Park et al. 2007), and theoretical and model analyses have been made by Anttila et al. (2006) and Riemer et al. (2009).

6.1.6 HNO_3

HNO_3 has a large affinity for H_2O (l), and it has been known that the accommodation coefficient rather than solubility or bulk reaction in the liquid phase, is the rate-determining step for the removal process from the gas phase.

As for the uptake coefficient of HNO_3 on H_2O (l), Van Doren et al. (1990) showed that γ has negative temperature dependence such that 0.19 ± 0.02 at 268 K and 0.071 ± 0.02 at 293 K, from the experiment of mono-disperse droplet train in a flow reactor. As seen in Sect. 6.1.1, this experimental method underestimates the accommodation coefficient due to the effect of gas phase diffusion (Garrett et al. 2006). Using the similar experimental method Ponche et al. (1993) obtained the values of 0.11 ± 0.01 at 298 K, and Schütze and Herrman (2002) reported the values of $\alpha \geq 0.03$ after the correction for the effect of gas phase diffusion. The recommended value of the NASA/JPL panel is $\alpha \geq 0.05$ (Sander et al. 2011). In general it is thought that the diffusion in the gas phase is rate determining for the uptake of gaseous molecules when $\alpha \geq 10^{-3}$.

6.1.7 SO_2

When SO_2 contacts with the surface of H_2O (l) and is taken into the aqueous phase, SO_2 react with water molecule as,

$$SO_2(g) + H_2O \rightleftarrows SO_2 \bullet H_2O \tag{6.8}$$

$$SO_2 \bullet H_2O \rightleftarrows HSO_3^- + H^+ \tag{6.9}$$

$$HSO_3^- \rightleftarrows SO_3^{2-} + H^+ \tag{6.10}$$

and reaches to the chemical equilibrium (Wallington et al. 2012). Thus, SO_2 in the interface can be quickly removed by reactions (6.9) and (6.10) when a water droplet is alkaline, but the uptake rate is dependent on contact time for the range of low pH. Jane et al. (1990) reported that the uptake is governed by the formation rate of HSO_3^- from the reaction of SO_2 with H_2O for pH > 5 and is determined by Henry's law solubility for the low pH. In either of the case, however, the uptake coefficients for the experiments changing the pH and the gas-liquid contact time by using the droplet flow method, are larger than those expected from the known constants. Donaldson et al. (1995) detected the chemisorbed SO_2•H_2O on water surface spectroscopically which was assumed by Jane et al. (1990).

The values of γ have been obtained by the experiment of Boniface et al. (2000) as 0.43 ± 0.01, and 0.175 ± 0.015 at 263 and 291 K, respectively showing the negative temperature dependence. Other values are $\gamma = 0.13 \pm 0.01$ with the low pressure reactor by Ponche et al. (1993), and $\gamma = 0.2$ with the liquid impinging method by Shimono and Koda (1996). The recommended value of the NASA/JPL evaluation No. 17 is $\alpha \geq 0.12$ (Sander et al. 2011). The IUPAC subcommittee proposed the bulk accommodation coefficient α_b and recommends $\alpha_b, = 0.11$ (Wallington et al. 2012).

A series of multiphase reactions after SO_2 is taken into fog and rain droplets followed by liquid phase oxidation processes by O_3 and H_2O_2 are described in Chap. 7, Sect. 7.6.2.

6.2 Uptake and Surface Reactions on Sea Salt and Alkaline Halides

The interface reactions of O_3, N_2O_5, etc. on sea salt is important heterogeneous reactions that have a possibility of releasing inorganic molecules containing Cl, Br, etc. However, those reaction processes have not been fully elucidated and are still important themes of research experimentally and theoretically. Furthermore, uptake of HO_2 radicals on sea salt is an important process having a large effect on the efficiency of HO_x cycle and ozone formation in the troposphere, together with the uptake on soil and mineral dust which will be described in the next section. In addition, when nitric acid is converted to $NaNO_3$ by the reaction with sea salt, it can be transported to far longer distance than HNO_3 and has a large influence on nitrogen budget in the troposphere, since the surface deposition rate of $NaNO_3$ is much smaller than HNO_3. The reaction of $ClONO_2$ with sea salt, which will be described in the last paragraph of this section, is important as a process in the tropospheric halogen cycle in the troposphere. As for the uptake and reactions of atmospheric species on sea salt and alkali halide salts, reviews have been given by Finlayson-Pitts (2003) and Rossi (2003).

The efflorescence and deliquescence point of NaCl is relative humidity of 43 % and 75 %, respectively. It is thought that water is adsorbed on the solid surface at above the efflorescence point, and salt is liquefied above the deliquescence point. In this section, uptake and heterogeneous reactions of atmospheric molecules on the surface of sea salts and halogenated alkali salts as their surrogate compounds are described. In general the surfaces of inorganic salts are not porous so that the adsorption surface area by the BET method agree with those of geometrical surface area.

6.2.1 O_3

The uptake coefficients of O_3 on the solid surface of pure NaCl, NaBr, etc. are very small ($\gamma \approx 10^{-6}$) and the decay of O_3 cannot be observed experimentally (Alebic-Juretic et al. 1997; Mochida et al. 2000). However, when water vapor is adsorbed or other ions coexist as in the case of actual sea salts, it has been known that the value of γ increases largely and Cl_2 and/or Br_2 are released.

Sadanaga et al. (2001) reported that the values of γ increases greatly from $<10^{-5}$ to 3.5×10^{-2} when Fe^{3+} coexists in NaCl more than 0.1 %, and under the condition that the ratio of Fe^{3+} increased to 0.5–1 %, the release of Cl_2 is observed in the dark reaction. In the experiment using synthetic sea salt, release of only Br_2 is observed and suggested that Cl_2 could be released after Br_2 is depleted in the reaction on actual sea salt. Hirokawa et al. (1998) reported that the uptake of O_3 and the formation of Br_2 were observed on the surface of NaBr only when the relative humidity was increased to near the deliquescence point. Similar uptake of O_3 and the release of Br_2 have been confirmed on Pyrex glass surfaces (Anastasio and Mozurkewich 2002) and in aerosol chambers (Hunt et al. 2004). Furthermore, Oum et al. (1998b) reported that Br_2 is released from the dark reaction of O_3 on sea ice surface, and Mochida et al. (2000) reported that the formation of Br_2 with $\gamma{\sim}10^{-3}–10^{-2}$ from synthetic and natural sea salt in the Knudsen cell experiment, which are three orders of magnitude larger than those estimated in the reaction in the bulk solution of bromides.

Several studies have been made experimentally and theoretically seeking for the reason of preferential release of Br_2 in the reaction of O_3 with sea salt in spite of the fact that the ratio of Cl: Br is 660:1. Ghosal et al. (2000) showed experimentally by X-ray photoelectron spectroscopy that Br^- is aggregated on the surface of NaCl doped with small amount of Br^- under the exposure of water vapor. Zangmeister et al. (2001) found by use of a spectroscopic method, that the ratio of Br at the surface of mixed salt which has the same Br: Cl ratio as the sea salt is 4–5 %, that is higher than the ratio in the bulk crystal by a factor of 35. Similarly, it has been shown that Br aggregation occurs on the surface of NaCl single crystal doped with Br when the relative humidity increased to 50–65 % by Hess et al. (2007) using the Rutherford backward scattering spectroscopy. It is suggested that the Br/Cl ratio is increased to 0.2 if the results are applied to natural sea salt. On the other hand, from the theoretical molecular dynamics calculation, it has been shown that polar ions such as Br^- and I^- float up to the surface in the presence of water molecules (Jungwirth and Tobias 2002).

6.2.2 OH

Although the uptake coefficient of the OH radical on sea salt has not been measured directly, it has been reported that active chlorine species are formed by the irradiation of O_3 under the coexistence of sea salt particles in a chamber (Behnke et al. 1995). Oum et al. (1998a) confirmed that Cl_2 is released to the gas phase by the irradiation of UV at 254 nm to the deliquescent sea salt in the coexistence of O_3, and proposed the formation of Cl_2 by the surface reaction of sea salt and OH produced by the photolysis of O_3 in the presence of water vapor (reaction 4.3).

Detailed analysis has been made for a chamber experiment with the use of molecular dynamics calculation by Knipping et al. (2000), and this reaction has been confirmed to occur on pure NaCl and sea salt without the existence of other metal ions. From the molecular dynamics calculations, it has become clear that Na^+ is buried into bulk water by solvation through hydrogen bonds on the deliquescent NaCl surface while Cl^- with large molecular diameter is pushed to the liquid surface resulting higher probability of surface reaction, and the affinity energy between OH and Cl is 4.0 kJ mol^- which is much larger than the value 1.2 kJ mol^- for H_2O to allow easier formation of an $OH\bullet\ Cl^-$ complex. Based on these results, Knipping et al. (2000) proposed the formation mechanism of Cl_2 from OH and sea salt as,

$$OH(g) + Cl^-(int) \rightarrow OH \bullet Cl^-(int) \tag{6.11}$$

$$2\ OH \bullet Cl^-(int) \rightarrow Cl_2 + 2\ OH^-. \tag{6.12}$$

Here, Cl^- (int) and $OH\bullet\ Cl^-$ (int) stands for the species on interface surface. Furthermore, it has been shown from kinetic analysis that the experimental results can be reproduced by this mechanism, while they cannot be explained by the known liquid phase reactions unless a strong acidic solution is postulated.

6.2.3 HO_2

Uptake of HO_2 radicals to sea salt has attracted interest as a loss process of HO_2 in the marine boundary layer. Experimentally, the formation of H_2O_2 has been reported by a self-reaction of HO_2 on the surface under the conditions with the radical concentration of a few 100 times higher than in the atmosphere. Recently, Taketani et al. (2008, 2009) reported the experimental uptake coefficient γ under the condition of comparable concentration of HO_2 to the atmosphere by direct measurement of HO_2 using a chemical conversion LIF method used in the atmospheric detection of the radicals. According to their results, the values of γ for dry NaCl and $(NH_4)_2SO_4$ are small, 0.04–0.05 (RH = 20–45 %) and 0.01–0.02 (RH = 20–55 %), respectively, but they increase to 0.11–0.19 (RH = 44–75 %) and 0.09–0.11 (RH = 53–75 %) under increased humidity higher than deliquescence point, and they further increase distinctly to 0.53 ± 0.12 and 0.65 ± 0.17, respectively, with the dope of Cu(II). Thus, it is known that uptake coefficient of HO_2 to sea salt increases with the existence of H_2O(l) molecules and metal ions such as Cu(II) at the surface. Taketani et al. (2009) also reported that γ for synthetic sea salt and natural sea salt crystallized from sea water is 0.07–0.13 and 0.10–0.11, respectively, under the condition of RH = 35–75 %. Based on these results, the IUPAC subcommittee recommends 0.1 for the uptake coefficient for sea salt to be used in atmospheric models (Wallington et al. 2012). Negative temperature dependences for the reaction of HO_2 on sea salt, NaCl, NaBr and $MgCl_2 \cdot 6H_2O$ have been reported by Loukhovitskaya et al. (2009).

From these results, the uptake of HO_2 on sea salt does not have direct chemical interaction with Cl^- and Br^- being different from the case of O_3 and OH, and the following processes are thought if as processes at the interface with water solution (Wallington et al. 2012),

$$HO_2(g) \rightarrow O_2^-(aq) + H^+ \tag{6.13}$$

$$O_2^-(aq) + HO_2(aq) + H_2O(l) \rightarrow H_2O_2(aq) + O_2(aq) + OH^-(aq) \tag{6.14}$$

$$O_2^-(aq) + Me^{n+} \rightarrow \text{products.} \tag{6.15}$$

Here, Me^{n+} stands for metal ions such as Cu(II).

6.2.4 N_2O_5

The reactions of N_2O_5 with sea salt has attracted interest since they are important as a loss process of N_2O_5 in the marine boundary layer, and also as a trigger for tropospheric halogen chemistry by forming photochemically active $ClNO_2$ in the marine boundary layer in urban coasts.

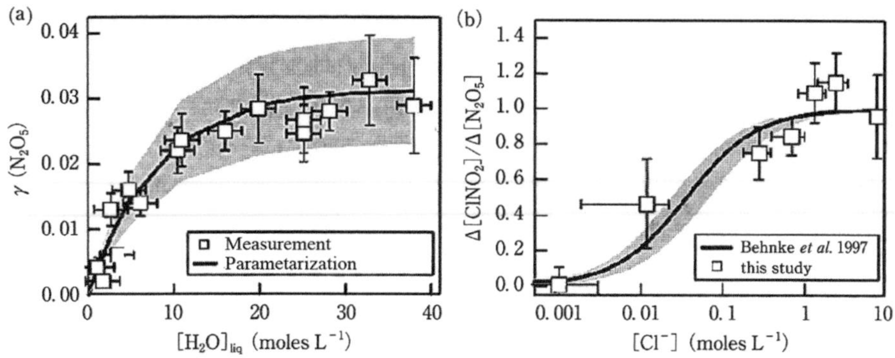

Fig. 6.1 Uptake of N_2O_5 on NaCl particles. (**a**) Dependence of $\gamma(N_2O_5)$ on surface water concentration, (**b**) Dependence of production yields of $ClNO_2$, ($\Delta[ClNO_2]/\Delta[N_2O_5]$), on surface concentration of Cl^- (Adapted from Bertram and Thornton 2009)

In the surface reactions of N_2O_5 with NaCl, a surrogate species for sea salt, the formation of $ClNO_2$ by the reaction with Cl^- and formation of HNO_3 by the hydrolysis reaction are known to occur simultaneously under the condition of the presence of water on the solid surface. Based on many measurements, Bertram and Thornton (2009) compiled the concentration dependence on $H_2O(l)$ and Cl^- for the surface reactions of N_2O_5 with dissolved NaCl, and showed that uptake coefficient γ increases with water concentration on the NaCl surface (Fig. 6.1a). The ratio of the formation yield of $ClNO_2$, $\Delta(ClNO_2)$, to the decreased amount of N_2O_5, $\Delta(N_2O_5)$, increases with the increase of the concentration of Cl^-, and the yield of $ClNO_2$, $\Delta(ClNO_2)/\Delta(N_2O_5)$ is unity for $[Cl^-] > 1$ M (Fig. 6.1b).

The experiments of Hoffman et al. (2003) gave $\gamma = 0.0029 \pm 0.0017(2\sigma)$ for the overall reactive uptake for the $ClNO_2$ and HNO_3 formation, and the branching ratio of $ClNO_2$ formation reaction is 0.73 ± 0.28. Thornton and Abbatt (2005) reported $\gamma = 0.005 \pm 0.004$ for NaCl crystal under the humidity lower than the deliquescence point, which agrees with those of Hoffman et al. (2003). For the deliquescent sea salt, γ is one order of magnitude larger, $\gamma = 0.03 \pm 0.008$ (RH =65 %), agreeing well with the result of Stewart et al. (2004), 0.025 (RH > 40 %). Recently, Roberts et al. (2009) obtained $0.2 < \Delta(ClNO_2)/\Delta(N_2O_5) < 0.8$ for $0.02 < [Cl^-] < 0.5$ M in the experiments of changing the concentration of NaCl in the aqueous solution. These values agree well with the uptake coefficient and the formation yield of $ClNO_2$ as a function of $[H_2O(l)]$ and $[Cl^-]$ shown in Fig. 6.1a, b, respectively.

It has been reported by Hoffman et al. (2003) that for the experiments using synthetic sea salt, the value of γ is one order of magnitude larger, 0.034 ± 0.08, and the formation yield of $ClNO_2$ is 100 %. The value of γ by Stewart et al. (2004) is $\gamma = 0.025$ is independent of humidity under the condition of relative humidity larger than 30 %, and those by Thornton and Abbatt (2005) is $\gamma = 0.02$–0.03 under the relative humidity of 43–70 %.

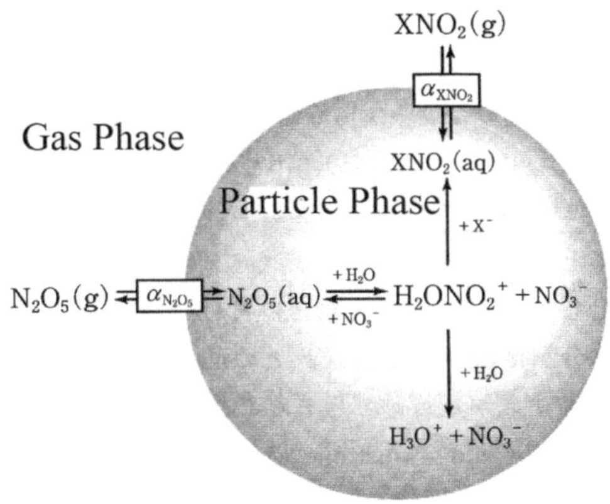

Fig. 6.2 Schematic diagram of uptake on an NaCl particle and reactions of N_2O_5 in liquid particles (Adapted from Bertram and Thornton 2009)

Based on these many experiments, the process for the reaction of N_2O_5 with the surface of dissolved sea salt has been proposed as,

$$N_2O_5(aq) + H_2O(l) \rightarrow H_2ONO_2^+(aq) + NO_3^-(aq) \tag{6.5}$$

$$H_2ONO_2^+(aq) + Cl^-(aq) \rightarrow ClNO_2(aq) + H_2O(l) \tag{6.16}$$

to form $ClNO_2$ (Bertram and Thornton 2009). Figure 6.2 shows schematic picture of the processes in a liquid droplet including these reactions.

6.2.5 *HNO_3*

The reaction of HNO_3 with sea salt is an important heterogeneous process in the troposphere for converting gaseous nitric acid to particulate sodium nitrate $NaNO_3$. Since the uptake coefficient of this reaction is large, most HNO_3 is transported from the coastal terrestrial boundary layer in polluted areas to the marine boundary layer is converted to $NaNO_3$ in a few hours. The deposition rate of HNO_3 to the ground surface is very large and the atmospheric lifetime in the boundary layer is short, while the deposition rate of $NaNO_3$ is much smaller. This means that NO_3^- is transported long distances by the conversion of HNO_3 to $NaNO_3$, and has a large impact on NO_3^- in remote areas. On the other hand, since there is a possibility that NaCl and other salts could be injected to the stratosphere, this heterogeneous reaction is also interested in the stratospheric chemistry.

Many studies have been conducted on the uptake of HNO_3 on NaCl particles and deliquescent salts (Rossi 2003), and the uptake is known to be facilitated by the substitution reaction on the NaCl surface,

$$HNO_3(g) + NaCl(aq) \rightarrow HCl(g) + NaNO_3(aq). \tag{6.17}$$

This reaction is dependent strongly on the adsorbed amount of water, and Ghosal and Hemminger (2004), and Davies and Cox (1998) showed adsorbed water at the surface of NaCl increases mobility of ions and increases the uptake coefficient even below the deliquescent point. The NASA/JPL panel (Sander et al. 2011) recommends the initial uptake coefficient γ_0 as 0.002, based on the above measurements and those of Hoffman et al. (2003) and Leu et al. (1995).

The value of the uptake coefficient of HNO_3 on NaCl differs largely by the surface conditions and γ_0 is as large as 0.5 ± 0.2 for deliquescent salt under RH of 60 %, but the uptake coefficient decreases by a factor of 5–50 when the surface is coated by long-chain fatty acid (Stemmler et al. 2008). Saul et al. (2006) reported that the uptake coefficient becomes as large as >0.1 under the co-existence of $MgCl_2$ even under the low humidity of RH = 10 %.

De Haan and Finlayson-Pitts (1997) reported $\gamma_0 = 0.07$–0.75 and the value at the stationary state $\gamma_{ss} = 0.03$–0.2 by the experiments of using the synthetic sea salt, showing that the uptake coefficient becomes large for sea salt as compared with pure salt similar to the reaction with O_3 mentioned in Sect. 6.2.1. Similarly, the uptake coefficient is very large, 0.5 ± 0.2 for the deliquescent sea salt (RH 55 %) in the experiment of Guimbaud et al. (2002).

6.2.6 $ClONO_2$

In the tropospheric halogen chemistry, $ClONO_2$ is a quasi-stable compound formed in the chain termination reaction of ClO in the urban coastal area where NO_x concentration is relatively high. The reaction of $ClONO_2$ with sea salt results in photochemically active Cl_2 as in the process,

$$ClONO_2(aq) + NaCl(aq) \rightarrow Cl_2(aq) + NaNO_3(aq) \qquad (6.18)$$

and facilitates halogen chain reaction in the troposphere (Finlayson-Pitts et al. 1989) (see Sect. 7.5.2). This reaction may have a role as a source of the Cl_2 which is episodically observed in the marine boundary layer.

Several measured values have been reported for the uptake coefficient of $ClONO_2$ on NaCl crystals. Very large values such as $\gamma_0 = 0.23 \pm 0.06$, (Caloz et al. 1996), 0.10 ± 0.05 (Aguzzi and Rossi 1999), 0.23 ± 0.06 (Gebel and Finlayson-Pitts 2001) have been reported for the initial uptake coefficient, while Hoffman et al. (2003) reported the stationary state value as $\gamma_{ss} = 0.024 \pm 0.012$ after the correction by a model. Deiber et al. (2004) obtained an accommodation coefficient $\alpha = 0.108 \pm 0.03$ from an experiment using the droplet train method with NaCl aqueous solution. As for the product, well-agreed results that the yield of Cl_2 is unity in the reaction (6.18) have been obtained in these experiments.

Similar experiments for NaBr and KBr have been conducted, and large uptake coefficients similar to NaCl described above, and the formation of BrCl as a product has been reported (Caloz et al. 1996; Aguzzi and Rossi 1999; Deiber et al. 2004). Furthermore, Gebel and Finlayson-Pitts (2001) conducted an experiment using synthetic sea salt, and obtained the initial uptake coefficient of 0.42, stationary state uptake coefficient of 0.16, and the yield of Cl_2 as 0.78 ± 0.13.

6.3 Uptake and Surface Reactions on Soil Dust and Mineral Particles

Soil dust blown up from the desert of Sahara, Gobi, etc. is an important constituent of the troposphere, and uptake of O_3, N_2O_5, HNO_3, SO_2, etc. on the particle surface can have a large impact on tropospheric chemistry (Bauer et al. 2004). The soil dust consists of chemical components of SiO_2, Al_2O_3, Fe_2O_3, $CaCO_3$, NaCl, $MgCO_3$, etc. The main components of typical aeolian dust are oxides of silicone and aluminum e.g. 60 % SiO_2, 10–15 % Al_2O_3 and the ratio of minerals containing other elements varies widely by location (Usher et al. 2003). Therefore, SiO_2, Al_2O_3, Fe_2O_3, etc. are often used as surrogates of dust particles in laboratory experiments. Since the surfaces of mineral dust are in general porous, different from inorganic salt crystals described in the previous section, it should be noted that the value of uptake coefficient differs largely before and after the correction for BET surface.

Chemical reactions on mineral dust have been reviewed by Usher et al. (2003), and the impacts on global tropospheric chemistry have been discussed using global chemical transport models by Bian and Zender (2003), Bauer et al. (2004), Evans and Jacob (2005), and other studies.

6.3.1 *O_3*

The decrease of O_3 concentration under aeolian dust has been found in field observations, and the causes have been discussed as due to the decomposition of O_3 on the dust surface, uptake of N_2O_5, the removal of NO_x from gas phase due to the conversion to HNO_3, etc. (Usher et al. 2003).

Many measurements of the uptake coefficient of O_3 on the surface of actual soil particles such as Saharan dust, and surrogates such as SiO_2 have been made studying the dependences on exposure time and on water molecules.

From these results, the following scheme for surface decomposition process has been proposed (Li et al. 1998),

$$O_3 + SS \rightarrow SS - O + O_2 \tag{6.19}$$

$$O_3 + SS - O \rightarrow SS - O_2 + O_2. \tag{6.20}$$

Here, SS expresses reactive sites on the surface of e.g. Al_2O_3 and Fe_2O_3 particles. The uptake of O_3 on mineral surface is irreversible, and the desorption of O_3 from the surface has not been observed (Crowley et al. 2010). From the above scheme, it is expected that the uptake of O_3 is large initially, decreases later and reach to a stationary state after all of the SS are oxidized to SS-O_2. The presence of water molecule is supposed to either inhibit the uptake of O_3 by covering SS competitively or facilitate by reactivating SS-O or SS-O_2, so that the uptake coefficient would have positive or negative humidity effect depending on which process is predominate (Sullivan et al. 2004; Mogili et al. 2006). The presence of SS-O on Al_2O_3 surface has been confirmed by the experiment of Roscoe and Abbatt (2005) using infrared spectroscopy.

Initial uptake coefficients γ_0 have been reported by Michel et al. (2003) in the experiments with a Knudsen cell at 296 K as $(1.2 \pm 0.4) \times 10^{-4}$, $(2.0 \pm 0.3) \times 10^{-4}$ and $(6.3 \pm 0.9) \times 10^{-5}$ for α-Al_2O_3, α-Fe_2O_3 and SiO_2, and $(2.7 \pm 0.8) \times 10^{-5}$ and $(6 \pm 2) \times 10^{-5}$ for yellow dust and Saharan dust, respectively, after the BET surface correction. Meanwhile, stationary state uptake coefficients are $\gamma_{ss,\,BET} = 2.2 \times 10^{-5}$ for α-Fe_2O_3 and 6×10^{-6} for Saharan dust. Sullivan et al. (2004) obtained $\gamma_{0,\,BET} = 1.0 \times 10^{-5}$ and 6×10^{-6} for α-Al_2O_3 and Saharan dust at the low mixing ratio (ppmv) region of O_3 by using a coated wall flow reactor. Hanisch and Crowley (2003) showed that the uptake coefficients for Saharan dust are dependent on the mixing ratio of O_3, and gave $\gamma_{0,\,BET} \sim 3 \times 10^{-5}$, $\gamma_{ss,\,BET} = \sim 7 \times 10^{-6}$ at 30 ppbv of O_3. From these measurements, it has been revealed that $\gamma_{0,\,BET}$ and $\gamma_{ss,\,BET}$ are $\approx 10^{-5}$ and $\approx 10^{-6}$, respectively, for actual soil dust and SiO_2, and the values for pure Al_2O_3 and α-Fe_2O_3 are one order of magnitude larger than these.

6.3.2 HO_2

A measurement of uptake coefficient of HO_2 on soil dust particles has been made recently and its temperature dependence has been reported. Bendjamian et al. (2013) obtained the reactive uptake coefficient of HO_2 on the membrane coated with Arizona desert dust by using the combined method of low-pressure flow system and molecular beam mass spectrometer. They reported $\gamma_0 = 1.2/(18.7 + RH^{1.1})$ at RH = 0.02–94 % independent of temperature (275–320 K) and photo-irradiation, and the upper limit of the formation yield of H_2O_2 on the surface is 5 %. Taketani et al. (2012) measured the uptake coefficients of HO_2 on the particles field-sampled at Mt. Tai (Shandong Province) and Mt. Meng (Beijing City) in China by using a LIF method, and reported the values of 0.13–0.34 and 0.09–0.40, respectively. These values are very large compared to those for single component particles and sea salt, and they suggest the enhancement of uptake coefficient of HO_2 on atmospheric aerosols due to the effect of co-existent trace metal ions and/or organic materials.

6.3.3 N_2O_5

The uptake of N_2O_5 on soil dust is thought to have the largest impact on the decrease of tropospheric ozone by aeolian dust. Although several experimental studies in laboratories have been made, there are still large discrepancies in the measured uptake coefficients. A recent review on heterogeneous reactions of N_2O_5 is given by Chang et al. (2011), including field observations and model calculations.

In the experiments using a Knudsen cell for Saharan dust, Seisel et al. (2005) and Karagulian et al. (2006) reported γ_{ss} (RH = 0 %) as 0.008 ± 0.003 and 0.2 ± 0.05, respectively, which are different by more than an order of magnitude. The reason is thought to be the difference in the concentration of N_2O_5 (the uptake coefficient increases at lower concentration) and in the estimated surface area of particles. Recently, Wagner et al. (2008, 2009) obtained the uptake coefficient (RH = 0) as $\gamma = 0.013 \pm 0.002$ for Saharan dust by using a particle flow reactor simultaneously obtained by Knudsen cell experiments which gave the upper limit, $\gamma_O = \gamma_{ss} = 0.037 \pm 0.012$ and 0.022 ± 0.00 for Saharan and Arizona dust, respectively.

As for the reaction products after N_2O_5 is taken up on particles, Karagulian et al. (2006) reported that the yield of HNO_3 is large for Arizona test dust and Kaolinite (silicate mineral including Al), but the yield is small for Saharan dust and $CaCO_3$, and CO_2 is produced with the yield of 42–50 % particularly for $CaCO_3$.

Seisel et al. (2005) proposed a mechanism for the reaction of N_2O_5 with mineral dust particles as,

$$N_2O_5(g) \bullet S - OH \rightarrow N_2O_5 \bullet S - OH \tag{6.21}$$

$$N_2O_5{*}S - OH \rightarrow HNO_3(ads) + S - NO_3 \tag{6.22}$$

$$HNO_3(ads) + S - OH \rightarrow H_2O(ads) + S - NO_3 \tag{6.23}$$

to convert N_2O_5 (g) to HNO_3 (ads). Here, S-OH is the OH group existing on the surface site. When adsorbed water exists on the surface, it is thought that HNO_3 is formed directly as,

$$N_2O_5(g) + H_2O(ads) \rightarrow 2\ HNO_3(ads) \tag{6.24}$$

$$HNO_3(ads) + H_2O(ads) \rightarrow H_3O^+ + NO_3^-. \tag{6.25}$$

6.3.4 HNO_3

The uptake of gaseous nitric acid HNO_3 into atmospheric dust and mineral aerosols is thought to be an important process resulting in the removal of HNO_3 from the gas phase, and thus affects tropospheric O_3 formation by decreasing the supply of NO_x by the photolysis in the upper troposphere (Bian and Zender 2003; Bauer et al. 2004). On the other hand, the mineral dust coated by HNO_3 is interested from the viewpoint of climate impact since the optical property is changed and it could act as condensation nuclei by changing the particles to hydrophilic (Lohmann et al. 2004).

Although many studies has been reported on the measurements of uptake coefficients of HNO_3 on mineral dust, there are large scatter of values, whose cause is thought to be due to the uncertainty in estimation of surface area of samples. In general, the corrected values for diffusion into surface holes and the BET surface area are smaller than those for the geometric surface area by orders of magnitude. Hanisch and Crowley (2001a, 2001b) gave the uncorrected values 0.13 ± 0.033, 0.10 ± 0.025, 0.11 ± 0.03 for Al_2O_3, $CaCO_3$, and Sahara Dust, respectively, by use of a Knudsen cell. Also by using a Knudsen cell, Seisel et al. (2004) obtained $\gamma_0 \approx 0.1$ for Saharan, Chinese and Arizona dust. Vlasenko et al. (2006) reported the uptake coefficient increases from 0.022 ± 0.007 (RH 12 %) to 0.113 ± 0.017 (RH73 %) with the increase of humidity for the Arizona dust using a flow system experiment. Smaller values $2.5 \pm 0.1 \times 10^{-4}$ (Goodman et al. 2000) and $2 \pm 1 \times 10^{-3}$ (Johnson et al. 2005) are reported for $CaCO_3$ corrected with the BET surface area. The NASA/JPL panel evaluation No. 17 recommends $\gamma_0 < 0.2$ for α-Al_2O_3 base on the measurements by Hanisch and Crowley (2001a, b) and Seisel et al. (2004).

As for the uptake process of HNO_3 on mineral dust, Seisel et al. (2004) proposed the two step mechanism of initial surface adsorption of HNO_3 (g) succeeded by a surface reaction of HNO_3 (ads) based on the experimental evidence that initial uptake coefficient is as large as ~0.1 whereas γ for the formation of NO_3^- at the surface is much smaller, $8 \times 10^{-3} < \gamma < 5.4 \times 10^{-2}$.

Overall reaction of HNO_3 and $CaCO_3$ is expressed as

$$2\ HNO_3 + CaCO_3 \rightarrow Ca(NO_3)_2 + CO_2 + H_2O. \tag{6.26}$$

but the yields of CO_2/HNO_3 and H_2O/HNO_3 are much smaller than the stoichiometric values, which suggests that the actual process is more complex (Goodman et al. 2000).

6.3.5 SO_2

The uptake of SO_2 on soil dust has been interested in relating to the observational evidence that negative correlation is seen between the concentrations of SO_2, and Saharan and Chinese dust (yellow sand) in the field measurement (Hanke et al. 2003), surface of mineral particles is in general coated by nss-SO_4^{2-} (Kojima et al. 2006), and positive correlation is seen between the concentration of nss-SO_4^{2-} and mineral particles in aerosols (Carmichael et al. 1996). The existence of the oxidation process on aerosol surface in addition to the gas phase reaction and liquid phase oxidation in water droplets has also been suggested by Kasibhatla et al. (1997) since the overestimation of SO_2 and underestimation of SO_4^{2-} is seen systematically in the comparison between the global model calculation and observation.

The values of the uptake coefficient of SO_2 on mineral particle surfaces scatter largely as they are dependent on the preparation and processing methods of samples, method of estimation of surface area, humidity and SO_2 concentrations in the gas phase, and other factors. The values of $\gamma_{0,BET}$ are reported as $(9.5 \pm 0.3) \times 10^{-5}$ (Goodman et al. 2001), $(1.6 \pm 0.5) \times 10^{-4}$ (Usher et al. 2002) for Al_2O_3, and $\sim 10^{-6}$ (Ullerstam et al. 2003), $(6.6 \pm 0.8) \times 10^{-5}$ (Adams et al. 2005) for Saharan dust. Without BET correction, large values of γ_0 are reported by Seisel et al. (2006) as $(7.4 \pm 0.9) \times 10^{-3}$, $(8.8 \pm 0.4) \times 10^{-2}$, $(7.6 \pm 0.5) \times 10^{-2}$ for γ-Fe_2O_3, γ-Al_2O_3, and Saharan dust, respectively.

When SO_2 is taken into the surface of mineral dust, it is known to be oxidized to $SO_4{}^{2-}$ and $HOSO_3{}^{-}$ in the presence of O_3 (Usher et al. 2002; Ullerstam et al. 2003). Wu et al. (2011) studied the temperature dependence of the uptake coefficient of SO_2 to form $SO_4{}^{2-}$ on $CaCO_3$ in the presence of O_3, and showed that the formation rate of $SO_4{}^{2}$ increases with temperatures between 230 and 250 K, but decreases with temperatures between 250 and 298 K. The apparent activation energy between 230 and 245 K is relatively large, 14.6 ± 0.2 KJ mole^{-1}. The measured $\gamma_{0,\ \mathrm{BET}}$ value is 1.27×10^{-7} at 298 K, agreeing well with 1.4×10^{-7} on $CaCO_3$ previously obtained by Li et al. (2006). Meanwhile, the value using geometrical surface area instead of the BET area is $(7.7 \pm 1.6) \times 10^{-4}$, which is nearly four orders of magnitude larger. The value of $\gamma_{0,\ BET}$ for Saharan dust by Ullerstam et al. (2002) is 5×10^{-7}, three times larger than for $CaCO_3$, but the value of using the geometrical area is $\sim 10^{-3}$, which is in the same orders of magnitude as Li et al. (2006). The NASA/JPL panel recommends $\gamma_0 > 5 \times 10^{-3}$ for γ-Al_2O_3 (Sander et al. 2011).

From these results uptake of SO_2 in the presence of O_3 is thought to occur in two steps, adsorption on the mineral particle surface and succeeding oxidation (Ullerstam et al. 2002; Li et al. 2006). The rate-determining step is the adsorption on the mineral particle surface, and in the case of Al_2O_3, Fe_2O_3, etc., physisorption and $HSO_3{}^{-}$ and $SO_3{}^{2-}$ formation by dissolution to H_2O (l) are thought to occur in the presence of water. In case of $CaCO_3$ reactions,

$$SO_2 + Ca(OH)(HCO_3) \rightarrow Ca(OH)(HSO_3) + CO_2 \tag{6.27}$$

$$SO_2 + Ca(OH)(HCO_3) \rightarrow CaSO_3 + H_2CO_3 \tag{6.28}$$

have been proposed as reaction processes (Al-Hosney and Grassian 2004; Santschi and Rossi 2006).

6.4 Uptake and Surface Reactions on Soot

Soot is emitted by incomplete combustion of fossil fuel and biomass. It is also called black carbon since it strongly absorbs the solar and terrestrial radiation in all of ultraviolet, visible and infrared regions, and attention is paid as one of the radiatively active species from the climate change point of view. As reactions on soot surface, uptake and dissipation of O_3 by soot emitted from aircrafts into upper troposphere and lower stratosphere, impact on photochemical budget of O_3 by the reactive uptake of NO_2, N_2O_5, HNO_3 in the boundary layer have been discussed. Particularly, formation of HONO by the reaction of NO_2 on soot has been of much concern since it gives important contribution to the formation of OH in polluted atmosphere. Although the fresh surface of soot is hydrophobic, it becomes hydrophilic by taking up H_2SO_4 and HNO_3 in the gas phase, and by the formation of H_2SO_4 in the surface reaction of SO_2. It is thus activated as cloud condensation nuclei and is interested in from a viewpoint of cloud physics.

The components of soot is mainly amorphous carbon containing polycyclic aromatic hydrocarbons and oxygen containing polycyclic aromatic hydrocarbons giving the elemental composition of C: ~ 95 %, H: ~1 %, O: ~1–5 %, N: <1 % (Chughtai et al. 1998; Stadler and Rossi 2000). The existence of various functional groups such as the carbonyl group has been known by infrared spectroscopy (Kirchner et al. 2000; Liu et al. 2010), etc., and the types and amounts of the functional group have important effects on the uptake and reaction of gaseous molecules. The types of functional group on the surface is very much dependent on fuel types, combustion conditions, and sampling location of soot in the flame. As a soot for the use of laboratory experiments on uptake coefficients and surface reactions, samples taken from combustion flame of n-hexane and diesel fuel, and from the spark generators of diesel engines of automobiles are often used. International Steering Committee for Black Carbon Reference Materials recommends those from the n-hexane flame as the standard soot (Sander et al. 2011).

Even though the mass ratio of soot in aerosol is small, its surface area ratio is much larger than the mass ratio since soot has an amorphous and fractal structure, and it has the similar importance for the heterogeneous reactions in the atmosphere as other aerosols. The uptake coefficients of soot obtained so far has an uncertainty of a few orders of magnitude, and the causes are thought to be largely ascribed to the difference in the type and amount of surface functional groups due to the difference in formation methods, and the estimate of surface area. It has been known that there are at least two reactive sites on soot, and the fresh soot near the emission source has sites with large reaction rates, but changes to less reactive site for the aged soot in the atmosphere. The actual values of uptake coefficients of soot in the atmosphere is thus thought to be affected by the difference in the gas-solid contact time resulting in the chemical transformation of the soot surface.

6.4.1 O_3

It has been suggested that the uptake and decomposition of O_3 on soot particles contribute to the loss of O_3 due to the emission from aircraft in the lower stratosphere and upper troposphere (Lary et al. 1997) and nighttime loss of O_3 in the urban atmosphere (Berkowitz et al. 2001), and many measurement have been made (Sander et al. 2011).

The uptake process of O_3 by the new soot causes very large initial loss of O_3, but the uptake rate decreases drastically as the reaction proceeds or for the pre-treated soot. The initial uptake coefficients are reported as $\gamma_O = 10^{-4}$–10^{-3} (Fendel et al. 1995; Rogaski et al. 1997), while they decrease to 10^{-6}–10^{-7} with the progress of the time (Kamm et al. 1999; Pöschl et al. 2001). Longfellow et al. (2000) showed that the value of γ_O gets smaller by a factor of 30 depending on the estimate of the surface area. From these values, discussion has been made that the importance of the process for O_3 loss on soot is limited.

6.4.2 NO_2

The uptake and reaction of NO_2 on soot is very important as a process to release nitrous acid (HONO) into the gas phase. The formation of HONO by the heterogeneous surface reaction and its enhancement by light irradiation has been found for the first time by Akimoto et al. (1987) relevant to the "unknown radical source" in a smog chamber (see column on p.278). The uptake of NO_2 and photocatalytic reaction on soot has been interested in as a model reaction of such heterogeneous process to elucidate the characteristics of HONO formation in the atmosphere.

In the polluted atmosphere, HONO has been measured at far higher concentrations than expected from the production rate by the homogeneous gas phase reaction, $OH + NO + M \rightarrow HONO + M$ (Sect. 5.2.4), and the discussions on the contribution other than the homogeneous reaction have been made in model analysis studies (Gonçalves et al. 2012; Elshorbany et al. 2012).

The uptake of NO_2 is composed of two processes, fast initial uptake and succeeding slow process as in the case of O_3, and many measurements have been made for γ_O and γ_{aged}. The NASA/JPL panel compiled a large range of values, $\gamma_O \approx 10^{-1}$–10^{-4} and $\gamma_{aged} \approx 10^{-4}$–$10^{-6}$ for soot prepared by various methods (Sander et al. 2011). Recently reported values are $\gamma_{O,\ BET} \sim 10^{-5}$ (Lelievre et al. 2004) and $\gamma_{aged,\ BET} \sim 10^{-8}$ (Prince et al. 2002), which are smaller than before and the discussion was made that the process cannot be a source of high concentration of HONO in the polluted atmosphere. However, it has also been pointed out that there is enough possibility to contribute if the photo-enhancement effect on the heterogeneous HONO formation process is considered.

As for the reaction products of NO_2 uptake on soot, the release of HONO into the gas phase is well known. The formation of NO is reported depending on the conditions, but the formation of HNO_3 has not been found (Arens et al. 2001). The yield of HONO is about 100 % for the soot from the flame of high fuel/oxygen ratio, while the yield is lower for soot from lean flame (Stadler and Rossi 2000; Khalizov et al. 2010). Also the uptake coefficients of NO_2 and the yields of HONO are known to be affected by the coating of the surface of soot by H_2SO_4 or organics (Aubin and Abbatt 2007; Khalizov et al. 2010).

Recently, Monge et al. (2010) found that the formation of HONO by the reaction of NO_2 on soot is accelerated by the irradiation of light in the range of 300–420 nm,

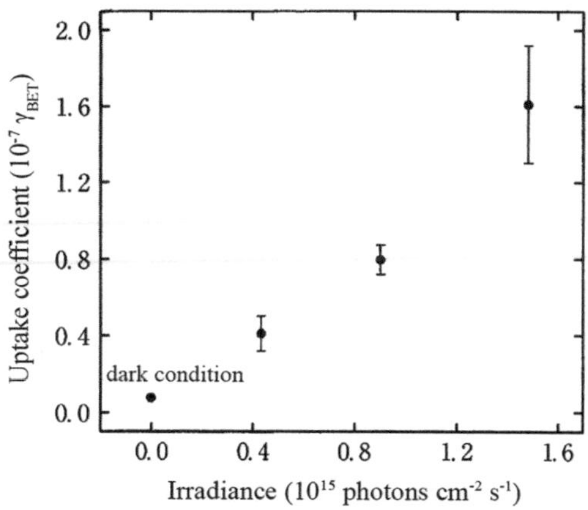

Fig. 6.3 Dependence on the irradiated light intensity of the uptake coefficients γ of NO_2 on soot (Adapted from Monge et al. 2010)

the HONO formation rate does not decrease with time under the irradiation, and the uptake coefficient of NO_2 keeps the value of $(2.0 \pm 0.6) \times 10^{-6}$. Figure 6.3 shows the dependence of γ on light intensity of irradiation. As shown in the figure, the values of γ is proportional to the light intensity. This is thought to be an important experimental results suggesting the importance of the photocatalytic reaction of NO_2 as a source of high concentration of HONO at daytime. The photo-enhancement of HONO formation by the heterogeneous surface reaction has been found in the smog chamber (Akimoto et al. 1987; Rohrer et al. 2005) as noted above and on the thin layer of PAH, pyrene (Brignite et al. 2008), so that the reaction is supposed to occur on various surfaces.

6.4.3 N_2O_5

The reaction of N_2O_5 on soot has been studied considering the possibility of heterogeneous reaction to convert N_2O_5 to HNO_3 effectively alternative to the very slow gas phase bimolecular reaction, $N_2O_5 + H_2O \rightarrow 2\ HNO_3$. According to the measurement by Longfellow et al. (2000), release of NO_2 into gas phase has been found, and the upper limit of uptake coefficient using the geometrical surface area, $\gamma = 0.016$ has been reported. The measurement by Saathoff et al. (2001) gave $\gamma = (4 \pm 2) \times 10^{-5}$ for the $N_2O_5 + \text{soot} \rightarrow 2\ HNO_3$ reaction under dry conditions, $(2 \pm 1) \times 10^{-4}$ for RH 50 %, and $\gamma = (4 \pm 2) \times 10^{-6}$ for the $N_2O_5 + \text{soot} \rightarrow NO + NO_2 + \text{prod.}$ reaction under dry conditions. Kargulian and Rossi (2007) reported the large reactive uptake coefficient for the $NO + NO_2$ formation, $\gamma_{r,\ ss} = 5.0 \times 10^{-3}$ by extrapolating N_2O_5 concentration to zero.

6.4.4 HNO_3

The reaction of HNO_3 with soot is expected to solve the discrepancy between the model and observation for the HNO_3/NO_2 ratio by converting HNO_3 to NO and NO_2 on the surface of soot emitted from aircrafts in the lower stratosphere and upper troposphere (Lary et al. 1997), and several measurements have been made.

According to the measurements of Kirchner et al. (2000), the uptake of HNO_3 on soot particles consists of fast and slow processes as in the uptake of O_3 and NO_2 on soot. For the soot prepared from spark generator, $\gamma = 10^{-3}$–10^{-6} and $\gamma = 10^{-6}$–10^{-8} have been reported for the fast and slow process, respectively. For the soot from diesel engine exhaust, one order of magnitude smaller values are reported. The functional group such as -C = O, -NO_2, -ONO_2, -ONO, etc. are known to be formed on the surface of soot which takes up HNO_3 by using Fourier transform infrared spectroscopy (FTIR), which suggests HNO_3 reacts on the soot surface. Longfellow et al. (2000) reported that the uptake of HNO_3 is reversible at 298 K, and the formation of NO_2 and NO was not observed by the chamber experiment using a flow tube/chemical ionization mass spectrometry (CIMS). On the other hand, Disselkamp et al. (2000) reported the formation of NO_2 by the chamber experiment with a long-path infrared spectrometer. They showed the yield of NO_2 to the decrease of HNO_3 is dependent on the type of soot, but the active points on the surface to form NO_2 are not reproduced leaving the unreacted HNO_3 on the surface. Similarly, Saathoff et al. (2001) reported the formation of NO_2 with $\gamma_{BET} \leq 3 \times 10^{-7}$ in a chamber experiment and Salgado-Muñoz and Rossi (2002) obtained the NO_2 forming uptake coefficients of $\gamma_0 = (2.0 \pm 0.1) \times 10^{-2}$ and $\gamma_{ss} = (4.6 \pm 1.6) \times 10^{-3}$ for geometric surface area using a Knudsen cell for the soot from lean decane flame.

Thus, the formation of NO_2 in the heterogeneous reaction on soot has been confirmed for the atmospheric concentration of HNO_3, but the uptake coefficient γ_r is different by the type and formation method of soot, and the uncertainty is still large as for the importance of this reaction in the atmosphere.

6.4.5 SO_2

The uptake and reaction of SO_2 on soot has been discussed as one of the heterogeneous oxidation process of S(IV) $\rightarrow$ S(VI) together with the oxidation of SO_2 to H_2SO_4 in liquid droplet and on mineral particles, but the measurements of the uptake coefficient are scarce.

Rogaski et al. (1997) and Koehler et al. (1999) reported the initial uptake coefficients of $\gamma \leq (3 \pm 1) \times 10^{-3}$ (298 K), and $(2 \pm 1) \times 10^{-3}$ (173 K), respectively, for geometric surface area using the combustion flame soot. The value decreases to one thirty-third when the roughness of the surface is considered (Koehler et al. 1999). However, since the uptake coefficient decreases with exposure time to nearly zero in a short time, the contribution of the reaction of SO_2 on soot to the formation of H_2SO_4 in the atmosphere is thought to be small. Meanwhile, it has been reported that when metal oxides such as Fe_2O_3, MnO_2, V_2O_5, etc. are present on the soot surface, the formation rate of H_2SO_4 increases significantly as compared to simple substance of soot or metal oxide (Chughtai et al. 1993).

6.5 Reactions on Polar Stratospheric Cloud (PSC)

While the many heterogeneous reactions in the troposphere so far described plays a complementary role to the homogeneous gas phase reactions, heterogeneous reactions on polar stratospheric cloud (PSC) are of primary importance for the formation of stratospheric ozone hole.

When the temperature decreases to lower than 200 K in the lower stratosphere of the polar regions, so-called PSC, clouds consisting of aerosols, appears (Brasseur and Solomon, 2005). The aerosols consist of H_2SO_4, HNO_3 and H_2O in general and it is known that their mixing ratios are known to change according to the thermodynamic stability (Koop et al. 1997). In the stratosphere there is sulfuric acid aerosol of 40–80 wt % H_2SO_4/H_2O (l) (SSA, stratospheric sulfate aerosols) as liquid droplet by nature (see Sect. 8.5), gas phase molecules of HNO_3 and H_2O are taken into the sulfuric acid droplet as temperature decreases, and becomes diluted acids consisting of H_2SO_4 and HNO_3 about 30 % each. When the temperature decreases further, sulfuric acid is further diluted by HNO_3 and H_2O, and becomes a liquid particle almost consisting of nitric acid and water at around 195 K. From this state nitric acid trihydrate (NAT) $HNO_3{\bullet}3H_2O$ freezes out as thermodynamically most stable form (Molina et al. 1993; Voight et al. 2000), and it is called type Ia particles from the viewpoint of LIDAR observation, which accompany depolarization. At the temperature 3–4 K still lower than the NAT formation, super-cooled ternary solution (STS) consisting of $H_2SO_4/HNO_3/H_2O$, or type Ib particles are formed, which does not depolarize LIDAR beam. Furthermore, at temperatures less than 188 K, the freezing temperature of H_2O, large particle of ice consisting of H_2O(s) is formed, which is called type II PSC. Other than this sulfuric acid tetrahydrate (SAT) $H_2SO_4{\bullet}4H_2O$ is known to be frozen out as solid particles from the $H_2SO_4/HNO_3/H_2O$ solution. Although still other particles with different constituents are thought as PSC, three solid particles, ice, NAT and SAT, and two liquid particles, SSA and STS are describe in this section as the most typical PSCs. Table 6.2 shows the characteristics of these main PSCs.

On the other hand, reactions between H_2O, N_2O_5, HCl, HOCl and $ClONO_2$ on PSC are described in this section. The reason that PSC plays a primary role in the formation of Antarctic and Arctic ozone hole is that visible light absorbing species like Cl_2, HOCl and $ClNO_2$ are formed by the reactions on PSC, and they are promptly photolyzed to generate Cl atoms right after the polar night to cause rapid ozone destruction (see Sect. 8.4). Although similar reactions are known for

Table 6.2 Typical particles constituting polar stratospheric clouds (PSC)

Particle names	Chemical composition	Shape · Phase	Particle diameter (μm)	Threshold emperature (K)
Stratospheric sulfate aerosols(SSA)	H_2SO_4/H_2O	Fine liquid droplet	0.1–5	T < 261
Sulfuric acid tetrahydrate(SAT)	$H_2SO_4 \cdot 4H_2O$	Fine solid crystals	<1	T < 213
Nitric acid trihydrate (NAT)	$HNO_3 \cdot 3H_2O$	Solid crystals	1–5	T < 196
Super cooled ternary solution(STS)	$H_2SO_4/HNO_3/H_2O$	Liquid droplet	<1	T < 192
Ice	H_2O	Large solid crystals	5–50	T < 189

Based on Brasseur and Solomon (2005)

Br-containing species (Finlayson-Pitts and Pitts 2000), only the Cl-containing species are described in this chapter as heterogeneous halogen reactions on PSC.

As for the heterogeneous reaction on PSC, the review have been made in the NASA/JPL panel evaluation No. 17 (Sander et al. 2011), and the IPAC subcommittee report Vol. V (Crowley et al. 2010) and data sheet (Wallington et al. 2012), and recommended values for many reactive uptake coefficients are given. Table 6.3 shows the values of reactive uptake coefficients γ_r for Cl-compounds on PSC based on these evaluation reports.

6.5.1 $N_2O_5 + H_2O$

The reaction of N_2O_5 and H_2O on PSC is the process,

$$N_2O_5 + H_2O \rightarrow 2\ HNO_3 \tag{6.29}$$

to form nitric acid on the particle. This reaction is important for the formation of the ozone hole, as it removes reactive NO_x from the gas phase during polar night and accelerates the ozone destruction in spring by negating the ClO_x chain termination reaction $ClO + NO_2 + M \rightarrow ClONO_2 + M$. Furthermore, when N_2O_5 is taken into big ice particles, nitric acid is removed from the stratosphere by gravitational sedimentation, which is called denitrification, and the ozone destruction proceeds even more effectively since NO_2 is no more supplied by the photolysis of HNO_3.

The measurements of uptake coefficients of N_2O_5 on ice particles at 188 K by Leu (1988), and Hanson and Ravishankara (1991b, 1993a) using a coated wall flow reactor, and by Quinlan et al. (1990) and Seisel et al. (1998) using a Knudsen cell agreed well. The NASA/JPL panel (Sander et al. 2011) and the IUPAC subcommittee (Wallington et al. 2012) both recommend $\gamma = 0.02$.

Table 6.3 Reactive uptake coefficients (γ_r) on polar stratospheric clouds under typical polar night conditions

Reactions	Polar stratospheric clouds (a)				
	Solid ice	Solid NAT nitrogen trihydrate	Solid SAT sulfuric acid tetrahydrate	Liquid SSA stratospheric sulfate aerosols	Liquid STS supercooled ternary solution
$N_2O_5 + H_2O \rightarrow 2HNO_3$	0.02 (b)	4×10^{-4} (b)	6×10^{-3} (b)	0.05–0.20 (b)	0.09 (218 K) (c)
					0.02–0.03 (195 K) (c)
$N_2O_5 + HCl \rightarrow ClNO_2 + HNO_3$	0.03 (b)	3×10^{-3} (b)	$<1 \times 10^{-4}$ (SAM) (b)	–	---
$HOCl + HCl \rightarrow Cl_2 + H_2O$	0.2 (b)	0.1 (b)	–	0.15 (58 wt % H_2SO_4	–
				1×10^{-8} atm HCl, 220 K) (d)	
				Decrease with H_2SO_4Concentration	
				Increase with HCl Partial Pressure (e)	
$ClONO_2 + H_2O \rightarrow HOCl + HNO_3$	0.3 (b)	0.004 (b)	0.01 (RH100 %) (f)	0.038 (45 wt % H_2SO_4, 230 K) (g)	0.019 (4.6 % HNO_3
			5×10^{-4} (RH 8 %) (f)	1.1×10^{-5} (75 wt % H_2SO_4, 230 K) (g)	44 % H_2SO_4, 205K) (g)
			Increase with RH in the second order (f)	Rapidly decrease with H_2SO_4 concentration (e)	Decrease with HNO_3 concentration (g)

(continued)

Table 6.3 (continued)

Reactions	Polar stratospheric clouds (a)				
	Solid ice	Solid NAT nitrogen trihydrate	Solid SAT sulfuric acid tetrahydrate	Liquid SSA stratospheric sulfate aerosols	Liquid STS supercooled ternary solution
$ClONO_2 + HCl \rightarrow Cl_2 + HNO_3$	0.3 (b)	0.2 (b)	$\geq$0.1 (RH100 %) (f)	0.6 (45 wt % H_2SO_4, HCl 10^{-8} atm) (g)	0.18 (4.4 % HNO_3
			0.0035 (RH 18 %) (f)	0.043 (55 wt % H_2SO_4, HCl 10^{-8} atm) (g)	44 % H_2SO_4, 205 K) (g)
			Increase with RH in the second order (f)	(both 203–205 K)	Decrease with HNO_3 Concentration (g)
				Rapidly decrease with H_2SO_4 concentration	
				Increase with HCl partial pressure (e)	

Temperature range is in general 180–220 K. Refer to the sources except noted

(a) ice = H_2O, NAT (nitric trihydrate) = HNO_3•$3H_2O$, SAT (sulfuric acid tetrahydrate) = H_2SO_2•$4H_2O$, SSA (stratospheric sulfate aerosols)/LBA (liquid binary sulfate aerosol) = H_2SO_4/H_2O, STS (Supercooled ternary solution) = $H_2SO_2/HNO_3/H_2O$, SAM (sulfuric acid monohydrate) = H_2SO_2•H_2O

(b) NASA/JPL panel evaluation No. 17 (Sander et al. 2011)

(c) Zhang et al. (1995)

(d) Donaldson et al. (1997)

(e) Shi et al. (2001)

(f) Zhang et al. (1994b)

(g) Hanson (1998)

As for the uptake coefficient of N_2O_5 on NAT, the NASA/JPL panel recommends $\gamma = 4 \times 10^{-4}$ based on the measurement of Hanson and Ravishankara (1993a), which is much smaller than for ice particles.

The uptake coefficients of N_2O_5 on SAT surface are recommended as $\gamma = 6 \times 10^{-3}$ by the NASA/JLP panel (Sander et al. 2011), and as $\gamma = 6.5 \times 10^{-3}$ for RH 22–100 % at 195–205 K by the IUPAC subcommittee (Wallington et al. 2012).

Compared to these reactions on solid PSCs, many more measurements have been conducted for the reaction of N_2O_5 on the liquid sulfuric acid H_2SO_4/H_2O surface in the wide temperature range of from 195 K to room temperature. The measurements at low temperature assuming PSC have been made with the use of an aerosol flow reactor (Fried et al. 1994; Hanson and Lovejoy 1994), a wet wall flow reactor (Zhang et al. 1995), Knudsen cell (Beichert and Finlayson Pitts 1996), droplet train method (Robinson et al. 1997), and chamber method (Wagner et al. 2005). The obtained values for 40–80 % H_2SO_4 agree fairly well in the range of $\gamma_r = 0.05$–0.20. Robinson et al. (1997) presumed that the uptake of N_2O_5 is controlled by the hydrolysis rate of N_2O_5 in the bulk liquid and proposed a parameterized formula for the temperature and concentration dependence. The IUPAC subcommittee report recommends $\gamma_r = [(7353/T)\text{-}24.83]^{-1}$ as an equation for the uptake coefficient of N_2O_5 in the temperature range of 210–300 K (Wallington et al. 2012). Figure 6.4 depicts the temperature dependence of the uptake coefficients of N_2O_5 on H_2SO_4/H_2O aerosols summarized in the report. In the figure, α is the physical accommodation coefficient, γ_r is the reactive uptake coefficient of $N_2O_5 + H_2O \rightarrow 2\ HNO_3$, and the solid line is the calculated values by the above equation.

As N_2O_5 is taken into H_2SO_4/H_2O, the sulfuric acid aqueous solution is diluted by nitric acid HNO_3 and the PSC changes to super-cooled ternary solution (STS), $H_2SO_4/HNO_3/H_2O$. The uptake coefficient of N_2O_5 on STS has been shown to decrease as the concentration of HNO_3 increases (Hanson 1997). From the experiments of Zhang et al. (1995) and Wagner et al. (2005), this nitrate effect gets larger

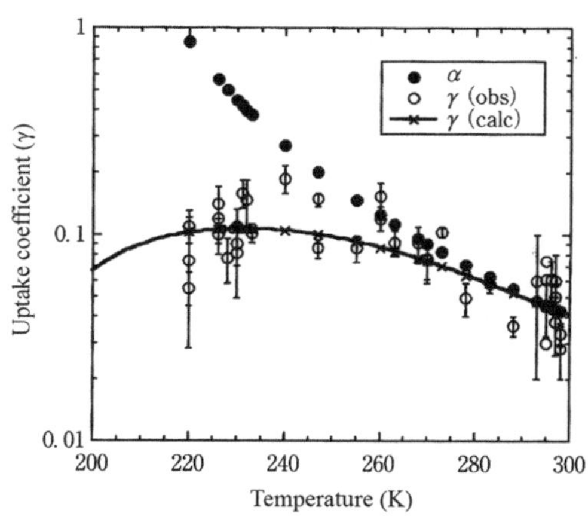

Fig. 6.4 Temperature dependence of uptake coefficient of N_2O_5 on SAA (H_2SO_4/H_2O) surface (Adapted from the IUPAC subcommittee evaluation data sheet, Wallington et al. 2012)

as the temperature decreases, and under the stratospheric condition of $P(H_2O) = 3.8 \times 10^{-4} - 1.0 \times 10^{-3}$ Torr, γ_r decreases from 0.09 at 195 K to 0.02–0.03 at 218 K by a factor of 2–5 (Zhang et al. 1995).

6.5.2 $N_2O_5 + HCl$

The reaction of $N_2O_5 + HCl$ on PSC proceeds as,

$$N_2O_5 + HCl \rightarrow ClNO_2 + HNO_3 \tag{6.30}$$

to convert metastable HCl to photochemically active $ClNO_2$ which is released to the gas phase, and to convert NO_x to HNO_3 to remove NO_2 from the gas phase. Both processes accelerate the destruction of polar stratospheric ozone in springtime.

For the reaction of $N_2O_5 + HCl$ on the surface of ice particles $H_2O(s)$, it has been reported by Seisel et al. (1998) that the uptake coefficient of N_2O_5 is $\gamma_r = 0.03$ with the $ClNO_2$ yield of 63 % for the consumed N_2O_5. The NASA/JPL panel recommends this value. Hanson and Ravishankara (1991a), however, pointed out that this reaction forms NAT layer on the ice surface, which decreases the uptake coefficient, and the measurement of reactive uptake coefficient for pure $H_2O(s)$ is difficult.

The uptake coefficient 3.2×10^{-3} of this reaction on NAT particle has been recommended by the NASA/JPL panel (Sander et al. 2011) based on the measurement by Hanson and Ravishankara (1991a). The measurement on SAT has not been made and a small value of $\gamma_r = 1 \times 10^{-4}$ has been reported for $H_2SO_4 \cdot H_2O$ (SAM, sulfuric acid monohydrate) by Zhang et al. (1995). The IUPAC subcommittee (Wallington et al. 2012) recommends the upper limit value of $\gamma_r < 1 \times 10^{-4}$ based on this value.

6.5.3 $HOCl + HCl$

The reaction of HOCl and HCl on PSC particles forms Cl_2 as follows.

$$HOCl + HCl \rightarrow Cl_2 + H_2O \tag{6.31}$$

This reaction and those of $ClONO_2$ described in the following Sects. 6.5.4 and 6.5.5 are called chlorine activation since they convert metastable HCl and $ClONO_2$ stored in polar night of the stratosphere to photochemically active Cl_2 and HOCl to be released into the gas phase. As shown in Fig. 4.38 in Sect. 4.4.3, HCl only has an absorption spectrum shorter than 200 nm, whereas HOCl has an absorption spectrum longer than 300 nm, as shown in Fig. 4.38 (Sect. 4.4.4), and can be photolyzed by the low altitude sunlight of early polar spring in the stratosphere. Cl_2

has much larger absorption cross-sections in the visible region as shown in Fig. 4.36 (Sect. 4.4.1) so that it can be photolyzed more effectively to release two Cl atoms.

It has been known that the reaction of HOCl and HCl has a large reactive uptake coefficient of $\gamma_r > 0.1$ on either PSCs of ice particle, NAT, SAT and SSA (Hanson and Ravishankara 1992; Abbatt and Molina 1992a). McNeill et al. (2006) deduced that the presence of HCl in the gas phase causes surface disorder by the dissolution of ice surface layer, and induces quasi liquid layer (QLL) at stratospheric temperature (188–203 K), which accelerate heterogeneous reaction of HOCl and $ClONO_2$ with HCl on ice surface. For this reason, the uptake coefficient of HOCl on the ice surface doped with HCl is very large, and the NASA/JPL panel recommends $\gamma_r = 0.2$ (uncertainty by a factor of 2) taking the average of the values reported by Hanson and Ravishankara (1992), Abbatt and Molina (1992a), and Chu et al. (1993). Meanwhile, this value is reported to decrease by a factor of 3-4 if the surface area of porosity is considered (Chu et al. 1993). The products of this reaction are reported to be Cl_2 and H_2O as seen in reaction (6.31). The production yield of Cl_2 for the loss of HOCl is reported to be 0.87 ± 0.20 by Abbatt and Molina (1992a), and it is generally thought this reaction forms Cl_2 and H_2O with the yields of unity (Sander et al. 2011).

The γ_r for the reaction of HOCl and HCl on NAT particles is known to increase with water vapor and reaches a constant value at more than a certain vapor pressure. The average of measured values by Hanson and Ravishankara (1992) and Abbatt and Molina (1992a) is $\gamma_r = 0.135 \pm 0.049$ without considering the porosity of the surface. The γ_r decreases to one tenth when water vapor is not present (Abbatt and Molina 1992a). The recommended values by the NASA/JPL panel is $\gamma = 0.1$ with an uncertainty by a factor of 2 (Sander et al. 2011).

Many measurements have been made for the reaction of HOCl and HCl on sulfuric acid liquid droplets $H_2SO_4 \cdot nH_2O$ (l). The uptake coefficient of this reaction is dependent on temperature, water vapor, and concentration ratio of H_2SO_4/H_2O. The reason is thought to be that the solubility of HOCl and HCl is affected largely by these parameters. The experimental results of the reaction of HCl with HOCl and with $ClONO_2$ in sulfuric acid droplets described in Sect. 6.5.5 has been proposed to be explained by the following acid-catalyzed protonation reaction pathways,

$$\mathrm{HOCl(g)} \rightleftarrows \mathrm{HOCl(l)} \tag{6.32}$$

$$\mathrm{HOCl(l)} + \mathrm{H^+} \rightleftarrows \mathrm{H_2OCl^+(l)} \tag{6.33}$$

$$\mathrm{H_2OCl^+(l)} + \mathrm{HCl(l)} \rightarrow \mathrm{H_3O^+(l)} + \mathrm{Cl_2(l)} \tag{6.34}$$

$$\mathrm{Cl_2(l)} \rightleftarrows \mathrm{Cl_2(g)}. \tag{6.35}$$

Shi et al. (2001) modeled the reaction considering this scheme in the temperature range of 185–260 K, and showed that the measured values of γ_r by Donaldson et al. (1997), Hanson and Lovejoy (1996) and Zhang et al. (1994a) can be reproduced well by the model as shown in Fig. 6.5. As seen in the figure, the value of γ_r increases with the increase of partial pressure of HCl, and H_2O ratio in sulfuric acid. It decreases with the increase of temperature, and Zhang et al. (1994a)

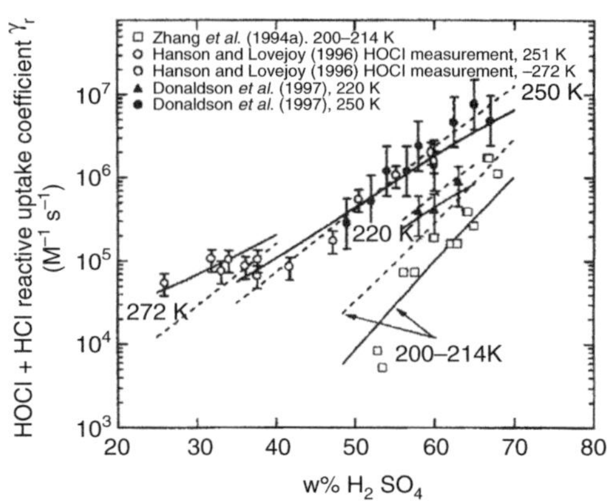

Fig. 6.5 Dependence of reactive uptake coefficients for the reaction of HOCl and HCl on sulfuric acid liquid droplet on temperature and sulfuric acid concentration (Adapted from Shi et al. 2001)

reported that γ_r decreases by a factor of 50 when the temperature increases from 198 k to 208 K. The rate of the reaction HOCl + HCl is very fast in the cold stratosphere at T < 199 K, and it decreases as HCl is depleted.

6.5.4 $ClONO_2 + H_2O$

The formation of HOCl in the reaction of $ClONO_2$ and H_2O on PSC,

$$ClONO_2 + H_2O \rightarrow HOCl + HNO_3 \tag{6.36}$$

is an important chlorine activation reaction (see the previous paragraph) together with the Cl_2 production in the reaction of $ClONO_2$ and HCl described in the following paragraph. In the polar night with no solar irradiation, most ClO radicals are converted to metastable chlorine nitrate $ClONO_2$ by the reaction with NO_2. The absorption spectrum of $ClONO_2$ has large absorption cross-sections only at wavelengths shorter than 300 nm, as seen in Fig. 4.37 (Sect. 4.4.2), so that if $ClONO_2$ stays as it is during the polar night, release of active chlorine by the photolysis does not occur effectively by the solar irradiation with relatively long wavelength in early spring at the lower altitude of sun, and the rapid ozone depletion producing the ozone hole cannot be not caused. Only in the presence of PSC at polar night, the conversion of $ClONO_2$ to HOCl occurs. Since the HOCl has absorption in the range of 300–350 nm as shown in Fig. 4.39a (Sect. 4.4.4), the release of Cl atoms by the low altitude sunlight is possible, and the formation of ozone hole results.

The uptake of $ClONO_2$ on ice surface followed by the reaction with H_2O is thought to form HOCl and HNO_3. At the stratospheric temperature, the product HOCl is released to the gas phase, and HNO_3 stays on the ice surface to form NAT. The accumulation of HNO_3 on the ice surface decreases the number of H_2O molecules that can participate the surface reaction, and interfere with the total reaction. Therefore, the uptake coefficient of this reaction decreases in general with time, and the evidence is more distinct for the experiments using high concentration of $ClONO_2$ (Sander et al. 2011; Wallington et al. 2012). The NASA/JPL panel (Sander et al. 2011) recommends the reactive uptake coefficient $\gamma_r = 0.3$ at 180–200 K for the geometrical surface area, based on the experiments by Hanson and Ravishankara (1991a, 1992), Oppliger et al. (1997) and Fernandez et al. (2005) using relatively low concentrations of $ClONO_2$. The IUPAC subcommittee (Wallington et al. 2012) gives the negative temperature dependence of this reaction as depicted in Fig. 6.6 and temperature parameters based on the result of Fernandez et al. (2005).

The reactive uptake coefficient of the reaction $ClONO_2$ and H_2O on NAT has been measured by Hanson and Ravishakara (1991a, 1992, 1993b), Abbatt and Molina (1992b), Zhang et al. (1994b), Barone et al. (1997), etc., and shown that γ_r is larger under the higher water vapor pressure (Wallington et al. 2012). Sander et al. (2011) presented the average value of $\gamma_r = 0.0043 \pm 0.0021$ for RH $\geq$ 90 %. This reaction is known to have positive temperature dependence as shown in Fig. 6.7 conversely to the reaction on ice particles, and $\gamma_r = 7.1 \times 10^{-3}$ exp (-2940/T) is given for the Arrhenius plot of γ_r at 100 % RH.

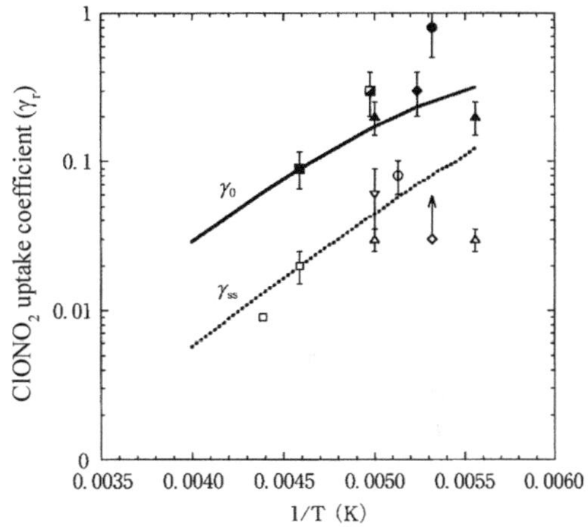

Fig. 6.6 Temperature dependence of reactive uptake coefficients for the reaction $ClONO_2$ and H_2O on PSC ice surface (Adapted from IUPAC subcommittee evaluation data sheet, Wallington et al. 2012). Refer to the original literature for the source of each experimental point. *Solid* and *dotted lines* are calculated values by the model for γ_0 and γ_{ss}, respectively

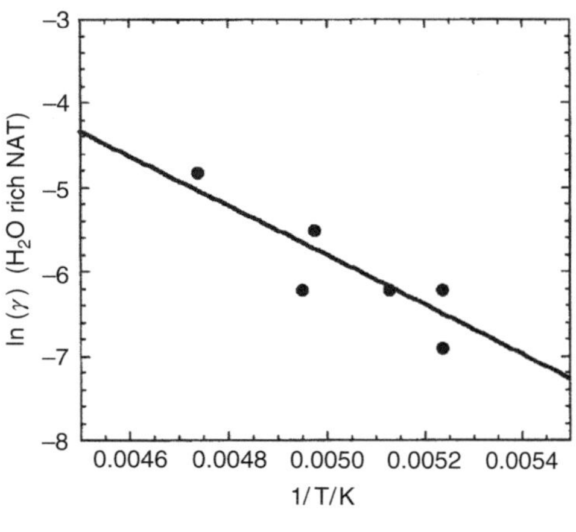

Fig. 6.7 Temperature dependence of reactive uptake coefficients for the reaction $ClONO_2$ and H_2O on NAT (Adapted from the IUPAC subcommittee evaluation data sheet, Wallington et al. 2012)

The relative humidity dependence of the reactive uptake coefficient for the reaction, $ClONO_2$ and H_2O, on SAT has been measured by Hanson and Ravishankara (1993b), and Zhang et al. (1994b) for the temperature range 192–205 K. From these results, γ_r of this reaction increases rapidly with the increase of RH. The measurement of Zhang et al. (1994b) gave $\gamma_r = 0.016$ at 195 K, which is in between ice particles and H_2O-rich NAT. It decreases rapidly with the decrease of RH to 5×10^{-4} at RH 8 %. Based on these data, the IUPAC subcommittee recommends the dependence of γ_r on relative humidity as $\gamma_r = 1 \times 10^{-4} + 1 \times 10^{-4}$ [RH] $+ 1 \times 10^{-7}$ [RH] ^{2} (192–205 K) (Crowley et al. 2010; Wallington et al. 2012).

Many measurements have been made for the reaction of $ClONO_2 + H_2O$ on SSA, sulfuric acid liquid droplets $H_2SO_4/H_2O(l)$, as compared to the solid PSCs mentioned above, and the dependence on temperature, humidity, and sulfuric acid composition has been investigated. The reaction products of $ClONO_2 + H_2O$ on the sulfuric acid liquid droplets are HOCl and HNO_3 that are the same for the solid PSCs, but HNO_3 are released to the gas phase being different from the cases of the solid surface. The reactive uptake coefficient is known to be strongly dependent on the H_2O content in sulfuric acid, and increases with the H_2O in the range of 20–70 % wt % H_2SO_4, and it increases with the decrease of temperature (Hanson and Ravishankara 1991b; Zhang et al. 1995; Ball et al. 1998; Hanson 1998). For example, γ_r is close to 0.1 for 40 % wt % H_2SO_4, but decreases to ~10^{-5} for 75 wt % H_2SO_4 (Hanson 1998) at around 200 K.

Shi et al. (2001) analyzed the reaction rate of this reaction based on the assumption that both of direct reaction (6.36) and the acid-catalyzed reactions proposed similarly for the reaction of HOCl and HCl (Sect. 6.5.3),

$$ClONO_2 + H^+ \rightleftarrows ClONO_2{}^+(l) \tag{6.37}$$

$$ClONO_2{}^+(l) + H_2O(l) \rightarrow H_3O^+(l) + HOCl(l) \tag{6.38}$$

occurs simultaneously. Figure 6.8 depicts the dependence of the reactive uptake coefficients of the $ClONO_2 + H_2O$ reaction on temperature and sulfuric acid concentration to compare the experimental data and model equation of Shi et al. (2001) based on the above mechanism. The detailed parameters of the model equation are given in the IUPAC subcommittee data sheet (Wallington et al. 2012).

The uptake coefficient of this reaction is smaller for the nitric acid co-existing sulfuric acid water droplets, $H_2SO_4/HNO_3/H_2O(l)$. The values of γ_r under the typical condition of gas phase concentration ~ 5ppb HNO_3 in the polar lower stratosphere are reported to be one half of the reaction on $H_2SO_4/H_2O(l)$ (Zhang et al. 1995; Hanson 1998).

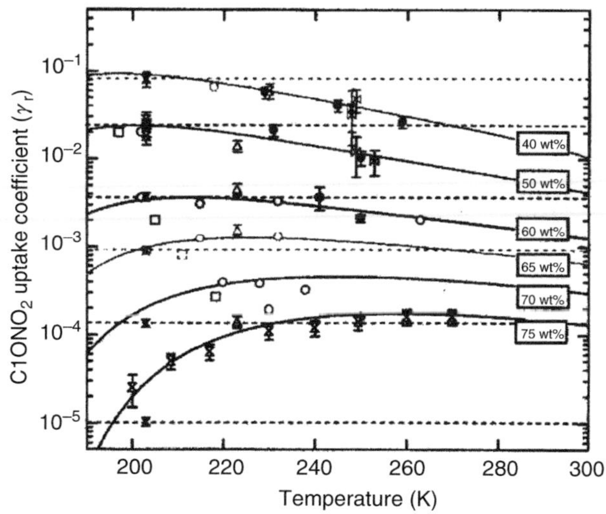

Fig. 6.8 Temperature and sulfuric acid concentration dependence of reactive uptake coefficients for the reaction of $ClONO_2$ and H_2O on sulfuric acid liquid droplet (Adapted from Shi et al. 2001). Refer to original literature for the source of each experimental point

6.5.5 *$ClONO_2$ + HCl*

The reaction of $ClONO_2$ and HCl on PSC,

$$ClONO_2 + HCl \rightarrow Cl_2 + HNO_3 \tag{6.39}$$

is a main reaction for the formation of the polar ozone hole together with the reaction of $ClONO_2$ and H_2O described in the previous paragraph. Particularly, reaction (6.39) forms photochemically active Cl_2 containing two chlorine atoms each from metastable $ClONO_2$ and HCl. Since the absorption spectrum of Cl_2 extends to the visible region, as seen in Fig. 4.35 (Sect. 4.4.1), so that Cl_2 can release Cl atoms more effectively than HOCl under the low altitude sun, and brings rapid ozone destruction in the polar stratosphere in early spring.

The heterogeneous reaction of $ClONO_2$ + HCl occurs rapidly on the ice H_2O (s) surface and the product Cl_2 is released to the gas phase immediately (Oppliger et al. 1997). According to the measurement by McNeill et al. (2006) using a Knudsen cell, the uptake coefficient of this reaction γ_r has large values of $\geq$0.3 being independent on p_{HCl} under the condition that partial pressure of HCl (p_{HCl}) is larger than $ClONO_2$ (p_{ClONO2}). The NASA/JPL panel (Sander et al. 2011) recommends $\gamma_r = 0.3$ at 180–200 K based on this value and other measurements (Leu 1988; Hanson and Ravishankara 1991a; Chu et al. 1993; Lee et al. 1999; Fernandez et al. 2005). The γ_r of this reaction decreases under the condition of $p_{HCl} \leq p_{ClONO2}$ (Oppliger et al. 1997), and also decreases as the ice surface is coated by HNO_3, the same as for the reaction of $ClONO_2$ and H_2O in the previous paragraph (Fernandez et al. 2005).

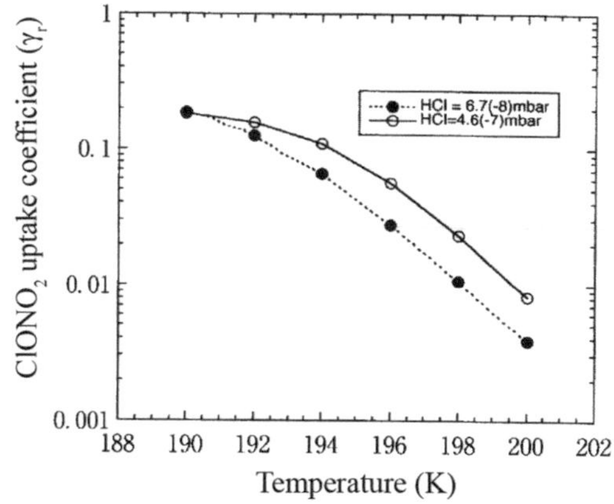

Fig. 6.9 Temperature dependence of reactive uptake coefficients for the reaction of $ClONO_2$ and HCl on NAT (adapted from the IUPAC subcommittee evaluation data sheet, Wallington et al. 2012)

The reactive uptake coefficients for the reaction of $ClONO_2$ and HCl on NAT have been measured by Hanson and Ravishankara (1991a, 1992, 1993b), Leu et al. (1991), and Abbatt and Molina(1992b), and agreed well with $\gamma_r > 0.1$. The recommended value by the NASA/JPL panel based on these is $\gamma_r = 0.2$ at 185–210 K. According to Abbatt and Molina (1992b), γ_r of this reaction is >0.2 at RH 90 %, about the same as on H_2O(s) mentioned above. It decreases rapidly to 0.002 at RH 20 %, and it is suggested that solvation is necessary for this reaction to occur. Figure 6.9 shows the reactive uptake coefficients on H_2O-rich NAT as a function of temperature as compiled by the data sheets of IPAC subcommittee (Wallington et al. 2012). demonstrating that they decrease with temperature in the range of 190–202 K. Meanwhile, the reactive uptake coefficient on the HNO_3-rich NAT increases with the partial pressure of HCl, and it has been shown that HCl is the limiting factor of this reaction (Abbatt and Molina 1992b). The HCl dependence of this reaction has been modeled by Carslaw and Peter (1997).

The reactive uptake coefficient of the reaction $ClONO_2$ and HCl on SAT has been measured by Hanson and Ravishankara (1993b), and Zhang et al. (1994b), and is known to be strongly dependent on temperature and water vapor pressure. The γ_r of this reaction has a large value of 0.12 at around RH 100 %, but decreases rapidly with the decrease of RH to give 0.0035 at RH 18 % in the similar way as on NAT. A parameterized equation of the RH-dependence of γ_r has been presented by Zhang et al. (1994b).

Many more measurements for the $ClONO_2 + HCl$ reaction on SSA, sulfuric acid liquid particle under the stratospheric condition of polar night, have been made as compared to those on solid PSCs similar to the case of the $ClONO_2 + H_2O$ reaction (Tolbert et al. 1988; Hanson and Ravishankara 1991b, 1994; Zhang et al. 1994a; Elrod et al. 1995; Hanson 1998). The reaction products are Cl_2 and HNO_3 as shown in reaction (6.39), both of which are released to the gas phase. Alike the reaction of $ClONO_2$ and H_2O described in a precedent paragraph, this reaction depends on temperature, humidity and composition of sulfuric acid, and has a complex dependence on the partial pressure of HCl. The γ_r of this reaction has a large value of

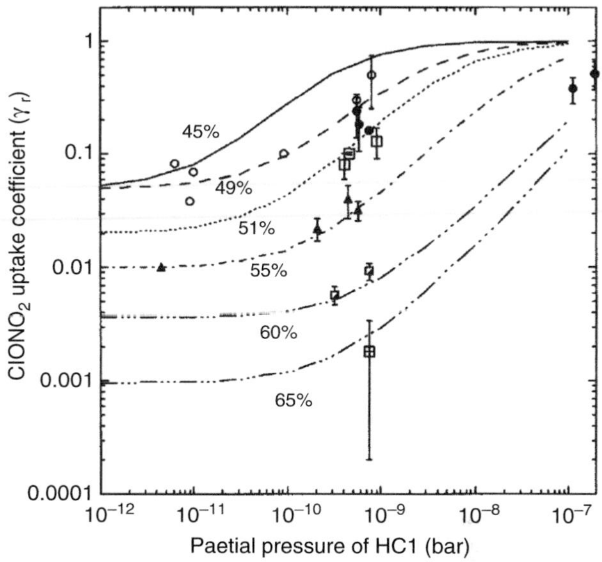

Fig. 6.10 Dependence of reactive uptake coefficients for the reaction of $ClONO_2$ and HCl on sulfuric acid liquid on gas phase HCl partial pressure for different H_2SO_4 wt % (Adapted from the IUPAC subcommittee evaluation data sheet, Wallington et al. 2012)

$\gamma_r = 0.6$ at 202 K when the HCl partial pressure is relatively high and decreases to 0.01 as p(HCl) decreases. The γ_r has a negative temperature dependence, which is thought to reflect the fact solubility of HCl to sulfuric acid decreases with the increase of temperature.

Shi et al. (2001) assumed the acid-catalyzed reaction by H^+ for this reaction similar to the $ClONO_2 + H_2O$ reaction in the preceding paragraph, and presented a modeling equation for the dependence of γ_r on temperature, humidity and composition of sulfuric acid by evaluating the available experimental data. The IUPAC subcommittee report sheet adopted the parameters given by Shi et al. (2001), and Fig. 6.10 depicts the dependence of γ_r on p_{HCl} at 200 K for the different sulfuric acid concentration wt % given in the sheet (Wallington et al. 2012). The γ_r of this reaction is also known to decrease with the increase of the concentration of HNO_3 in $H_2SO_4/H_2O(l)$ (Zhang et al. 1994b; Hanson 1998).

References

Abbatt, J.P.D., Molina, M.J.: The heterogeneous reaction of $HOCl + HCl \rightarrow Cl_2 + H_2O$ on ice and nitric acid trihydrate: reaction probabilities and stratospheric implications. Geophys. Res. Lett. **19**, 461–464 (1992a)

Abbatt, J.P.D., Molina, M.J.: Heterogeneous interactions of $ClONO_2$ and HCl on nitric acid trihydrate at 202 K. J. Phys. Chem. **96**, 7674–7679 (1992b)

Adams, J.W., Rodriguez, D., Cox, R.A.: The uptake of SO_2 on Saharan dust: a flow tube study. Atmos. Chem. Phys. **5**, 2679–2689 (2005)

Aguzzi, A., Rossi, M.J.: The kinetics of the heterogeneous reaction of $BrONO_2$ with solid alkali halides at ambient temperature. A comparison with the interaction of $ClONO_2$ on NaCl and KBr. Phys. Chem. Chem. Phys. **1**, 4337–4346 (1999)

Akimoto, H., Takagi, H., Sakamaki, F.: Photoenhancement of the nitrous acid formation in the surface reaction of nitrogen dioxide and water vapor: extra radical source in smog chamber experiments. Int. J. Chem. Kinet. **19**, 539–551 (1987)

Alebic-Juretic, A., Cvitas, T., Klasinc, L.: Ozone destruction on solid particles. Environ. Monit. Assess. **44**, 241–247 (1997)

Al-Hosney, H.A., Grassian, V.H.: Carbonic acid: an important intermediate in the surface chemistry of calcium carbonate. J. Am. Chem. Soc. **126**, 8068–8069 (2004)

Anastasio, C., Mozurkewich, M.: Laboratory studies of bromide oxidation in the presence of ozone: evidence for glass-surface mediated reaction. J. Atmos. Chem. **41**, 135–162 (2002)

Anttila, T., Kiendler-Scharr, A., Tillmann, R., Mentel, T.F.: On the reactive uptake of gaseous compounds by organic-coated aqueous aerosols: theoretical analysis and application to the heterogeneous hydrolysis of N_2O_5. J. Phys. Chem. A **110**, 10435–10443 (2006)

Arens, F., Gutzwiller, L., Baltensperger, U., Gäggeler, H.W., Ammann, M.: Heterogeneous reaction of NO_2 on diesel soot particles. Environ. Sci. Technol. **35**, 2191–2199 (2001)

Aubin, D.G., Abbatt, J.P.D.: Interaction of NO_2 with hydrocarbon soot: focus on HONO yield, surface modification, and mechanism. J. Phys. Chem. A **111**, 6263–6273 (2007)

Ball, S.M., Fried, A., Henry, B.E., Mozurkewich, M.: The hydrolysis of $ClONO_2$ on sub-micron liquid sulfuric acid aerosol. Geophys. Res. Lett. **25**, 3339–3342 (1998)

Barone, S.B., Zondlo, M.A., Tolber, M.A.: A Kinetic and product study of the hydrolysis of $ClONO_2$ on type Ia polar stratospheric cloud materials at 185 K. J. Phys. Chem. A **101**, 8643–8652 (1997)

Bauer, S.E., Balkanski, Y., Schulz, M., Hauglustaine, D.A., Dentener, F.: Global modeling of heterogeneous chemistry on mineral aerosol surfaces: influence on tropospheric ozone chemistry and comparison to observations. J. Geophys. Res. **109**(D02304), 17 (2004). doi:10.1029/2003JD003868

Bedjanian, Y., Romanias, M.N., El Zein, A.: Uptake of HO_2 radicals on Arizona test dust surface. Atmos. Chem. Phys. Discuss. **13**, 8873–8900 (2013)

Behnke, W., Zetzsch, C.: Production of a photolytic precursor of atomic Cl from aerosols and Cl^- in the presence of O_3. In: Grimvall, A., de Leer, E.W.B. (eds.) Naturally-Produced Organohalogens, pp. 375–384. Kluwer Acad, Norwell (1995)

Behnke, W., George, C., Scheer, V., Zetzsch, C.: Production and decay of ClNO2 from the reaction of gaseous N2O5 with NaCl solution: Bulk and aerosol experiments, J. Geophys. Res. 102, 3785-3804 (1997)

Beichert, P., Finlayson-Pitts, B.J.: Knudsen cell studies of the uptake of gaseous HNO3 and other oxides of nitrogen on solid NaCl: the role of surface-adsorbed water. J. Phys. Chem. **100**, 15218 (1996)

Berkowitz, C.M., Chapman, E.G., Zaveri, R.A., Laulainen, N.S., Disselkamp, R.S., Bian, X.: Evidence of nighttime ozone depletion through heterogeneous chemistry. Atmos. Environ. **35**, 2395–2404 (2001)

Bertram, T.H., Thornton, J.A.: Toward a general parameterization of N_2O_5 reactivity on aqueous particles: the competing effects of particle liquid water, nitrate and chloride. Atmos. Chem. Phys. **9**, 8351–8363 (2009)

Bian, H., Zender, C.S.: Mineral dust and global tropospheric chemistry: relative roles of photolysis and heterogeneous uptake. J. Geophys. Res. **108**(4672), 10 (2003). doi:10.1029/2002JD003143

Boniface, J., Shi, Q., Li, Y.Q., Chueng, J.L., Rattigan, O.V., Davidovits, P., Worsnop, D.R., Jayne, J.T., Kolb, C.E.: Uptake of gas-phase SO_2, H_2S, and CO_2 by aqueous solutions. J. Phys. Chem. A **104**, 7502–7510 (2000)

Brasseur, G.P., Solomon, S.: Aeronomy of the Middle Atmosphere: Chemistry and Physics of the Stratosphere and Mesosphere, 3rd edn. Springer, Dordrecht (2005)

Brigante, M., Cazoir, D., D'Anna, B., George, C., Donaldson, D.J.: Photoenhanced uptake of NO_2 by pyrene solid films. J. Phys. Chem. A **112**, 9503–9508 (2008)

Caloz, F., Fentner, F.F., Rossi, M.J.: Heterogeneous kinetics of the uptake of $ClONO_2$ on NaCl and KBr. J. Phys. Chem. **100**, 7494–7501 (1996)

Carmichael, G.R., Zhang, Y., Chen, L.L., Hong, M.S., Ueda, H.: Seasonal variation of aerosol composition at Cheju Island, Korea. Atmos. Environ. **30**, 2407–2416 (1996)

Carslaw, K.S., Peter, T.: Uncertainties in reactive uptake coefficients for solid stratospheric particles-1. Surface chemistry. Geophys. Res. Lett. **24**, 1743–1746 (1997)

Chakraborty, P., Zachariah, M.: On the structure of organic-coated water droplets: from "net water attractors" to "oily" drops. J. Geophys. Res. **116**(D21205), 8 (2011). doi:10.1029/2011JD015961

Chang, W.L., Bhave, P.V., Brown, S.S., Riemer, N., Stutz, J., Dabdub, D.: Heterogeneous atmospheric chemistry, ambient measurements, and model calculations of N_2O_5: a review. Aerosol Sci. Technol. **45**, 665–695 (2011)

Chu, L.T., Leu, M.-T., Keyser, L.F.: Heterogeneous reactions of hypochlorous acid + hydrogen chloride $\rightarrow Cl_2 + H_2O$ and chlorosyl nitrite + HCl $\rightarrow Cl_2 + HNO_3$ on ice surfaces at polar stratospheric conditions. J. Phys. Chem. **97**, 12798–12804 (1993)

Chughtai, A.R., Brooks, M.E., Smith, D.M.: Effect of metal oxides and black carbon (soot) on $SO_2/O_2/H_2O$ reaction systems. Aerosol Sci. Technol. **19**, 121–132 (1993)

Chughtai, A.R., Atteya, M.M.O., Kim, J., Konowalchuck, B.K., Smith, D.M.: Adsorption and adsorbate interaction at soot particle surfaces. Carbon **36**, 1573–1589 (1998)

Crowley, J.N., Ammann, M., Cox, R.A., Hynes, R.G., Jenkin, M.E., Mellouki, A., Rossi, M.J., Troe, J., Wallington, T.J.: Evaluated kinetic and photochemical data for atmospheric chemistry: volume V – heterogeneous reactions on solid substrates. Atmos. Chem. Phys. **10**, 9059–9223 (2010)

Davidovits, P., Worsnop, D.R., Jayne, J.T., Kolb, C.E., Winkler, P., Vrtala, A., Wagner, P.E., Kulmula, M., Lehtinen, K.E.J., Vessala, T., Mozurkewich, M.: Mass accommodation coefficient of water vapor on liquid water. Geophys. Res. Lett. **31**, L22111 (2004)

Davies, J.A., Cox, R.A.: Kinetics of the heterogeneous reaction of HNO3 with NaCl: effect of water vapor. J. Phys. Chem. A **102**, 7631–7642 (1998)

DeHaan, D.O., Finlayson-Pitts, B.J.: Knudsen cell studies of the reaction of gaseous nitric acid with synthetic sea salt at 298 K. J. Phys. Chem. A **101**, 9993–9999 (1997)

Deiber, G., George, C., Le Calve, S., Schweitzer, F., Mirabel, P.: Uptake study of $ClONO_2$ and $BrONO_2$ by Halide containing droplets. Atmos. Chem. Phys. **4**, 1291–1299 (2004)

Disselkamp, R.S., Carpenter, M.A., Cowin, J.P.: A chamber investigation of nitric acid-Soot aerosol chemistry at 298 K. J. Atmos. Chem. **37**, 113–123 (2000)

Donaldson, D.J., Guest, J.A., Goh, M.C.: Evidence for adsorbed SO_2 at the aqueous-air interface. J. Phys. Chem. **99**, 9313–9315 (1995)

Donaldson, D.J., Ravishankara, A.R., Hanson, D.R.: Detailed study of HOCl + HCl $\rightarrow Cl_2 + H_2O$ in sulfuric acid. J. Phys. Chem. A **101**, 4717–4725 (1997)

Elrod, M.J., Koch, R.E., Kim, J.E., Molina, M.S.: HCl vapour pressures and reaction probabilities for $ClONO_2$+ HCl on liquid H_2SO_4–HNO_3–HCl–H_2O solutions. Faraday Discuss. **100**, 269–278 (1995)

Elshorbany, Y.F., Steil, B., Bruhl, C., Lelieveld, J.: Impact of HONO on global atmospheric chemistry calculated with an empirical parameterization in the EMAC model. Atmos. Chem. Phys. **12**, 9977–10000 (2012)

Evans, M.J., Jacob, D.J.: Impact of new laboratory studies of N_2O_5 hydrolysis on global model budgets of tropospheric nitrogen oxides, ozone, and OH. Geophys. Res. Lett. **32**, L09813 (2005). doi:10.1029/2005GL022469

Fendel, W., Matter, D., Burtscher, H., Schimdt-Ott, A.: Interaction between carbon or iron aerosol particles and ozone. Atmos. Environ. **29**, 967–973 (1995)

Fernandez, M.A., Hynes, R.G., Cox, R.A.: Kinetics of $ClONO_2$ reactive uptake on ice surfaces at temperatures of the upper troposphere. J. Phys. Chem. A **109**, 9986–9996 (2005)

Finlayson-Pitts, B.J., Ezell, M.J., Pitts Jr., J.N.: Formation of chemically active chlorine compounds by reactions of atmospheric NaCl particles with gaseous N_2O_5 and $ClONO_2$. Nature **337**, 241–244 (1989)

Finlayson-Pitts, B.J., Pitts Jr., J.N.: Chemistry of the Upper and Lower Atmosphere. Academic Press, San Diego (2000)

Finlayson-Pitts, B.J.: The tropospheric chemistry of sea salt: a molecular-level view of the chemistry of NaCl and NaBr. Chem. Rev. **103**, 4801–4822 (2003)

Folkers, M., Mentel, T.F., Wahner, A.: Influence of an organic coating on the reactivity of aqueous aerosols probed by the heterogeneous hydrolysis of N_2O_5. Geophys. Res. Lett. **30**(12), 1644 (2003). doi:10.1029/2003GL017168

Fried, A., Henry, B.E., Calvert, J.G., Mozukewich, M.: The reaction probability of N_2O_5 with sulfuric acid aerosols at stratospheric temperatures and compositions. J. Geophys. Res. **99**, 3517–3532 (1994)

Garrett, B.C., Schenter, G.K., Morita, A.: Molecular simulations of the transport of molecules across the liquid/vapor interface of water. Chem. Rev. **106**, 1355–1374 (2006)

Gebel, M.E., Finlayson-Pitts, B.J.: Uptake and reaction of $ClONO_2$ on NaCl and synthetic sea salt. J. Phys. Chem. A **105**, 5178–5187 (2001)

George, C., Ponche, J.L., Mirabel, P., Behnke, W., Sheer, V., Zetzsch, C.: Study of the uptake of N_2O_5 by water and NaCl solutions. J. Phys. Chem. **98**, 8780–8784 (1994)

Ghosal, S., Shbeeb, A., Hemminger, J.C.: Surface segregation of bromine in bromide doped NaCl: implications for the seasonal variations in Arctic ozone. Geophys. Res. Lett. **27**, 1879–1882 (2000)

Ghosal, S., Hemminger, J.C.: Surface adsorbed water on NaCl and its effect on nitric acid reactivity with NaCl powders. J. Phys. Chem. B **108**, 14102–14108 (2004)

Gonçalves, M., Dabdub, D., Chang, W.L., Jorba, O., Baldasano, J.M.: Impact of HONO sources on the performance of mesoscale air quality models. Atmos. Environ. **54**, 168–176 (2012)

Goodman, A.L., Underwood, G.M., Grassian, V.H.: A laboratory study of the heterogeneous reaction of nitric acid on calcium carbonate particles. J. Geophys. Res. **105**, 29053–29064 (2000)

Goodman, A.L., Li, P., Usher, C.R., Grassian, V.H.: Heterogeneous uptake of sulfur dioxide on aluminum and magnesium oxide particles. J. Phys. Chem. A **105**, 6109–6120 (2001)

Griffiths, P.T., Badger, C.L., Cox, R.A., Folkers, M., Henk, H.H., Mentel, T.F.: Reactive uptake of N_2O_5 by aerosols containing dicarboxylic acids. Effect of particle phase, composition, and nitrate content. J. Phys. Chem. A **113**, 5082–5090 (2009)

Guimbaud, C., Arens, F., Gutzwiller, L., Gäggeler, H.W., Ammann, M.: Uptake of HNO_3 to deliquescent sea-salt particles: a study using the short-lived radioactive isotope tracer ^{13}N. Atmos. Chem. Phys. **2**, 249–257 (2002)

Hallquist, M., Stewart, D.J., Stephenson, S.K., Cox, R.A.: Hydrolysis of N_2O_5 on sub-micron sulfate aerosols. Phys. Chem. Chem. Phys. **5**, 3453–3463 (2003)

Hanisch, F., Crowley, J.N.: Heterogeneous reactivity of gaseous nitric acid on Al_2O_3, $CaCO_3$, and atmospheric dust samples: a Knudsen cell study. J. Phys. Chem. A **105**, 3096–3106 (2001a)

Hanisch, F., Crowley, J.N.: The heterogeneous reactivity of gaseous nitric acid on authentic mineral dust samples, and on individual mineral and clay mineral components. Phys. Chem. Chem. Phys. **3**, 2474–2482 (2001b)

Hanisch, F., Crowley, J.N.: Ozone decomposition on Saharan dust: an experimental investigation. Atmos. Chem. Phys. **3**, 119–130 (2003)

Hanke, M., Umann, B., Uecker, J., Arnold, F., Bunz, H.: Atmospheric measurements of gas-phase HNO_3 and SO_2 using chemical ionization mass spectrometry during the MINATROC field campaign 2000 on Monte Cimone. Atmos. Chem. Phys. **3**, 417–436 (2003)

Hanson, D.R.: Reaction of N_2O_5 with H_2O on bulk liquids and on particles and the effect of dissolved HNO_3. Geophys. Res. Lett. **24**, 1087–1090 (1997)

Hanson, D.R.: Reaction of $ClONO_2$ with H_2O and HCl in sulfuric acid and $HNO_3/H_2SO_4/H_2O$ mixtures. J. Phys. Chem. A **102**, 4794–4807 (1998)

Hanson, D.R., Burkholder, J.B., Howard, C.J., Ravishankara, A.R.: Measurement of OH and HO_2 radical uptake coefficients on water and sulfuric acid surfaces. J. Phys. Chem. **96**, 4979–4985 (1992)

Hanson, D.R., Lovejoy, E.R.: The uptake of N_2O_5 onto small sulfuric acid particle. Geophys. Res. Lett. **21**, 2401–2404 (1994)

Hanson, D.R., Lovejoy, E.R.: Heterogeneous reactions in liquid sulfuric acid: HOCl + HCl as a model system. J. Phys. Chem. **100**, 6397–6405 (1996)

Hanson, D.R., Ravishankara, A.R.: The reaction probabilities of $ClONO_2$ and N_2O_5 on polar stratospheric cloud materials. J. Geophys. Res. **96**, 5081–5090 (1991a)

Hanson, D.R., Ravishankara, A.R.: The reaction probabilities of $ClONO_2$ and N_2O_5 on 40 to 75 % sulfuric acid solutions. J. Geophys. Res. **96**, 17307–17314 (1991b)

Hanson, D.R., Ravishankara, A.R.: Investigation of the reactive and nonreactive processes involving $ClONO_2$ and HCl on water and nitric acid doped ice. J. Phys. Chem. **96**, 2682–2691 (1992)

Hanson, D.R., Ravishankara, A.R.: Response to "Comment on porosities of ice films used to simulate stratospheric cloud surfaces". J. Phys. Chem. **97**, 2802–2803 (1993a)

Hanson, D.R., Ravishankara, A.R.: Reaction of $ClONO_2$with HCl on NAT, NAD, and frozen sulfuric acid and hydrolysis of N_2O_5 and $ClONO_2$ on frozen sulfuric acid. J. Geophys. Res. **98**, 22931–22936 (1993b)

Hanson, D.R., Ravishankara, A.R.: Reactive uptake of $ClONO_2$ onto sulfuric acid due to reaction with HCl and H_2O. J. Phys. Chem. **98**, 5728–5735 (1994)

Hess, M., Krieger, U.K., Marcolli, C., Huthwelker, T., Ammann, M., Lanford, W.A., Peter, T.: Bromine enrichment in the near-surface region of Br-doped NaCl single crystals diagnosed by Rutherford backscattering spectrometry. J. Phys. Chem. A **111**, 4312–4321 (2007)

Hirokawa, J., Onaka, K., Kajii, Y., Akimoto, H.: Heterogeneous processes involving sodium halide particles and ozone: molecular bromine release in the marine boundary layer in the absence of nitrogen oxides. Geophys. Res. Lett. **25**, 2449–2452 (1998)

Hoffman, R.C., Gebel, M.E., Fox, B.S., Finlayson-Pitts, B.J.: Knudsen cell studies of the reactions of N_2O_5 and $ClONO_2$ with NaCl: development and application of a model for estimating available surface areas and corrected uptake coefficients. Phys. Chem. Chem. Phys. **5**, 1780–1789 (2003)

Hunt, S.W., Roesolová, M., Wang, W., Wingen, L.M., Knipping, E.M., Tobias, D.J., Dabdub, D., Finlayson-Pitts, B.J.: Formation of molecular bromine from the reaction of ozone with deliquesced NaBr aerosol: evidence for interface chemistry. J. Phys. Chem. A **108**, 11559–11572 (2004)

Jayne, J.T., Davidovits, P., Worsnop, D.R., Zahniser, M.S., Kolb, C.E.: Uptake of sulfur dioxide (g) by aqueous surfaces as a function of pH: the effect of chemical reaction at the interface. J. Phys. Chem. **94**, 6041–6048 (1990)

Johnson, E.R., Sciegienka, J., Carlos-Cuellar, S., Grassian, V.H.: Heterogeneous uptake of gaseous nitric acid on Dolomite ($CaMg(CO_3)_2$) and Calcite ($Ca\ CO_3$) particles: a Knudsen cell study using multiple, single, and fractional particle layers. J. Phys. Chem. A **109**, 6901–6911 (2005)

Jungwirth, P., Tobias, D.J.: Ions at the air/water interface. J. Phys. Chem. B **106**, 6361–6373 (2002)

Kamm, S., Mohler, O., Naumann, K.-H., Saathoff, H., Schurath, U.: The heterogeneous reaction of ozone with soot aerosol. Atmos. Environ. **33**, 4651–4661 (1999)

Karagulian, F., Santschi, C., Rossi, M.J.: The heterogeneous chemical kinetics of N_2O_5 on $CaCO_3$ and other atmospheric mineral dust surrogates. Atmos. Chem. Phys. **6**, 1373–1388 (2006)

Karagulian, F., Rossi, M.J.: Heterogeneous chemistry of the NO_3 free radical and N_2O_5 on decane flame soot at ambient temperature: reaction products and kinetics. J. Phys. Chem. A **111**, 1914–1926 (2007)

Kasibhatla, P., Chameides, W.L., John, J.S.: A three-dimensional global model investigation of seasonal variations in the atmospheric burden of anthropogenic sulfate aerosols. J. Geophys. Res. **102**, 3737–3759 (1997)

Khalizov, A.F., Cruz-Quinones, M., Zhang, R.: Heterogeneous reaction of NO_2 on fresh and coated soot surfaces. J. Phys. Chem. A **114**, 7516–7524 (2010)

Kirchner, U., Scheer, V., Vogt, R.: FTIR Spectroscopic investigation of the mechanism and kinetics of the heterogeneous reactions of NO_2 and HNO_3 with soot. J. Phys. Chem. A **104**, 8908–8915 (2000)

Knipping, E.M., Lakin, M.J., Foster, K.L., Jungwirth, P., Tobias, D.J., Gerber, R.B., Dabdub, D., Finlayson-Pitts, B.J.: Experiments and simulations of ion-enhanced interfacial chemistry on aqueous NaCl aerosols. Science **288**, 301–306 (2000)

Koehler, B.G., Nicholson, V.T., Roe, H.G., Whitney, E.S.: A Fourier transform infrared study of the adsorption of SO_2 on n-hexane soot from $-130°$ to -40 °C. J. Geophys. Res. **104**, 5507–5514 (1999)

Kojima, T., Buseck, P.R., Iwasaka, Y., Matsuki, A., Trochkine, D.: Sulfate-coated dust particles in the free troposphere over Japan. Atmos. Res. **82**, 698–708 (2006)

Koop, K., Carslaw, K.S., Peter, T.: Thermodynamic stability and phase transitions of PSC particles. Geophys. Res. Lett. **24**, 2199–2202 (1997)

Laaksonen, A., Vesala, T., Kulmala, M., Winkler, P.M., Wagner, P.E.: Commentary on cloud modelling and the mass accommodation coefficient of water. Atmos. Chem. Phys. **5**, 461–464 (2005)

Lary, D.J., Lee, A.M., Toumi, R., Newchurch, M.J., Pirre, M., Renard, J.B.: Carbon aerosols and atmospheric photochemistry. J. Geophys. Res. **102**, 3671–3682 (1997)

Lee, S.H., Leard, D.C., Zhang, R., Molina, L.T., Molina, M.J.: The $HCl + ClONO_2$ reaction rate on various water ice surfaces. Chem. Phys. Lett. **315**, 7 (1999)

Lelievre, S., Bedjanian, Y., Laverdet, G., Le Bras, G.: Heterogeneous reaction of NO_2 with hydrocarbon flame soot. J. Phys. Chem. A **108**, 10807–10817 (2004)

Leu, M.-T.: Laboratory studies of sticking coefficients and heterogeneous reactions important in the Antarctic stratosphere. Geophys. Res. Lett. **15**, 17–20 (1988)

Leu, M.-T., Moore, S.B., Keyser, L.F.: Heterogeneous reactions of chlorine nitrate and hydrogen chloride on type I polar stratospheric clouds. J. Phys. Chem. **95**, 7763–7771 (1991)

Leu, M.-T., Timonen, R.S., Keyser, L.F., Yung, Y.L.: Heterogeneous reactions of HNO_3 (g) + NaCl (s) $\rightarrow$ HCl(g) + $NaNO_3$ (s) and N_2O_5 (g) + NaCl(s) $\rightarrow$ $ClNO_2$(g) + $NaNO_3$(s). J. Phys. Chem. **99**, 13203–13212 (1995)

Li, L., Chen, Z.M., Zhang, Y.H., Zhu, T., Li, J.L., Ding, J.: Kinetics and mechanism of heterogeneous oxidation of sulfur dioxide by ozone on surface of calcium carbonate. Atmos. Chem. Phys. **6**, 2453–2464 (2006)

Li, Y.Q., Davidovits, P., Shi, Q., Jayne, J.T., Kolb, C.E., Worsnop, D.R.: Mass and thermal accommodation coefficients of H_2O(g) on liquid water as a function of temperature. J. Phys. Chem. A **105**, 10627–10634 (2001)

Li, W., Gibbs, G.V., Oyama, S.T.: Mechanism of ozone decomposition on a manganese oxide catalyst. 1. in situ Raman spectroscopy and ab initio molecular orbital calculations. J. Am. Chem. Soc. **120**, 9041–9046 (1998)

Liu, Y., Liu, C., Ma, J., Ma, Q., He, H.: Structural and hygroscopic changes of soot during heterogeneous reaction with O_3. Phys. Chem. Chem. Phys. **12**, 10896–10903 (2010)

Lohmann, U., Karcher, B., Hendricks, J.: Sensitivity studies of cirrus clouds formed by heterogeneous freezing in the ECHAM GCM. J. Geophys. Res. **109**, D16204 (2004). doi:10.1029/2003JD004443

Longfellow, C.A., Ravishankara, A.R., Hanson, D.R.: Reactive and nonreactive uptake on hydrocarbon soot: HNO_3, O_3, and N_2O_5. J. Geophys. Res. **105**, 24,345–24,350 (2000)

Loukhovitskaya, E., Bedjanian, Y., Morozov, I., Le Bras, G.: Laboratory study of the interaction of HO_2 radicals with the NaCl, NaBr, $MgCl_2 \cdot 6H_2O$ and sea salt surfaces. Phys. Chem. Chem. Phys. **11**, 7896–7905 (2009)

Magi, L., Schweitzer, F., Pallares, C., Cherif, S., Mirabel, P., George, C.: Investigation of the uptake rate of ozone and methyl hydroperoxide by water surfaces. J. Phys. Chem. A **101**, 4943–4948 (1997)

McNeill, V.F., Loerting, T., Geiger, F.M., Trout, B.L., Molina, M.J.: Hydrogen chloride-induced surface disordering on ice. Proc. Nat. Acad. Sci. **103**, 9422–9427 (2006)

Michel, A.E., Usher, C.R., Grassian, V.H.: Reactive uptake of ozone on mineral oxides and mineral dusts. Atmos. Environ. **37**, 3201–3211 (2003)

Mochida, M., Hirokawa, J., Akimoto, H.: Unexpected large uptake of O_3 on sea salts and the observed Br_2 formation. Geophys. Res. Lett. **27**, 2629–2632 (2000)

Mogili, P.K., Kleiber, P.D., Young, M.A., Grassian, V.H.: Heterogeneous uptake of ozone on reactive components of mineral dust aerosol: an environmental aerosol reaction chamber study. J. Phys. Chem. A **110**, 13799–13807 (2006)

Molina, M.J., Zhang, R., Wooldridge, P.J., McMahon, J.R., Kim, J.E., Chang, H.Y., Beyer, K.D.: Physical chemistry of the $H_2SO_4/HNO_3/H_2O$ system: implications for polar stratospheric cloud. Science **261**, 1418–1423 (1993)

Monge, M.E., D'Anna, B., Mazri, L., Giroir-Fendler, A., Ammann, M., Donaldson, D.J., George, C.: Light changes the atmospheric reactivity of soot. Proc. Natl. Acad. Sci. **107**, 6605–6609 (2010)

Morita, A., Sugiyama, M., Koda, S.: Gas-phase flow and diffusion analysis of the droplet train/flow-reactor technique for the mass accommodation process. J. Phys. Chem. A **107**, 1749–1759 (2003)

Morita, A., Kanaya, Y., Francisco, J.S.: Uptake of the HO_2 radical by water: molecular dynamics calculations and their implications for atmospheric modeling. J. Geophys. Res. **109**(D09201), 10 (2004a). doi:10.1029/2003JD004240

Morita, A., Sugiyama, M., Kameda, H., Koda, S., Hanson, D.R.: Mass accommodation coefficient of water: molecular dynamics simulation and revised analysis of droplet train/flow experiment. J. Phys. Chem. B **108**, 9111–9120 (2004b)

Mozurkewich, M., McMurray, P.H., Gupta, A., Calvert, J.G.: Mass accommodation coefficient for HO_2 radicals on aqueous particles. J. Geophys. Res. **92**, 4163–4170 (1987)

Mozurkewich, M., Calvert, J.: Reaction probability of N_2O_5 on aqueous aerosols. J. Geophys. Res. **93**, 15882–15896 (1988)

Müller, B., Heal, M.R.: The mass accommodation coefficient of ozone on an aqueous surface. Phys. Chem. Chem. Phys. **4**, 3365–3369 (2002)

Oppliger, R., Allanic, A., Rossi, M.J.: Real-time kinetics of the uptake of $ClONO_2$ on ice and in the presence of HCl in the temperature range 160 $K \leq T \leq 200$ K. J. Phys. Chem. A **101**, 1903–1911 (1997)

Oum, K.W., Lakin, M.J., DeHaan, D.O., Brauers, T., Finlayson-Pitts, B.J.: Formation of molecular chlorine from the photolysis of ozone and aqueous sea-salt particles. Science **279**, 74–76 (1998a)

Oum, K.W., Lakin, M.J., Finlayson-Pitts, B.J.: Bromine activation in the troposphere by the dark reaction of O_3 with seawater ice. Geophys. Res. Lett. **25**, 3923–3926 (1998b)

Park, S.-C., Burden, D.K., Nathanson, G.M.: The inhibition of N_2O_5 hydrolysis in sulfuric acid by 1-Butanol and 1-Hexanol surfactant coatings. J. Phys. Chem. A **111**, 2921–2929 (2007)

Ponche, J.L., George, C., Mirabel, P.: Mass transfer at the air/water interface: mass accommodation coefficients of SO_2, HNO_3, NO_2 and NH_3. J. Atmos. Chem. **16**, 1–21 (1993)

Pöschl, U., Letzel, T., Schauer, C., Niessner, R.: Interaction of ozone and water vapor with spark discharge soot aerosol particles coated with benzo[*a*]pyrene: O_3 and H_2O adsorption, benzo[*a*]pyrene degradation, and atmospheric implications. J. Phys. Chem. A **105**, 4029–4041 (2001)

Prince, A.P., Wade, J.L., Grassian, V.H., Kleiber, K.D., Young, M.A.: Heterogeneous reactions of soot aerosols with nitrogen dioxide and nitric acid: atmospheric chamber and Knudsen cell studies. Atmos. Environ. **36**, 5729–5740 (2002)

Quinlan, M.A., Reihs, C.M., Golden, D.M., Tolbert, M.A.: Heterogeneous reactions on model polar stratospheric cloud surfaces: reaction of dinitrogen pentoxide on ice and nitric acid trihydrate. J. Phys. Chem. **94**, 3255–3260 (1990)

Riemer, N., Vogel, B., Anttila, T., Kiendler-Scharr, A., Mentel, T.F.: Relative importance of organic coatings for the heterogeneous hydrolysis of N_2O_5 during summer in Europe. J. Geophys. Res. **114**(D17307), 14 (2009)

Roberts, J.M., Osthoff, H.D., Brown, S.S., Ravishankara, A.R., Coffman, D., Quinn, P., Bates, T.: Laboratory studies of products of N_2O_5 uptake on Cl^- containing substrates. Geophys. Res. Lett. **36**, L20808 (2009). doi:10.1029/2009GL040448

Robinson, G.N., Worsnop, D.R., Jayne, J.T., Kolb, C.E., Davidovits, P.: Heterogeneous uptake of $ClONO_2$ and N_2O_5 by sulfuric acid solutions. J. Geophys. Res. **102**, 3583–3601 (1997)

Roeselová, M., Jungwirth, P., Tobias, D.J., Gerber, R.B.: Impact, trapping, and accommodation of hydroxyl radical and ozone at aqueous salt Aerosol surfaces. A molecular dynamics study. J. Phys. Chem. B **107**, 12690–12699 (2003)

Roeselová, M., Vieceli, J., Dang, L.X., Garrett, B.C., Tobias, D.J.: Hydroxyl radical at the air-water interface. J. Am. Chem. Soc. **126**, 16308–16309 (2004)

Rogaski, C.A., Golden, D.M., Williams, L.R.: Reactive uptake and hydration experiments on amorphous carbon treated with NO_2, SO_2, O_3, HNO_3, and H_2SO_4. Geophys. Res. Lett. **24**, 381–384 (1997)

Rohrer, F., Bohn, B., Brauers, T., Brüning, D., Johnen, F.-J., Wahner, A., Kleffmann, J.: Characterisation of the photolytic HONO-source in the atmosphere simulation chamber SAPHIR. Atmos. Chem. Phys. **5**, 2189–2201 (2005)

Roscoe, J.M., Abbatt, J.P.D.: Diffuse reflectance FTIR study of the interaction of alumina surfaces with ozone and water vapor. J. Phys. Chem. A **109**, 9028–9034 (2005)

Rossi, M.J.: Heterogeneous reactions on salts. Chem. Rev. **103**, 4823–4882 (2003)

Saathoff, H., Naumann, K.-H., Riemer, N., Kamm, S., Möhler, O., Schurath, U., Vogel, H., Vogel, B.: The loss of NO_2, HNO_3, NO_3/N_2O_5, and $HO_2/HOONO_2$ on soot aerosol: a chamber and modeling study. Geophys. Res. Lett. **28**, 1957–1960 (2001)

Sadanaga, Y., Hirokawa, J., Akimoto, H.: Formation of molecular chlorine in dark condition: heterogeneous reaction of ozone with sea salt in the presence of ferric ion. Geophys. Res. Lett. **28**, 4433–4436 (2001)

Sakaguchi, S., Morita, A.: Mass accommodation mechanism of water through monolayer films at water/vapor interface. J. Chem. Phys. **137**(064701), 9 (2012). doi:10.1063/1.4740240

Salgado-Muñoz, M.S., Rossi, M.J.: Heterogeneous reactions of HNO_3 with flame soot generated under different combustion conditions. Reaction mechanism and kinetics. Phys. Chem. Chem. Phys. **4**, 5110–5118 (2002)

Sander, S.P., Baker, R., Golden, D.M., Kurylo, M.J., Wine, P.H., Abbatt J.P.D., Burkholder, J.B., Kolb, C.E., Moortgat, G.K., Huie, R.E., Orkin, V.L.: Chemical Kinetics and Photochemical Data for Use in Atmospheric Studies, Evaluation Number 17, JPL Publication 10-6, Pasadena, California. http://jpldataeval.jpl.nasa.gov/ (2011)

Santschi, C., Rossi, M.J.: Uptake of CO2, SO2, HNO3 and HCl on calcite (CaCO3) at 300 K: mechanism and the role of adsorbed water. J. Phys. Chem. A **110**, 6789–6802 (2006)

Saul, T.D., Tolocka, M.P., Johnston, M.V.: Reactive uptake of nitric acid onto sodium chloride aerosols across a wide range of relative humidities. J. Phys. Chem. A **110**, 7614–7620 (2006)

Schütze, M., Herrmann, H.: Determination of phase transfer parameters for the uptake of HNO_3, N_2O_5 and O_3 on single aqueous drops. Phys. Chem. Chem. Phys. **4**, 60–67 (2002)

Seisel, S., Flückiger, B., Rossi, M.J.: The heterogeneous reaction of N_2O_5 with HBr on ice comparison with N_2O_5 + HCl. Ber. Bunsen. Phys. Chem. **102**, 811–820 (1998)

Seisel, S., Börensen, C., Vogt, R., Zellner, R.: The heterogeneous reaction of HNO_3 on mineral dust and γ-alumina surfaces: a combined Knudsen cell and DRIFTS study. Phys. Chem. Chem. Phys. **6**, 5498–5508 (2004)

Seisel, S., Börensen, C., Vogt, R., Zellner, R.: Kinetics and mechanism of the uptake of N_2O_5 on mineral dust at 298 K. Atmos. Chem. Phys. **5**, 3423–3432 (2005)

Seisel, S., Keil, T., Lian, Y., Zellner, R.: Kinetics of the uptake of SO_2 on mineral oxides: Improved initial uptake coefficients at 298 K from pulsed Knudsen cell experiments. Int. J. Chem. Kinet. **38**, 242–249 (2006)

Shi, Q., Jayne, J.T., Kolb, C.E., Worsnop, D.R.: Kinetic model for reaction of $ClONO_2$ with H_2O and HCl and HOCl with HCl in sulfuric acid solutions. J. Geophys. Res. **106**, 24259–24274 (2001)

Shimono, A., Koda, S.: Laser-spectroscopic measurements of uptake coefficients of SO_2 on aqueous surfaces. J. Phys. Chem. **100**, 10,269–10,276 (1996)

Smith, J.D., Cappa, C.D., Drisdell, W.S., Cohen, R.C., Saykally, R.J.: Raman thermometry measurements of free evaporation from liquid water droplets. J. Am. Chem. Soc. **128**, 12892–12898 (2006)

Stadler, D., Rossi, M.J.: The reactivity of NO_2 and HONO on flame soot at ambient temperature: the influence of combustion conditions. Phys. Chem. Chem. Phys. **2**, 5420–5429 (2000)

Stemmler, K., Vlasenko, A., Guimbaud, C., Ammann, M.: The effect of fatty acid surfactants on the uptake of nitric acid to deliquesced NaCl aerosol. Atmos. Chem. Phys. **8**, 5127–5141 (2008)

Stewart, D.J., Griffiths, P.T., Cox, R.A.: Reactive uptake coefficients for heterogeneous reaction of N_2O_5 with submicron aerosols of NaCl and natural sea salt. Atmos. Chem. Phys. **4**, 1381–1388 (2004)

Sullivan, R.C., Thornberry, T., Abbatt, J.P.D.: Ozone decomposition kinetics on alumina: effects of ozone partial pressure, relative humidity and repeated oxidation cycles. Atmos. Chem. Phys. **4**, 1301–1310 (2004)

Takahama, S., Russell, L.M.: A molecular dynamics study of water mass accommodation on condensed phase water coated by fatty acid monolayers. J. Geophys. Res. **116**(D02203), 14 (2011). doi:10.1029/2010JD014842

Takami, A., Kato, S., Shimono, A., Koda, S.: Uptake coefficient of OH radical on aqueous surface. Chem. Phys. **231**, 215–227 (1998)

Taketani, F., Kanaya, Y., Akimoto, H.: Kinetics of heterogeneous reactions of HO_2 radical at ambient concentration levels with $(NH_4)\ _2SO_4$ and NaCl aerosol particles. J. Phys. Chem. A **112**, 2370–2377 (2008)

Taketani, F., Kanaya, Y., Akimoto, H.: Heterogeneous loss of HO_2 by KCl, synthetic sea salt, and natural seawater aerosol particles. Atmos. Environ. **43**, 1660–1665 (2009)

Taketani, F., Kanaya, Y., Pochanart, P., Liu, Y., Li, J., Okuzawa, K., Kawamura, K., Wang, Z., Akimoto, H.: Measurement of overall uptake coefficients for HO_2 radicals by aerosol particles sampled from ambient air at Mts. Tai and Mang (China). Atmos. Chem. Phys. **12**, 11907–11916 (2012)

Thornton, J.A., Abbatt, J.P.D.: N_2O_5 reaction on submicron sea salt aerosol: kinetics, products, and the effect of surface active organics. J. Phys. Chem. A **109**, 10004–10012 (2005)

Tolbert, M.A., Rossi, M.J., Golden, D.M.: Heterogeneous interactions of chlorine nitrate, hydrogen chloride, and nitric acid with sulfuric acid surfaces at stratospheric temperatures. Geophys. Res. Lett. **15**, 847–850 (1988)

Ullerstam, M., Vogt, R., Langer, S., Ljungström, E.: The kinetics and mechanism of SO_2 oxidation by O_3 on mineral dust. Phys. Chem. Chem. Phys. **4**, 4694–4699 (2002)

Ullerstam, M., Johnson, M.S., Vogt, R., Ljungström, E.: DRIFTS and Knudsen cell study of the heterogeneous reactivity of SO_2 and NO_2 on mineral dust. Atmos. Chem. Phys. **3**, 2043–2051 (2003)

Usher, C.R., Al-Hosney, H., Carlos-Cuellar, S., Grassian, V.H.: A laboratory study of the heterogeneous uptake and oxidation of sulfur dioxide on mineral dust particles. J. Geophys. Res. **107**, 4713 (2002)

Usher, C.R., Michel, A.E., Grassian, V.H.: Reactions on mineral dust. Chem. Rev. **103**, 4883–4940 (2003)

Van Doren, J.M., Watson, L.R., Davidovits, P., Worsnop, D.R., Zahniser, M.S., Kolb, C.E.: Temperature dependence of the uptake coefficients of nitric acid, hydrochloric acid and nitrogen oxide (N_2O_5) by water droplets. J. Phys. Chem. **94**, 3265–3269 (1990)

Vieceli, J., Roeselov, M., Tobias, D.J.: Accommodation coefficients for water vapor at the air/water interface. Chem. Phys. Lett. **393**, 249–255 (2004)

Vieceli, J., Roeselová, M., Potter, N., Dang, L.X., Garrett, B.C., Tobias, D.J.: Molecular dynamics simulations of atmospheric oxidants at the air-water interface: solvation and accommodation of OH and O3. J. Phys. Chem. B **109**, 15876–15892 (2005)

Vlasenko, A., Sjogren, S., Weingartner, E., Stemmler, K., Gäggeler, H.W., Ammann, M.: Effect of humidity on nitric acid uptake to mineral dust aerosol particles. Atmos. Chem. Phys. **6**, 2147–2160 (2006)

Voight, C., Schreiner, J., Kohlmann, A., Zink, P., Mauersberger, K., Larsen, N., Deshler, T., Kröger, C., Rosen, J., Adriani, A., Cairo, F., Donfrancesco, G.D., Viterbini, M., Ovarlez, J., Ovarlez, H., David, C., Dörnbrack, A.: Nitric acid trihydrate (NAT) in polar stratospheric clouds. Science **290**, 1756–1758 (2000)

Voigtländer, J., Stratmann, F., Niedermeier, D., Wex, H., Kiselev, A.: Mass accommodation coefficient of water: a combined computational fluid dynamics and experimental data analysis. J. Geophys. Res. **112**(D20208), 8 (2007). doi:10.1029/2007JD008604

Wagner, R., Naumann, K.-H., Mangold, A., Möhler, O., Saathoff, H., Schurath, U.: Aerosol chamber study of optical constants and N_2O_5 uptake on supercooled $H_2SO_4/H_2O/HNO_3$ solution droplets at polar stratospheric cloud temperatures. J. Phys. Chem. A **109**, 8140–8148 (2005)

Wagner, C., Hanisch, F., Holmes, N.S., de Coninck, H.C., Schuster, G., Crowley, J.N.: The interaction of N_2O_5 with mineral dust: aerosol flow tube and Knudsen reactor studies. Atmos. Chem. Phys. **8**, 91–109 (2008)

Wagner, C., Schuster, G., Crowley, J.N.: An aerosol flow tube study of the interaction of N_2O_5 with calcite, Arizona dust and quartz. Atmos. Environ. **43**, 5001–5008 (2009)

Wallington, T., Ammann, M., Atkinson, R., Cox, R.A., Crowley, J., Hynes, R., Jenkin, M.E., Mellouki, W., Rossi, M.J., Troe, J.: Evaluated kinetic and photochemical data for atmospheric chemistry – Data Sheet V, VI, IUPAC Subcommittee for Gas Kinetic Data Evaluation for Atmospheric Chemistry http://www.iupac-kinetic.ch.cam.ac.uk/members.html (2012)

Winkler, P.M., Vrtala, A., Rudolf, R., Wagner, P.E., Riipinen, I., Vesala, T., Lehtinen, K.E.J., Viisanen, Y., Kulmala, M.: Condensation of water vapor: experimental determination of mass and thermal accommodation coefficients. J. Geophys. Res. **111**(D19202), 12 (2006). doi:10.1029/2006JD007194

Wu, L.Y., Tong, S.R., Wang, W.G., Ge, M.F.: Effects of temperature on the heterogeneous oxidation of sulfur dioxide by ozone on calcium carbonate. Atmos. Chem. Phys. **11**, 6593–6605 (2011)

Zangmeister, C.D., Turner, J.A., Pemberton, J.E.: Segregation of NaBr in NaBr/NaCl crystals grown from aqueous solutions: implications for sea salt surface chemistry. Geophys. Res. Lett. **28**, 995–998 (2001)

Zhang, R., Leu, M.-T., Keyser, L.F.: Heterogeneous reactions of $ClONO_2$, HCl, and HOCl on liquid sulfuric acid surfaces. J. Phys. Chem. **98**, 13,563–13,574 (1994a)

Zhang, R., Jayne, J.T., Molina, M.J.: Heterogeneous interactions of $ClONO_2$ and HCl with sulfuric acid tetrahydrate: implications for the stratosphere. J. Phys. Chem. **98**, 867–874 (1994b)

Zhang, R., Leu, M.-T., Keyser, L.F.: Hydrolysis of N_2O_5 and $ClONO_2$ on the $H_2SO_4/HNO_3/H_2O$ ternary solutions under stratospheric conditions. Geophys. Res. Lett. **22**, 1493–1496 (1995)

Chapter 7
Tropospheric Reaction Chemistry

About 90 % of the main constituents of earth's atmosphere (nitrogen and oxygen), and most atmospheric trace constituents exists in the troposphere. Almost all trace species found in the troposphere are substances emitted from anthropogenic and/or natural sources on the ground, and from volcanoes and aircraft into the free troposphere. The exceptions are O_3 and other secondary products formed by the chemical reactions in the atmosphere, or produced by lightning such as NO. Tropospheric chemistry is a research field for studying a series of processes as a system, including the identification and quantification of emission sources, chemical reactions and transport in the atmosphere, transformation from the gas phase molecules to liquid and/or solid particles, and deposition to clouds and fog, raindrops, and earth surface. Tropospheric chemistry is the most important fundamental discipline for various air pollution issues on the global, regional and urban scales that affect the social life of human beings. For the integrated management of air pollution global warming/climate change, atmospheric chemistry, particularly tropospheric chemistry, provides important scientific knowledge together with atmospheric physics and meteorology. As for atmospheric /tropospheric chemistry as a holistic system science, many textbooks have been published in recent decades (for example, Jacob 1999; Finlayson-Pitts and Pitts 2000; Wayne 2000; Akimoto et al. 2002; Brasseur et al. 1999, 2003; Seinfeld and Pandis 2006). This chapter provides specialized descriptions of the chemical reaction system in the troposphere, one of the main elements of tropospheric chemistry.

7.1 Oxidation of Methane in the Natural Atmosphere and OH Radical Chain Reaction

As described in Chap. 1, in contrast to stratospheric chemistry that can be traced back to the 1930s when the chemical theory of ozone layer formation was published by Chapman (1930), research on chemical reaction system in the tropospheric

H. Akimoto, *Atmospheric Reaction Chemistry*, Springer Atmospheric Sciences,
DOI 10.1007/978-4-431-55870-5_7

chemistry started only after the end of the 1960s to the early 1970s when the OH/HO_2 radical chain reaction system was proposed (Crutzen 1973). This era overlaps with the era of elucidation of the chain reactions of HO_x, NO_x and ClO_x, which modified the Chapman theory in the stratospheric chemistry (see Chap. 8). This section describes the oxidation reaction mechanism of methane in the natural atmosphere, which provided the opportunity for the proposal of the chain mechanism with OH and HO_2 radicals as chain carriers. The column 1 on page 288 provides a historical story when the OH chain reaction mechanism was proposed.

As for the primary trace chemical species in the remote natural atmosphere where there is no influence of anthropogenic activities, methane (CH_4) emitted from lakes and marshes, biogenic volatile organic compounds (BVOCs), nitric oxide (NO) from natural soils and lightning, dimethyl sulfide (DMS) from marine organisms, O_3 descended from the stratosphere can be conceived. Among them, the most important chemical species that is subjected to photolysis by the actinic flux (see Sect. 3.5) in the troposphere is O_3 and NO_2 (see Sects. 4.2.1 and 4.2.2).

The OH radicals formed by the reaction of water vapor H_2O with excited oxygen atoms $O(^1D)$ produced by the photolysis of O_3 (Sects. 4.2.1 and 5.1.4), play an unequivocally important role in tropospheric chemistry.

$$O_3 + h\nu \rightarrow O(^1D) + O_2 \tag{7.1}$$

$$O(^1D) + H_2O \rightarrow 2OH \tag{7.2}$$

$$O(^1D) + N_2 \rightarrow O(^3P) + N_2 \tag{7.3}$$

$$O(^1D) + O_2 \rightarrow O(^3P) + O_2 \tag{7.4}$$

Among the $O(^1D)$ formed in the photolysis (7.1), the ratio of reacting with H_2O to form OH depends on the humidity. For example, under the condition of RH = 50 % at 298 K, it is about 10 % against the deactivation (also called quenching) by reactions (7.3) and (7.4) referring to the rate constants given in Table 5.1. The ratio of OH formation by this reaction decreases with the altitude due to the decrease of humidity. Meanwhile, the ground state oxygen atoms $O(^3P)$ formed by the deactivation, react with O_2 returning to O_3 (Sect. 5.1.1),

$$O(^3P) + O_2 + M \rightarrow O_3 + M, \tag{7.5}$$

so that they hardly play a role in tropospheric chemistry.

The OH radicals formed in reaction (7.2), react with CH_4 (Sect. 5.2.7) in the natural atmosphere followed by a series of reactions under the condition of low concentration of NO (Levy 1971; Warneck 1988),

$$OH + CH_4 \rightarrow CH_3 + H_2O \tag{7.6}$$

$$CH_3 + O_2 + M \rightarrow CH_3O_2 + M \tag{7.7}$$

$$CH_3O_2 + NO \rightarrow CH_3O + NO_2 \tag{7.8}$$

$$CH_3O + O_2 \rightarrow HCHO + HO_2 \tag{7.9}$$

$$HO_2 + NO \rightarrow OH + NO_2. \tag{7.10}$$

A fundamentally important processes in tropospheric chemistry is the formation of HO_2 in a series of reactions initiated by the reaction of OH with CH_4, and the regeneration of OH by reaction (7.10) (Sect. 5.3.2). Thus, this series of reactions (7.6, 7.7, 7.8, 7.9 and 7.10), forms a chain reaction with OH and HO_2 as chain carriers. This chain reaction is called the HO_x chain cycle, but is often called the OH radical chain reaction since the reaction of OH with CH_4 is the rate-determining step. Including stratospheric chemistry, described in the following chapter, the fundamental aspect of atmospheric chemical reaction system is chain reactions where ultra-trace levels of radicals act as catalysis, which enables loss and formation of trace species whose concentrations are much higher than these radicals.

In the oxidation processes of CH_4 in the troposphere, formaldehyde (HCHO) is formed by reaction (7.9). The HCHO further reacts with OH (Sect. 5.2.11) or photolyses (Sect. 4.2.5 to form CO, H_2O and H_2,

$$HCHO + OH \rightarrow HCO + H_2O \tag{7.11}$$

$$HCHO + h\nu \rightarrow HCO + H \tag{7.12}$$

$$\rightarrow CO + H_2 \tag{7.13}$$

$$HCO + O_2 \rightarrow CO + HO_2. \tag{7.14}$$

The CO reacts with OH (Sect. 5.2.3) to give CO_2 as a final product,

$$OH + CO \rightarrow CO_2 + H \tag{7.15}$$

$$H + O_2 + M \rightarrow HO_2 + M. \tag{7.16}$$

Since these reactions of HCHO and CO with OH forms HO_2 by reactions (7.14) and (7.16) and OH is regenerated by reaction (7.10), they form HO_x chain cycle, just the same as the case of CH_4.

Here, referring to CO, the HO_x chain cycle expressed by reactions (7.6, 7.7, 7.8, 7.9 and 7.10) can be represented more simply as,

$$OH + CO \rightarrow CO_2 + H \tag{7.15}$$

$$H + O_2 + M \rightarrow HO_2 + M \tag{7.16}$$

$$HO_2 + NO \rightarrow OH + NO_2. \tag{7.10}$$

Incidentally, the typical mixing ratios of CH_4 and CO in the northern hemisphere mid-latitude clean atmosphere are ca. 1.8 ppmv and 120 ppbv, respectively, (Brasseur et al. 1999; Finlayson-Pitts and Pitts 2000). Using these values, the ratios of OH radicals to react with CH_4 and CO can be estimated as ca. 30 and 70 %, respectively. Therefore, through a series of these reactions, CH_4 is oxidized finally to CO_2 and H_2O via HCHO and CO.

In any chain cycle, there exist chain termination reactions other than the chain propagation reactions presented above. In the case of the HO_x chain cycle in the natural atmosphere,

$$HO_2 + HO_2 \rightarrow H_2O_2 + O_2 \tag{7.17}$$

$$HO_2 + CH_3O_2 \rightarrow CH_3OOH + O_2 \tag{7.18}$$

are the main termination reactions to form peroxides such as hydrogen peroxide (H_2O_2) and methyl hydroperoxide (CH_3OOH) (see Sects. 5.3.5 and 5.3.6). Therefore, assuming the trace amount of NO and NO_2, the existence of HCHO, CO, H_2O_2 and CH_3OOH in the natural troposphere can be explained by the reactions of O_3 and CH_4 in the troposphere as presented by Levy (1971). As for the termination reaction of the HO_x chain cycle other than reactions (7.17) and (7.18),

$$OH + HO_2 \rightarrow H_2O + O_2 \tag{7.19}$$

can also be conceived. Although the atmospheric concentration of OH is in general lower than HO_2 by two orders of magnitude (see Sect. 7.3.1), this reaction (7.19) is also important as chain termination reaction, since the rate constant of this reaction (Table 5.2) is two orders of magnitude larger than that of reaction (7.17) (Table 5.4) at room temperature.

7.2 Column 1 "Discovery" of the OH Radical Chain Reaction

The chain reaction mechanism with OH radical as a carrier, which is now accepted as the most fundamental reaction process in tropospheric chemistry, was proposed at the beginning of 1970s. The proposals were made in two research communities independently. One is out of the scientists who were interested in the mechanism of formation and dissipation of CO and HCHO in the natural atmosphere. In 1971 Levy (1971) presented the OH chain reaction scheme shown in Fig. 7.1 in Science for the purpose of explaining the formation of HCHO in the clean atmosphere. The chain reaction mechanism initiated by the reaction of OH and methane CH_4,

$$OH + CH_4 \rightarrow CH_3 + H_2O \tag{1}$$

and propagated by the regeneration of OH in the presence of NO,

$$HO_2 + NO \rightarrow OH + NO_2, \tag{2}$$

was the essence of the proposed scheme. One of the point of this proposal was the assumption that the chain mechanism will be completed if then unknown rate

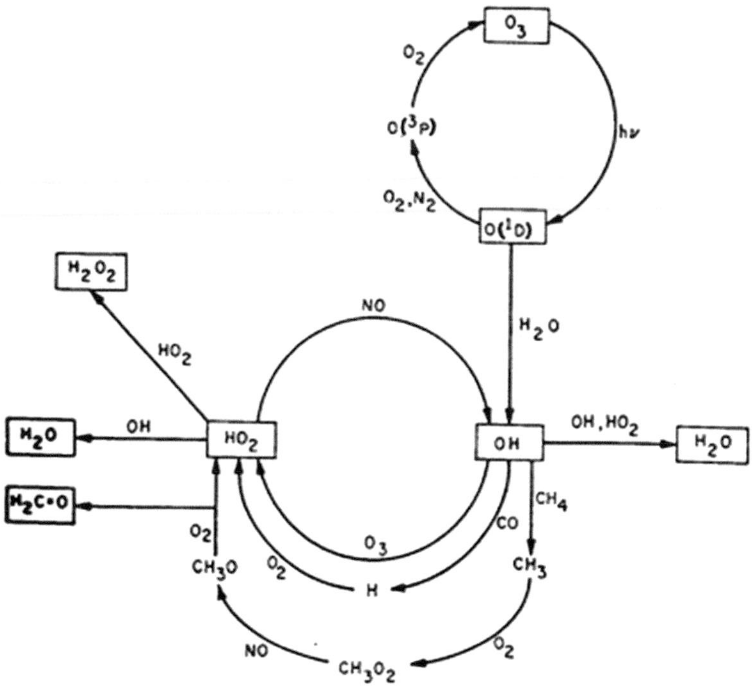

Fig. 7.1 Photochemical reaction scheme in the clean troposphere (Levy 1971)

constant of the reaction, $HO_2 + NO$, is larger than 10^{-12} cm^3 $molecule^{-1}$ s^{-1}. Later the rate constant of this reaction was found to be as large as 8.0×10^{-12} cm^3 $molecule^{-1}$ s^{-1} (298 K) as given in Table 5.4, and the Levy's hypothesis was verified to hold. Another point of this proposal is the initial supply of OH by the reaction of $O(^1D)$ and H_2O in which $O(^1D)$ is formed in the photolysis of O_3. By combining the OH generation by the photolysis of O_3 in the clean atmosphere and the above chain reaction, the Levy's hypothesis realized the breakthrough to tropospheric chemistry, and now describe in all textbooks.

On the other hand, smog chamber experiments (see Column 2 on page 317) irradiating the mixtures of NO_x and NMHC in ppm-range with simulated sunlight (($\lambda \geq 300$ nm) were then conducted for studying the chemistry of the polluted atmosphere. In those experiments, the evidence of NO to NO_2 conversion and O_3 formation accompanying the decay of NMHC had been confirmed, and what chemical species react with NMHC and convert NO to NO_2 was a big puzzle. For example, in the case of propylene C_3H_6, the reaction with then known O atoms and O_3 could not explain even a half of the decay rate as shown in Fig. 7.2, strongly suggesting some other reactive species was necessary. In this research community, Weinstock (1971) and Heicklen (1971) discussed the possibility of OH radicals to

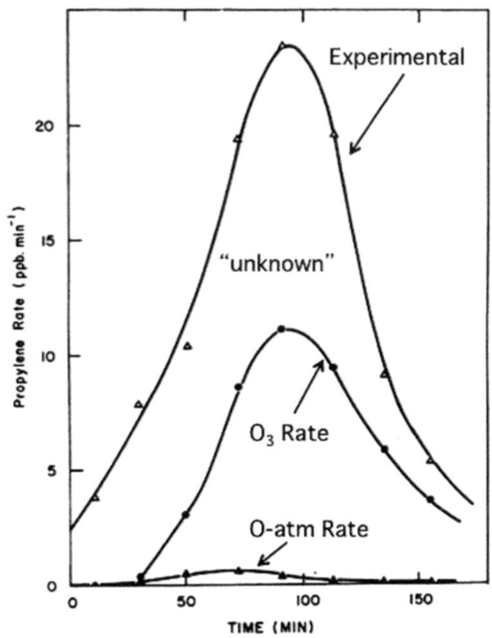

Fig. 7.2 Comparison of experimental loss rate of C_3H_6 in the irradiation of a mixture of C_3H_6 and NO_x in air in a smog chamber with the calculated loss rate by the reaction with O atom and O_3 (Niki et al. 1972)

act a reactive species, and Heicklen (1971) proposed the OH chain reaction mechanism for reactive hydrocarbons, for example i-butene,

$$C_4H_{10} + OH \rightarrow C_4H_9 + H_2O \tag{3}$$

$$C_4H_9 + O_2 \rightarrow C_4H_9O_2 \tag{4}$$

$$C_4H_9O_2 + NO \rightarrow C_4H_9O + NO_2 \tag{5}$$

$$C_4H_9O + O_2 \rightarrow C_4H_8O + HO_2 \tag{6}$$

$$HO_2 + NO \rightarrow OH + NO_2 \tag{2}$$

This mechanism is equivalent to the chain reaction for CH_4 proposed by Levy, and was proposed independently. This reaction mechanism was originally presented in 1969 in the bulletin of Pennsylvania University (Heicklen et al. 1969). However, it was never published as a peer-reviewed journal paper, and followed the unhappy history that it is not referred in most of the present textbooks on atmospheric chemistry.

7.3 Oxidation Reaction Mechanisms of VOCs in Polluted Atmosphere

As the influence of human activities to the atmosphere is enhanced, the atmospheric concentrations of anthropogenic species such as nitrogen oxides ($NO_x = NO + NO_2$), volatile organic compounds (VOCs), etc. increase over a certain level. Tropospheric chemistry then exceeds the range of perturbation to the natural atmosphere, a chemical reaction system characteristic to the polluted atmosphere, sometimes called "smog reactions," is brought about. Oxidation reaction mechanisms for the NO_x-VOC mixtures directly related to the OH chain reaction are described in this section.

In contrast to the NO_x mixing ratios 10–100 pptv in the clean lower atmosphere, those of NO_x in the urban polluted air are typically one to tens of ppbv, more than 100 times higher than the former (Finlayson-Pitts and Pitts 2000). Similarly, the mixing ratios of components of non-methane volatile organic compounds (NMVOCs) or non-methane hydrocarbons (NMHCs) in the polluted atmosphere, are also 0.1–100 ppbv, typically 100 times higher than 1–1000 pptv in the clean atmosphere (Finlayson-Pitts and Pitts 2000). In such a polluted atmosphere, most of the OH radicals formed by reactions (7.1) and (7.2) react with anthropogenic NHVOCs rather than CH_4 and CO.

The oxidation reaction processes of VOC in the presence of NO_x are different for alkanes (saturated hydrocarbons), alkenes (unsaturated hydrocarbons with a double bond, also called olefins), alkynes (unsaturated hydrocarbons with a triple bond), and aromatic hydrocarbons (unsaturated hydrocarbons with a benzene ring). However, all NMVOCs react with OH, and the HO_x chain cycle can formally be expressed as,

$$OH + RH + O_2 \rightarrow RO_2 + \text{products} \tag{7.20}$$

$$RO_2 + NO \rightarrow RO + NO_2 \tag{7.21}$$

$$RO + O_2 \rightarrow R'CHO(R'COR'') + HO_2 \tag{7.22}$$

$$HO_2 + NO \rightarrow OH + NO_2, \tag{7.10}$$

in common to most NMVOCs in a similar way as CH_4. As for aldehydes, nearly the same form of the chain reaction mechanism can also be applied. The oxidation reactions with O_3 and NO_3 for alkenes and the reaction with NO_3 for aldehydes are also important. In this section, the oxidation reaction mechanisms of hydrocarbons and aldehydes with OH, O_3 and NO_3 are summarized.

The oxidation reactions of alkanes, alkenes and aromatic hydrocarbons treated in this section are described in detail in monographs by Calvert et al. (2000, 2002, 2008), and reaction mechanisms for air quality models are summarized by Stockwell et al. (2012). Also, Calvert et al. (2011) and Mellouki et al. (2003) reviewed the oxidation reactions of oxygenated volatile organic compounds (OVOCs) which are not treated in this book. Detailed models for the photolysis and the OH, O_3 and NO_3 reactions called MCM (Master Chemical Mechanism) have been reported in v3 (Part B) for aromatic hydrocarbons (Jenkin et al. 2003), and in v3 (Part A) for other than aromatic hydrocarbons (Saunders et al. 2003). Bloss et al. (2005) further reported the improved version v3.1 for aromatic VOCs.

7.3.1 *Reaction Rate Constants of OH, O_3, NO_3 with Hydrocarbons and Aldehydes*

The rate constants for the elementary reactions of C_1-C_3 hydrocarbons and aldehyde with OH, O_3 and NO_3 have been cited in Table 5.2 in Chap. 5, Table 7.1 summarizes the OH rate constants at 298 K for the alkanes, alkynes, and aromatic hydrocarbons with higher carbon numbers that are observed in polluted atmospheres. Table 7.2 shows rate constants at 298 K for C_2-C_6 alkenes, and isoprene and α-, β-pinene, the most important biogenic hydrocarbons, with OH, O_3 and NO_3. Although the rate constants for the C_1-C_3 compounds have already been given in Table 5.2, they are cited again in Table 7.1 and 7.2 for the purpose of comparing with other $> C_3$ NMVOCs. The temperature parameters of the rate constants given in Table 7.1 and 7.2 are referenced to Atkinson (1989), Atkinson and Arey (2003), and the IUPAC subcommittee data sheet (Wallington et al. 2012).

The rate constants of OH at 298 K for the $\geq C_3$ hydrocarbons given in Table 7.1 and 7.2 are several dozen times as large even for the most slowly reacting species, ethane, and a factor of three orders of magnitude is large for most hydrocarbons as compared to methane (Table 5.2). Assuming the daytime peak concentration of OH radical in urban atmosphere in summer as 4×10^6 molecules cm^{-3} (see Sect. 7.4), the atmospheric lifetime of the species with the OH rate constants larger than 1×10^{-11} cm^3 molecule^{-1} s^{-1} is estimated to be within a few hours, and is understood to play an important role in the urban photochemical air pollution. Although the OH rate constants for the small molecules such as propane, n-butane, acetylene, benzene, ethylene, etc. are one order of magnitude smaller than this, these species are also important as other VOCs since their concentrations are generally relatively high in the urban atmosphere. From these reasons, in the discussion of the urban atmosphere, methane and other hydrocarbons are separated, and NMVOCs or NMHCs are more focused on discussion. Incidentally, the rate constants of ethane and OH, 2.4×10^{-13} cm^3 molecule^{-1} s^{-1} at 298 K is nearly the same as for CO (2.4×10^{-13} cm^3 molecule^{-1} s^{-1} at 298 K, 1 atm) (Table 5.2), and corresponds to the atmospheric lifetime of more than 1 month. Therefore, ethane does not directly contribute to urban photochemical air pollution, but can be a good tracer of anthropogenic pollution on the hemispherical scale.

As seen in Table 7.2, O_3 and NO_3 have large rate constants for alkenes, and they increase with the number of carbons. It is characteristic that the difference in rate constants are largely dependent on molecular structure, even though the carbon number is the same. As discussed in Sects. 5.4.3 and 5.5.3 in Chap. 5, these

Table 7.1 Rate constants of reactions of alkanes, alkynes and aromatic hydrocarbons with OH at 298 K

Compounds	Chemical formula	Rate constants (298 K) (10^{-11} cm^3 molecule^{-1} s^1)	Ref.
Alkanes			
Methane	CH_4	0.00064	(a)
Ethane	CH_3CH_3	0.024	(a)
Propane	CH_3 CH_2CH_3	0.11	(a)
n-Butane	$CH_3CH_2CH_2CH_3$	0.24	(b)
2-methyl propane	$CH_3CH(CH_3)CH_3$	2.1	(b)
n -Pentane	$CH_3CH_2CH_2CH_2CH_3$	3.8	(b)
2-Methylbutane	$CH_3CH(CH_3)CH_2CH_3$	3.6	(b)
2,2-Dimethylpropane	$CH_3C(CH_3)_2CH_3$	0.83	(b)
Cyclopentane		5.0	(b)
n-Hexane	$CH_3CH_2CH_2CH_2CH_2CH_3$	5.2	(b)
2-Methylpentane	$CH_3CH(CH_3)CH_2CH_2CH_3$	5.2	(b)
3- Methylpentane	$CH_3CH_2CH(CH_3)CH_2$ CH_3	5.2	(b)
2,2-Dimentylbutane	$CH_3C(CH_3)_2CH_2CH_3$	2.2	(b)
2,3-Dimethylbutane	$CH_3CH(CH_3)CH(CH_3)CH_3$	5.8	(b)
Cyclohexane		7.0	(b)
Alkynes			
Acetylene	$CH{\equiv}CH$	0.078[a]	(a)
Propyne	$CH{\equiv}CCH_3$	0.59	(c)
1-butyne	$CH{\equiv}CCH_2CH_3$	0.80	(c)
2-butyne	$CH_3C{\equiv}CCH_3$	2.7	(c)
Aromatic hydrocarbons			
Benzene		0.12	(b)
Toluene	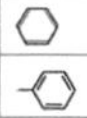	0.56	(b)
Ethylbenzene	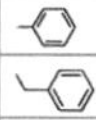	0.70	(b)
o-Xylene	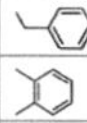	1.4	(b)
m-Xylene	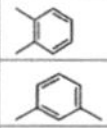	2.3	(b)
p-Xylene	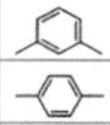	1.4	(b)
1,2,3-Trimethylbenzene	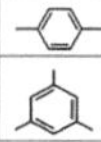	5.8	(b)

[a] 1 atm
(a) IUPAC Subcommittee Report Vol. II (Atkinson et al. 2006), (b) Atkinson and Arey (2003), (c) Atkinson (1989)

reactions are electrophilic addition reactions so that the rate constants for alkenes with a double bond adjacent to multiple methyl groups that pushes out electrons are very fast. Such alkenes are also called internal olefins. Although not shown in Table 7.2, it is known that NO_3 reacts with $\geq C_4$ alkanes, and $\geq C_8$ aromatic hydrocarbons with the rate constants with the order of ~10^{-16} cm^3 molecule^{-1} s^{-1} (Atkinson and Arey 2003).

Aldehydes do not react with O_3, but they have relatively large rate constants with OH and NO_3, and their rate constants are in between the C_2H_4 and C_3H_6, as seen in

Table 7.2 Rate constants for the reactions of alkenes, biogenic VOCs, and aldehydes with OH, O_3, and NO_3 at 298 K

		Rate constants (298 K) (cm^3 $molecule^{-1}$ s^1)		
		OH (Ref.)	O_3 (Ref.)	NO_3 (Ref.)
Compounds	Chemical formula	$\times 10^{-11}$	$\times 10^{-17}$	$\times 10^{-14}$
Ethylene	$CH_2=CH_2$	0.79[a] (a)	0.16 (a)	0.0021 (a)
Propylene	$CH_2=CHCH$	2.9[a] (a)	1.0 (a)	0.095 (a)
1-Butene	$CH_2=CHCH_2CH_3$	3.1 (b)	0.96 (b)	0.13 (b)
cis-2-Butene	$CH_3CH=CHCH_3$	5.6 (b)	12.5 (b)	3.5 (b)
trans-2-Butene	$CH_3CH=CHCH_3$	6.4 (b)	19.0 (b)	3.9 (b)
2-Methypropane	$CH_2=C(CH_3)CH_3$	5.1 (b)	1.1 (b)	3.4 (b)
1-Pentene	$CH_3CH=CHCH_2CH_3$	3.1 (b)	1.1 (b)	0.15 (b)
cis -2-Pentene	$CH_3CH=CHCH_2CH_3$	6.5 (b)	13 (b)	n/a[b]
trans -2-Pentene	$CH_3CH=CHCH_2CH_3$	6.7 (b)	16 (b)	n/a
Cyclopentene		6.7 (b)	57 (b)	4.2 (b)
2-Methy-1-butene	$CH_2=C(CH_3)\ CH_2CH_3$	6.1 b)	1.4 (b)	n/a
3-Methy-1-butene	$CH_2=CHCH(CH_3)CH_3$	3.2 (b)	0.95[c] (b)	n/a
2-Methyl-2-butene	$CH_3C(CH_3)=CHCH_3$	8.7 (b)	40 (b)	94 (b)
1-Hexene	$CH_3CH=CHCH_2CH_2CH_3$	3.7 (b)	1.1 (b)	0.18 (b)
2-Methyl-1-pentene	$CH_2=C(CH_3)\ CH_2CH_2CH_3$	6.3 (b)	1.6 (b)	n/a
2-Methyl-2-pentene	$CH_3C(CH_3)=CHCH_2CH_3$	8.9 (b)	n/a	n/a
Cyclohexene		6.8 (b)	8.1 (b)	5.1 (b)
1,3-Butadiene	$CH_2=CH=CH=CH_2$	6.7 (b)	0.63 (b)	1.0 (b)
Isoprene	$CH_2=C(CH_3)=CH=CH_2$	10 (a)	1.3 (a)	7.0 (a)
α-Pinene	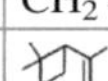	5.3 (a)	9.0 (a)	62 (a)
β-Pinene	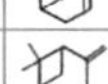	7.4 (b)	1.5 (b)	25 (b)
Formaldehyde	HCHO	0.85 (a)	n/r[d]	0.0056 (a)
Acetaldehyde	CH_3CHO	1.5 (a)	n/r	0.027 (a)
Propanal	CH_3CH_2CHO	2.0 (a)	n/r	0.064 (a)
Butanal	$CH_3CH_2CH_2CHO$	2.4 (a)	−n/r	0.11 (a)
2-Methylpropanal	$CH_3CH(CH_3)CHO$	2.6 (b)	−n/r	0.11 (b)

[a] High pressure limit, [b] Not available, [c] 293 K, [d] Not reactive
(a) IUPAC Subcommittee Report II (Atkinson et al. 2006, (b) Atkinson and Arey (2003)

Table 7.2. The reactions of aldehydes with NO_3 are the main OH and HO_2 radical source in the nighttime polluted atmosphere together with the reactions of O_3 and NO_3 with alkenes (see Sect. 7.4.1).

As for the correlation between the rate constants of VOCs with OH, O_3 and NO_3 and molecular parameters, it has been reported that the rate constants of OH and alkene reactions have good correlation with highest occupied molecular orbitals (HOMO) of the reacting molecules as mentioned in Sect. 5.2.8 (King et al. 1999). Pfrang et al. (2006) gives the correlation formulas between the reaction rate constants for each of OH, O_3 and NO_3 and HOMO energy based on recently obtained quantum chemical calculations.

7.3.2 Oxidation Reaction Mechanism of Alkanes by OH

The initial process of the alkane reaction with OH is the hydrogen abstraction, the same as for CH_4, and the abstraction of hydrogen from either the primary, secondary and tertiary carbon atom is possible for alkane with the carbon number $\geq C_3$ (see Sect. 5.2.7). As an example of alkanes, the oxidation reaction mechanism of n-butane (n-C_4H_{10}) by OH in the presence of NO_x is shown in Reaction Scheme 7.1. In the case of n-butane, 2-butyl and 1-butyl radical is formed by the abstraction process (a) from the secondary carbon and (b) from the primary carbon, respectively. In general, the reaction probability gets larger for primary < secondary < tertiary carbon reflecting the decrease of bond energy as ~420, ~ 410, ~ 400 kJ $mole^{-1}$ for the primary, secondary and tertiary C-H, respectively (Haynes 2012–2013) as described in Sect. 5.2.7. The ratio of the process (a) and (b) is ca. 85 and 15 % in the case of n-butane at 298 K (Atkinson et al. 2006).

The alkyl radicals formed by the abstraction reaction react exclusively with O_2 in the atmosphere to form alkylperoxy radicals (pathways (c) (h)), and most of them are then transformed to alkoxy radicals by oxidizing NO to NO_2 in the presence of NO_x (pathways (d) and (i)). However, some alkylperoxy radicals react with NO by the recombination isomerization reactions (e) and (j), to form alkyl nitrate (see Sect. 5.3.3). Although the yields of 2-butyl and 1-butyl nitrate are not large in the case of 2-butyl and 1-butyl radicals, 0.083 and $\leq$0.04, respectively, the production yields of alkyl nitrates increase with the increase of carbon number of alkyl radicals as described in Sect. 5.3.3, and it is as large as >0.2 for the C_6, and C_7 secondary n-alkyl radicals (2-hexyl, 3-hexyl, 2-heptyl, and 3-heptyl radicals) (Lightfoot et al. 1992; Arey et al. 2001). Since these reactions act as termination reactions for the OH chain reaction, they are important as parameters in determining the ozone formation efficiency in model calculations. The production yields of $RONO_2$

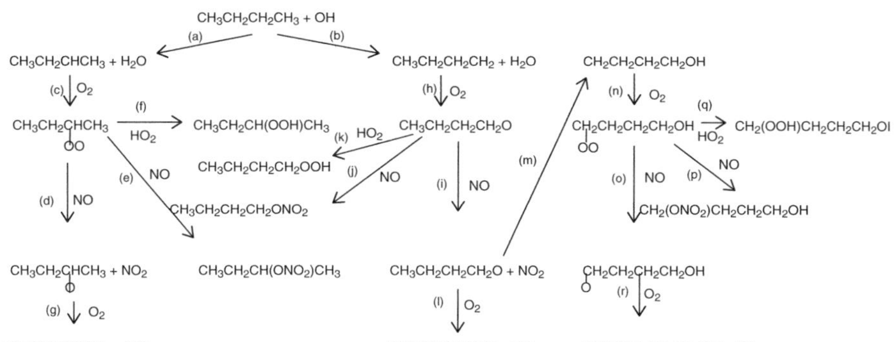

Reaction Scheme 7.1 Oxidation reaction mechanism of n-butane with OH in the presence of NO_x

in various alkylperoxy radicals (RO_2) with NO are summarized by Arey et al. (2001).

The 2-butoxy, and 1-butoxy radicals formed in the reactions of 2-butyl and 1-butyl peroxy radicals with NO (path (d) and (j)), react with O_2 to produce HO_2 together with carbonyl compounds such as methyl ethyl ketone and butanal (path (g) and (l)), and complete the HO_x cycle. Under the typical NO_x concentrations in the polluted atmosphere, the formation rate of alkyl nitrates by the reactions of alkoxy radicals with NO_2 is negligible, the recombination isomerization reaction by the alkylperoxy radicals with NO is the major pathway as the formation process of alkyl nitrate.

It is known that for straight-chain alkoxy radicals with a carbon number ≥ 4, such as 1-butoxy radicals, isomerization forms alkoxy radicals to alcohol radicals though six-members rings can occur (Atkinson 1997a, b) as follows.

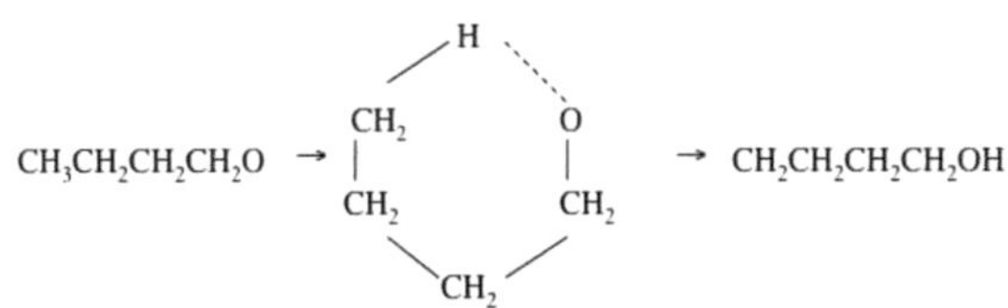

(7.23)

In the above Reaction Scheme 7.1, pathway (m) corresponds this process, and the formed n-butanol radicals produce hydroxyl butanal and 1-hydroxy butyl nitrate following the similar process as butyl radicals mentioned above (pathways (r) and (p), respectively). The rate constant for the isomerization reaction of alkoxy to alcohol radicals is in general $10^5\ s^{-1}$ while the rate constants for the production of aldehydes or ketones by hydrogen abstraction by O_2,

$$CH_3CH_2CH_2CH_2O + O_2 \rightarrow CH_3CH_2CH_2CHO + HO_2 \quad (7.24)$$

is $\sim 1 \times 10^{-15}\ cm^3\ molecule^{-1}\ s^{-1}$ (Atkinson and Arey 2003), so that the isomerization and the abstraction reactions occur with nearly the same rates in the atmosphere.

Furthermore, as for alkoxy radicals with a carbon number $\geq C_4$, it is known that the following type of unimolecular decomposition to rupture C-C bond between the carbon attached to carboxyl group and the adjacent carbon atom occurs.

$$RCH(O)R' \rightarrow RCHO + R' \quad (7.25)$$

Since the unimolecular decomposition rate constants for the secondary alkoxy radicals with $\geq C_4$ carbons, are $\sim 10^4\ s^{-1}$ (Atkinson and Arey 2003), the reactions of the types (7.23), (7.24) and (7.25) can occur in parallel to give hydroxyl aldehyde, aldehyde with the same carbon number as the reactant alkanes, and aldehydes with one carbon less than the original alkane, respectively. The rate constants of for the isomerization reaction, reaction with O_2, and the unimolecular decomposition reaction of typical alkoxy radicals are given by Finlayson-Pitts and Pitts (2000).

Under relatively low NO_x conditions in the atmosphere, a part of alkyl peroxy radicals, and hydroxyalkyl peroxy radicals react with HO_2 to give hydroperoxy butane (pathways (f), (k)), and hydroxyhydroperoxy butane (pathway (q)). Thus, in oxidation reactions of alkane in the atmosphere, hydroperoxides, hydoxyhydor-peroxides, and hydroxyalkyl nitrate, could also be produced in addition to the normal aldehydes, ketones and alkyl nitrates.

7.3.3 *Oxidation Reaction Mechanism of Alkenes by OH*

The initial reactions of alkenes and OH are the addition, and they are mostly in the high-pressure limit under atmospheric conditions including ethylene as seen in Chap. 5, Sect. 5.2.8. The addition reaction forms β-hydroxyalkyl radicals which have an OH group on the carbon adjacent to the carbon atom with an unpaired electron.

$$\mathrm{OH} + \mathrm{RCH} = \mathrm{CH_2} \rightarrow \mathrm{RCH} - \mathrm{CH_2OH} \quad (7.26)$$

$$\rightarrow \mathrm{RCH(OH)} - \mathrm{CH_2} \quad (7.27)$$

For asymmetric alkenes with carbon number $\geq C_3$, there is a possibility that the OH adds to either end of the double bond as shown above. It is known that the formation of a secondary radical by the addition of OH to the end carbon like reaction (7.26) predominates, and the ratios of the reaction type (7.26) and (7.27) is ca. 65 and 35 %, respectively in case of propylene (Finlayson-Pitts and Pitts 2000; Calvert et al. 2000).

Reaction Scheme 7.2 summarizes the reaction mechanism for 1-butene (1-C_4H_8) as an example of alkenes. The hydroxyalkyl radicals formed by the pathways (a) and (b) is a kind of alkyl radicals mentioned in the previous Sect. (7.2.2), and exclusively forms hydroxyperoxy radicals by the reaction with O_2 in the atmosphere. From the hydroxyperoxy radicals, oxyradicals (hydroxybutoxy radicals) and NO_2 (pathways (d), (k)), and partially hydroxybutyl nitrate (pathways (e), (l)) are produced by the reaction with NO as in the case of alkylperoxy radicals described in the previous paragraph. The yields of hydroxylalkyl nitrates are 2–6 % for C_4-C_6 alkenes (O'Brien et al. 1998), which is about half of those for alkyl nitrates from the alkoxy radicals. Hydroxyalkoxy radicals formed in pathways (d) and (k) are known to follow the three reaction pathways, unimolecular decomposition ((g), (n)), H-atom abstraction by O_2 ((h), (o)), and dihydroxyl radical formation by isomerization (p), corresponding to reactions (7.25), (7.24) and

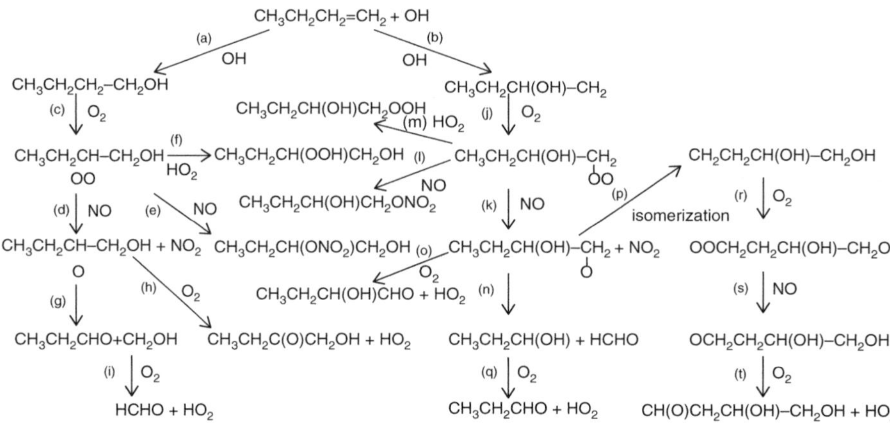

Reaction Scheme 7.2 Oxidation reaction mechanism of the 1-butene by OH in the presence of NO_x

(7.23), in the case of oxyradicals, respectively (Atkinson 1997b). In the unimolecular decomposition pathway, carbonyl compounds (CH_3CH_2CHO, HCHO) with less carbon atoms than a reactant, and hydroxyl alkyl radicals (CH_2OH, CH_3CHOH) are formed. From the hydroxyl alkyl radicals, HO_2 is produced by the reaction of O_2. From the H-atom abstraction reaction by O_2, hydoxyketone, hydoxyaldehyde, and HO_2 radicals are formed. From the HO_2 radicals, OH is reproduced by the reaction of HO_2 and NO (reaction (7.2)) so that HO_x cycle is completed.

The hydroxyalkoxy radicals formed in pathway (k) can take place isomerization by intramolecular shift of H-atom via a six-membered ring as in the case of the alkoxy radical in the previous paragraph. Dihydroxyl aldehyde is then formed in pathways (q), (r), (s) though dihydroxyl radicals which has two OH groups in a molecule. The formation of dihydroxyl aldehyde has been confirmed in the laboratory experiment, and the yield of dihydroxyl aldehyde (3,4-dihydroxyl butanal) is 0.04 for 1-butene but it is as high as 0.6 for 1-octene (Kwok et al. 1996b). Under the low NO_x concentrations, a part of hydroxyperoxy radicals react with HO_2, and is known to produce hydroxyhydroperoxy butane (pathways (f), (m)) (Hatakeyama et al. 1995; Tuazon et al. 1998).

Thus, oxidation reactions of alkenes with OH in the polluted atmosphere are thought to result in various organic nitrates and hydroperoxides such as hydroxy nitrate, dihydroxyl nitrate, hydroxyhydroperoxide, and dihydroxyl hydroperoxyide. This suggests that there are many undetected and unidentified organic peroxides, and organic nitrates in the polluted atmosphere together with the OH oxidation products of alkanes discussed in the previous paragraph.

7.3.4 Oxidation Reaction Mechanism of Alkenes by O_3

As for the pathways of O_3 and alkene reactions, detailed explanation has been made for C_2H_4 in Chap. 5, Sect. 5.4.3, and the similar reaction mechanisms are basically considered for other alkenes. In the case of asymmetric chain alkenes, two reaction pathways can be considered depending on either end of double bond forms carbonyl oxide (Criegee intermediate) when the primary ozonides rupture corresponding to reaction (5.56).

$$O_3 + RCH = CH_2 \rightarrow RCHO + [CH_2OO]^{\dagger} \quad (0.50) \tag{7.28}$$

$$\rightarrow [RCHOO]^{\dagger} + HCHO \quad (0.50) \tag{7.29}$$

$$O_3 + R_1R_2C = CH_2 \rightarrow R_1C(O)R_2 + [CH_2OO]^{\dagger} \quad (0.35) \tag{7.30}$$

$$\rightarrow [R_1R_2C\ OO]^{\dagger} + HCHO \quad (0.65) \tag{7.31}$$

$$O_3 + R_1R_2C = CHR_3 \rightarrow R_1C(O)R_2 + [R_3CHOO]^{\dagger} \quad (0.35) \tag{7.32}$$

$$\rightarrow [R_1R_2C\,OO]^{\dagger} + R_3CHO \quad (0.65) \tag{7.33}$$

As for the ratios of the two alternative pathways, carbonyl oxides containing more alkyl groups are attached are known to be formed preferentially, and the ratios are obtained experimentally to be 0.50, 0.65 and 0.65 for reactions (7.29), (7.31) and (7.33), respectively (Atkinson et al. 1997, 2006).

As for the reactions of vibrationally excited carbonyl oxides formed above, the formation of OH radicals by unimolecular decomposition and stabilized carbonyl oxides by collisional deactivation can be conceived.

$$[RCH_3C\ OO]^{\dagger} + M \rightarrow RCH_3COO + M \tag{7.34}$$

→ 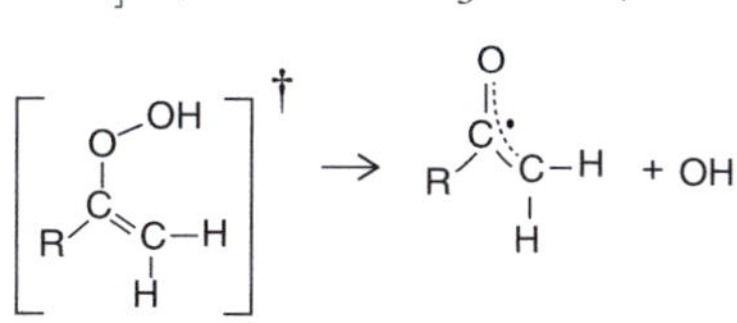 (7.35)

The formation of OH radicals by the reaction of alkenes and O_3 in the polluted atmosphere is very important in atmospheric chemistry of ozone formation described in the next section. The yields of OH radicals under the atmospheric pressure are compiled by the IUPAC subcommittee (Atkinson et al. 2006) and summarized in Table 7.3. These values are those obtained by the batch type experiments with OH scavengers, while Kroll et al. (2001a) investigated pressure dependence (1–400 torr) of the initial OH yields in the reaction of O_3 and alkene using a high-pressure flow system coupled with the direct detection of OH by LIF. From this experiment, strong pressure dependence was found for the OH formation yields from alkenes other than ethylene with the OH yields higher than unity at low

Table 7.3 The yields of OH in the O_3-alkene reactions under the atmospheric pressure

Alkenes	OH yields	Ref.
Ethene	0.14–0.20	(a)
Propene	0.32–0.35	(a)
1-Butene	0.41	(b)
1-Pentene	0.37	(b)
1-Hexene	0.32	(b)
cis-2-Butene	0.33	(a)
trans-2-Butene	0.54–0.75	(a)
2-Methylpropene	0.60–0.72	(a)
2-Methy-1-butene	0.83	(b)
2-Methyl-2-butene	0.80–0.98	(a)
2,3-Dimethyl-2-butene	0.80–1.00	(a)
Cyclopentene	0.61	(b)
Cyclohexene	0.68	(b)
Isoprene	0.25	(a)
α-Pinene	0.70–1.00	(a)
β-Pinene	0.35	(b)

(a) IUPAC Subcommittee Report Vol. II (Atkinson et al. 2006)
(b) Finlayson-Pitts and Pitts (2000)

pressure. This implies that other than reaction (7.35) the OH formation by the process,

$$H + O_3 \rightarrow OH + O_2 \tag{7.36}$$

due to H atoms formed in the reaction (5.60c) in the case of C_2H_4 is important at low pressure. Meanwhile, the OH yields at around 1 atm is much smaller than pervious values shown in Table 7.3 except for C_2H_4. In order to solve the problem, Kroll et al. (2001b) obtained unimolecular decomposition rate of carbonyl oxides formed in the O_3-alkene reactions by the RRKM calculation. As a result, it was confirmed that OH radicals are produced, other than those from the vibrationally excited carbonyl oxides, also from the stabilized carbonyl oxides formed by collisional stabilization in reaction (7.34) in a much longer time scale. Thus, it is thought that the total OH yields including this slow process has been obtained by the previous experiments with OH scavengers, and the values in Table 7.3 would be appropriate for use in atmospheric chemistry models. In the case of C_2H_4, the pressure dependence was not observed in the above experiment with the flow system, and the obtained value, 0.14 agreed well with the batch type experiments, suggesting a different reaction mechanism from other alkenes (Kroll et al. 2001c).

Johnson and Marston (2008) reviewed on the reactions of unsaturated VOC with O_3 in the troposphere, and Donahue et al. (2011) introduced new data and interpretation on the pressure dependence of the OH yields.

7.3.5 *Oxidation Reaction Mechanism of Alkenes by NO_3*

The initial reactions of NO_3 and alkenes are the addition to double bonds as seen in Chap. 5, Sect. (5.5.3). As shown in Table 7.2, the reaction rate constants increase with carbon number, but for the same carbon number species those for the internal alkenes are much larger. The reaction mechanisms of the NO_3-alkene reactions are similar to those with OH and alkenes and an example for cis-2-butene can be illustrated as below.

As seen in Reaction Scheme 7.3, in the NO_3 and alkene reactions epoxy alkanes (also called epoxides or oxiranes), hydroperoxy nitrates, carbonyl nitrates and dinitrates are produced characteristically (Bandow et al. 1980; Kwok et al. 1996a; Calvert et al. 2000), and some of them have been detected in the polluted atmosphere (Schneider et al. 1998). Recently, reactions of NO_3 with alkenes are interesting from the view point of organic aerosol formation (Gong et al. 2005; Ng et al. 2008).

As for the initial pathways of the NO_3-alkene reactions, it has been shown that ring-opening addition reaction as seen in Reaction Scheme 7.3 occurs for C_2H_4 and C_3H_6 by quantum chemical calculations (Pérez-Casany et al. 2000; Nguyen et al. 2011), and the possibility of cycloaddition is also suggested for some alkenes (Cartas-Rosado et al. 2004).

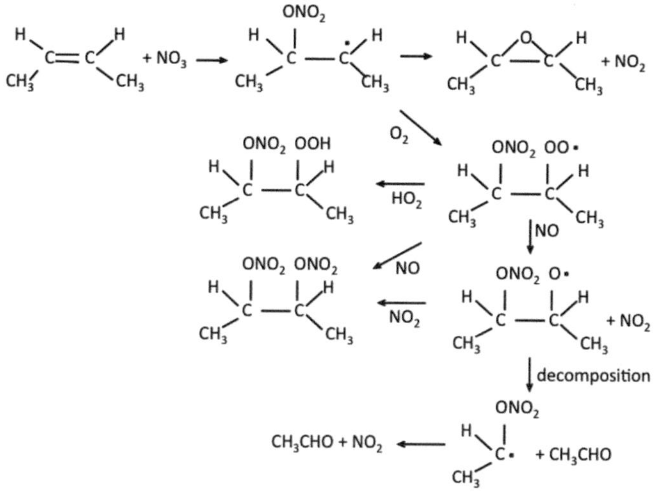

Reaction Scheme 7.3 Oxidation reaction mechanism of cis-2-butene by NO_3

7.3.6 *Oxidation Reaction Mechanism of Isoprene by OH, O_3, and NO_3*

Isoprene ($CH_2=C(CH_3)$-$CH=CH_2$) is called 2-methyl-1,3-butadiene by the nomenclature of IUPAC, a compound with two double bonds in a molecule, and is the most important biogenic hydrocarbons, which account for 50 % of the global emissions (Guenther et al. 2006). The rate constants of isoprene are large, 1.0×10^{-11}, 1.3×10^{-17} and 6.8×10^{-13} cm^3 $molecule^{-1}$ s^{-1} at 298 K for OH, O_3 and NO_3, respectively, as shown in Table 7.2. The reactions of either of these species are important in the atmosphere, particularly those with OH and O_3 in daytime and with NO_3 in nighttime (Calvert et al. 2000).

The oxidation mechanisms of isoprene with OH, O_3, and NO_3 are addition reaction similar to alkenes and can be thought as an application of reactions described in Sects. 7.2.3, 7.2.4, and 7.2.5. However, since isoprene has asymmetric two double bonds, four reaction pathways have to be considered depending on the addition of active species to either of double bonds and either side of carbons. Many experimental and theoretical studies have been conducted as for the oxidation mechanism of isoprene (Finlayson-Pitts and Pitts 2000; Seinfeld and Pandis 2006), and Fan and Zhang (2004) presented the reaction scheme for each of OH, O_3 and NO_3 by summarizing those studies. Reaction Schemes 7.4, 7.5 and 7.6 illustrate the oxidation reaction mechanism of isoprene initiated by OH, O_3 and NO_3, respectively, adapted from Fan and Zhang (2004).

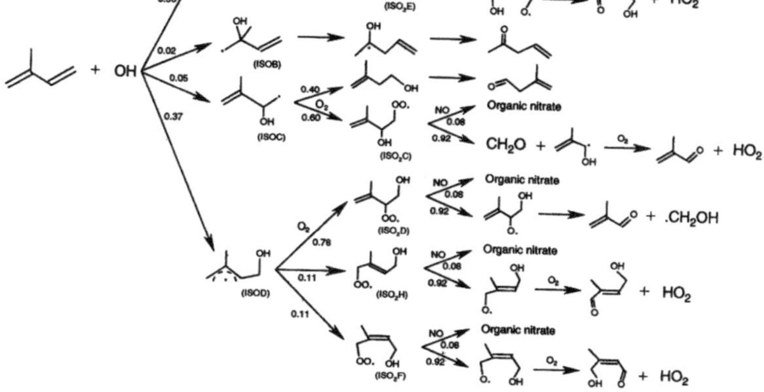

Reaction Scheme 7.4 Oxidation reaction mechanism of isoprene by OH in the presence of NO_x (Adapted from Fan and Zhang 2004)

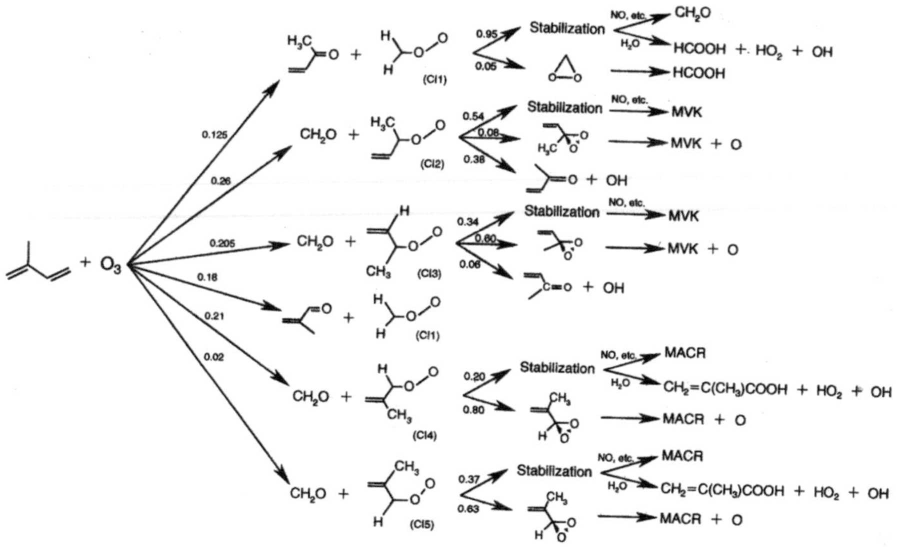

Reaction Scheme 7.5 Oxidation reaction mechanism of isoprene by O_3 in the presence of NO_x (Adapted from Fan and Zhang 2004)

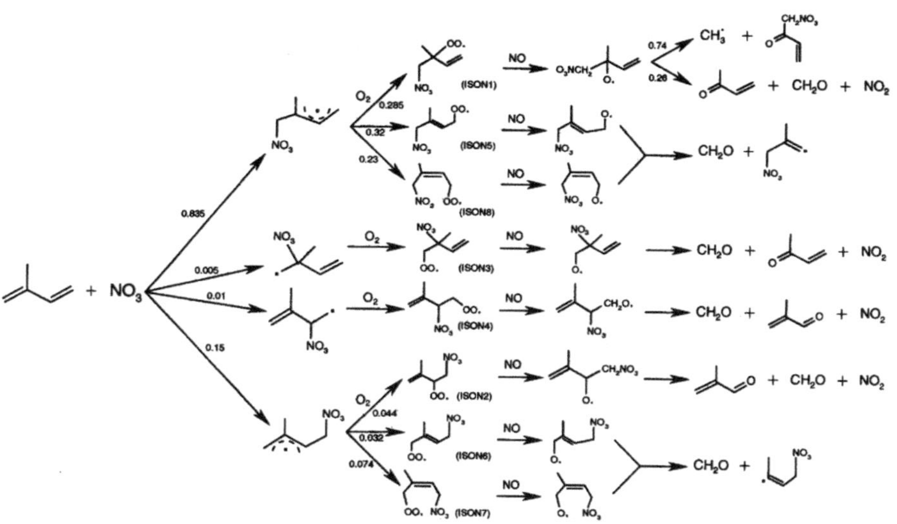

Reaction Scheme 7.6 Oxidation reaction mechanism of isoprene by NO_3 (Adapted from Fan and Zhang 2004)

As can be seen in Reaction Scheme 7.4, the main products of the oxidation reaction of isoprene by OH are methacrolein ($CH_2=CH(CH_3)CHO$, MACR) and methyl vinyl ketone ($CH_3C(O)CH=CH_2$, MVK) (Karl et al. 2006). Besides them unsaturated hydroxycarbonyl compounds and hydroxy nitrate are formed as characteristic products.

In the reaction of isoprene and O_3, via five kinds of carbonyl oxides (Criegee intermediates), MACR and MVK are formed as main products as in the case of oxidation reaction by OH, as well as HCHO, HCOOH and unsaturated carboxylic acids. In this reaction similar to the reaction of alkene and O_3, OH radicals are formed, which is important as a nighttime OH source. Zhang and Zhang (2002) showed by the quantum chemical calculation that the formation of cyclic primary ozonides occur to either of double bonds in the same ratio, and OH is formed by unimolecular decomposition of carbonyl oxides formed by the decomposition of primary ozonides. There are two kinds of OH formed by the unimolecular decomposition of carbonyl oxides, one generated from the vibrationally excited carbonyl oxides in a short timescale, and the another generated from collisionally stabilized carbonyl oxides with a time delay as described in the above Sect. 7.2.4 (Kroll et al. 2001b). The summed OH yields of these processes are important in atmospheric reaction chemistry. Malkin et al. (2010) reported the average OH yield as 0.26 ± 0.02 obtained by the product analysis with scavenger and the direct measurement with FAGE/LIF (see Sect. 7.4.1), and the yield of HO_2 by the FAGE method as 0.26 ± 0.03 at 298 K and 1 atm. Meanwhile, the summed OH formation yield obtained by theoretical calculation is 0.24, agreeing well with the experiments (Zhang and Zhang 2002; Zhang et al. 2002). Although not taken up in this book, the reaction mechanisms of O_3 with α-pinene and β-pinene, which are important biogenic hydrocarbons next to isoprene, are summarized by Zhang and Zhang (2005).

In the reaction of isoprene with NO_3 (Reaction Scheme 7.6), the reactions proceed through eight kinds of nitrooxyalkyl peroxy radicals (ISON1-ISON8) and 8 kinds of nitrooxyalkoxy radicals (ISN1-ISN8) formed from the formers by the reaction with NO. The main final products are formaldehyde, unsaturated aldehydes, ketones and organic nitrates, and the yields of MACR, MVK are small different form the reactions with OH and O_3 (Barnes et al. 1990; Skov et al. 1992; Kwok et al. 1996a). From recent chamber experiments using proton-transfer mass spectrometer (PTR-MS) and thermal decomposition laser induced fluorescence (TD-LIF) as analytical methods, the yields of each of MACR and MVK are ~10 %, and those of organic nitrates are 0.65 ± 0.12 (Perring et al. 2009). Furthermore, hydroxy nitrate (C), nitoxyaldehyde (D), nitroxyhydroperoxide (E), etc. are observed among the products. Their reaction pathways, which are not included in Reaction Scheme 7.6, are shown in Reaction Scheme 7.7.

In recent field observations in the foothills of Sierra Nevada, United States, using a chemical ionization mass spectrometer (CIMS) and TD-LIF method, many organic nitrates, including hydroxy nitrates from isoprene and other BVOCs, are detected, and their ratio is reported amounting to two thirds of total organic nitrates (Beaver et al. 2012). Oxidation reactions of isoprene and other BVOCs in the

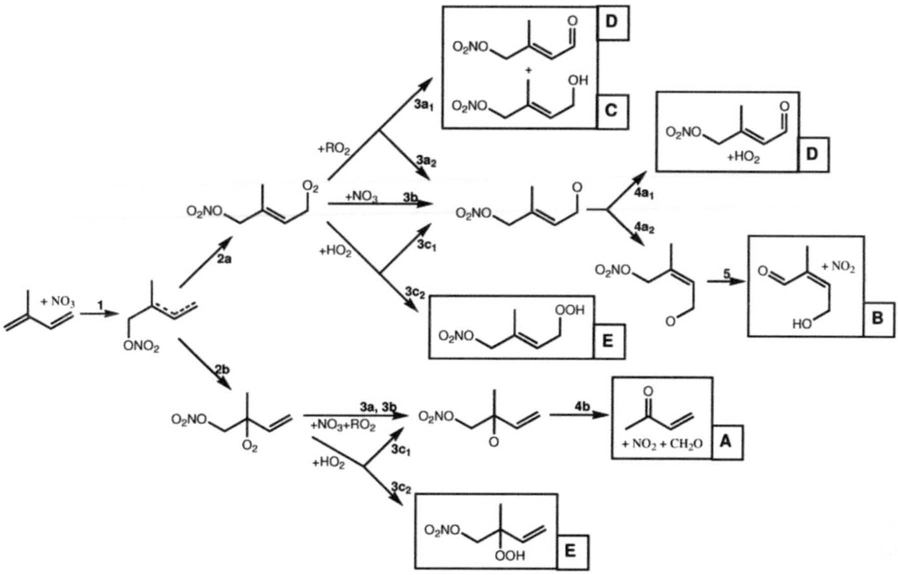

Reaction Scheme 7.7 Pathways of the products found in the reaction of isoprene and NO_3 for (**A**) C_4-carbonyl, (**B**) C_5-hydroxy carbonyl, (**C**) hydroxy nitrate, (**D**) nitroxy aldehyde, and (**E**) nitroxy hydroperoxide (Perring et al. 2009)

atmosphere are also interested form the viewpoint of organic aerosols formation (Claeys et al. 2004; Kroll et al. 2006; Zhang et al. 2007; Ng et al. 2008; Rollins et al. 2009).

7.3.7 Oxidation Reaction Mechanism of Alkynes by OH

As for the atmospheric dissipation of alkynes, the reaction with OH is the sole process. In addition, the initial reaction of OH with alkynes is similar to alkenes, and the reaction is in the high-pressure limit at 1 atm for alkynes with carbon number more than 3, although several atm is necessary to reach the high-pressure limit for acetylene (C_2H_2) (see Sect. 5.2.9) (see Table 7.1). The main alkyne in the polluted atmosphere is C_2H_2, and its oxidation mechanism by OH is shown in Reaction Scheme 7.8.

As seen in Reaction Scheme 7.8, glyoxal (CHO-CHO) is the main product from the OH-added radical in the reaction of OH and C_2H_2 through peroxy and oxy radicals (Hatakeyama et al. 1986; Galano et al. 2008). The main reaction pathway yields HO_2 in the last step (e), and forms HO_x chain cycle. Similarly, methyl glyoxal (CH_3COCHO) and biacetyl CH_3COCH_3CO) are formed as main products from propyne (C_3H_4), and 2-butyne (2-C_4H_6), respectively (Hatakeyama et al. 1986).

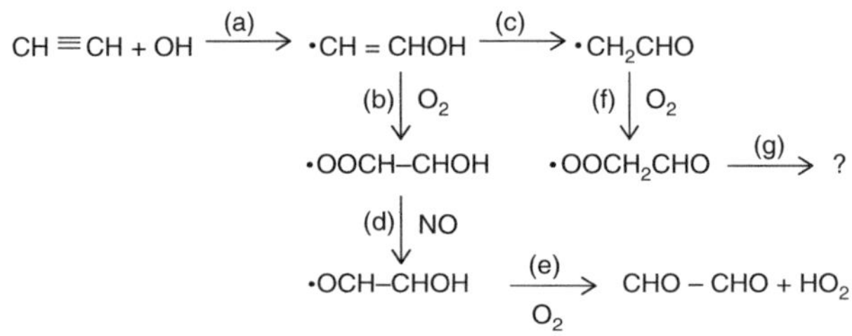

Reaction Scheme 7.8 Oxidation reaction scheme of acetylene by OH in the presence of NO_x

A part of the OH-added radical isomerizes as in pathway (c), and forms vinoxy radical (CH_2CHO) (Schmidt et al., 1985), but its succeeding reactions are not well known yet.

7.3.8 Oxidation Reaction Mechanism of Aromatics by OH

The reactions of OH and aromatic hydrocarbons under the atmospheric conditions consists of two processes, the addition to benzene ring and the H-atom abstraction from alkyl group as seen in Chap. 5, Sect. 5.2.10. Here, taking toluene (CH_3-C_6H_5) whose concentration is the highest in general among aromatic hydrocarbons in the polluted atmosphere, as a representative example, the H-atom abstraction from the side chain methyl group gives benzyl radical as follows (Calvert et al. 2002; Atkinson and Arey 2003) (Sect. 5.2.10).

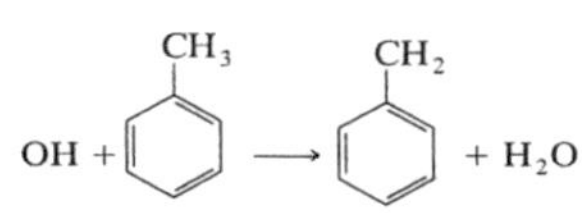

(7.37)

The main products from the benzyl radical in the presence of NO_x are known to be benzaldehyde and benzyl nitrate (Akimoto et al. 1978; Klotz et al. 1998; Calvert et al. 2002; Atkinson and Arey 2003), and their reaction mechanism can be represented as,

According to this reaction scheme, HO_2 radicals are formed similar to the processes of H-atom abstraction forming alkanes seen in Sect. 7.2.2. Since OH radicals are regenerated from HO_2, HO_x chain reaction cycle is completed, and NO is oxidized to NO_2 simultaneously. However, the ratio of H-atom abstraction is only a few percent in the case of toluene, and it has been know from the product analysis that the main pathway is addition reaction described below (Klotz et al. 1998; Calvert et al. 2002; Atkinson and Arey 2003) (Reaction Scheme 7.9).

The first step of the addition reaction of OH to the benzene ring is the formation of hydroxymethyl cyclohexadienyl radical (A) as shown in Reaction Scheme 7.10 (see Sect. 5.2.10).

The direct measurements of the rate constants of the reaction of hydroxymethyl cyclohexadienyl radical with O_2 has been made using the UV absorption method (Sect. 5.2.10), and the values of 2.5×10^{-16} cm^3 molecule^{-1} s^{-1} for benzene (Bohn and Zetzsch 1999; Grebenkin and Krasnoperov 2004; Raoult et al. 2004; Nehr et al. 2011), 6.0×10^{-16} cm^3 molecule^{-1} s^{-1} for toluene (Knispel et al. 1990; Bohn 2001) are reported. Therefore, most hydroxymethyl cyclohexadienyl radicals are thought to react solely with O_2. Also, the unimolecular decomposition rate of cyclohexadienyl radical from benzene back to $C_6H_6 + OH$ has been reported as (3.9 ± 1.3) s^{-1} at 298 K (Nehr et al. 2011).

In the case of toluene, as shown in Reaction Scheme 7.10, three reaction pathways of hydroxymethyl cyclohexadienyl radical with O_2 are conceived; the formation of (1) cresol (B) by the H-atom abstraction from the OH-added carbon (pathway a), (2) toluene 1,2-epoxide (C) and the 2-methyl oxepin (D) equilibrated with it, and (3) hydroxymethyl cyclohexadienyl peroxy radical (E) formed by the O_2 addition to the benzene ring (Suh et al. 2003; Cartas-Rosado and Castro 2007; Baltaretu et al. 2009). The formation of cresol by pathway (a) is well known, and o-cresol is produced predominantly followed by p- and m-cresol reflecting the ortho- and para-orientation of the OH addition to benzene ring (see Sect. 5.2.10). However, the ratio of the H-atom abstraction from the benzene ring of hydroxy

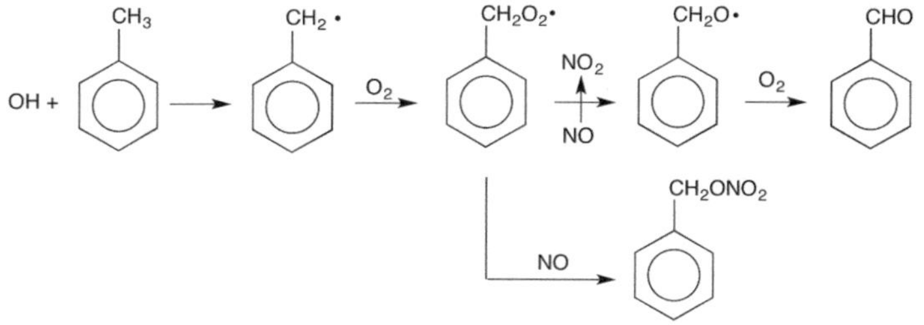

Reaction Scheme 7.9 Oxidation reaction mechanism of toluene by OH through H-atom abstraction process in the presence of NO_x

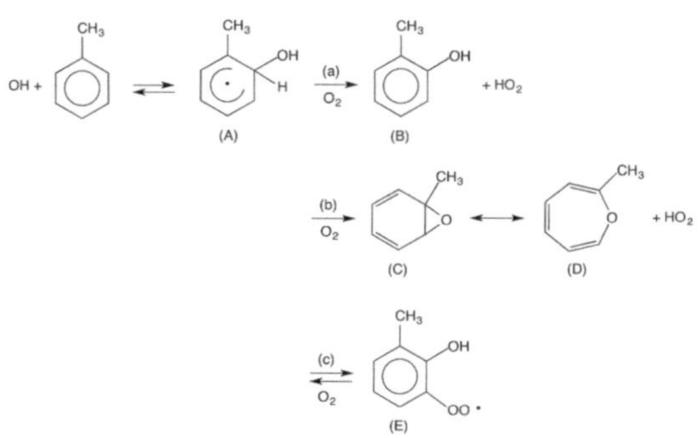

Reaction Scheme 7.10 Formation of hydroxymethyl cyclohexadienyl radical in the addition reaction of OH to toluene and its pathway for the reaction with O_2

methyl cyclohexadienyl radical by O_2 is c.a. 20 % for the reactions of OH with toluene, xylene and trimethyl benzene, and the pathways (b) and (c) are known to occur preferentially based on the product analysis (Klotz et al. 1998; Calvert et al. 2002; Atkinson and Arey 2003).

The production of toluene 1,2-epoxide (C) and 2-methy oxepin (D) by the pathway (b) was proposed by Klotz et al. (2000). Their reaction rates with OH was found to be fast experimentally and theoretically, and they are thought to be one of the formation pathways to the open-ring compounds described below (Cartas-Rosado and Castro 2007). In the case of benzene, it has been reported that phenol is formed from the photolysis of benzene oxide and oxepin (Klotz et al. 1997), but the cresols are not formed from toluene 1.2-epoxide or 2-methyl oxepin (Klotz et al. 2000).

The formation of hydroxymethyl cyclohexadienyl peroxy radical (E) by pathway (c) in Reaction Scheme 7.10 is thought to be the most important as a process to produce ring-opened compounds, which are well known reaction products of OH and aromatic hydrocarbons. The O_2 addition reaction to the OH-added benzene ring radical is reversible, and in the case of benzene, the equilibrium constant and the enthalpy change is given as $K(298\ \mathrm{K}) = (8.0 \pm 0.6) \times 10^{-20}\ \mathrm{cm^3\ molecule^{-1}}$ (Johnson et al. 2002), and $= (-44 \pm 2)\ \mathrm{kJ\ mol^{-1}}$ (Grebenkin and Krasnoperov 2004) for the hydroxycyclohexadienyl radical. Therefore, hydroxycyclohexadienyl radicals formed in pathway (c) are in general stable in the atmosphere, and play an important role as intermediates in the oxidation reaction of aromatic compounds.

A peculiar feature of oxidation reaction of aromatic hydrocarbons is the production of dicarbonyl compounds by opening of the benzene ring. For example, glyoxal (CHOCHO) are known to be formed from benzene, glyoxal and methyl glyoxal ($CHOC(CH_3)O$) from toluene, glyoxal, methyl glyoxal and biacetyl ($C(CH_3)OC(CH_3)O$) from o-xylene (Calvert et al. 2002; Atkinson and Arey 2003). Also, in the case of toluene C_5, C_6 and C_7 dialdehyde, and hydroxy dicarbonyl compounds are formed. Further, formation of epoxide such as 2,3-epoxy-2-methyl hexenedial, has been confirmed (Yu and Jeffries 1997; Baltaretu et al. 2009) as mentioned below. The production yields and formation mechanism of these ring-opened compounds are not well established yet and the many studies are still going on.

Two reaction pathways to form ring-opened compounds in the oxidation reaction of toluene and OH are thought to occur, one through hydroxymethyl cyclohexadienyl peroxy radical (E) and the other through toluene-1,2-epoxide (C) or 2-metyl oxepin (D) in Reaction Scheme 7.10. Reaction Scheme 7.11 summarizes former pathways for peroxy radical (E) shown in Reaction Scheme in 7.10 based on recent findings.

From hydroxymethyl cyclohexadienyl peroxy radical (E), formation of an intermediate in which O-O bridges to a benzene ring, bicyclooxy radical (F) is predicted theoretically (Andino et al. 1996; García-Cruz et al. 2000; Suh et al. 2003, 2006). Most of ring-opened products are thought to be formed through this bicyclo radical, and in the case of toluene it is assumed that C_7 epoxide, epoxymethyl hexenedial (L), is thought to be formed by the isomerization of bicyclo-peroxy radical to

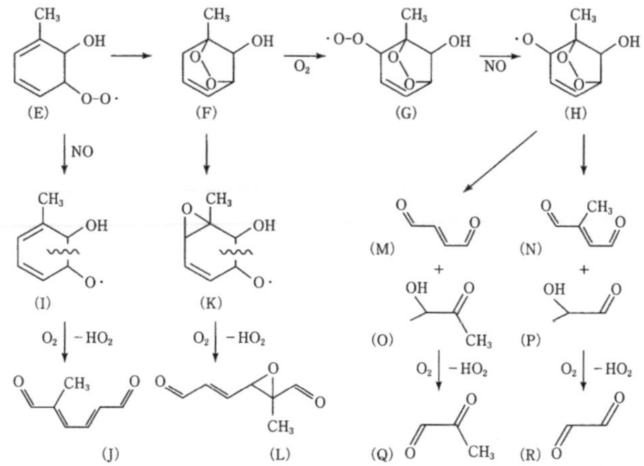

Reaction Scheme 7.11 Formation pathways of ring-opened compounds through hydroxymethyl cyclohexadienyl peroxy radical (E) in the oxidation reaction of toluene with OH

hydroxy methyl epoxyoxy radical (K) (Bartolotti and Edney 1995; Baltaretu et al. 2009). Furthermore, Baltaretu et al. (2009) found methyl hexadienedial (J) containing all the carbon atoms of toluene is formed in high yields in the oxidation reaction of toluene with OH using a flow system chemical ionization mass spectrometer (Birdsall et al., 2010), and proposed a pathway through a oxy radical (I) formed from the reaction of cyclohexadienyl peroxy radical with NO without passing though the bicyclo intermediate (F).

Bicycloperoxy peroxy radical (G) formed by the further O_2 addition to bicycloperoxy radicals has been detected experimentally by use of a chemical ionization mass spectrometer, and the production pathway of butanedial (M) and methyl glyoxal (Q), and 2-methyl butanedial (N) and glyoxal (R) are proposed through the unimolecular decomposition of bicyclo peroxy oxy radical (H) which is formed by the peroxyradical and NO reaction as shown in Reaction Scheme 7.11 (Volkamer et al. 2001; Suh et al. 2003; Baltaretu et al. 2009).

Meanwhile, reactions of toluene 1,2-epoxide (C) and 2-methy oxepin (D) have not been studied well. According to quantum chemical calculations, the reaction rate constants of the former epoxide with OH largely at $\approx 1 \times 10^{-10}$ cm^3 molecule^{-1} s^{-1} (Cartas-Rosado and Castro 2007) reproducing well the experimental value of $\approx 2 \times 10^{-10}$ cm^3 molecule^{-1} s^{-1} (Klotz et al. 2000), the theoretical value of the rate constant of OH with the oxepin is very small as $\approx 1 \times 10^{-14}$ cm^3 molecule^{-1} s^{-1}. The formation of the reaction product 6-oxohepta-2,4-dienal obtained experimentally by Klotz et al. (2000) is also predicted by theoretical calculation, and the reaction pathways is proposed by Cartas-Rosado and Castro (2007), as shown in Reaction Scheme 7.12.

The oxidation reaction mechanism of aromatic hydrocarbons in the atmosphere has not been established yet. Although the validation for the reaction model MCM

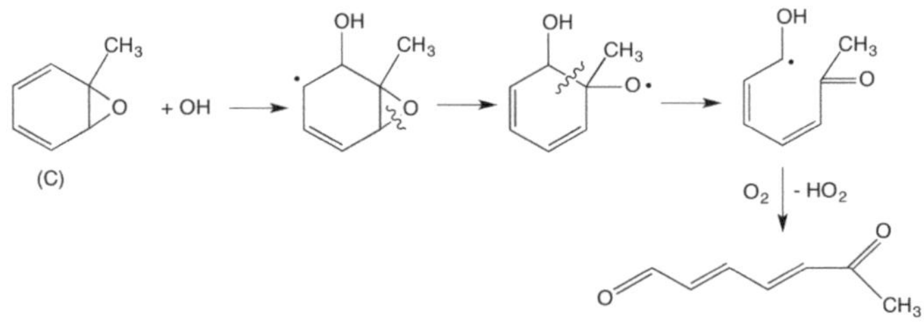

Reaction Scheme 7.12 The formation pathway of 6-oxohepta-2,4-dial from the OH oxidation reaction of toluene 1,2-opoxide (C)

(master chemical reaction) based on the reaction schemes mentioned here, has been made by the experimental data using the outdoor smog chamber of European Union (EURPHORE) (see Column 2 on p. 317), still large discrepancies are reported regarding the concentrations of formed ozone, OH and HO_2, etc. (Wagner et al. 2003; Bloss et al. 2005). In particular, experimental evidence on the NO_x dependence of the products and the identification of nitrogen containing products are still poor.

As for the reaction products, many compounds not mentioned above are also reported and their production pathways are not known well. Furthermore, many of the products mentioned above such as dienes containing two double bonds and dials containing two aldehyde groups in a molecule, have large reaction rate constants with OH, and there is a possibility that the compounds formed secondarily from them may play an important role in the polluted atmosphere. It is easily anticipated that they would share the large fraction of unidentified VOCs in the atmosphere (see Sect. 7.4.2) and constitutes a large uncertainty factor in the predictive models for photochemical ozone and oxi dants. Also, there are many carcinogenic and mutagenic substances among them, such as dials and epoxides, which are concerning from the aspect of health impact of photochemical air pollution.

7.3.9 *Oxidation Reaction Mechanism of Aldehydes by OH and NO_3*

The oxidation reactions of aldehydes with OH in the presence of NO_x are very important from the point of forming peculiar compounds with strong biological toxicity called peroxy acyl nitrates (PANs, $RC(O)OONO_2$). The initial reaction of OH and aldehydes are H-atom abstraction forming the aldehyde group to form acyl radicals as seen in Chap. 5, Sect. 5.2.11. For example, in the case of acetaldehyde, the reaction mechanism is,

$$CH_3CHO + OH \rightarrow CH_3CO + H_2O \tag{7.38}$$

$$CH_3CO + O_2 + M \rightarrow CH_3C(O)OO + M \tag{7.39}$$

$$CH_3C(O)OO + NO \rightarrow CH_3C(O)O + NO_2 \tag{7.40}$$

$$CH_3C(O)OO + NO_2 + M \rightarrow CH_3C(O)OONO_2 + M \tag{7.41}$$

$$CH_3C(O)OO + HO_2 \rightarrow CH_3C(O)OOH + O_2 \tag{7.42}$$

$$CH_3C(O)O \rightarrow CH_3 + CO_2 \tag{7.43}$$

$$CH_3 + O_2 + M \rightarrow CH_3O_2 + M \tag{7.7}$$

$$CH_3O_2 + NO \rightarrow CH_3O + NO_2 \tag{7.8}$$

$$CH_3O + O_2 \rightarrow HCHO + HO_2. \tag{7.9}$$

These reaction pathways are in parallel with those for alkanes mentioned in Sect. 7.2.2, and the reactions (7.7, 7.8 and 7.9) after CH_3 radicals are formed in reaction (7.43) are the same as those in the oxidation processes of methane described in Sect. 7.1. The specific feature of oxidation reactions of aldehydes is the formation of a metastable peroxy acyl nitrates from the reaction of peroxy acyl radicals with NO_2 by reaction (7.41). In the case of acetaldehyde, peroxy acetyl nitrate, $CH_3C(O)OONO_2$, is formed. This compound is called PAN (Peroxy Acetyl Nitrate), and is known to have much stronger toxicity to plants than ozone. A group of peroxy acyl nitrates are collectively called PANs.

Since the bond energy D°(O-NO_2) in $CH_3C(O)OONO_2$ is as small as 92 kJ mol^{-1}, this molecule is easily thermal decomposed, and it is in thermal equilibrium in the troposphere,

$$CH_3C(O)OO + NO_2 + M \quad CH_3C(O)OONO_2 + M. \tag{7.44}$$

The thermal decomposition rate constants of PAN is given by the IUPAC Subcommittee Report Vol. II as $k_{\infty,7.44}$ (298 K) = $3.8 \times 10^{-4}\ s^{-1}$ (Atkinson et al. 2006), and the atmospheric lifetime of PAN is calculated as 43 min at 298 K. Thus, PAN is lost by the thermal decomposition and is not transported in a long range in the lower troposphere. However, the lifetime is much longer in the upper troposphere where the temperature is low, and it is transported in a long range as a NO_x reservoir, and serves as a slow regeneration source of NO_x subject to the reaction with OH or photolysis.

The reaction of aldehyde with NO_3 is the H-atom abstraction (Mora-Diez and Boyd, 2002), the same as with OH, and for example, in the case of CH_3CHO,

$$CH_3CHO + NO_3 \rightarrow CH_3CO + HONO_2. \tag{7.45}$$

Therefore, the succeeding reactions are the same as reaction (7.39) and the following described above, and HO_2 is formed by reaction (7.9). Thus the reaction of aldehyde and NO_3 is important as a nighttime source of HO_x radicals.

7.4 Formation and Loss of O_3 by OH Radical Chain Reaction

The OH radical chain reaction described in Sect. 7.1 plays an essential role in tropospheric chemistry relating to the removal of almost all atmospheric organic compounds, the formation of secondary pollutants including organic aerosols, and the formation and destruction of ozone. Under the low NO_x concentrations where the anthropogenic influence is small, OH radicals generated from the reaction of water vapor and the excited oxygen atom $O(^1D)$ formed in the photolysis of O_3 causes the loss of O_3 by reacting O_3 itself competing with the reactions with CH_4 and CO. On the other hand, under the condition of NO_x level higher than a certain level, the OH radical chain reaction coupled with the photolysis of NO_2 causes the net formation of O_3. Particularly in the polluted atmosphere with high concentrations of NO_x and VOC, it produces a high concentration of O_3 causing photochemical air pollution. Therefore, the distribution of NO_x and the resultant formation and destruction of O_3 is fundamentally important in determining the spatial distribution of O_3. Further, quantitative elucidation of the relationships between the O_3 formation and the concentrations of NO_x and VOC in the polluted atmosphere is important from the control strategy of photochemical air pollution. This section describes the HO_x chain reaction mechanism which is the basis of such relationships between the NO_x and VOC with O_3 formation.

7.4.1 Formation and Loss of O_3 in the Clean Atmosphere

When the atmospheric concentration of NO_x is very low, the OH radicals formed in the reactions (7.1) and (7.2) triggered by the photolysis of O_3, and the HO_2 radicals formed in the reactions of OH with CO and CH_4, react with O_3 in addition to the radical-radical reactions (7.17) and (7.19) (Sects. 5.2.1 and 5.3.1),

$$OH + O_3 \rightarrow HO_2 + O_2 \tag{7.46}$$

$$HO_2 + O_3 \rightarrow HO + 2O_2 \tag{7.47}$$

which constitute a chain reaction to destruct O_3. Also among the $O(^1D)$ formed in the photolysis of O_3, the reactions with H_2O to form OH itself,

$$O_3 + h\nu \rightarrow O(^1D) + O_2 \tag{7.1}$$

$$O(^1D) + H_2O \rightarrow 2OH \tag{7.2}$$

causes the loss of O_3. From these reactions, the in situ loss rate of O_3 can be expressed as,

$$L(O_3) = (j_{7.1}f_{7.2} + k_{7.46}[OH] + k_{7.47}[HO_2])[O_3], \quad (7.48)$$

where $j_{7.1}$ is the photolysis rate by reaction (7.1), $f_{7.2}$ is the yield of OH formation from $O(^1D)$ produced in this reaction, $k_{7.46}$ and $k_{7.47}$ are the rate constants of reactions (7.46) and (7.47), respectively. The value of $j_{7.1}$ can be calculated from the absorption cross sections of O_3 (Table 4.1) and the actinic flux given in Table 3.5, for example, for the earth surface,

$$f_{7.2} = \frac{k_{7.2}[H_2O]}{k_{7.2}[H_2O] + k_{7.3}[N_2] + k_{7.4}[O_2]}. \quad (7.49)$$

Here, $k_{7.2}$, $k_{7.3}$ *and* $k_{7.4}$ are the rate constants of reaction (7.2), (7.3) and (7.4).

On the other hand, when the atmospheric NO_x concentration increases, the HO_x chain reaction of OH with CH_4 and CO operating under the low NO_x conditions described in Sect. (7.1), is coupled to the O_3 formation reactions $O(^3P)$ from the photolysis of NO_2 (Sect. 4.2.2) and O_2, to bring O_3 production.

$$HO_2 + NO \rightarrow OH + NO_2 \quad (7.10)$$

$$RO_2 + NO \rightarrow RO + NO_2 \quad (7.21)$$

$$NO_2 + h\nu \rightarrow NO + O(^3P) \quad (7.50)$$

$$O(^3P) + O_2 + M \rightarrow O_3 + M. \quad (7.51)$$

From these equations, in situ O_3 production rate can be expressed as,

$$P(O_3) = (k_{7.10}[HO_2] + k_{7.21}[RO_2])[NO], \quad (7.52)$$

where $k_{7.10}$ and $k_{7.21}$ are the rate constants of reactions (7.10) and (7.21), and RO_2 stands for organic peroxy radicals. In the remote clean atmosphere over the open ocean, it is enough to consider only CH_3O_2, many other organic peroxide radicals seen in Sect. 7.2 has to be taken into account as natural and anthropogenic VOCs are added. From Eqs. (7.52) and (7.48), net in situ production rate of O_3, $N(O_3)$ can be expressed as,

$$N(O_3) = P(O_3) - L(O_3). \quad (7.53)$$

Figure 7.3 is an example of plot of the calculated formation and loss rate of O_3 as function of NO_x in the clean atmosphere (Liu et al. 1992). The calculations are made for the diurnal averaged values for the atmospheric condition of Mauna Loa with the 1–100 pptv of NO_x. As shown in Fig. 7.3, the loss rate of O_3 is almost constant independent of NO_x concentration whereas the formation rate is nearly proportional to NO_x concentration. This means that the daytime and nighttime averaged concentrations [OH], $[HO_2]$ and $[RO_2]$ in Eqs. (7.48) and (7.52) are almost constant being independent of NO_x concentration.

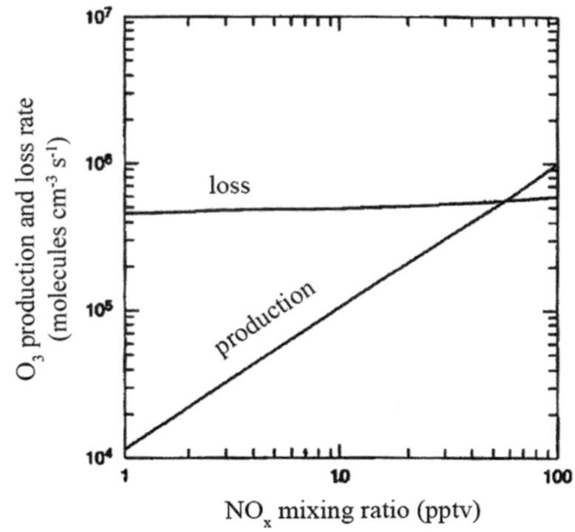

Fig. 7.3 Production and loss rate of O_3 as a function of NO_x mixing ratio in the clean troposphere calculated by a model simulation (Adapted from Liu et al. 1992)

From Fig. 7.3, it can be seen that there is a threshold of NO_x concentration under which net O_3 destruction and above which net O_3 production occurs in the clean troposphere. Putting $P(O_3) = L(O_3)$ in Eq. (7.53), the threshold value of NO concentration, $[NO]_{th}$, is calculated as,

$$[NO]_{th} = \frac{(j_{7.1}f_{7.2} + k_{7.46}[OH] + k_{7.47}[HO_2])[O_3]}{k_{7.10}[HO_2] + k_{7.21}[RO_2]}. \tag{7.54}$$

Here, substituting the typical values, $j_{7.1} = 4 \times 10^{-5}$ s^{-1}, $f_{7.2} = 0.1$; $[HO_2] = 6 \times 10^8$, $[OH] = 4 \times 10^6$, $[CH_3O_2] = 3 \times 10^8$ molecules cm^{-3}; $k_{7.46} = 1.9 \times 10^{-15}$, $k_{7.47} = 7.3 \times 10^{-14}$, $k_{7.10} = 8.0 \times 10^{-12}$, $k_{7.21}(CH_3O_2) = 7.7 \times 10^{-12}$ cm^3 molecule^{-1} s^{-1}, and $[O_3] = 40$ ppb, the value of $[NO]_{th} = \sim 25$ ppt is obtained.

Under the solar irradiation in daytime, the concentration ratio of NO and NO_2 can be determined by the ratio of NO production rate by the photolysis of NO_2 in reaction (7.50) and the conversion rate of NO to NO_2 by O_3 (Sect. 5.4.1),

$$O_3 + NO \rightarrow NO_2 + O_2, \tag{7.55}$$

so that,

$$\frac{[NO]}{[NO_2]} = \frac{j_{7.50}}{k_{7.55}[O_3] + k_{7.10}[HO_2] + k_{7.21}[RO_2]}. \tag{7.56}$$

Typical O_3 mixing ratios are ~10 ppbv and 30–50 ppbv over the clean ocean and the clean terrestrial, respectively, while those of peroxy radicals are ~10 pptv, three orders of magnitude lower than O_3. However, the rate constants of the reaction of NO with the peroxy radicals, e.g. HO_2, CH_3O_2, are ~ 8×10^{-12} cm^3 molecule^{-1} s^{-1}, which is nearly three orders of magnitude larger than that of NO and O_3,

1.8×10^{-14} cm^3 $molecule^{-1}$ s^{-1} (see Table 5.4), the contributions of these peroxy radicals and O_3 are comparable to determine the equilibrium concentration ratio of NO and NO_2 in the daytime. The calculated ratio from Eq. (7.56) is $[NO]/[NO_2] \approx 0.3$. Using this value, the threshold value of NO_x calculated from Eq. (7.53) is $[NOx]_{th} \approx 80$ pptv. Form Fig. 7.3, the threshold value of NO_x can be read as ~60 pptv, and thus the threshold mixing ratio of NO_x to determine whether production or destruction of net O_3 occurs in the clean atmosphere is in general thought to be a few 10 s to 100 pptv.

In the actual remote atmosphere, O_3 concentrations are determined mostly by long-range transport, and do not necessarily reflect the in situ production and destruction directly. However, the production and loss of O_3 described above are reflected in the regional scale distribution of O_3, and are important for the consideration of the global budget of tropospheric ozone.

7.4.2 Formation of O_3 in the Polluted Atmosphere

The dependence of production and loss of O_3 on NO_x mixing ratio shows totally different features in the polluted atmosphere where the NO_x mixing ratio exceeds 1 ppbv. Fig. 7.4 shows the plot of the OH and HO_2 concentrations, and net O_3 production rate against NO_x mixing ratio from the clean regime of 1–100 pptv to the polluted regime of over 10 ppbv by a model calculation (Brune 2000). The calculated results for the two cases of planetary boundary layer (PBL) and upper troposphere (UP) are shown in the figure. As seen in Fig. 7.4a for the boundary layer, the OH concentrations are almost constant up to the NO_x mixing ratio of a

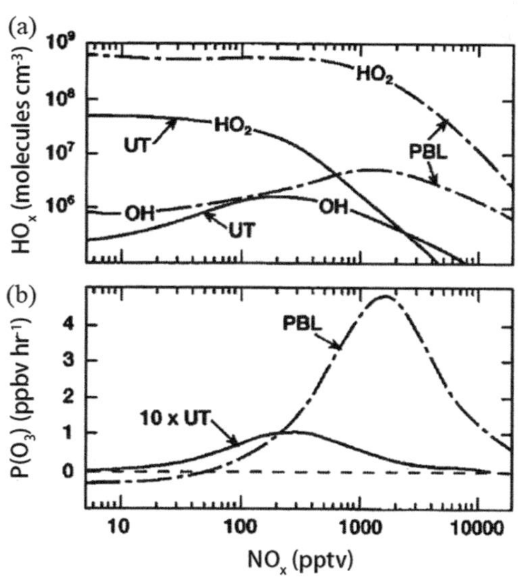

Fig. 7.4 Model-calculated dependence on NO_x mixing ratio of (**a**) OH and HO_2 concentrations, (**b**) net ozone production rate. *PBL* planetary boundary layer, *UT* upper troposphere (Adapted from Brune 2000)

few 10s of pptv, and it increases when the NO_x mixing ratio exceeds this value, while the concentrations of OH and HO_2 decreases rapidly when the NO_x mixing ratio exceeds ca. 1 ppbv. In the upper troposphere, these changes occur nearly one order of magnitude lower mixing ratios of NO_x, and the increase of OH is seen over ca. 10 pptv of NO_x and the decrease of OH and HO_2 is seen over ca. 200 pptv of NO_x. On the other hand, as seen in Fig. 7.4b, the net production rate of O_3, $N(O_3)$, in the boundary layer turns to positive over a few 10s of pptv being consistent with Fig. 7.3, rapidly increases over 100 pptv, and maximizes at 1–2 ppbv of NO_x and then rapidly decreases over the higher mixing ratios of NO_x.

These NO_x dependence of HO_x concentrations and $N(O_3)$ can be explained as follows. In the low NO_x region of less than 100 pptv (lower than the threshold value of net O_3 production) in the planetary boundary layer, OH formed in the photolysis of O_3 and the succeeding HO_2 are controlled by loss processes by the reaction with O_3, reactions (7.46) and (7.47), and radical-radical reactions, reactions (7.17, 7.18 and 7.19). Since the HO_2 + NO, reaction (7.10), cannot compete with them, the concentrations of OH and HO_2 are nearly constant being independent on NO_x mixing ratio. As the increase of NO_x, the contribution of the reaction, HO_2 + NO, increases to control the O_3 production, and simultaneously facilitate the conversion of HO_2 to OH resulting the increase of OH concentrations. When the NO_x mixing ratio further increases over 1 ppbv, the reaction of OH and NO_2 (see Sect. 5.2.4),

$$OH + NO_2 + M \rightarrow HONO_2 + M \tag{7.57}$$

becomes effective to reduce OH and HO_2 concentration as the NO_x mixing ratio increases. Thus, in the regime where NO_x exceeds 1–2 ppbv, HO_2 concentration and net O_3 production rate $N(O_3)$ starts to decrease with NO_x. The NO_x mixing ratio of 10–100 ppbv can causes photochemical air pollution with high O_3 pollution over 100 ppbv. The production rate and concentration of O_3 shows strong non-linearity toward NO_x mixing ratios in this range as described in the next paragraph.

In the upper troposphere, the formation rate of OH by the reaction of $O(^1D)$ and H_2O is small due to the low humidity, so that the contribution of the photolysis of H_2O_2 (Sect. 4.2.8), HCHO (Sect. 4.2.5) and $CH_3C(O)CH_3$ (Sect. 4.2.7) as a source of HO_x gets larger compared to the PBL. As seen in Fig. 7.4, the concentrations of OH and HO_2 in the low NO_x range in the upper troposphere are small, and about one third for OH and about one tenth for HO_2 as compared to the boundary layer. Therefore, the self-reactions of HO_x as a loss process of HO_x are not efficient there. On the other hand, the reaction of HO_2 and NO_2 to form HO_2NO_2 (see Sect. 5.3.4),

$$HO_2 + NO_2 + M \rightarrow HO_2NO_2 + M \tag{7.58}$$

becomes important with the increase of NO_x in the upper troposphere. This is because the thermal decomposition reaction rate of

$$HO_2NO_2 + M \rightarrow HO_2 + NO_2 + M \quad (7.59)$$

gets small at the low temperatures in the upper troposphere, and reaction (7.58) becomes an in situ substantial loss process of NO_x. From the reason that the reaction with NO_x is more effective compared to self-reactions of HO_x, the OH and HO_2 concentrations and the net O_3 production rate starts to decrease at much lower NO_x mixing ratio than in PBL as shown in Fig. 7.4. Thus, the maximum of OH concentration and $N(O_3)$ appears around 200 pptv of NO_x, and the HO_2 concentration starts to decrease rapidly over 200 pptv, both happens for one order of magnitude lower NO_x mixing ratios as compared to the PBL. The mixing ratio of NO_x more than 100 pptv has been observed in the upper troposphere (Brasseur et al. 1999), and it should be noted that the production of O_3 is in the non-linear for NO_x in this range.

7.4.3 Column 2 Photochemical Smog Chamber

The large reactive vessel designed to study the photochemical reactions in the polluted atmosphere is called "photochemical smog chamber" or simply "smog chamber" (also called photochemical chamber or environmental chamber). The prototype of the smog chamber is those fabricated in 1960s, which was a metal-framed plastic film box of a few m^3 volume with fluorescent lights (called black lights) outside for irradiating light of longer wavelength than 300 nm (for example, Rose and Brandt 1960). The large volume was required for smog chambers from the reasons of analysis and reaction. For the analysis of reactants and products in the chamber, air pollution monitoring instruments were conventionally used, which required to suck substantial amount of air, so that large volume was necessary for the reaction vessel. From the point of reaction involving low concentration gases, adsorption and desorption of sample and product gases to and from the vessel wall cannot be neglected, and a large reaction vessel was required in order to reduce surface/volume ratio.

In order to fulfill these two requirement, the smog chamber has been developed to two directions. One is the construction of evacuable smog chamber, which enabled to remove the adsorbed species by heating the wall under high vacuum and erase the memory of previous reactions to improve experimental accuracy. The idea is the extension of physical chemistry experiments on photochemistry using classical glass cells. The first evacuable smog chamber was built at University of California at Riverside (UCR) (Winer, et al. 1980), and the second one was built at National Institute for Environment (NIES) in Japan (Akimoto et al. 1979a). Figure 7.5 illustrates the picture of the NIES chamber. The volume of the chamber is 6.3 m^3 and heating oil is circulated through the aperture of the double wall realizing the baking at 200 °C and temperature control during experiments. Irradiation is made co-axially by the solar simulator consisting of xenon arc lamps. The chamber is

Fig. 7.5 Evacuable photochemical smog chamber at National Institute for Environmental Studies, Tsukuba, Japan

equipped with the long-path (133.5 m) Fourier transform infrared spectrometer in the short-axial direction, and the multi-reflection mirrors are supported on the base of concrete optical benches mechanically separated from the chamber wall which suffers the thermal strain of heating. This chamber was successful to give reproducible experimental data, and provided many useful results from the end of 1970s–1980s (see Sect. 7.3.3). Recent evacuable chamber has been built at Universit'es Paris-Est-Cr'eteil et Paris Diderot in France (Wang et al. 2011).

Another direction of the smog chamber is the development of outdoor chambers. The advantage of the outdoor chamber is the usage of the natural sunlight as a light source, which enables much larger volume for the chamber. The first outdoor chamber with 156 m^3 was constructed at University of North Carolina (Fox et al. 1975). Later many outdoor chambers have been built. Among them European Commission (EC) built an outdoor chamber (photoreactor) with the volume of ca. 200 m^3 (EUPHORE) in Valencia, Spain (Becker 1996) as shown in Fig. 7.6, and it has been utilized for the validation of model simulation and the studies on atmospheric reaction mechanisms. A more recent one with the volume of 270 m^3 (SAPHIR) was built in Jülich, Germany, which is equipped with the long-path absorption cell for the measurement of OH radicals (Karl et al. 2004).

Thus, the smog chamber whose prototype was born in 1960s is the rare experimental equipment continuing to be constructed and utilized for more than 50 years. Recalling of early days when the smog chamber experiments were criticized that the scientific merit is low due to their low reproducibility, these seem to belong to another age.

Fig. 7.6 Outdoor photoreactor (EUPHORE) in Valencia, Spain

7.4.4 Dependence of O_3 Formation on NO_x and VOC, and Ozone Isopleths

In this paragraph, dependence of O_3 production on NO_x and VOC concentrations under the conditions of polluted urban atmosphere is summarized referring to the photochemical smog chamber experiments (see Column 2 p. 317). The almost sole reaction of direct O_3 formation in the troposphere is the reaction of O_2 with the ground state oxygen atom $O(^3P)$ from the photolysis of NO_2,

$$NO_2 + h\nu \rightarrow NO + O(^3P) \tag{7.50}$$

$$O(^3P) + O_2 + M \rightarrow O_3 + M \tag{7.51}$$

If only these reactions proceed, all NO_2 molecules in the atmosphere are photolyzed, and the same amount of O_3 as the initial concentration of NO_2 will be formed. Thus, for example, if the NO_2 mixing ratio in the polluted atmosphere is 100 ppb, the formed O_3 should also be 100 ppb. Figure 7.7 depicts the experimental plot of the mixing ratios of NO_2, NO and O_3 when about 100 ppb of NO_2 is irradiated by the solar simulator in the smog chamber (Akimoto et al. 1979a). As seen in the figure, NO_2, NO and O_3 reach to photo-equilibrium within a couple of minutes so that only about 30 % of the initial concentration of NO_2 converts to the equivalent amounts of NO and O_3, and no further conversion of NO_2 to O_3 can be seen. The k_1 value shown in Fig. 7.7 is the experimental value of photolysis rate constant of NO_2 calculated from the concentrations of NO, NO_2 and O_3 by putting HO_2 and RO_2 is equal to zero in the photo-equilibrium Eq. (7.56). The k_1 value, 0.24 min^{-1}, after reaching to the photo-equilibrium corresponds to $j_{7.50}$ in the text.

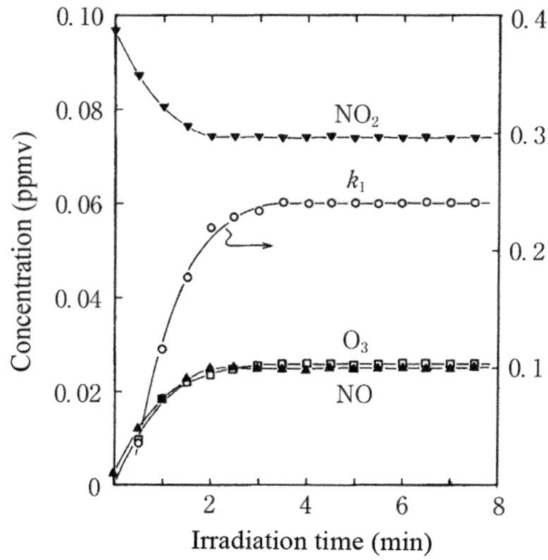

Fig. 7.7 Time profiles of NO, NO_2 and O_3 mixing ratios under the irradiation of NO_2 in the smog chamber. The values of k_1 in the figure is an experimentally obtained photolysis rate constant of NO_2 from the stationary state equation (see text) (Adapted from Akimoto et al. 1979a)

The reason of establishment of the photo-equilibrium is due to the reverse reaction between O_3 and NO (Sect. 5.4.1),

$$O_3 + NO \rightarrow NO_2 + O_2, \tag{7.55}$$

which brings back NO to NO_2. From reactions (7.50), (7.51) and (7.50), the concentration of O_3 produced in the photostationary state when only NO_x is present in the atmosphere can be approximated by,

$$[O_3]_{ps} = [NO]_{ps} = \frac{-j_{7.50} + \sqrt{j_{7.50}^2 + j_{7.50}k_{7.55}[NO_2]_0}}{2k_{7.55}} \approx \sqrt{\frac{j_{7.50}}{k_{7.55}}[NO_2]_0} \tag{7.60}$$

(Akimoto et al. 1979a). Here, $[NO_2]_0$ is the initial concentration of NO_2, and $[O_3]_{ps}$ and $[NO]_{ps}$ are the photostationary concentrations of O_3 and NO, respectively, and $j_{7.50}$ and $k_{7.50}$ are NO_2 photolysis rate (4×10^{-3} s^{-1} in the experiment of Fig. 7.7) and the rate constant of reaction (7.50) (1.8×10^{-14} cm^3 molecule^{-1} s^{-1}, see Table 5.5), respectively. Incidentally, reaction (7.50) is important as causing in situ dissipation of O_3 by NO emitted from sources in urban atmosphere, and is called "NO titration reaction" since it decreases O_3 concentration stoichiometrically equal to the amount of NO concentration.

When VOC is added to the NO_2-air mixture, the HO_x chain reaction (7.20), (7.21), (7.22) and (7.10) mentioned in Sect. 7.2 are combined to the above reactions (7.50), (7.51) and (7.55) to constitute the following chain reaction system.

$$OH + RH + O_2 \rightarrow RO_2 + \text{product} \tag{7.20}$$

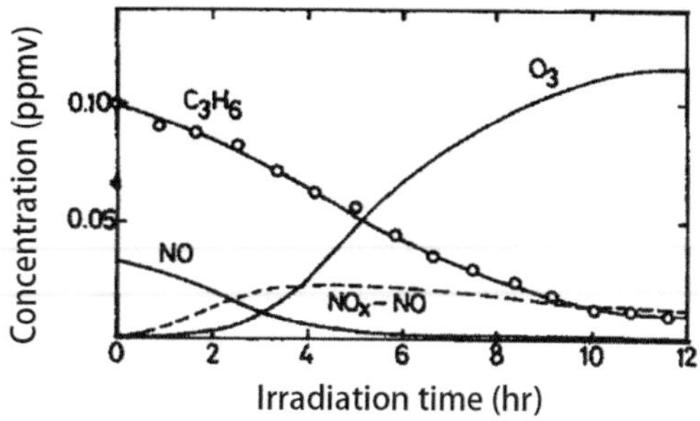

Fig. 7.8 Time profiles of the mixing ratios of C_3H_6, NO, $NO_x - NO$ and O_3 in the irradiation of the mixture of C_3H_6,and NO_x in air in the smog chamber (Adapted from Akimoto et al. 1979b)

$$RO_2 + NO \rightarrow RO + NO_2 \tag{7.21}$$

$$RO + O_2 \rightarrow R'CHO(R'COR'') + HO_2 \tag{7.22}$$

$$HO_2 + NO \rightarrow OH + NO_2 \tag{7.10}$$

$$NO_2 + h\nu \rightarrow NO + O(^3P) \tag{7.50}$$

$$O(^3P) + O_2 + M \rightarrow O_3 + M \tag{7.51}$$

$$O_3 + NO \rightarrow NO_2 + O_2 \tag{7.55}$$

In this reaction system, NO formed in the photolytic reaction (7.50) returns to NO_2 preferentially reacting with HO_2 and RO_2 by reactions (7.10) and (7.21) without consuming O_3 by reaction (7.55). From the reproduced NO_2, O_3 is produced again together with oxidation of NO to NO_2, so that O_3 is accumulated to result in high concentration as the cycle proceeds in many times.

This feature is demonstrated by the chamber experiment in Fig. 7.8. This figure shows the change of concentrations of each compound when 34 ppbv of NO_x (NO 33 ppbv, NO_2 1 ppbv) and 100 ppbv of propylene (C_3H_6) are irradiated by the solar simulator (Akimoto et al. 1979b). From the figure, the features of NO to NO_2 oxidation, formation of O_3 in much higher concentration than the initial concentration of NO_x and the decay of C_3H_6 can be seen. The [NO_x-NO] in the figure stands for the concentration of nitrogeneous compounds other than NO, which is almost identical to the concentration of NO_2 until reaching the peak of [NO_x-NO], but reflects the sum of nitrogeneous compounds including PAN and partially nitric acid as the irradiation time gets longer. This is because the commercial chemiluminescent instrument with a molybdenum convertor used in the experiment reduces many of the secondary nitrogeneous compounds other than NO_2 to NO, and their sum is represented as NO_x. As shown in Fig. 7.8, O_3 formation rate decreases in the latter half of the experiment, and reaches the maximum because the reactant NO_x is removed from the reaction system by the formation of nitric acid (reaction (7.57)), PAN (reaction (7.44)), and hydroxy alkyl nitrate (reaction scheme 7.2), etc. Meanwhile, the decay of C_3H_6 shown in Fig. 7.8 is due to the reaction of OH and O_3 as seen in Sect. 7.2.

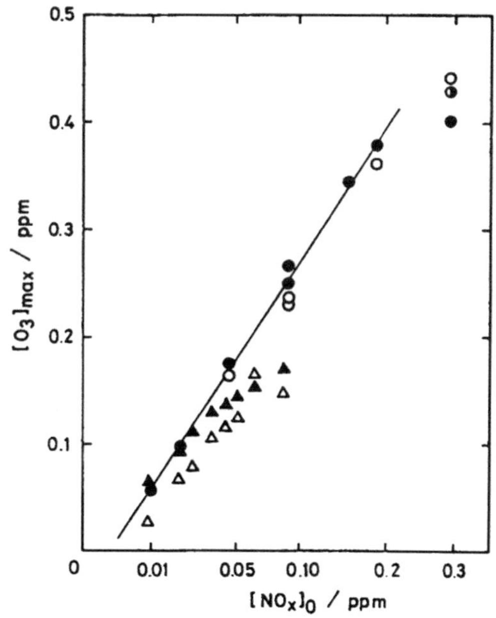

Fig. 7.9 The dependence of the maximum ozone mixing ratio $[O_3]_{max}$ as a function of the square root of initial mixing ratio of NO_x, $[NO_x]_O$, in the irradiation of the mixture of C_3H_6 and NO_x in air in the smog chamber. The abscissa axis is in a square root scale. The initial concentration of C_3H_6 is 500 ppbv (○, ●) and 100 ppbv (△, ▲). Experimental (*open symbols*) and model-calculated (*filled symbols*) values are compared (Sakamaki et al. 1982)

In Fig. 7.8, it took nearly 10 hours for O_3 to reach its maximum, but it is generally known that the O_3 formation rate is roughly proportional to VOC concentration (Akimoto and Sakamaki 1983). According to Eq. (7.52), the O_3 formation rate $P(O_3)$ is proportional to the concentration of HO_2 and RO_2, and the reaction of OH with RH, reaction (7.20), is the rate-determining step for the formation rate of the peroxy radicals. As deduced by the chamber experiment, OH concentration is nearly constant when the VOC/NO_x ratio is higher than a certain value (VOC-excess or NO_x-limited), so that the reaction rate of OH and VOC is nearly proportional to the VOC concentration (Akimoto et al. 1980).

Figure 7.9 shows a comparison between experiment and reaction model calculation for the relationship between the initial concentrations of NO_x and VOC and maximum O_3 concentration $[O_3]_{max}$ obtained by the chamber experiment under the VOC-excess condition. In this experiment, C_3H_6 is used as VOC, and prolonged irradiation was made for the C_3H_6 initial concentration of 100 and 500 ppbv by changing the initial concentrations of NOx in the range of 10–300 ppbv. As shown in Fig. 7.9, the O_3 concentrations finally reached are almost independent of C_3H_6 and is approximately proportional to the square root of the initial concentration of NO_x (Akimoto et al. 1979b), which is well reproduce by the model calculation (Sakamaki et al. 1982). Similarly, it was confirmed that $[O_3]_{max}$ is proportional to square root of photolysis rate constant of NO_2, $j_{7.44}$, by the experiments and the model calculations (Sakamaki et al. 1982). From these results, they defined $[O_3]_{ps}$ as,

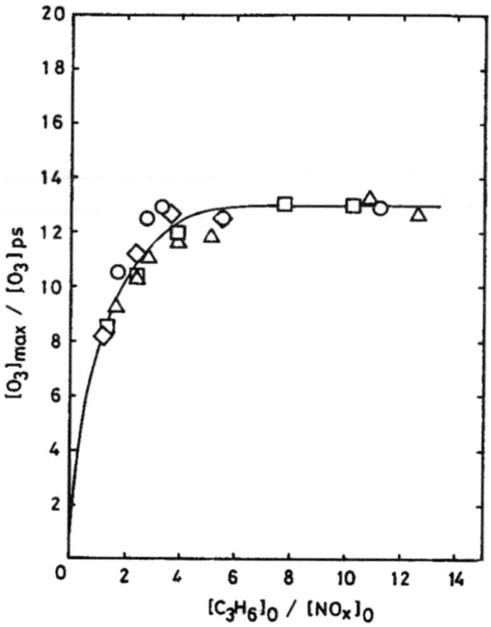

Fig. 7.10 Dependence of $[O_3]_{max}/[O_3]_{ps}$ on $[C_3H_6]_0/[NO_x]_0$ in the irradiation of C_3H_6-NO_x mixtures in air in the smog chamber (Sakamaki et al. 1982)

$$[O_3]_{\max} \propto [O_3]_{ps} \equiv \sqrt{\frac{j_{7.44}}{k_{7.50}}[NO_x]_0} \tag{7.61}$$

by modifying Eq. (7.60), and showed that $[O_3]_{max}$ is proportional to $[O_3]_{ps}$ for the long-enough irradiation under the VOC-excess conditions. (Sakamaki et al. 1982).

It is seen in Fig. 7.9, both experimental and calculated $[O_3]_{max}$ values for $[C_3H_6]_0 = 100$, 500 ppbv are deviated downward from the linear line against $[NOx]_0$ of higher than a certain value for each $[C_3H_6]_0$. This means that the maximum concentration of O_3 gets dependent more strongly on the initial concentration of $[VOC]_0$ rather than $[NO_x]_0$ as $[VOC]_0/[NO_x]_0$ decreases. In such NO_x-excess it has been shown that $[O_3]_{max}$ is proportional approximately to the square root of the initial concentration of VOC (Sakamaki et al. 1982). Figure 7.10 shows their plot of the calculated values of $[O_3]_{max}/[O_3]_{ps}$ by the reaction model against various combinations of $[C_3H_6]_0/[NOx]_0$ (Sakamaki et al. 1982). It is shown in the figure that the maximum concentrations of O_3 are primarily determined by the ratios of $[VOC]_0/[NO_x]_0$ under the concentration range met usually in the polluted urban atmosphere.

The two-dimensional plot showing the dependence of maximum ozone concentration on the initial concentrations of NO_x and VOC is called "ozone isopleths". Usually, VOC concentration is taken as abscissa and NOx concentration as ordinate as in Fig. 7.11a, b. Figure 7.10a is the plot of the data of Fig. 7.9 obtained in the

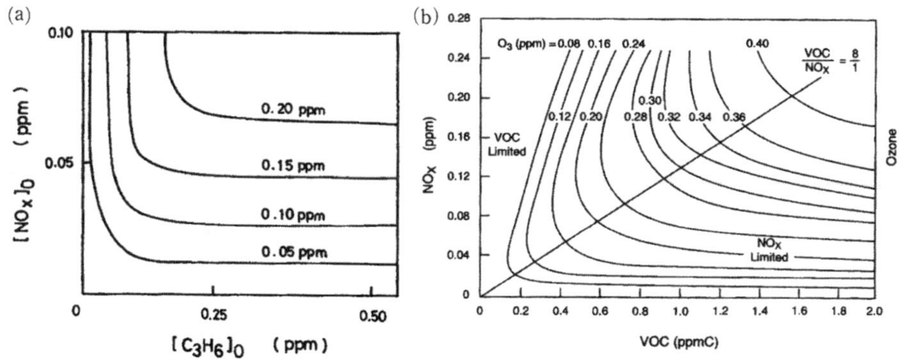

Fig. 7.11 Ozone isopleths for the maximum concentration of O_3 obtained by (**a**) a smog chamber experiment after long-time irradiation to reach ultimate maximum (Sakamaki et al. 1982), and (**b**) EKMA model with limited-time irradiation to simulate daily ozone (Dodge 1977)

chamber experiment in the form of isopleths, while Fig. 7.10b shows the isopleths calculated for the mixture of VOC obtained by the EKMA(Empirical Kinetic Modeling Approach) model calculation (Dodge 1977; Finlayson-Pitts and Pitts 2000). It is apparent that the shapes of the both curves look different, but the cause of difference is due to the difference in irradiation time. In Fig. 7.11b maximum ozone concentrations reached after a fixed few hours irradiation are plotted, while in Fig. 7.11a ultimate maximum ozone concentrations reached after the irradiation of more than 10 h. In general, when the irradiation is terminated in a few hours, ozone has not yet reached the ultimate maximum, which tends to give a larger positive slope of the isopleths in the high NO_x side and to give a plot showing that ozone increases with the decrease of NO_x. In Fig. 7.11b, the straight line drawn at $VOC/NO_x = 8$ represents the ridge of the isopleths, and the right lower and the left upper region are usually called the NO_x-limited and VOC-limited, respectively. Such plots are often discussed as a basis of ozone (oxidant) control strategy, claiming that an emission control of VOC is more effective when the VOC/NO_x ratio is in the VOC-limited, and NOx control is more effective if the ratio is in the NO_x-limited in a particular city. However, the following caution is necessary for the application of ozone isopleths to the policy proposal.

The first issue is the problem of uncertainty of the model, which arises from the situation that the chemical accuracy of the model is not fully validated in the actual ambient atmosphere. The second problem is the temporal and spatial expanse of photochemical air pollution, i.e. the relationship between the O_3 and the initial concentrations of NO_x and VOC in the source area and down-wind area where O_3 is reached after a few hours of transport is quite different. As for the chemical accuracy of the model, in the case of chamber experiments where the reactant VOC is well defined, the isopleths based on the model which can reproduce the experiment are recognized as accurate within the limit of validation, and they are useful for understanding the relationships between O_3 and initial concentrations of NO_x and VOC conceptually and quantitatively. However, in the actual ambient atmosphere, VOCs are not captured fully well in the conventional chemical analysis in the most cases as will be discussed in the next section (Sect. 7.4.2). In such a case, the calculation that considers only the captured VOC tends to lead to the VOC-limited than the reality. Furthermore, although the accuracy of the reaction model has to be validated by the field observation of the concentrations of chain carriers, OH and HO_2, previous measurements tend of underestimate of HO_2 in the higher NO_x region. This means there is a missing process which facilitates O_3 production under the higher NO_x concentrations, and the O_3 forming potential based on a such model calculation also leads to VOC-limited tendency.

In the air pollutant emission area of the central part of a city, the results of model calculation tend to give VOC-limited since the concentration of NOx is high relative to VOC. Since the O_3 formation rate is proportional to VOC concentration in the morning, it is generally shown that the reduction of VOC is more effective for the O_3 control. However, under the geographical and meteorological conditions where the polluted plume is transported to downwind areas in the afternoon, the O_3 concentration tends to depend more on NO_x as the reaction proceeds, and it tends to switch to the NO_x-limited regime. Then, the concentration of O_3 gets more sensitive to the further addition of NO_x. Such a situation has been analyzed by three-dimensional models, and the conclusion of the preference of either NO_x or VOC control is largely different depending on which area affected by the transport is targeted for the control. Such an example has been clearly shown in the Los Angeles basin. According to the ozone isopleths, VOC control is more effective in the central part of Los Angeles city, but the 100 km downwind area NO_x control is more effective (Milford et al. 1989; Finlayson-Pitts and Pitts 2000). Also in the calculation targeting still wider area in the eastern part of United States, it has been shown that the VOC control is more effective within cities, but NO_x control is more effective in rural areas (Sillman et al. 1990).

7.5 Atmospheric Measurements of OH and HO_2 Radicals, and Model Validation

It is a prerequisite for proving the validity of the chemical transport model to compare the reproducibility of calculated results with the observational values of the concentrations of O_3 and its precursors, NO_x, VOC, CO etc. at various sites. However, since the concentrations of chemical species in the atmosphere are determined by the combination of in situ production and loss, and inflow and outflow by transport, the agreement between the calculated and observational values does not necessarily assure the accuracy of chemical reaction mechanism included in the model. Particularly, in the case of using the chemical transport model for ozone control policy to discuss how NO_x and VOC should be controlled for the reduction of O_3 in future, the conclusion can largely be dependent on if the model accurately describes important reaction pathways as well as on if the emissions of NO_x and VOCs are accurately input.

In order to validate the reaction mechanism related to the formation and loss of O_3, comparisons between the direct measurements of the chain carriers, OH and HO_2 radicals in the ambient atmosphere and the model calculations have been made. Since the atmospheric lifetime of OH and HO_2 is as short as less than 1 s and a few tens of seconds, respectively, their concentrations are determined in situ and the effect of transport can be excluded. Therefore, the comparison of the calculated OH and HO_2 concentrations by a box model (zero-dimensional model without including transport) with constraints of using simultaneously measured concentrations of many chemical species and photolysis rate with observed values, can be one-level higher method of validation of the chemistry model than the comparison of the concentrations of O_3 and its precursors. The measurements of OH, HO_2 and RO_2 radicals in the ambient atmosphere are themselves challenging scientific subjects and have become possible through many technical developments. In this section, OH measured by the direct measurements by optical and mass spectroscopic methods, and HO_2 measured by converting to OH are addressed. However, it has recently been noted that in the measurement of HO_2 by adding NO to convert to OH, OH is also produced by the reaction of RO_2 and interfere the measurement of HO_2 (Fuchs et al. 2011). This interference is known to be large when the isoprene, alkene and aromatic hydrocarbons are present in high concentrations, and RO_2 from alkanes with small number of carbon atoms does not produce OH. Thus, previous measurements in the free troposphere and remote sites may not be affected, but the previously reported HO_2 values in the urban and particularly in the forest atmosphere where the concentrations of BVOC is high would have been overestimated. Although the qualitative conclusion of the comparison with the model would not be affected, caution should be used for quantitative discussions, and the values may be revised in future.

The agreement between the measured and calculated values of these radicals is, however, still a necessary but not a sufficient condition of the accuracy of the chemical reaction mechanism. As for these radicals, their concentrations are determined by the ratio of formation and loss rate, and the possibility cannot be excluded that if the error of formation and loss rates are in the similar magnitude, agreement is obtained when these factors are compensated. For further validation of the HO_x chemical reaction processes in the atmosphere, a method of checking of the unknown loss processes of OH has been developed, which produces OH in pulse in the air and measures its time decay to compare with the decay rate calculated from the simultaneously measured VOCs and NO_x concentrations. This method is called the measurement of OH reactivity and it is an effective method for checking the OH loss process relating to the OH budget directly.

In this section, validation of the HO_x chain reaction mechanism by the comparison of OH and HO_2 concentrations obtained by recent field measurements and box model calculations, and the grasp of unknown OH-removing species by the field measurements of the OH reactivity are described. Meanwhile, although the comparison between the output of three-dimensional models with aircraft measurements of OH and HO_2 concentrations is important for the validation of quantitative estimate of O_3 formation and loss in the free troposphere. It is mentioned rather briefly in the next paragraph since it is difficult to incorporate detailed chemical species and reaction mechanism in 3-D models. As for the comparison between the field observations and models of OH and HO_2 radicals, a recent review has been made by Stone et al. (2012).

7.5.1 *Measurements of Concentrations of OH and HO_2, and Comparison with Models*

The typical atmospheric concentrations of OH and HO_2 radicals in the daytime are ~10^6 and ~10^8 molecules cm^{-3} (~0.1 pptv and a few pptv), respectively. The detection techniques whose accuracy is thought to be satisfactory for their measurements are laser-induced fluorescence (LIF) at low pressure which is called Fluorescence Assay by Gas Expansion (FAGE), and chemical ionization mass spectrometry (CIMS), and they have recently been used widely. In either of the methods, directly measured is OH, and HO_2 is measured by converting to OH utilizing the reaction of $HO_2 + NO \rightarrow OH + NO_2$ (reaction (7.10)) adding NO to the atmospheric samples just before the introduction to the detector. Other than these methods, differential optical absorption spectroscopy (DOAS) is also used for the measurements of OH in fields and smog chambers.

Measurements of OH and HO_2 have been made in the upper troposphere using aircrafts, and in the marine, forest and urban boundary layers. From the point of validation of HO_x reaction mechanism, it implies verification for more complex systems with more variety of chemical species in this order.

Upper Troposphere The measurements of HO_x in the upper troposphere over the ocean correspond to the simplest reaction system with the lowest VOCs concentrations. The measurements of this atmospheric domain have been made by many missions of the United States NASA during the latter half of 1990s–2000s, and comparisons with models have been made (Stone et al. 2012). From these investigations, the primary source of HO_x in the upper troposphere under less than 100 ppmv of H_2O at the altitude of higher than 8 km, it has been analyzed that in addition to the process,

$$O(^1D) + H_2O \rightarrow 2OH, \tag{7.2}$$

the processes (Sects. 4.2.5 and 4.2.7),

$$HCHO + h\nu \rightarrow HCO + H \tag{7.62}$$

$$H_2O_2 + h\nu \rightarrow 2OH \tag{7.63}$$

$$CH_3COOH + h\nu \rightarrow CH_3CO + OH \tag{7.64}$$

contribute substantially (Tan et al. 2001b; Ren et al. 2008). Photolysis of acetone (Sect. 4.2.6),

$$CH_3COCH_3 + h\nu \rightarrow CH_3CO + CH_3 \tag{7.65}$$

was once thought to be a major source of HO_x in the upper troposphere, its contribution is now estimated to be lower since it has been found that the photolysis quantum yield has temperature dependence and it is much lowered at low temperature (Arnold et al. 2004; Blitz et al. 2004). Meanwhile, it is estimated that the reaction of $O(^1D) + H_2O$ is dominant at lower than 7 km (Tan et al. 2001b).

The main loss processes of HO_x in the upper troposphere are,

$$OH + CH_4 \rightarrow CH_3 + H_2O \tag{7.6}$$

$$HO_2 + HO_2 \rightarrow H_2O_2 + O_2 \tag{7.17}$$

$$HO_2 + CH_3O_2 \rightarrow CH_3OOH + O_2 \tag{7.18}$$

$$OH + HO_2 \rightarrow H_2O + O_2. \tag{7.19}$$

Further, the reaction,

$$OH + NO_2 + M \rightarrow HONO_2 + M \tag{7.57}$$

gains importance as a sink of HO_x at the altitude of higher than 8 km as the concentration of NO_x increases with the altitude due to the influence of stratospheric NO_x. Ren et al. (2008) obtained the plot of the comparison of the measured values obtained in INTEX, TRACE-P, and PEM-Tropics-B with the box-model calculations as shown in Fig. 7.12. The results show a good agreement of the calculated and the observed values for OH within observational errors over the whole domain in the upper troposphere. As for HO_2, the ratios of the observed and

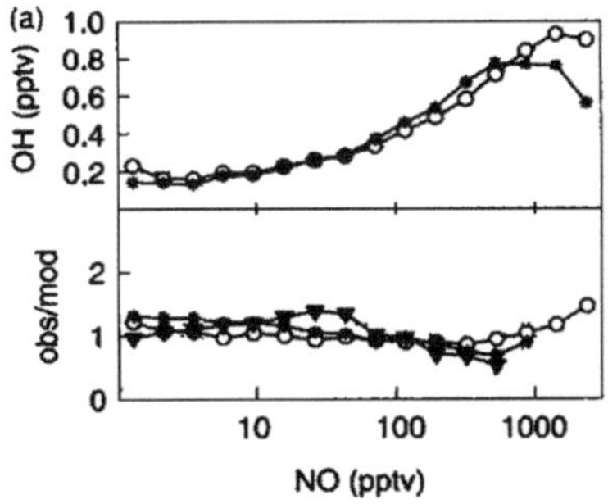

Fig. 7.12 Comparisons of the NO dependence of (**a**) OH and (**b**) HO_2 in the upper troposphere by aircraft campaigns. *Top*: measured (○), modeled (★), *Bottom*: measured to modeled ratios in INTEX-A (○), TRACE-P (★), PEM-Tropics B (▼) (Adapted from Ren et al. 2008)

calculated values agree within 1.2 at the altitude of lower than 8 km where NO mixing ratios are less than 100 pptv, the ratio exceeds 3 at higher than 11 km where the NO mixing ratio is over 1000 pptv. From these analysis it is generally thought that the calculated values of the concentrations of OH and HO_2 agrees well with the observed one at the upper troposphere where the NO_x mixing ratio is low and the HO_x chain cycle can be well explained by the known reaction processes. On the other hand, the underestimate of the calculated concentrations of HO_2 is the same situation as in the polluted atmosphere, and suggests the presence of unknown reaction pathways from OH to HO_2 as will be discussed later.

The three-dimensional simulations are also performed for such aircraft measurements of OH and HO_2, to make comparison in the free troposphere (Sudo et al. 2002; Zhang et al. 2008; Regelin et al. 2012). Figure 7.13 is an example by Sudo et al. (2002). In the figure, the comparisons are made for the observation over Hawaii, Fiji, and Easter islands by the CHASER model showing reasonably good agreement in general.

Marine Boundary Layer The measurements of OH and HO_2 in the remote marine boundary layer are useful for the validation of the model of the atmospheric reaction mechanism under the condition without the influence of anthropogenic activities, and many measurements have been performed (Stone et al. 2012). From these measurements, diurnal maximum concentrations of OH and HO_2 have been obtained typically as (3–10) $\times 10^6$ molecules cm^{-3} and (1–6) $\times 10^8$ molecules cm^{-3} (4–25 pptv), respectively. In the comparison between the observational and calculated values, it has been reported that the calculated concentrations of HO_2 often overestimate the observed one by a factor of up to 3 in the field campaigns, OKIPEX 1998 in Oki Island, and RISOTTO 2000 and 2003 in Rishiri Island in Japan (Kanaya et al. 2000, 2002, 2007a). On the other hand, the differences between the calculated and observed values for OH concentrations are within 10–30 %, much smaller than for HO_2. Figure 7.14 shows the comparison between the box-model calculation using the RACM reaction model (Stockwell et al. 1997) for HO_2 and OH with the observation in the RISOTTO campaign (Kanaya and Akimoto 2002). Halogen chemistry is not included in this model, and Kanaya et al. (2002), and Kanaya and Akimoto (2002) suggested the possibility that the chain reaction involving halogens such as IO which have been detected in the marine boundary layer (Alicke et al. 1999),

$$HO_2 + XO \rightarrow HOX + O_2 \quad (7.66)$$

$$HOX + h\nu \rightarrow OH + X \quad (7.67)$$

$$X + O_3 \rightarrow XO + O_2 \quad (7.68)$$

$$(X = I, Br)$$

may convert more HO_2 to OH. Other suggested possibilities include the heterogeneous loss of HO_2 on aerosols, net loss of HO_x radicals by the heterogeneous

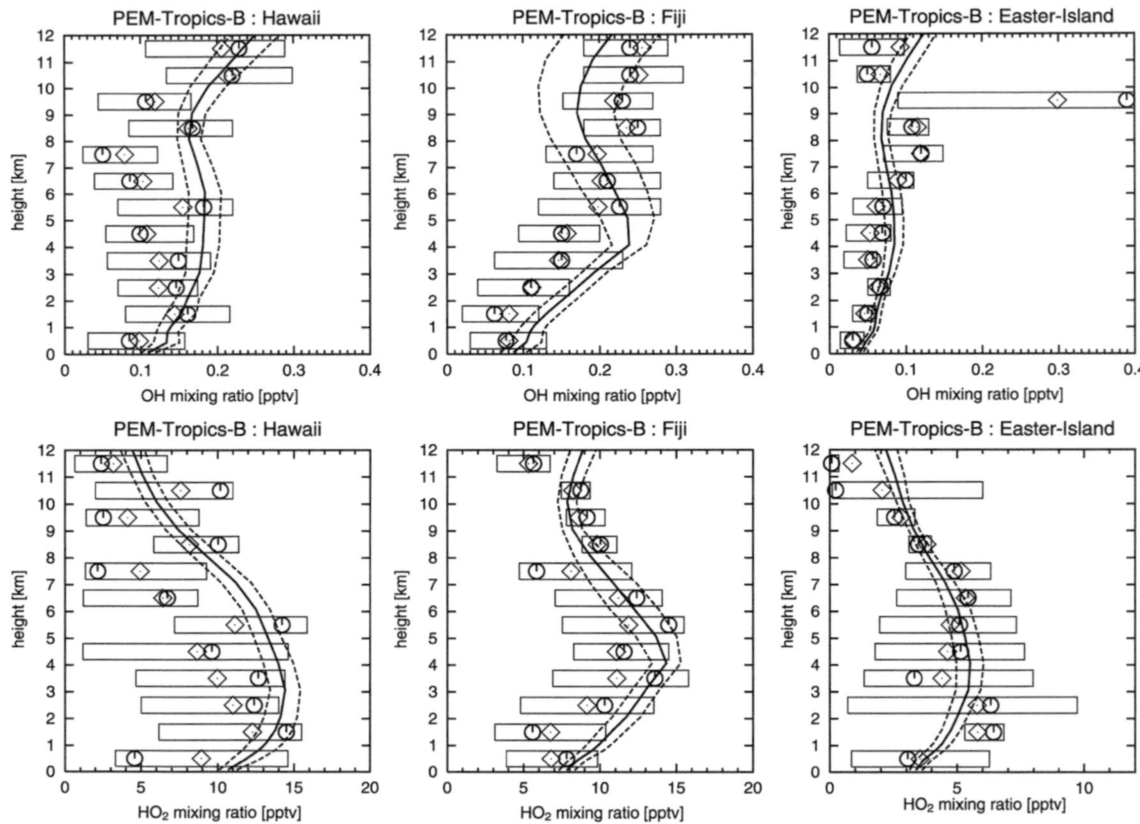

Fig. 7.13 Comparison of the vertical profiles of OH and HO_2 mixing ratios between aircraft observation (PEM-Tropics-B) and three-dimensional model (CHASER). *Solid* and *dotted lines* are the average values and the values of $\pm 1\sigma$, respectively, by the model calculation. ◇ and ⊙ are the average and median values of the observation, respectively, and the *boxes* represent the data within ±50 % (Adapted from Sudo et al. 2002)

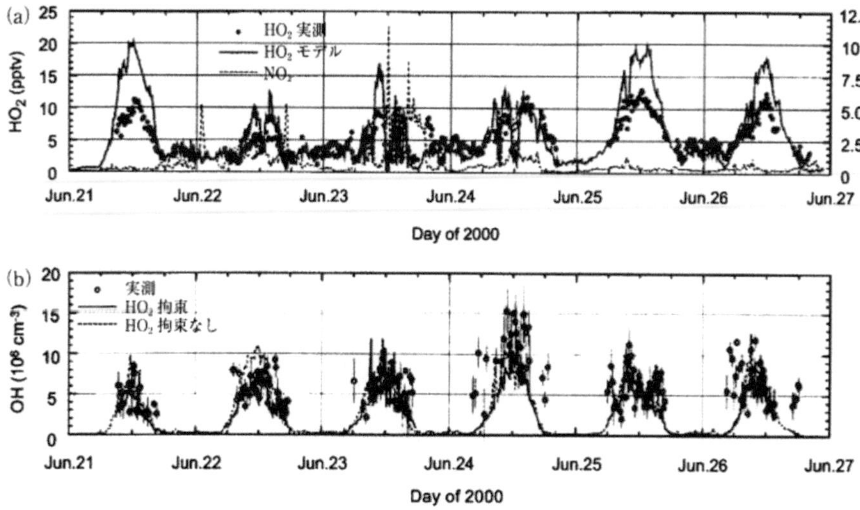

Fig. 7.14 Comparison of HO$_2$ and OH in the RISOTTO campaign at Rishiri, Japan, between the box-model calculation and observation (Adapted from Kanaya and Akimoto 2002)

reactions of HOX (HOI, etc.), and underestimates of some of the unknown rate constants of the HO$_2$ + RO$_2$ reactions.

Simultaneous measurements of OH and HO$_2$ with BrO, IO, OIO and I$_2$ were made in the NAMBLEX campaign at Mace Head, Island in 2002 (Saiz-Lopez et al. 2006). Observed maximum mixing ratios of IO and BrO are 4 and 6.5 pptv, respectively, and the comparison between the measurement and calculation by the MCM reaction model (Bloss et al. 2005) based on these observed values are shown in Fig. 7.15 for HO$_2$ (Sommariva et al. 2006). In this model calculation, a wide range of VOC reactions the calculations are included, and the comparisons between with and without including IO reactions, is shown in Fig. 7.15a, and the those between with and without uptake of HO$_2$ and HOI on aerosols are shown in Fig. 7.15b. As shown in Fig. 7.15a, the overestimate of HO$_2$ is improved as much as maximum 30 % when the IO reactions are included, but the improvement is not enough. When the uptake of HO$_2$ on aerosols with the maximum coefficient of $\gamma = 1$, overestimation of the calculation improves by as much as 40 %, for example on August 15 and 20, and the calculated values are reversed to underestimate on August 16. The uptake coefficient of HO$_2$ for the actual aerosols sampled at Mt. Tai and Mt. Mang in China has been given to be as large as $\gamma = 0.1 - 0.4$ by Taketani et al. (2012) (Sect. 6.2.3), which implies that the heterogeneous uptake of HO$_2$ is important, as well as IO, in the marine boundary layer.

Furthermore, in the RHaMBLe campaign in 2007 at Cape Verde in the tropical Atlantic Ocean, IO and BrO has been measured by DOAS together with HO$_2$. The calculation by MCM model based on these observed values agrees well with observed values within the measurement error of 20 % when halogen chemistry and heterogeneous loss process of HO$_2$ with $\gamma(HO_2) = 0.1$ are taken into account

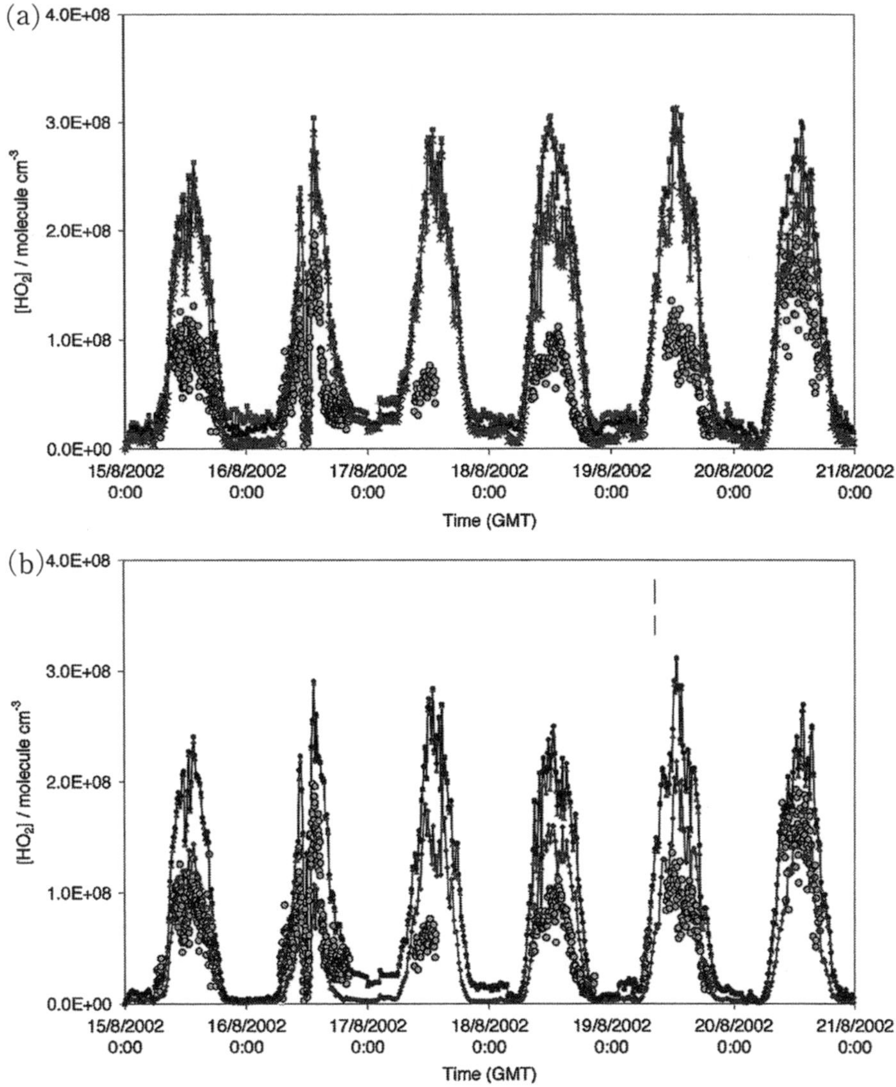

Fig. 7.15 Comparison of observed HO_2 concentration in the NAMBLEX campaign at Mace Head and the model calculation considering IO. (**a**) comparison of calculation without IO (■), with IO (⯎) and observation (◎), (**b**) comparison of calculation with IO and uptake of HOI (●), with IO and uptake of HO_2 (◆) and observation (◎) (Adapted from Sommariva et al. 2006)

(Whalley et al. 2010). Figure 7.16 depicts the contribution ratios of each chemical process for the formation and loss of OH and HO_2 obtained by the model calculation (Whalley et al. 2010). As shown in the figure, the contribution of halogens combining the photolysis of HOI and HOBr is 13 % whereas the $O(^1D)$ reaction by the photolysis of O_3 is 76 % for the formation of OH. As for the loss processes of

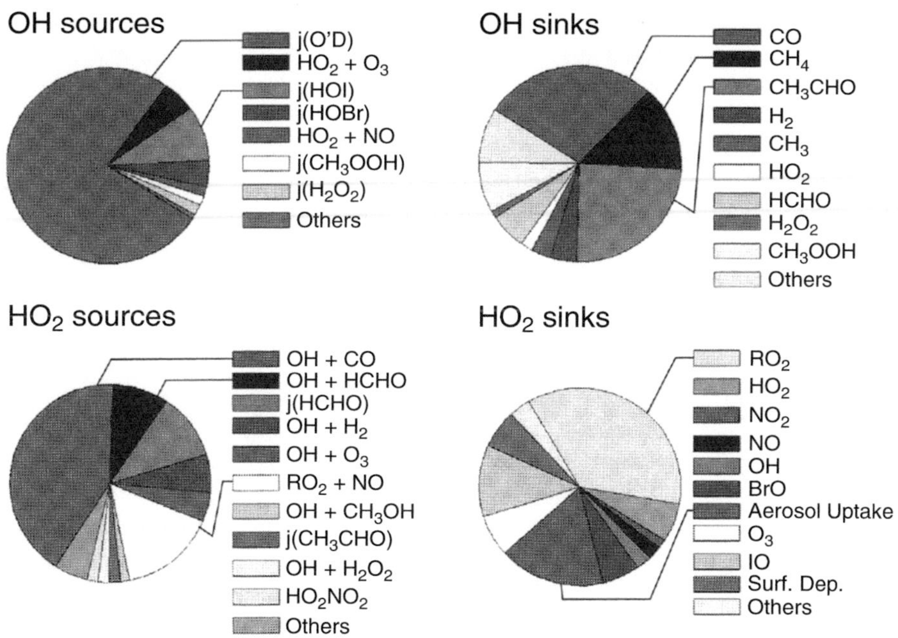

Fig. 7.16 The contribution of each chemical process for the production and loss of OH と HO_2 obtained by the simulation for the RHaMBLe campaign at Cape Verde (Adapted from Whalley et al. 2010)

OH, the contribution of the reactions with CO and CH_3CHO is the largest, 28 % and 25 %, respectively. Meanwhile, the contributions of OH + CO reaction is 41 %, and those of the OH + HCHO and the photolysis of HCHO are ca. 10 % each for the formation of HO_2. As the loss processes of HO_2, contribution of $RO_2 + HO_2$ is the largest, ca. 40 %, uptake on aerosol surface is 23 %, and the reaction with IO and BrO is 19 % total.

In the marine boundary layer, it is known that HO_2 exists in a few ppbv even at nighttime. They are thought to be due to the O_3 reaction with isoprene, monoterpenes, internal alkenes, etc. influenced by the emissions from surrounding lands (Kanaya et al. 2007a; Smith et al. 2006; Whalley et al. 2010). In general, the nighttime OH in the marine boundary layer is under the detection limit.

Forest Air The measurements of HO and HO_2 in the forest air influenced by biogenic hydrocarbons such as isoprene and monoterpenes and comparison with model calculations have also been made on many occasions (Stone et al. 2012). These studies have implications for the validation how well the reaction mechanism of for example isoprene described in Sect. 7.2.6 reflects the role of HO_x radical chain mechanism. In general, it has been known that the model calculation underestimates the observed values of OH to a large extent for the sites where the isoprene, etc. have significant impact. For example, the calculation by the RACM model for the PROPHET-98 campaign conducted in Michigan state, United States, where isoprene concentration is high, underestimates the observation of OH by a factor of 2.7, and the disagreement reached a factor of 6 when the NO concentration is low (Tan et al. 2001a). On the other hand, the calculated values for HO_2 agree well with the observation within 15 %. This means that unknown reactions converting HO_2 to OH exist in the HO_x chain reaction system when the concentration of NO is low. In the succeeding campaign, PROPHET-2000, the measurement of the OH reactivity was also made, and it was revealed that there are many unknown BVOCs are present (Di Carlo et al. 2004).

Substantial underestimate of calculated OH and HO_2 has also been reported for the measurements in GABRIEL in Amazonian tropical forest in 2005 (Sander et al. 2005), and in Borneo Sabah in 2008 (Whalley et al. 2011). However, observational values of HO_2 are thought to suffer the large positive interference by RO_2 originated from BVOC as described above so that the quantitative discussions would have to be revised in future.

Particularly under the low concentration of NO, several reactions of peroxy radicals that may produce excessive OH has been proposed. Dillon and Crowley (2008) reported that in the reaction of HO_2 and peroxy radicals with carbonyl group, e.g. acetyl peroxy radical,

$$HO_2 + CH_3C(O)O_2 \rightarrow CH_3C(O)O_2H + O_2 \quad (7.69)$$

$$\rightarrow CH_3C(O)OH + O_3 \quad (7.70)$$

$$\rightarrow CH_3C(O)O + O_2 + OH, \quad (7.71)$$

the yield of OH-forming pathway (7.71) is $\alpha(298\ K) = 0.5 \pm 0.2$. It is also suggested that in the case of peroxy radicals with carbonyl group formed in the reaction of OH and isoprene, the reactions with HO_2 produce OH radicals, which could contribute excess OH in the forest air under low NO concentrations. Meanwhile, Peeters and Müller (2010) suggested that the photolysis of hydroperoxy methyl butanal formed by the intramolecular hydrogen transfer in hydroperoxy radical in the reaction of OH with isoprene, gives OH with quantum yield of unity, and may contribute to the excess OH in forest air.

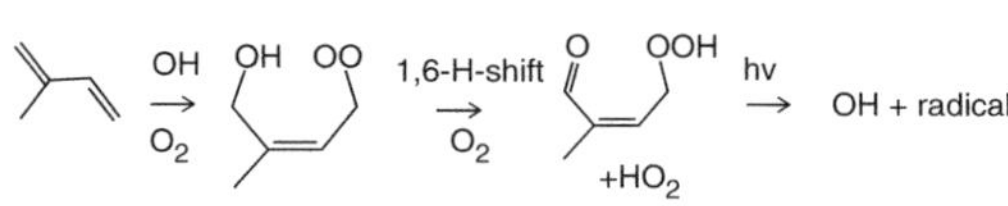

(7.72)

Therefore, the oxidation reaction mechanism of isoprene and monoterpenes under low NO concentration is still shown to have many uncertainties from the direct measurement of HO_x.

On the other hand, in the PRIDE-PRD campaign in the suburbs of Guangzhou, China in 2006, very high daytime concentrations of OH and HO_2, $(15–26) \times 10^6$ and $(3–25) \times 10^8$ molecules cm^{-3} (80 pptv), respectively, has been reported under the conditions of high VOC containing 20 % of isoprene and low NO (200 ppt in the daytime) (Hofzumahaus et al. 2009; Lu et al. 2012). The model calculations reproduce reasonably well for HO_2 concentrations, whereas those for the OH concentrations are underestimated by a factor of 3–5. These results imply that OH sources are in shortage in the known reaction mechanism. The difference cannot be explained by the OH forming reaction from $HO_2 + RO_2$ with taking $\alpha = 1$, and the presence of an unknown reaction cycle to form OH from HO_2 other than the reaction with NO has been suggested (Hofzumahaus et al. 2009; Lu et al. 2012).

The nighttime concentrations of OH and HO_2 in the PROPHET-98 in Michigan, United States are 1×10^6 molecules cm^{-3} and 2 pptv, respectively, showing very high OH but relatively low HO_2 concentrations. The model calculation using the observed monoterpenes and O_3 underpredicts the OH concentration by a factor of 2 (Faloona et al. 2001).

Urban Polluted Air Urban polluted air contains a few hundred species of hydrocarbons and oxygen containing volatile organic compounds (OVOCs) (Lewis et al. 2000), and the observation of OH and HO_2 concentrations there validates the reaction model with the whole set of these VOCs. Comparisons between the model calculation and observation of OH and HO_2 concentrations in urban air are also interesting from the point of checking the dependence of O_3 formation rate on NO_x and VOC in the polluted atmosphere, and are important as validation of a reaction model for the discussion of the oxidant control strategy as described in the previous section. From these viewpoints, many measurements of and the comparison with model calculations have been made in urban air, and the values of $(3–20) \times 10^6$ molecules cm^{-3} for OH, and $(1–12) \times 10^8$ molecules cm^{-3} (4–50 pptv) for HO_2 concentration, which are similar or higher than in the marine boundary layer, have been reported (Stone et al. 2012).

As an example of the comparison between the measurements and calculations by the RACM model for OH and HO_2 in urban air, Fig. 7.17 shows the results for the IMPACT campaign in Tokyo, Japan in the summer of 2004 by Kanaya et al. (2007b). In this measurement, the maximum concentration of OH 13×10^6 molecules cm^{-3} and the daytime median concentration 6.3×10^6 molecules cm^{-3} with relatively small day-to-day variation, and the maximum mixing ratio of HO_2 50 pptv and the daytime median value 5.7 pptv with large day-by-day variation have been obtained. The observational maximum and median values of OH and HO_2 at the same site in 2004 winter are 1.5×10^6 molecules cm^{-3} and 1.1 pptv, respectively, both are about one fifth of the summer values. As seen in the figure, the calculated and measured median values of OH and HO_2 are in reasonably good agreement within the error range for summer, and the ratios of the calculated to the measured values of OH and HO_2 are 0.81 and 1.21, i.e. about 20 % under- and overestimation, respectively. Meanwhile, the calculated/measured ratios of the median value are 0.93 for OH showing good agreement, but 0.48 for HO_2 resulting in underestimation about a factor of 2.

The comparisons of measurements and calculations for OH and HO_2 in urban air have been made in various cities, Los Angeles, Nashville, Birmingham, Houston, New York, Essex, Mexico City, etc. Although in most cases, the ratios of measured

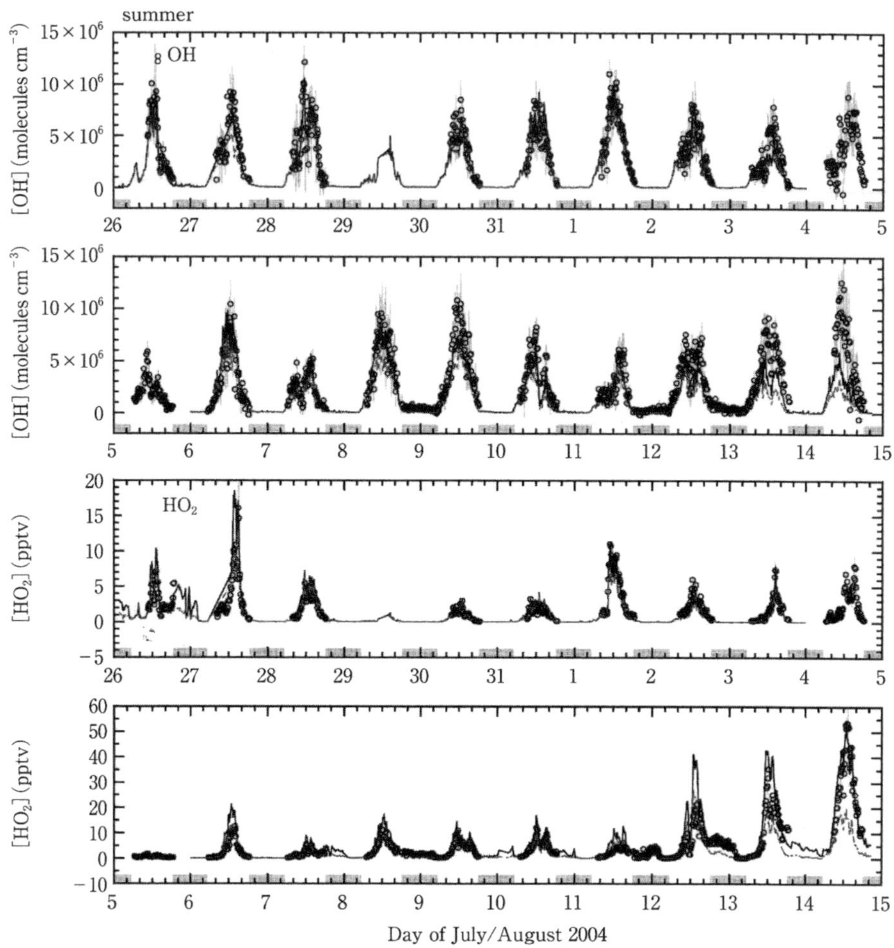

Fig. 7.17 Comparison of the observed and calculated OH and HO_2 concentrations in IMPACT campaign in Tokyo (Kanaya et al. 2007b). ○ observation, — base-case calculation, --- calculation including the heterogeneous loss process of HO_2

and calculated values agree within 50 %, the models tend to overestimate OH in general, and to underestimate HO_2 when the NO_x concentration is high (Stone et al. 2012). However, for example in Mexico City, the calculated values of HO_2 underestimated the observation by a factor of 5 (around at 10 am), and those of OH overestimated by a factor of 1.7 (around at noon), which are ascribed to the reason that many hydrocarbons e.g. aromatics are not captured in the VOC measurement (Dusanter et al. 2009). Furthermore, in the PMTACS campaign in New York City in winter, it has been reported that the calculated daytime HO_2 concentration gives underestimation by a factor of 6, and particularly the disagreement is large when NO concentration is high (Ren et al. 2006).

From these many measurements, it has been revealed in general that the photolysis of HONO, alkene-ozone reaction, and photolysis of aldehyde contribute to a large extent as a source of total HO_x in the urban air, and the contribution of the $O(^1D) + H_2O$ reaction due to the photolysis of O_3 is only a few %. The major photolytic species for the HO_x source is understandably dependent on cities, and for example, it is reported that the contribution of photolysis of dialdehyde (glyoxal and methyl glyoxal) in addition to formaldehyde, is large in Mexico City (Dusanter et al. 2009). Incidentally, HONO observed in high concentration in the daytime urban air is thought to be formed by the photocatalytic reaction of NO_2 near the ground surface as described in Chap. 6, Sect. 6.4.2. From the study of the model inter-comparison for the RACM, MCM, etc., the differences between the reaction models are relatively small for urban air, which suggests that the effect of differences in the VOC reaction mechanism is not large, and the effect of NO_x chemistry is more important as an uncertainty factor for the urban air as described below (Chen et al. 2010).

Figure 7.18 depicts the dependence of the concentrations of OH, HO_2, and the ratio of HO_2/OH on NO concentration obtained in winter and summer in Tokyo (Kanaya et al. 2007b). As seen in the figure, the concentrations of OH and HO_2, and the ratio of HO_2/OH decreases with the increase of NO concentration in the range of 1–100 ppbv, and this tendency is well reproduced by the model calculation. However, both of the calculated values OH and HO_2 tend to overestimate in the lower concentration range of NO, and underestimate in the higher concentration range of NO. This tendency is more apparent for HO_2 in winter, and also for the HO_2/OH ratio. The expanded underestimation of HO_2 concentration in the high concentration range of NO is also seen apparently in the observation in Mexico City in April (Sheehy et al. 2010; Volkamer et al. 2010), and in New York in winter, and this tendency is a commonality in many observations (Stone et al. 2012). A similar trend is also seen in the upper troposphere by the aircraft campaign (Ren et al. 2008), which suggests strongly that unknown process exists in the reaction mechanism related to NO_x used in the chemistry model.

As seen in Sect. 7.3.2, the net formation rate of O_3 in the atmosphere is expressed as $P(O_3)$-$L(O_3)$ where

$$P(O_3) = (k_{7.10}[HO_2] + k_{7.21}[RO_2])[NO] \tag{7.52}$$

$$L(O_3) = (j_{7.1}f_{7.2} + k_{7.47}[HO_2] + k_{7.46}[OH])[O_3], \tag{7.48}$$

so that it is found that the concentrations of OH, HO_2 and RO_2 are the key parameters of net O_3 formation. Therefore, the proper reproduction of the observed NO_x and VOC dependence of OH and HO_2 by the reaction mechanism used in the model should be prerequisite for the discussion of the NO_x-limited and VOC-limited using the isopleths described in Sect. 7.3.2. Kanaya et al. (2008) plotted the $P(O_3)$-$L(O_3)$ based on the measurement of OH and HO_2 in Tokyo on the isopleths calculated by the model as a function of NO_x and VOC concentrations,

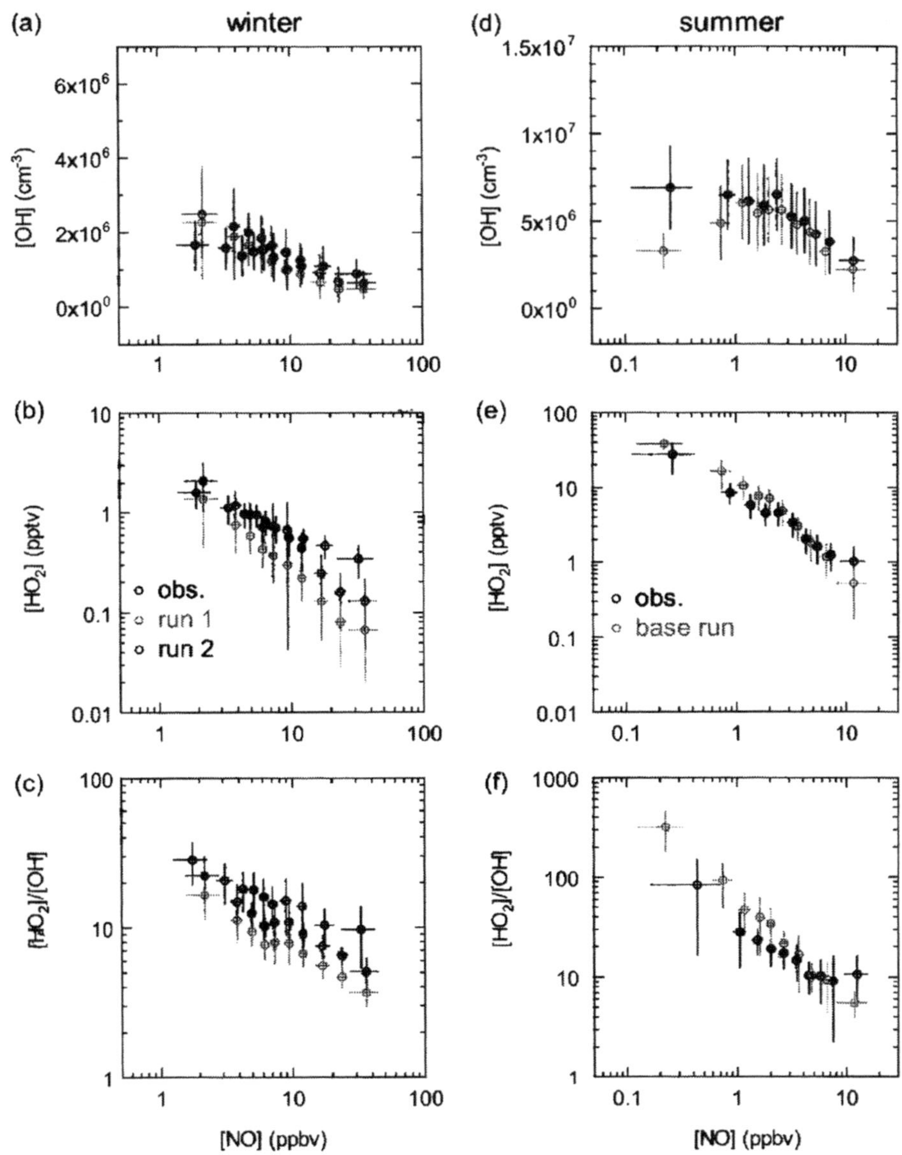

Fig. 7.18 Observed NO-dependence of the OH and HO_2 concentrations, and the HO_2/OH ratio obtained by the measurement in Tokyo in winter and summer (Adapted from (Kanaya et al. 2007b). Run 2 is the calculation when the internal olefin (OL1) and reactive alkene (HC8) in RACM was increased by a factor of 3.5

and showed that the behavior of the observed and calculated tendency does not necessarily agree, particularly in winter.

Nighttime OH and HO_2 are in general observed in the urban and suburban polluted atmosphere. For example, 1.9×10^5 molecules cm^{-3} of OH and 4 ppt of HO_2 are observed in the BERLIOZ campaign in Germany in summer 1998, and the analyses are made to show the contribution of 36 % of NO_3 reaction and 64 % of O_3 reaction for OH, and 53 and 47 % for HO_2, respectively (Geyer et al. 2003). The average concentrations of 2.6×10^5 molecules cm^{-3} of OH, and 2.9×10^7 molecules cm^{-3} (1.2 pptv) of HO_2 are observed in the TORCH campaign in Essex, UK in summer 2003. The contribution of 66 % of O_3 reaction, and 33 % of NO_3 reaction are estimated as radical sources, but the calculated values are 41, and 16 % underestimate for OH and HO_2, respectively (Emmerson and Carslaw 2009).

7.5.2 Measurements of OH Reactivity and Missing Reactivity

As described in the previous paragraph, since the atmospheric concentrations of OH and HO_2 are determined by the balance of formation and loss of these radicals, the reproduction of measured values by the calculated value by the model does not necessarily guarantee the accuracy of the terms of source and sink of these radicals. Among the formation and loss rates that determine the atmospheric concentration of OH, the measurement technique for the loss rate called "OH reactivity" has been developed by producing pulsed OH artificially in the ambient air to observe its time decay directly (Sadanaga et al. 2004; Sinha et al. 2008; Ingham et al. 2009).

The loss rate of atmospheric OH can be expressed by,

$$L(\mathrm{OH}) = \Sigma_i k_{\mathrm{VOCi}}[\mathrm{VOC_i}] + k_{\mathrm{CO}}[\mathrm{CO}] + k_{\mathrm{NO2}}[\mathrm{NO_2}] + k_{\mathrm{NO}}[\mathrm{NO}] + k_{\mathrm{SO2}}[\mathrm{SO_2}] + \mathrm{k_{O3}}[\mathrm{O_3}] + k_{\mathrm{CH4}}[\mathrm{CH_4}] + \text{---} \tag{7.73}$$

as the sum of the products of each reactant concentration and corresponding OH rate constant. In the measurement of OH reactivity, the loss rate L(OH) is observed directly by,

$$[\mathrm{OH}]_\mathrm{t} = [\mathrm{OH}]_0 \exp\{-L(\mathrm{OH})t\} \tag{7.74}$$

$$\ln([\mathrm{OH}]_\mathrm{t}/[\mathrm{OH}]_0) = -L(\mathrm{OH})t. \tag{7.75}$$

From such a measurement, the missing loss rate can be estimated quantitatively by comparing the observed with the computed value based on the simultaneously observed individual VOCs, NO_x and other species and their OH reaction rate constants. In the urban atmosphere few hundreds of VOCs are estimated to exist (Lewis et al. 2000), and furthermore if the biogenic VOCs are included, the numbers are enormous (Goldstein and Galbally, 2007) and the bottom-up description in the model is in general unrealistic. Since the OH reactivity measurements

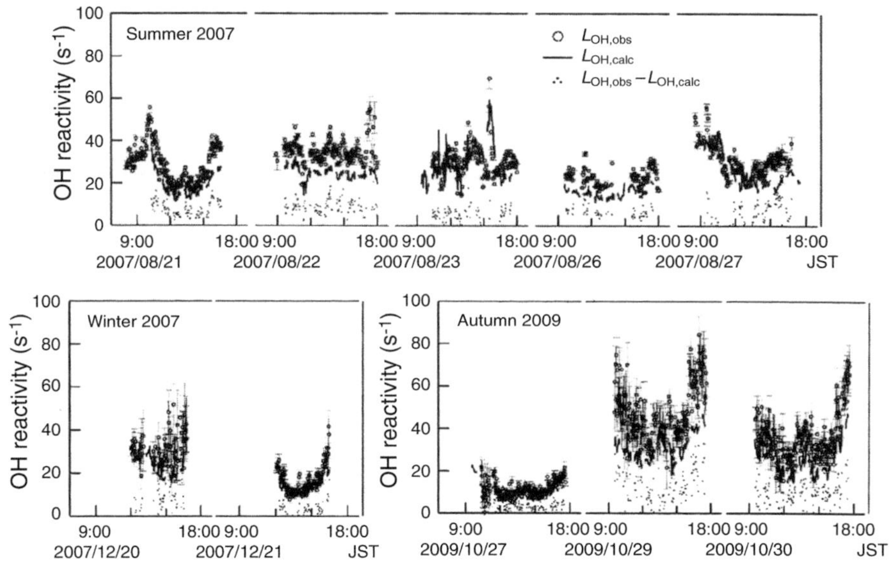

Fig. 7.19 Comparison of the observed ($L_{OH,\ obs}$) and calculated ($L_{OH,\ calc}$) values of the OH reactivity in urban air (Adapted from Yoshino et al. 2012)

can make their overall evaluation, it is very useful for providing the boundary condition to the model.

Measurements of the OH reactivity in forest air have been made in the temperate forest in Michigan United States, tropical forest in Amazon, and boreal forest in Finland. In Michigan, the missing reactivity that is not explained by the measured VOC is considerably high. Since its magnitude has the same temperature dependence as other terpenes, it is deduced that BVOCs and OVOCs formed from them are thought to be the unknowns (Di Carlo et al. 2004). Furthermore, in the tropical forest, the missing reactivity reached ten times of the calculated reactivity based on the measured VOCs (Lelieveld et al. 2008). In the forest in Finland, 30 species of BVOCs were quantified by using a proton transfer mass spectrometer (PTR-MS), but the calculated OH reactivity captured only 50 % of the measured one (Sinha et al. 2010).

Meanwhile the OH reactivity in urban air, has been measured in Tokyo in August, and December in 2007, and in October, 2009 by Yoshino et al. (2012). In these observations, more than 60 species of VOCs by GC-MS and several OVOCs e.g. HCHO, CH_3CHO, CH_3COCH_3, CH_3OH, etc. by PTR-MS have been measured simultaneously, and the comparison between the measured and calculated OH reactivity was made, as shown in Fig. 7.19. As seen in the figure, the measured OH reactivity was 17–70 s^{-1} in August, and 10–80 s^{-1} in October, which is substantially higher than the calculated one, suggesting that there are still many unanalyzed VOCs and OVOCs. The missing reactivity amounted to 27 % in August and 35 % in October. The contributions of the each category of measured species to

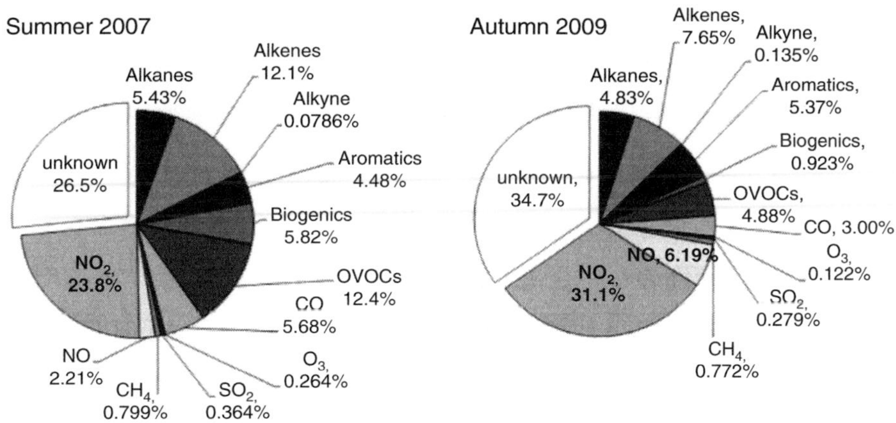

Fig. 7.20 The contribution percentage of each observed species to the OH reactivity observed in Tokyo in summer 2007, and autumn 2009 (Yoshino et al. 2012)

the OH reactivity are shown in Fig. 7.20. Thus, in summer, NO_x ($NO + NO_2$) is 26.0 %, anthropogenic VOC, 22.1 %, OVOC, 12.4 %, and biogenic VOC, 5.8 %, and in autumn, NO_x is 37.3 %, anthropogenic VOC, 18.0 %, OVOC, 4.9 %, and biogenic VOC, 0.9 %.

It should be noted that by adding unknown VOCs and OVOCs deduced from the measurements of OH reactivity in the model calculation, the calculated results of ozone formation tends to shift from the VOC-limited to NO_x-limited regime.

7.6 Tropospheric Halogen Chemistry

The important roles of halogens, Cl, Br, and I, in the chemical reaction system in the troposphere was recognized rather recently in 1980s. The trigger was the findings of the phenomenon that the mixing ratio of surface ozone in the Arctic Circle in spring decreases from a normal value of ca. 30 ppbv to nearly zero within a few hours depending on the day, which was then discovered to be due to the O_3 destruction by BrO by Barrie et al. (1988). Later the similar phenomenon was also found in the Antarctica. Furthermore, active halogen species containing Cl, Br and I, have been widely observed in the marine boundary layers in the mid- and high-latitudes and are found to affect the concentrations of OH and HO_2 as mentioned in the previous section to cause the decrease of O_3 in situ. The chemical reactions in which halogens are related have also been found out around salt lakes and in volcanic plumes (Saiz-Lopez and von Glasow 2012). Although volcanic halogens are not mentioned in this chapter, they are found not only in intermittent eruption but also in stationary plumes, and it is tend to be treated in general within tropospheric chemistry (Bobrowski, et al. 2003, 2007; Lee et al. 2005; Aiuppa et al. 2009; Vance et al. 2010).

The overall tropospheric halogen chemistry has recently been reviewed by Saiz-Lopez and von Glasow (2012). Also, reviews on heterogeneous reactions on sea salt (Finlayson-Pitts 2003; Rossi 2003) and other specific topics of tropospheric halogen chemistry (Carpenter 2003; Sander et al. 2003; Simpson et al. 2007; von Glasow 2010; Abbatt et al. 2012; Saiz-Lopez et al. 2012) are available.

7.6.1 *Initial Source of Halogens*

One of the big issues on tropospheric halogen chemistry is that the initial process of releasing halogens to the atmosphere has not been well established in spite that the succeeding chain reactions relevant to Cl, Br and I are relatively well known. So far, heterogeneous reactions on sea salt and biogenic emissions have been identified as the main initial source of inorganic halogens. However, their detailed processes and the contributing ratios are very much different by region and halogen species, and in many cases, the parameters relevant to the release have not been quantified well. In this section, present knowledge on these processes are described for each of Cl, Br and I.

7.6.1.1 Chlorine

In spite of the large molar ratio of 660:1 of Cl over Br, the ratio of Cl released to the atmosphere by heterogeneous reactions of gaseous oxidizing species on sea salt is much smaller than Br. Among the Cl releasing processes from sea salt as photochemically active species, the formation of nitryl chloride $ClNO_2$ by the reaction of N_2O_5 and sea-salt particles in polluted urban plumes has been well established (see Sect. 6.2.4) (Rossi 2003; Finlayson-Pitts 2003). The overall reaction combining reactions (6.5), (6.16) in Sect. 6.2.4 can be expressed as,

$$N_2O_5(g) + NaCl(aq) \rightarrow ClNO_2(g) + NaNO_3(aq), \tag{7.76}$$

and the product $ClNO_2$ is easily photolyzed in the troposphere to release Cl atoms (see Sect. 4.4.8).

$$ClNO_2 + h\nu \rightarrow Cl + NO_2 \tag{7.77}$$

Although the release of Cl by this reaction is limited to the areas under direct influence of polluted air, such as coastal zone of large cities and harbors where ship emissions prevail, the contribution of this reaction is thought to be large even globally since the condition applies to many coastlines worldwide. Recently, $ClNO_2$ has been observed not only in the coastal zone (Osthoff et al. 2008), but also in inland cities where sea salt does not exist (Thornton et al. 2010; Mielke et al. 2011). In this case, $ClNO_2$ is thought to be formed from the reaction of N_2O_5 on the aerosol surface containing anthropogenic chlorine compounds. The reaction of N_2O_5 on sea-salt particles also release Br_2 other than $ClNO_2$, and the ratio has been found to depend on temperature (Lopez-Hilfiker et al. 2012).

On the other hand, the reaction of HNO_3 and sea salt (Sect. 6.2.5)

$$HNO_3(g) + NaCl(aq) \rightarrow HCl(g) + NaNO_3(aq) \tag{7.78}$$

is important over the ocean apart from coastal lines since the atmospheric lifetime of HNO_3 is much longer than N_2O_5. However, the product, HCl is not photolyzed in the troposphere (see Sect. 4.4.3), and the rate constants of the reaction with OH is not large, 7.8×10^{-13} cm^3 molecule^{-1} s^{-1} at 298 K (see Table 5.2), so that reaction (7.70) is not important as a source of Cl atoms in the troposphere.

As mentioned in Sect. 7.5.4, Cl_2 has been observed sporadically in the marine boundary layer. As for the process of the release of Cl_2, the surface reaction of OH on deliquescent sea salt has been confirmed experimentally (Oum et al. 1998a; Knipping et al. 2000). While the reaction of O_3 with sea salt (see Sect. 6.2.1) is known to be the most important process of the release of Br_2 as will be described later, it has been reported that the reaction of O_3 with Fe^{3+}-doped NaCl releases Cl_2 (Sadanaga et al. 2001), and the possibility of Cl_2 release after the depletion of Br_2 has been suggested. Since rather ubiquitous distribution of Cl_2 in the marine boundary layer has been known recently, further studies are necessary for the processes of the Cl_2 release into the atmosphere. Since the absorption spectrum of Cl_2 extends to visible region (Fig. 4.36) and it is easily photolyzed to release Cl atoms (see Sect. 4.4.1), it is an important active species in the troposphere.

7.6.1.2 Bromine

Among the tropospheric halogen species, active bromine species such as BrO are best known and widely measured not only in the polar region but also in the low- and mid-latitude marine boundary layers. As a source of active bromine to the atmosphere, the reaction of O_3 on sea salt surface has been studied well and thought to be the most important (Hirokawa et al. 1998; Mochida et al. 2000; Anastasio and Mozurkewich 2002) (see Sect. 6.2.1). Furthermore, the reactions of O_3 with snow surface on which sea ice and sea salt are co-deposited, are also known as Br_2 source in the polar region (Tang and McConnell 1996; Oum et al. 1998b). Br_2 has absorption spectrum extending to still longer wavelength in the visible region than Cl_2, and thus is photolyzed in shorter periods of time to release Br atoms, and thus play the most important role in the tropospheric halogen chemistry.

On the other hand, since the absorption cross section of tribromomethane (bromoform, $CHBr_3$) released from seaweed is small in the spectrum region of actinic flux (Gillotay et al. 1989), it is not a very important source of photoactive bromine species in the troposphere.

7.6.1.3 Iodine

Since the ratio of contents of iodine in sea salt is very small, 5×10^{-8}, the release of I from sea salt is negligible, and the source of IO widely observed in the marine boundary layer is thought to be the photolysis of organoiodine compounds e.g. iodomethane (methyl iodide, CH_3I), iodoethane (ethyl iodide, C_2H_5I), 1-iodopropane (1-propyl iodide, 1-C_3H_7I), chloroiodomethane (CH_2ICl), bromoiodomethane (CH_2IBr) and diiodo methane (CH_2I_2) etc., and inorganic iodine, I_2, from marine organisms (Saiz-Lopez et al. 2012). These organoiodine compounds are emitted from seaweeds in the coastal zone and from microalgae in the ocean. Recently, non-biological emissions from sea surface reactions between ozone and organic compounds in sea surface water are also considered (Martino et al. 2009). Among the organoiodine compounds, the emissions of CH_3I and CH_2I_2 are estimated to be the largest globally (Saiz-Lopez et al. 2012). Meanwhile, I_2 is known to be emitted from seaweed in coastal zones, but the sources over the ocean are not well known yet.

The absorption spectrum of iodomethane is shown in Fig. 4.32 in Chap. 4. Organoiodine compounds including iodomethane have strong absorption in the tropospheric actinic flux region so that they are easily photolyzed to release I atoms (see Sect. 4.3.8 for CH_3I).

$$CH_3I + h\nu \rightarrow CH_3 + I \tag{7.79}$$

$$CH_2ICl + h\nu \rightarrow CH_2 + I + Cl \tag{7.80}$$

$$CH_2I_2 + h\nu \rightarrow CH_2 + 2I \tag{7.81}$$

The calculated photolytic lifetimes are few days, a few hours, and a few minutes for CH_3I, CH_2ICl, CH_2I_2, respectively (Saiz-Lopez et al. 2012). As seen in Sect. 4.4.1 and Fig. 4.36, I_2 is easily photolyzed by the visible light and forms I atoms.

7.6.2 Gas Phase Halogen Chain Reaction

The halogen atoms formed in the atmosphere by the photolysis of Cl_2, $ClNO_2$, Br_2, I_2, and organoiodine compounds reacts with O_3 to produce ClO, BrO, and IO radicals, which play a role in reducing the concentration of tropospheric O_3 by the chain reactions described below. Representing these halogen atoms by X and Y (= Cl, Br, I), the main chain cycle in e.g. in the polar region where the concentrations of active halogen are high, can be expressed as,

$$X + O_3 \rightarrow XO + O_2 \tag{7.82}$$

$$Y + O_3 \rightarrow YO + O2 \tag{7.83}$$

$$XO + YO \rightarrow X + Y + O_2. \quad (7.84)$$

These reactions constitute chain cycle to cause net O_3 dissipation,

$$2\ O_3 \rightarrow 3\ O_2. \quad (7.85)$$

However, the actual reaction of XO + YO follow the pathways other than reaction (7.84) depending on different kinds of halogens, and they have to be considered for the detailed analysis of reaction mechanism.

$$XO + YO \rightarrow OXO + Y \quad (7.86)$$

$$\rightarrow XY + O_2 \quad (7.87)$$

$$+M \rightarrow XOOY + M \quad (7.88)$$

For example, the reaction of ClO + ClO under the atmospheric pressure proceeds almost solely as,

$$ClO + ClO + M \rightarrow ClOOCl + M \quad (7.89)$$

(see Sect. 5.6.6), while BrO and IO, the following reaction pathways and branching ratios are given at 298 K (Sander et al. 2011; Atkinson et al. 2007; Saiz-Lopez et al. 2012).

$$BrO + ClO \rightarrow Br + OClO \quad (0.48) \quad (7.90)$$

$$\rightarrow Br + ClOO \quad (0.44) \quad (7.91)$$

$$\rightarrow BrCl + O_2 \quad (0.08) \quad (7.92)$$

$$BrO + BrO \rightarrow 2\ Br + O_2 \quad (0.85) \quad (7.93)$$

$$\rightarrow Br_2 + O_2 \quad (0.15) \quad (7.94)$$

$$BrO + IO \rightarrow OIO + Br \quad (0.80) \quad (7.95)$$

$$\rightarrow IBr + O_2 \quad (0.20) \quad (7.96)$$

$$IO + IO \rightarrow 2I + O_2 \quad (7.97)$$

$$\rightarrow I + OIO \quad (7.98)$$

The ClOOCl (Sects. 4.4.6, 5.6.6), BrCl (Sect. 4.4.1), Br_2 (Sect. 4.4.1), IBr, and OIO formed by these reactions will be susceptible to thermal decomposition or photolysis (see Sects. 4.4.1 and 4.4.6).

$$ClOOCl + M \rightarrow ClO + ClO + M \quad (7.99)$$

$$ClOOCl + h\nu \rightarrow 2\ Cl + O_2 \quad (0.80) \quad (7.100)$$

$$\rightarrow \mathrm{ClO} + \mathrm{ClO} \quad (0.20) \tag{7.101}$$

$$\mathrm{BrCl} + h\nu \rightarrow \mathrm{Br} + \mathrm{Cl} \tag{7.119}$$

$$\mathrm{Br_2} + h\nu \rightarrow 2\ \mathrm{Br} \tag{7.103}$$

$$\mathrm{IBr} + h\nu \rightarrow \mathrm{I} + \mathrm{Br} \tag{7.104}$$

$$\mathrm{OIO} + h\nu \rightarrow \mathrm{I} + \mathrm{O_2} \tag{7.105}$$

As for the photolysis of OIO in reaction (7.105), the quantum yield of unity is reported in the experiment (Gómez Martín et al. 2009). The absorption spectrum and photolytic processes of the chain carriers, ClO, BrO, IO, which result in the depletion of O_3 by the reactions (7.82), (7.83) and (7.84), have been described in Chap. 4, Sect. 4.4.5.

For such a gas phase chain reaction, radical-radical reactions (7.84, 7.86–7.88) predominate when the mixing ratios of halogen radicals (XO) exceed e.g. 3 pptv. However, as their mixing ratios decrease, the formation of HOX by the reaction of XO with HO_2,

$$\mathrm{XO} + \mathrm{HO_2} \rightarrow \mathrm{HOX} + \mathrm{O_2} \tag{7.106}$$

becomes more important. As for the successive reaction of HOX, other than the photolysis (see Chap. 4, Sect. 4.4.4),

$$\mathrm{HOX} + h\nu \rightarrow \mathrm{X} + \mathrm{OH}, \tag{7.107}$$

the multiphase reaction described in the following Sect. 7.5.3 is important particularly for HOBr.

In the environment affected by polluted air with high NO_x concentration, $XONO_2$ forming reaction by XO and NO_2 gets important as the chain termination reaction (see Sect. 5.6.5 for $ClONO_2$),

$$\mathrm{XO} + \mathrm{NO_2} + \mathrm{M} \rightarrow \mathrm{XONO_2} + \mathrm{M}. \tag{7.108}$$

The produced $XONO_2$ will be photolyzed as,

$$\mathrm{XONO_2} + h\nu \rightarrow \mathrm{X} + \mathrm{NO_3} \tag{7.109}$$

to form X and NO_3 (see Sect. 4.4.2), although the photolysis rate of $ClONO_2$ in the troposphere is relatively small.

7.6.3 Multiphase Halogen Chain Reaction

As for a chain reaction involving halogens in the troposphere, other than the gas phase reactions, multiphase reactions through the surface of sea salt solution is known particularly for bromine. From HOBr formed by the type of reaction (7.106) in the previous paragraph,

$$BrO + HO_2 \rightarrow HOBr + O_2, \quad (7.110)$$

Mozukewich (1995) suggested that Br_2 is released to the gas phase on the acidic sea salt surface in the polar region, and Sander and Crutzen (1996) proposed that a similar processes occurs in the polluted marine boundary layer in the mid-latitude.

$$HOBr(g) \rightarrow HOBr(aq) \quad (7.111)$$

$$HOBr(aq) + Br^- + H^+ \rightarrow Br_2(aq) + H_2O \quad (7.112)$$

$$Br_2(aq) \rightarrow Br_2(g) \quad (7.113)$$

In such a heterogeneous reactions on sea salt particles, Cl^- also participate in the reaction to form BrCl,

$$HOBr(aq) + Cl^- + H^+ \rightarrow BrCl(aq) + H_2O, \quad (7.114)$$

but BrCl is transformed to Br_2 in solution,

$$BrCl(aq) + Br^- + H^+ \rightarrow Br_2(aq) + HCl \quad (7.115)$$

as far as Br^- remains in the sea salt surface solution. Thus, Br_2 is released preferentially, and BrCl could be released into the atmosphere only after Br^- is depleted (Vogt et al. 1996),

$$BrCl(aq) \rightarrow BrCl(g). \quad (7.116)$$

Br_2 and BrCl thus emitted to the gas phase, returns to HOBr by the reactions,

$$Br_2 + h\nu \rightarrow 2Br \quad (7.103)$$

$$BrCl + h\nu \rightarrow Br + Cl \quad (7.102)$$

$$Br + O_3 \rightarrow BrO + O_2 \quad (7.117)$$

$$BrO + HO_2 \rightarrow HOBr + O_2, \quad (7.110)$$

and a multiphase chain reaction are constituted as far as the sea salt particle surface is available. The chain cycle composed of a series of above reactions are also called autocatalytic reactions. The rate determining step of this multiphase chain reaction is the heterogeneous reaction of HOBr(g) with NaBr/NaCl. The uptake coefficient of this reaction is $\gamma > 10^{-2}$ at 253 K (−20 °C), and increases as the temperature goes down (Adams et al. 2002).

The net reactions of this multiphase chain cycle can be expressed as,

$$Br^- + H^+ + O_3 + HO_2 + h\nu \rightarrow Br + 2O_2 + H_2O \tag{7.118}$$

$$Cl^- + H^+ + O_3 + HO_2 + h\nu \rightarrow Cl + 2O_2 + H_2O, \tag{7.119}$$

showing the loss of O_3 occurs by a series of this reactions. As seen in the next paragraph, it has been known that a sudden decrease of near surface O_3 occurs in the polar zone in springtime by high concentration of Br. Since the phenomenon cannot be explained by the gas phase chain reaction mentioned in the previous paragraph, it is thought that multiphase chain reaction occurs on the sea salt on ice, and the abrupt increase of Br in the polar region is called "bromine explosion" (Wennberg 1999).

In polluted marine boundary layer, a multiphase chain reaction involving the reactions of $BrONO_2$,

$$BrO + NO_2 + M \rightarrow BrONO_2 + M \tag{7.120}$$

$$BrONO_2 + Br^- \rightarrow NO_3{}^- + Br_2 \tag{7.121}$$

$$BrONO_2 + Cl^- \rightarrow NO_3{}^- + BrCl \tag{7.122}$$

has been analyzed by Sander et al. (1999).

7.6.4 Atmospheric Measurements of Active Halogens and Comparison with Models

The observed mixing ratios of active halogen species in the troposphere is in general 1–100 ppt, and the direct measurements of these species are usually made by using the methods, Long-Path Differential Absorption Spectroscopy (LP-DOAS), Multi-Axis Differential Optical Absorption Spectroscopy (MAX-DOAS), Chemical Ionization Mass Spectrometer (CIMS) and Laser Induced Fluorescence (LIF). As for BrO and IO, satellite observation gives spatial distribution of its tropospheric column.

In this paragraph, in situ measurements of active halogen species and their impact on ozone in the marine boundary layer and around salt lakes, and the satellite observation of BrO and IO are described comparing with the model calculation.

7.6.4.1 Polar Region

It has been known from the long-term monitoring data from Barrow, Alaska, that surface ozone concentration often decreases regularly in springtime (Oltmans and Komhyr 1986). Barrie et al. (1988) found that surface ozone rapidly drops in spring, from around 40 ppbv to a few ppb also at Alert in the polar zone of Canada, and

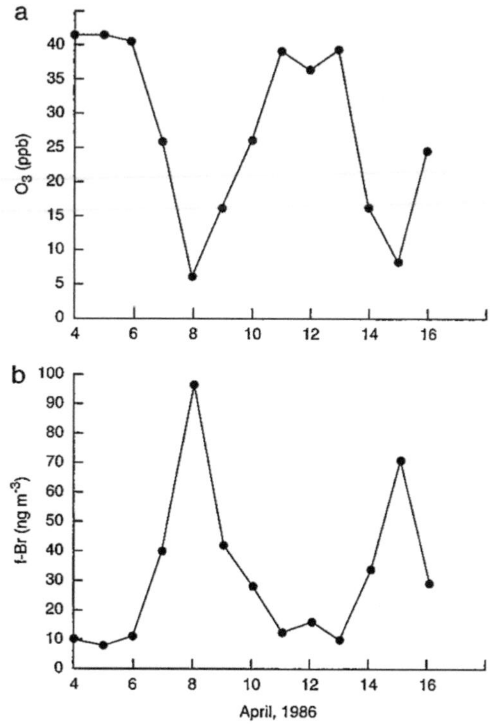

Fig. 7.21 Correlation between the (**a**) surface ozone and (**b**) filterable bromine ion (f-Br) concentration at Alert (Barrie et al. 1988)

reported a clear anti-correlation between the concentrations of O_3 and bromine in the aerosols sampled on filters as shown in Fig. 7.21. This finding brought the recognition of the importance of halogen chain reaction in the troposphere. Later, measurements of BrO in the polar sunrise at Alert were made by LP-DOAS, and the 24-h averaged mixing ratio, 4–17 pptv, was reported (Hausmann and Platt 1994). This concentration level of BrO has been shown to be enough for O_3 to decrease to near zero within a day by the reactions,

$$Br + O_3 \rightarrow BrO + O_2 \qquad (7.117)$$

$$BrO + BrO \rightarrow 2Br + O_2. \qquad (7.93)$$

Later, real-time measurement of Br_2 and BrCl were made by CIMS in ALERT2000 campaign, and their maximum mixing ratios were obtained to be 27 and 35 pptv, respectively, while Cl_2 could not be observed over the detection limit of 2 pptv (Foster et al. 2001; Spicer et al. 2002). On the other hand, the evidence of the major role of Cl atoms instead of OH radicals in the consumption of hydrocarbons in the period of low O_3 was found by Jobson et al. (1994), and the presence of Cl atoms in the low O_3 plume has been proved. Figure 7.22 is the plot of i-butane/n-butane against i-butane/propane ratio for the low ozone and normal ozone period (Jobson et al. 1994). The rate constants of i-butane and propane for

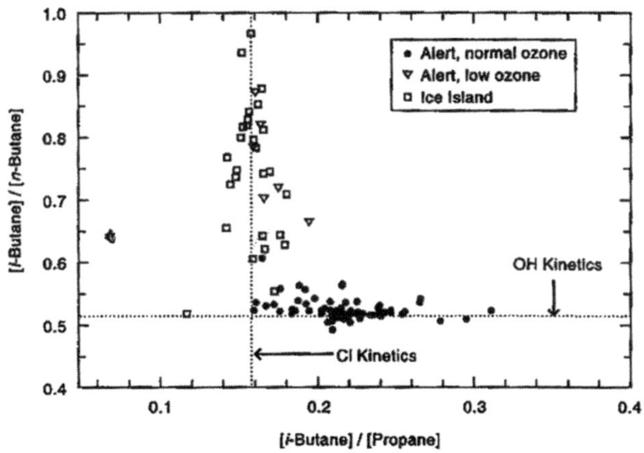

Fig. 7.22 Plot of the ratios of i-C_4H_{10}/n-C_4H_{10} vs. i-C_4H10/C_3H_8 at Alert when surface ozone is depleted (*open symbols*) and ozone is in normal (*filled symbols*) (Jobson et al. 1994)

Cl atoms are nearly the same and those for OH are different whereas the rate constants of i-butane and n-butane for OH are nearly the same and those for Cl atoms are different. Based on this, nearly constant ratio of i-butane/propane ratio for the low O_3 period and nearly constant ratio of i-butane/n-butane for the normal O_3 period as shown in Fig. 7.22, implies that hydrocarbons are consumed by Cl and OH, respectively, for each period.

Regarding Arctic Cycle sites other than Alert, at Ny Ålesund in Spitsbergen Island, maximum BrO, 30 pptv, ClO, 1–2 pptv, and filterable aerosols Br, 120 ng m^{-3} have been measured in a low ozone period by using DOAS in 1995–1996 (Tuckermann et al. 1997; Lehrer et al. 1997). On this occasion, NO_2 and SO_2 were less than 50 pptv indicating that that the surface ozone depletion in the polar region is a natural phenomenon which have no relation with anthropogenic pollution. Furthermore, in the recent observation at Barrow using CIMS, the average daytime HOBr mixing ratio was measured at 10 pptv with a maximum value of 26 pptv, and the average nighttime Br_2 was at 13 pptv with a maximum of 46 pptv which is strongly anti-correlated with ozone as shown in Fig. 7.23 (Liao et al. 2012). It has been reported that the box model calculation including the heterogeneous reactions reproduces well the diurnal variation of the BrO, HOBr, and Br_2 (Liao et al. 2012).

The surface ozone depletion by halogens in the polar region is also observed in Antarctica (Kreher et al. 1997). The daytime maximum mixing ratio of BrO was 13 pptv, and nighttime Br_2 and BrCl were 46 and 6 pptv, respectively, by the CIMS measurement at Hallay (Buys et al. 2013). Meanwhile, the maximum mixing ratio of IO at 10 pptv was measured at Neumayer, which is very different from Antarctica where IO is low (Frieß et al. 2001). In observations around Antarctica, these active halogens have good correlations with the contact time with sea ice by trajectory analysis, which supports the presumption that the phenomena is caused by the increase of the concentrations of active halogens due to the heterogeneous chain reaction on sea salt surface as described in the previous Sect. (Wagner et al. 2007). The high concentration of iodine over Antarctica as compared to the Arctic Circle may be ascribed to the difference in the emission strength of biogenic

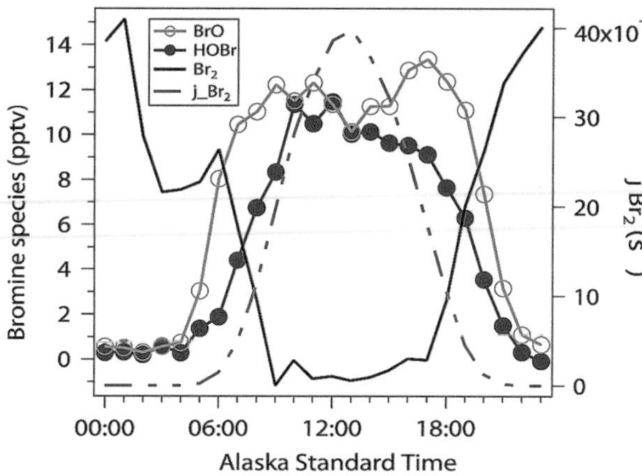

Fig. 7.23 Observed mixing ratios of BrO, HOBr and Br_2, and photolysis rate of Br_2 (j_{Br2}) at Barrow, Alaska (Adapted from Liao et al. 2012)

photoactive organoiodine compounds from the sea surface around the ice bound sea (Frieß et al. 2001). The satellite observation of IO in the Antarctica has been made by the SCIAMACHY, which shows the maximum column density in Spring (October) and relatively high in summer and autumn, but the increase of IO is not observed in winter (Schönhardt et al. 2008). It has been confirmed that such an increase of IO column density is not observed in the Arctic Circle, and the reason is thought to be the difference in the distribution of marine biota emitting iodine compounds.

The polar ozone depletion is confirmed to occur within the boundary layer, about 1 km from the ground surface by the slant column measurements of MAX-DOAS, and also by ozone sondes, while the active halogens are known to spread to the free troposphere as observed by aircrafts (Hönninger and Platt 2002; Choi et al. 2012). Furthermore, the BrO column density has been widely observed by satellite sensors, GOME, etc., and it is concluded that the high density in the polar region is due to the tropospheric column, and those of other regions are due to the stratospheric column (Platt and Wagner 1998). According to the satellite data, tropospheric BrO in the polar region in spring spreads in the timescale of 2–4 days and the spatial scale of 2000 km as "bromine clouds", which proves that tropospheric halogen chemistry is not limited to the local phenomena only around the sites where the surface ozone depletion is observed.

The three dimensional modelling studies on the BrO formation and boundary layer ozone depletion in the polar region have been conducted recently (Zhao et al. 2008; Toyota et al. 2011). Toyota et al. (2011) assumed the reaction of sea salt contained in the snow on the sea ice with O_3 is the sole initial process of Br_2 release into the atmosphere, and calculated the BrO formation and O_3 depletion for the Arctic Circle using the three dimensional model based on the autocatalytic chain reaction described in the previous paragraph. Figure 7.24 compares the GOME satellite data and model calculation of tropospheric column of BrO in the Arctic region in April 2001. On April 16–17, and 20–22, large bromine clouds are

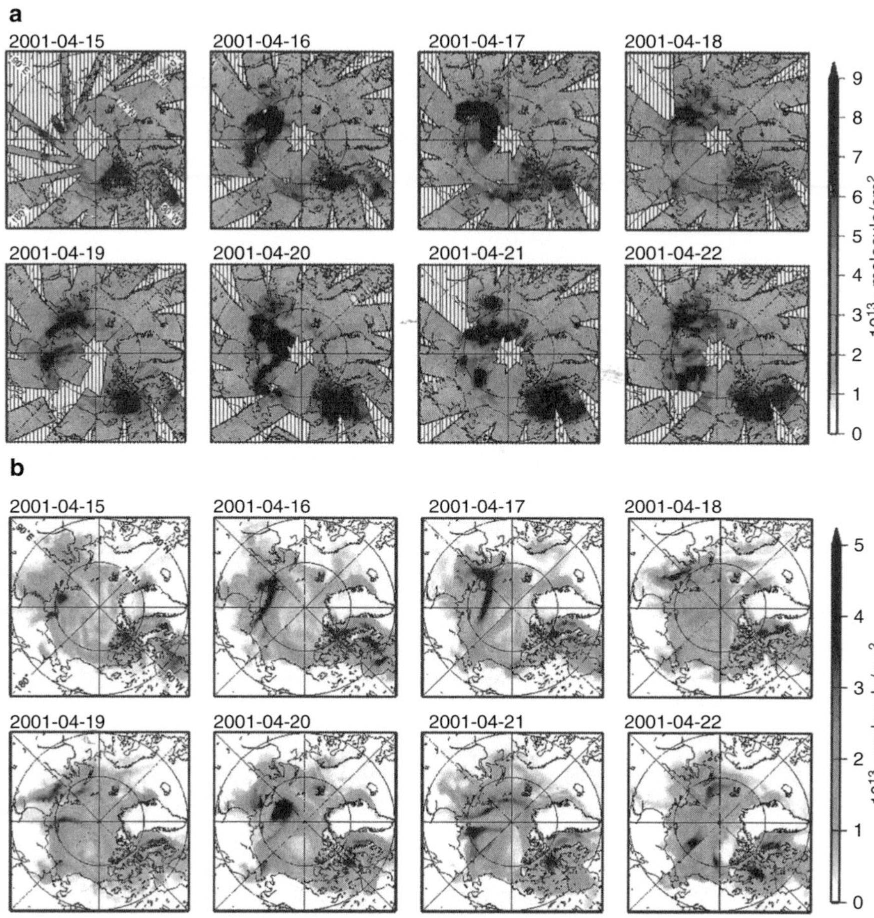

Fig. 7.24 Comparison of tropospheric BrO column density in "bromine explosion" events in Arctic Circle. (**a**) satellite observation and (**b**) model calculation (Adapted from Toyota et al. 2011)

seen over 1000 km over the Siberian fringe and Canadian Arctic region, respectively. Shown in Fig. 7.24b is the result of RUN 3 model in which critical temperature for the Br_2 from the reactions of O_3, HOBr and $BrONO_2$ with sea salt on snow cover was set as $T_c = -10$ °C. The RUN 3 model reproduced reasonably well the temporal variation and spatial distribution of BrO. In this model study, it is reported that the emerging timing and place of the bromine clouds do not match with the satellite data if T_c is set as -15 °C, and the bromine clouds do not appear at all when they set $T_c = -20$ °C. Meanwhile, the experimentally obtained uptake coefficient of HOBr on NaCl-covered ice is $\gamma \approx 1 \times 10^{-2}$ at -15 °C, and increase with decrease of temperature but the temperature dependence is not very strong (Adams et al. 2002). Figure 7.25 depicts the comparison between the observation and model calculation for surface ozone at Alert, Barrow, Zeppelin

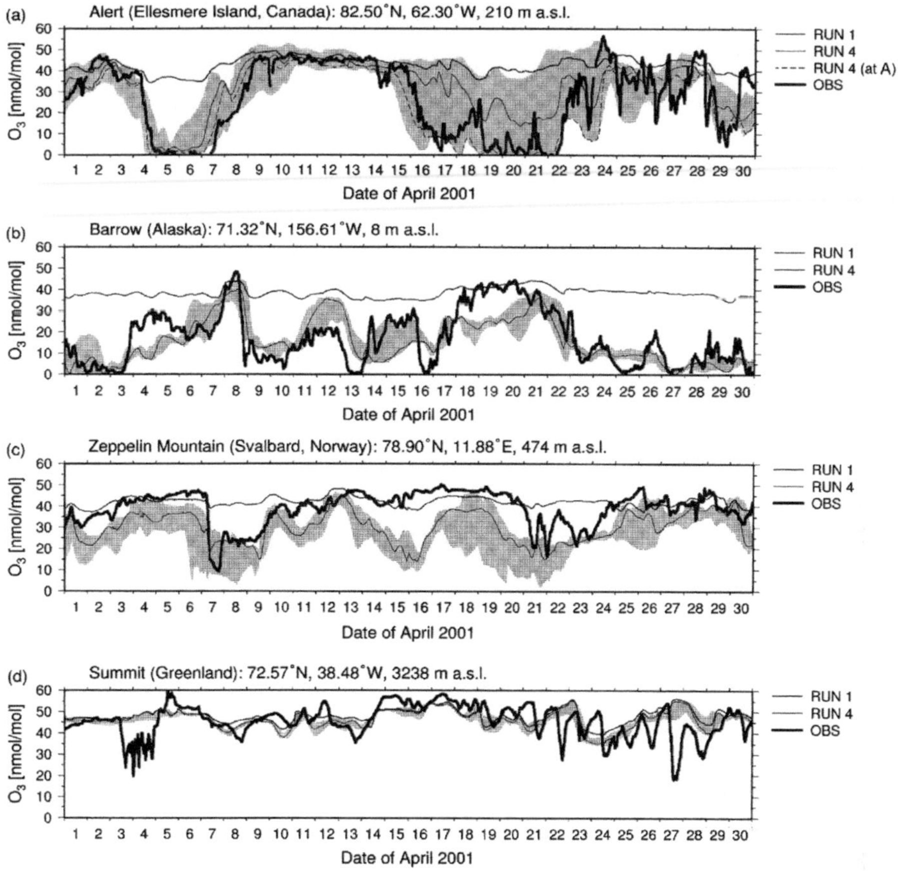

Fig. 7.25 Comparison of surface O_3 mixing ratios between the observation and model calculation at (**a**) Alert, Canada, (**b**) Barrows, Alaska (**c**) Zeppelin Mountain, Swalbard, Norway, and (**d**) Summit, Greenland (Adapted from Toyota et al. 2011)

Mountain (Spitsbergen Island, Norway), and Greenland Summit in April 2001 (Toyota et al. 2011). As shown in the figure, the O_3 depletion does not appear in RUN 1 in which the release of Br is not considered, the episodic O_3 depletions are reproduced well by RUN 4 in which T_c is set at −15 °C. From this model investigation, as for the cause for the sporadic bromine explosion, meteorological conditions is suggested to influence the contact time of O_3 with sea salt on snow cover by transport and turbulent diffusion. As for the bromine release, the first year fresh snow is thought to be important and it is sensitive to the critical temperature mentioned above.

7.6.4.2 Low- and Mid-Latitude Marine Boundary Layer

Inorganic halogens have been observed, not only in the polar region, but widely in the marine boundary layer. The concentrations of halogens in the marine boundary layer are apparently different in the polluted and clean air, and in general the concentration is much higher in the air influenced by pollution. The difference is marked for Cl, and for example at Cape Verde Island in the tropical eastern Atlantic Ocean in summer in 2009, mixing ratios of HOCl and Cl_2 in the air mass from the open ocean are maximum 60 pptv in the daytime, and 10 pptv in the nighttime, whereas those in the air mass from European continent are over 100 ppt and 35 pptv in maximum, respectively (Lawler et al. 2011). Lawler et al. (2011) analyzed the data by using a box model including the multiphase photochemical chain reaction,

$$HCl + OH \rightarrow Cl + H_2O \tag{7.123}$$

$$Cl + O_3 \rightarrow ClO + O_2 \tag{7.124}$$

$$ClO + HO_2 \rightarrow HOCl + O_2 \tag{7.125}$$

$$HOCl + Cl^- + H^+ \rightarrow Cl_2 + H_2O \tag{7.126}$$

$$Cl_2 + h\nu \rightarrow 2Cl \tag{7.127}$$

considering the presence of observed 0.6 pptv of HCl. The model reproduced 10 pptv of the nighttime Cl_2 well, but could not reproduce a few tens of pptv of the daytime HOCl, suggesting the existence of a missing source of HOCl. From the analysis for this observational data by a one-dimensional model including Br, Sommariva and von Glasow (2012) concluded that 5–11 % of CH_4 loss is due to Cl atoms, and 35–40 % of O_3 loss is due to halogens (mainly Br).

At Cape Verde, year-around observation of BrO and IO has been conducted, and the averaged mixing ratios of BrO 2.5 ± 1.1 (1σ) pptv and IO 1.4 ± 0.8 pptv are reported for November 2006–June 2007. Figure 7.26 shows the observed diurnal and seasonal variation of BrO and IO. The measured value of NO at this site is very low, 3.0 ± 1.0 pptv, and photochemical loss of O_3 occurs in the daytime, as mentioned in Sect. 7.3.1, and it is deduced that 30–50 % of the O_3 loss is due to the halogens (Read et al. 2008). Also, Dickerson et al. (1999) estimated by a model calculation for the observation in the tropical Indian Ocean that against the observed 32 % loss of the daytime O_3, only 12 % is explained by the HO_x chain reaction and loss by the multiphase chain reaction including BrO/HOBr/HBr amounts to 22 %. The sources of iodine at the sites where seaweed is not present, such as at Cape Verde and over the mid-latitude open ocean, are thought to be CH_2I_2, $CHClI_2$, and CH_3I formed by the sea surface reaction of organic compounds with iodine under the sunlight (Martino et al. 2009; Mahajan et al. 2010). As shown in Fig. 7.27, McFiggans et al. (2000) obtained a good agreement for the diurnal variation of IO with a box model calculation using the mixing ratios of CH_2I_2, CH_2BrI and CH_3I (0.03, 0.3, 3 pptv, respectively) measured in the remote Atlantic Ocean as a boundary condition. From this calculation, the ratios of initial I atom

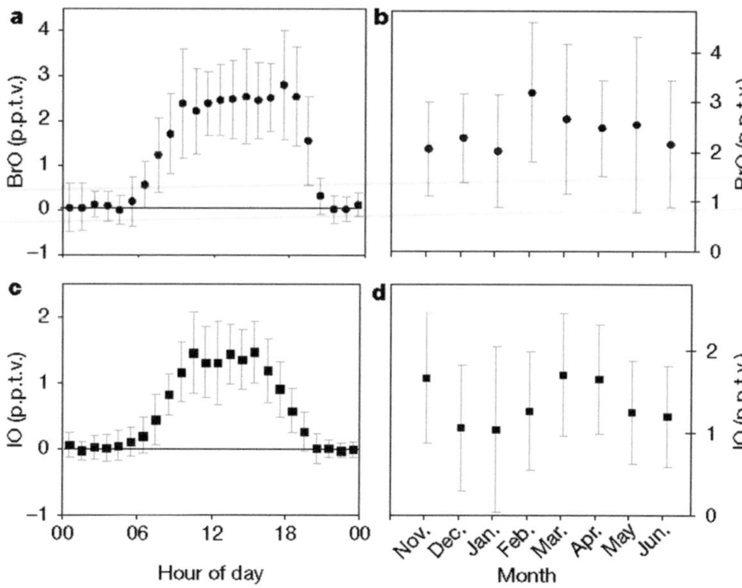

Fig. 7.26 (**a**) Diurnal variation of BrO, (**b**) seasonal variation of BrO, (**c**) diurnal variation of IO, and (**d**) seasonal variation of IO at Cape Verde. Designated error is 1σ. (Adapted from Read et al. 2008)

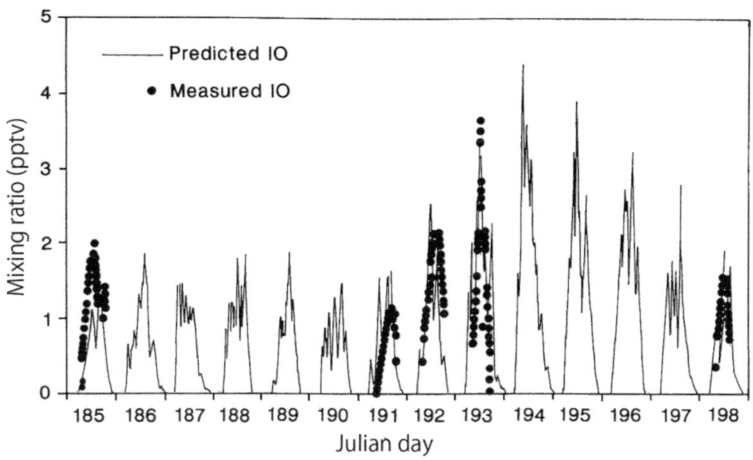

Fig. 7.27 Comparison between the observed mixing ratio of IO and the estimated values from the observed concentrations of CH_2I_2, CH_2BrI and CH_3I over Atlantic Ocean (Adapted from McFiggans et al. 2000)

release are calculated as CH_2I_2 64 %, CH_2BrI 34 % and CH_3I 4 %, and the observed maximum IO mixing ratio of 3 pptv is well reproduced.

In the relatively clean marine boundary layer at Mace Head, Ireland, maximum mixing ratios of BrO at 6.5 pptv, the daytime averaged value of 2.3 pptv, and the maximum values of I_2,IO, OIO at 93, 2.5 and 10.8 pptv, respectively, were observed by LP-DOAS (Saiz-Lopez et al. 2004, 2006). The source of high concentration of iodine at Mace Head is thought to be due to seaweed, based on the correlation with low tide.

On the other hand, in marine boundary layers under the strong influence of polluted air, very high concentrations of $ClNO_2$ and Cl_2 have been observed. For example, the maximum mixing ratios of $ClNO_2$ and Cl_2 at 2100 and 200 ppt (Riedel et al. 2012) in the harbor near Los Angeles, and nighttime maximum Cl_2 at 150 pptv in Long Island, United States (Spicer et al. 1998). These high concentration of halogens are thought to be $ClNO_2$, formed by reaction (7.76) between N_2O_5 with sea salt in the polluted air from cities and harbor ships, and Cl_2 formed by the self-catalytic reaction described in Sect. 7.5.3. However, as for Cl_2 in the suburbs of Los Angeles, either of the cases where it has high correlation or no correlation with $ClNO_2$ has been reported, and the direct emission of Cl_2 from urban pollution sources are suggested in the latter case (Riedel et al. 2012).

7.6.4.3 Salt Lakes

Reactive halogens in the troposphere have been known to exist not only over the marine boundary layer, but also in the surroundings of terrestrial salt lakes. As for the halogen species at the inland salt lakes, the Dead Sea in Israel is best investigated, and maximum mixing ratio of 100–200 pptv of BrO has been observed regardless of season (Matveev et al. 2001; Tas et al. 2005). Other than BrO, maximum 10 pptv of IO has also been measured (Zingler and Platt 2005). As for the sources of halogens at Dead Sea, discussions have been made that iodine source is not biogenic, but heterogeneous photochemical reactions between the halogen ions in water with gas-phase O_3 at the surface of water has been discussed (Zingler and Platt 2005; Tas et al. 2006; Smoydzin and von Glasow 2009). Furthermore, ClO and BrO have been observed at Great Salt Lake, United States and BrO at Salar de Uyuni, Bolivia (Stutz et al., 2002; Hönninger et al., 2004), and it has been established that the presence of active halogen species in the atmosphere around the inland salt lakes is a general phenomenon.

7.7 Tropospheric Sulfur Chemistry

Historically, sulfur compounds in the troposphere have been the cause of concern from the viewpoint of air pollution since the industrial revolution; sulfur dioxide directly emitted from fossil fuel combustion and sulfuric acid aerosols formed by its oxidation in the atmosphere have been studied as the most important air pollutants, and the acid rain caused by their uptake into clouds and fog followed by liquid phase reactions have also been studied intensively (Finlayson-Pitts and Pitts 2000). On the other hand, sulfur compounds of natural biogenic origin such as dimethyl sulfide (DMS) are oxidized in the atmosphere to form sulfuric acid aerosols over the open ocean. They have been attracting much attention from the viewpoint that they act as cloud condensation nuclei (CCN) to form clouds, which influences radiation budget of the earth and has a possibility to generate negative feedback to control the emission of DMS (Charlson et al. 1987). Sulfuric acid aerosols formed from the sulfur dioxide of anthropogenic sources also affect the radiation budget, and are important from the point of climate change caused by human activities (IPCC 2013).

Sulfur compounds in the atmosphere include hydrogen sulfide (H_2S), methanethiol (CH_3SH), dimethyl sulfide (DMS, CH_3SCH_3), dimethyl disulfide (CH_3SSCH_3), and carbonyl sulfide (COS) which are mainly marine biogenic origin, carbon disulfide (CS_2) which is both marine and anthropogenic origin, and sulfur dioxide (SO_2) which is mainly anthropogenic but partly volcanic origin. Among the natural origin sulfur compounds, DMS is the most important (Bates et al. 1992), and its gas phase oxidation mechanism is described in this section. The rate constant of COS with OH is small and its tropospheric lifetime is more than 10 years, so that most of them reach to stratosphere causing the stratospheric aerosols as described in Chap. 8, Sect. 8.5.

As for SO_2, after the gas phase oxidation reactions are described in Sect. 7.6.1, multiphase heterogeneous reactions (see Sect. 2.4) in fog water and rain droplet are covered in Sect. 7.6.2.

7.7.1 *Gas Phase Homogeneous Oxidation Reaction Mechanism*

Most of oxidation reactions of sulfur compounds in the troposphere are initiated by OH radicals, but the reactions with NO, and active halogen species such as BrO and IO has to be considered in the polluted air and in the marine boundary layer, respectively. As for SO_2, other than the gas phase reaction with OH, oxidation reaction in cloud and fog droplet is important which will be described later.

7.7.1.1 SO_2

The main emission source of sulfur dioxide SO_2 is anthropogenic fossil fuel combustion, and the emissions from volcanoes and biomass burning are added to it (Bates et al. 1992). Atmospheric SO_2 reacts with OH,

$$SO_2 + OH + M \rightarrow HOSO_2 + M \tag{7.128}$$

to form $HOSO_2$ radicals (see Sect. 5.2.6). The formation of sulfur trioxide (SO_3) and HO_2 from $HOSO_2$,

$$HOSO_2 + O_2 \rightarrow SO_3 + HO_2 \tag{7.129}$$

was first proposed by Stockwell and Calvert (1983), and an important consequence of this process is that it produces HO_2 so that the reaction of SO_2 and OH does not serve as the termination reaction of OH chain reaction (see Sect. 7.3). The reaction of SO_3 to form sulfuric acid is second order to H_2O, and it is known that the adduct of SO_3 and H_2O formed by

$$SO_3 + H_2O \rightarrow H_2OSO_3 \tag{7.130}$$

reacts with another H_2O molecule to form H_2SO_4 (Lovejoy et al. 1996).

$$H_2OSO_3 + H_2O \rightarrow H_2SO_4 + H_2O \tag{7.131}$$

The generated H_2SO_4 is a gaseous molecule but since the vapor pressure is low, it condense with water vapor in the atmosphere to yield $H_2SO_4(aq)$ and forms sulfuric acid aerosols. They further react with ammonia molecules to yield solid ammonium sulfate,

$$H_2SO_4(aq) + 2NH_3 \rightarrow (NH_4)_2SO_4(s). \tag{7.132}$$

From the rate constant of gaseous reaction of OH and SO_2 at 298 K and 1 atm given in Table 5.2, average atmospheric lifetime of SO_2 is more than 1 week when the average concentration of OH is assumed to be 8×10^5 molecules cm^{-3}, but it is generally much shorter when the multiphase reaction is considered.

7.7.1.2 CS_2

Carbon disulfide CS_2 is known to be mainly emitted from industries and partially form natural biota in the ocean (Chin and Davis 1993). The reaction of CS_2 and OH is the addition reaction,

$$CS_2 + OH + M \ CS_2OH + M. \tag{7.133}$$

The CS_2OH radical reacts with O_2 to generate OCS and SO_2 with the yield of ca. 0.83–0.84 (Stickel et al. 1993), and the reaction pathways,

$$CS_2OH + O_2 \rightarrow OCS + HOSO \tag{7.134}$$

$$HOSO + O_2 \rightarrow SO_2 + HO_2 \tag{7.135}$$

have been proposed by quantum chemical calculation (Zhang and Qin 2000; McKee and Wine 2001). Carbonyl sulfide, COS formed in the atmospheric reaction of CS_2 has a long lifetime in the troposphere as mentioned below, and it is mostly transported to the stratosphere. Although other products, CO and SO_2 are formed in this reaction with the yields of ca. 0.16 and 0.30, respectively, their detailed production pathways are not well known.

The reaction rate constant of CS_2 and OH under the atmospheric conditions is about the same as of SO_2 (see Table 5.2), and its atmospheric lifetime is 1–2 weeks.

7.7.1.3 COS

Carbonyl sulfide COS is a sulfur compound present in the troposphere at ca. 0.5 ppbv. It is emitted from volcanic activities and also formed in the atmospheric reaction of CS_2 and OH as mentioned above. Among the global emission of COS, the ratio of the secondary formation from CS_2 is estimated to be ca. 30 % (Chin and Davis 1993). The rate constant of the reaction of COS an OH is very small as seen in Table 5.2 (2×10^{-15} cm^3 molecule^{-1} s^{-1} at 298 K), and the atmospheric lifetime calculated from the average concentration of OH assumed to be 8×10^5 molecules cm^{-3} is about 20 years. Therefore, most of COS emitted and formed in the troposphere is transported to the stratosphere, where it is photolyzed to yield H_2SO_4, which causes the stratospheric aerosols (see Chap. 8, Sect. 8.5).

7.7.1.4 H_2S

Hydrogen sulfide H_2S is emitted to the atmosphere either from natural sources, e.g. volcanos, soils, biomass burning, and marine biology, or from anthropogenic sources, mainly industries. The reaction of H_2S with OH yields HS (sulfanyl radical) by abstraction reaction,

$$H_2S + OH \rightarrow HS + H_2O. \tag{7.136}$$

The formed HS radicals react with O_3 or NO_2 in the atmosphere,

$$HS + O_3 \rightarrow HSO + O_2 \tag{7.137}$$

$$HS + NO_2 \rightarrow HSO + NO \tag{7.138}$$

to yield HSO (sulfhydryl) radicals, from which SO_2 is produced.

$$HSO + O_3 \rightarrow HSO_2 + O_2 \tag{7.139}$$

$$HSO_2 + O_2 \rightarrow HO_2 + SO_2 \tag{7.140}$$

The rate constant of the OH and H_2S reaction is 4.7×10^{-12} cm^3 molecule^{-1} s^{-1} at 298 K as given in Table 5.2, and the tropospheric lifetime is calculated to be ca. 70 h assuming the OH concentration of 8×10^5 molecules cm^{-3}.

7.7.1.5 CH_3SH

Methanetiol, CH_3SH is also called methyl mercaptan, and its sources are mainly marine biology. The reaction with OH radicals is thought to be the hydrogen abstraction (Atkinson et al. 2004; Sander et al. 2011),

$$CH_3SH + OH \rightarrow CH_3S + H_2O. \tag{7.141}$$

The product, CH_3S (methylthiyl) radicals are very important as the intermediate of the atmospheric oxidation reaction of DMS (CH_3SCH_3), and will be described later in detail in this paragraph.

The rate constant of the reaction, CH_3SH and OH (see Table 5.2) is 7 times higher than H_2S, and the tropospheric lifetime of CH_3SH is calculated to be a few hours.

7.7.1.6 CH_3SCH_3

Dimethyl sulfide (DMS), CH_3SCH_3, has the highest emission rate among the natural origin sulfur compounds (Bates et al. 1992), and is concerned as the precursor of the natural origin SO_2, particulate methane sulfonic acid (MSA), CH_3SO_3H, and dimethyl sulfoxide (DMSO), CH_3SOCH_3. Other than OH, the rate constant of DMS with NO_3 is also large, and the reaction is important in the coastal area with strong influence of anthropogenic activities. DMS is known also to react with active halogen species such as ClO and BrO. Many studies have been reported on the atmospheric oxidation processes of DMS, and a review is given by Barnes et al. (2006).

It has been known that the initial reaction of DMS and OH is both hydrogen abstraction and addition to S atom (Albu et al. 2006).

$$CH_3SCH_3 + OH \rightarrow CH_3SCH_2 + H_2O \tag{7.142}$$

$$+M \rightarrow CH_3S(OH)CH_3 + M \tag{7.143}$$

As seen in Table 5.2, the abstraction reaction has positive activation energy and the reaction rate constant increases with temperature, while the addition reaction has negative temperature dependence and the rate constant increases with the decrease of temperature. The pressure dependence is also seen for the addition reaction as a three-body reaction, so that the branching ratio of reaction (7.142) and (7.143) changes depending on temperature and pressure. For example, about 80 % of the reaction is abstraction at 298 K and 1 atm, the ratio of the addition reaction increases with the decrease of temperature (Albu et al. 2006). The rate constant for the abstraction reaction at 298 K is 4.7×10^{-12} cm^3 molecule^{-1} s^{-1} (see Table 5.2), which is nearly the same as H_2S and the atmospheric lifetime is several tens of hours.

The CH_3SCH_2 radical formed in reaction (7.142) react with O_2, and is thought to follow the same pathways as alkyl radicals in the presence of NO (Barnes et al. 2006).

$$CH_3SCH_2 + O_2 + M \rightarrow CH_3SCH_2O_2 + M \quad (7.144)$$

$$CH_3SCH_2O_2 + NO \rightarrow CH_3SCH_2O + NO_2 \quad (7.145)$$

$$CH_3SCH_2O \rightarrow CH_3S + HCHO \quad (7.146)$$

$$CH_3SCH_2O_2 + HO_2 \rightarrow CH_3SCH_2OOH + O_2 \quad (7.147)$$

Many studies have been conducted as for the reaction pathways of CH_3S radicals formed in the above reaction (7.146), and reaction (7.141) of CH_3SH and OH, most of them are based on the analysis of final products, and the precise reaction mechanism has not been established yet. In the atmosphere, CH_3S is thought to react with NO_2 and O_3,

$$CH_3S + NO_2 \rightarrow CH_3SO + NO \quad (7.148)$$

$$CH_3S + O_3 \rightarrow CH_3SO + O_2 \quad (7.149)$$

to form CH_3SO, which would further reacts as,

$$CH_3SO + O_2 + M \rightarrow CH_3S(O)O_2 + M \quad (7.150)$$

$$CH_3S(O)O_2 + NO \rightarrow CH_3SO_2 + NO_2 \quad (7.151)$$

$$CH_3SO + O_3 \rightarrow CH_3SO_2 + O_2 \quad (7.152)$$

to produce CH_3SO_2. The reaction of CH_3SO_2 with O_3,

$$CH_3SO_2 + O_3 \rightarrow CH_3SO_3 + O_2 \quad (7.153)$$

would form CH_3SO_3, and a mechanism to produce methane sulfonic acid (MSA, CH_3SO_3H) as a final product,

$$CH_3SO_3 + RH \rightarrow CH_3SO_3H + R \quad (7.154)$$

has been considered (Barnes et al. 2006).

The production of SO_2 as well as CH_3SO_3H in the OH oxidation reaction of DMS under the atmospheric conditions has been known experimentally (Hatakeyama et al. 1982). As for the pathways to form SO_2, a thermal decomposition of CH_3SO_2 formed in the above reaction (7.152),

$$CH_3SO_2 \rightarrow CH_3 + SO_2 \tag{7.155}$$

has been proposed, but the reaction pathway has not been established.

Meanwhile, from $CH_3S(OH)CH_3$ formed in the addition reaction (7.143) of DMS and OH, dimethyl sulfoxide (DMSO), CH_3SOCH_3, and methanesulfenic acid, CH_3SOH, are thought to be produced (Gross et al. 2004; Ramírez-Anguita et al. 2009) as,

$$CH_3S(OH)CH_3 + O_2 \rightarrow CH_3SOCH_3 + HO_2 \tag{7.156}$$

$$\rightarrow CH_3SOH + CH_3O_2. \tag{7.157}$$

The reaction of DMS with NO_3 is assumed to give initial adduction from the observed negative temperature dependence, but is considered to be basically an abstraction reaction based on the experimentally observed H/D isotope effect,

$$CH_3SCH_3 + NO_3 \rightarrow [CH_3S(ONO_2)CH_3]^\dagger \rightarrow CH_3SCH_2 + HNO_3 \tag{7.158}$$

to give CH_3SCH_2 radicals, the same as in the reaction with OH (Atkinson et al. 2004; Sander et al. 2011).

As for the atmospheric reactions of DMS, the reactions with the active halogen species, as seen in Sect. 7.5 are also known. For example, the reaction with BrO,

$$CH_3SCH_3 + BrO \rightarrow [CH_3S(OBr)CH_3]^\dagger \rightarrow CH_3S(O)CH_3 + Br \tag{7.159}$$

has a negative temperature effect, and gives DMSO with a reaction yield of unity. The reactions of ClO and IO are also known to produce DMSO directly (Atkinson et al. 2004; Sander et al. 2011).

7.7.1.7 CH_3SSCH_3

Dimethyl disulfide (DMDS), CH_3SSCH_3, is released mainly from marine biology and biomass burning. The rate constant of the OH reaction is very fast, 2.3×10^{-10} cm^3 $molecule^{-1}$ s^{-1} at 298 K, and the atmospheric lifetime is very short within a few hours. The reaction is thought to give methane sulfonic acid (MSA) directly (Barnes et al. 1994) as,

$$CH_3SSCH_3 + OH \rightarrow CH_3S + CH_3SOH \tag{7.160}$$

$$CH_3SOH + O_2 \rightarrow CH_3SO_3H. \quad (7.161)$$

7.7.2 *Multiphase Reactions and Acidification of Clouds and Fog*

The studies on multiphase reactions in the troposphere have been developed through atmospheric environmental problem of acid rain/wet deposition. Most of oxidizing processes in rainwater relevant to sulfate, nitrate, and carbonate ions have been well established, as mentioned in this paragraph. However, multiphase reactions in the troposphere involving uptake and reaction on the surface of aerosol surface for many atmospheric species, and reactions in the aerosol aqueous solution relevant to organic aerosols are the research fields studied intensively at the present time, as they are relevant to secondary organic aerosol formation, and further development in future is expected. For the content of this paragraph, textbooks by Finlayson-Pitts and Pitts (2000), Seinfeld and Pandis (2006), and McElroy (2002) are of use as references. Since the actual field observations of acid rain are not covered in this book, the above textbooks should be referred for this aspect and also Cowling (1982) for the historical aspect.

Here, the uptake of SO_2 to clouds and fog droplet, liquid phase oxidation processes to sulfuric acid, and the acidification of water droplets are considered. As seen in Chap. 2, Sect. 2.4.1, multiphase reaction processes of atmospheric molecules consist of (1) transport and diffusion of gaseous molecules to the gas-liquid surface, (2) uptake of gaseous molecules to liquid phase, (3) diffusion in the bulk liquid, and (4) chemical reactions in the bulk liquid phase (Schwartz and Freiberg 1981).

7.7.2.1 Multiphase Processes in Water Droplets

First, assuming that the transport and diffusion of gaseous molecules occurs by molecular diffusion to spherical water particles, transport rate constant β_{gd} (s^{-1}) is equal to the heterogeneous rate constant k_{het} in Chap. 2 Sect. 2.4.1, it can be expressed as,

$$\beta_{gd} = k_{het} = \frac{1}{\tau_g} = \frac{1}{4}\Gamma_g u_{av} A \quad (7.162)$$

by putting $\gamma = \Gamma_g$ in Eq. (2.81). Here, τ_g is the characteristic time for the gas phase diffusion to the liquid droplets, u_{av} is the average thermal kinetic velocity, A is the surface area density of droplets contained in the unit volume of gas. Considering here water droplets with radius r, A can be expressed as,

$$A = 4\pi r^2 N_p \tag{7.163}$$

where N_p (particles cm^{-3}) is the number of particles in a unit gas volume. Meanwhile, if we define liquid water content (LWC) as the weight of water contained in a unit volume of atmosphere (Wallace and Hobbs 2006; Seinfeld and Pandis 2006),

$$L = \frac{4}{3}\pi r^3 N_p \tag{7.164}$$

from the definition, and

$$A = \frac{3L}{r} \tag{7.165}$$

By substituting this equation and Eq. (2.84) in Sect. 2.4.1 for Γ_g into Eq. (7.143),

$$\beta_{gd} = \frac{3D_g L}{r^2} \tag{7.166}$$

is obtained. When typical values of diffusion constant, $D_g = 0.1$ cm^2 s^{-1}, clouds and fog radius, $r = 7 \times 10^{-4}$ cm, and liquid water content, $L = 5 \times 10^{-7}$ (l(aq)/l(air)) (0.5 g m^{-3}) are assumed, β_{gd} is calculated to be ≈ 0.3 s^{-1}, and the characteristic time for diffusion is the order of ~ 1 s by talking the reciprocal.

On the other hand, transport rate constant for the accommodation process at the gas-liquid surface, β_i (s^{-1}), can be expressed similarly by $\gamma = \alpha$ putting in Eq. (2.81),

$$\beta_i = k_{het} = \frac{1}{4}\alpha u_{av} A = \frac{3u_{av}\alpha L}{4r} \tag{7.167}$$

If we use the accommodation coefficient of SO_2 molecule to the water particle surface given in Table 6.1, $\alpha \geq 0.12$, $u_{av} = 4.7 \times 10^4$ cm s^{-1}, and liquid water content, $L = 5 \times 10^{-7}$ (l(aq)/l(air)), and particle radius, $r = 7 \times 10^{-4}$ cm, β_i is calculated to be 3.0 s^{-1}, and the characteristic time necessary for the uptake from the gas phase to the liquid surface is the order of ~ 0.1 s by taking the reciprocal. Furthermore, SO_2 molecules taken into water droplets react with H_2O molecules at the liquid surface, and reach to chemical equilibrium with three sulfur species, $SO_2{\bullet}H_2O$, HSO_3 and $SO_3{}^{2-}$ as mentioned later. The time for reaching to the chemical equilibrium is as short as 10^{-3} s.

Next, the diffusion of the taken-up molecule in the water droplets can be treated by converting the one-dimensional diffusion Eq. (2.93) in Chap. 2, Sect. 2.4.3, to the three-dimensional polar coordinate considering diffusion in a spherical liquid droplet,

$$\frac{\partial N_{aq}}{\partial t} = D_{aq}\left(\frac{\partial^2 N_{aq}}{\partial r^2} + \frac{2}{r}\frac{\partial N_{aq}}{\partial r}\right) \tag{7.168}$$

Here, r is the position coordinate in radial direction, N_{aq} is the molecular density (molecules cm^{-3}) of molecule contained in the unit volume of aqueous solution, and D_{aq} is the diffusion constant of molecules in the aqueous solution. After solving this equation by assuming the initial and boundary conditions, the diffusion rate constant, β_{ad} (s^{-1}) of the molecules inside of the liquid phase can be obtained as,

$$\beta_{ad} = \frac{\pi^2 D_{aq}}{r^2} \tag{7.169}$$

The characteristic diffusion time of molecules in water droplets is smaller than the gas phase diffusion by four orders of magnitude if liquid phase molecular diffusion is assumed. Finally, reactions in the liquid layer is considered. For the particles with radius smaller than 10 μm such as clouds and fog droplets, the time for the gas phase diffusion to the liquid surface, transfer from the gas to the liquid phase, and diffusion within the liquid droplet is less than 1 s, and they cannot be a rate-limiting step for the multiphase oxidation reaction process. Thus, the liquid phase reaction within the water droplet as described below is the rate-limiting step.

7.7.2.2 Dissolution of SO_2 Into Water

The SO_2 molecule dissolved in water forms a molecular complex $SO_2{\bullet}H_2O$, and then generates bisulfate ion $HSO_3{}^-$ and sulfite ion $SO_3{}^{2-}$ though the following two step ionic dissociation,

$$SO_2(g) + H_2O\ \ SO_2\cdot H_2O \tag{7.170}$$

$$SO_2\cdot H_2O\ \ H^+ + HSO_3{}^- \tag{7.171}$$

$$HSO_3{}^-\ \ H^+ + SO_3{}^{2-} \tag{7.172}$$

Then,

$$K_{H,SO_2} = \frac{[SO_2\cdot H_2O]}{p_{SO_2}} \tag{7.173}$$

$$K_{S1} = \frac{[H^+][HSO_3^-]}{[SO_2\cdot H_2O]} \tag{7.174}$$

$$K_{S2} = \frac{[H^+][SO_3^{2-}]}{[HSO_3^-]}, \tag{7.175}$$

where $K_{H,SO2}$ is the Henry's law constant of SO_2 for water, K_{S1}, K_{S2} are the dissociation constants for reactions (152) and (153), is the partial pressure of SO_2

in the atmosphere. Each values at 298 K are given as $K_{H,SO2} = 1.4$ M atm^{-1} (Table 2.6), $K_{SI} = 1.3 \times 10^{-2}$ M, $K_{S2} = 6.6 \times 10^{-8}$ M (Seinfeld and Pandis 2006). As shown in Table 2.6, Henry's law constant of SO_2 is in the medium level, much smaller than those for nitric acid, aldehydes and organic acids, but much larger than for CO_2, NO and NO_2. Meanwhile, the dissociation constant K_{SI} for SO_2•H_2O to H^+ + HSO_3^- is considerably large but K_{S2} for HSO_3^- to H^+ + SO_3^{2-} is very small. Using these values of the constants, concentrations of SO_2•H_2O, HSO_3^-, and SO_3^{2-} are given by,

$$[SO_2 \cdot H_2O] = K_{H,SO_2} p_{SO_2} \tag{7.176}$$

$$\left[HSO_3^-\right] = \frac{K_{S1}[SO_2 \cdot H_2O]}{[H^+]} = \frac{K_{H,SO_2} K_{S1} p_{SO_2}}{[H^+]} \tag{7.177}$$

$$\left[SO_3^{2-}\right] = \frac{K_{S2}\left[HSO_3^-\right]}{[H^+]} = \frac{K_{H,SO_2} K_{S1} K_{S2} p_{SO_2}}{[H^+]^2}, \tag{7.178}$$

respectively.

Thus, SO_2 molecules taken up to aqueous solution exists in the form of three chemical species, SO_2•H_2O, HSO_3^- and SO_3^{2-}. Since they are all sulfur compounds with the oxidation number (valence) of 4, they are expressed as S(IV) in total,

$$\mathrm{S(IV)} = \mathrm{SO_2 \cdot H_2O} + \mathrm{HSO_3}^- + \mathrm{SO_3}^{2-}. \tag{7.179}$$

By using Eqs. (7.176), (7.177) and (7.178), the concentration of total sulfur S (IV) can be expressed as,

$$[S(IV)] = K_{H,SO_2} p_{SO_2} \left[1 + \frac{K_{S1}}{[H^+]} + \frac{K_{S1} K_{S2}}{[H^+]^2}\right]. \tag{7.180}$$

Here, if we define an effective Henry's law constant for S(IV),, as,

$$K^*_{H,S(IV)} = K_{H,SO_2} \left[1 + \frac{K_{S1}}{[H^+]} + \frac{K_{S1} K_{S2}}{[H^+]^2}\right], \tag{7.181}$$

the Henry equilibrium equation for the total amount of dissolving sulfur dioxide can be expressed as,

$$[\mathrm{S(IV)}] = K^*_{H,S(IV)} p_{SO_2}. \tag{7.182}$$

From these equations, mole concentrations (M) for SO_2•H_2O, HSO_3^-, SO_3^{2-} and S(IV) in the aqueous solution for 1 ppbv of SO_2 are shown in Fig. 7.28 as a

Fig. 7.28 Concentrations of $SO_2 \bullet H_2O$, HSO_3^-, SO_3^{2-} and S(IV) as a function of solution pH for gas phase SO_2 mixing ratio of 1 ppbv at 298 K (Seinfeld and Pandis 2006)

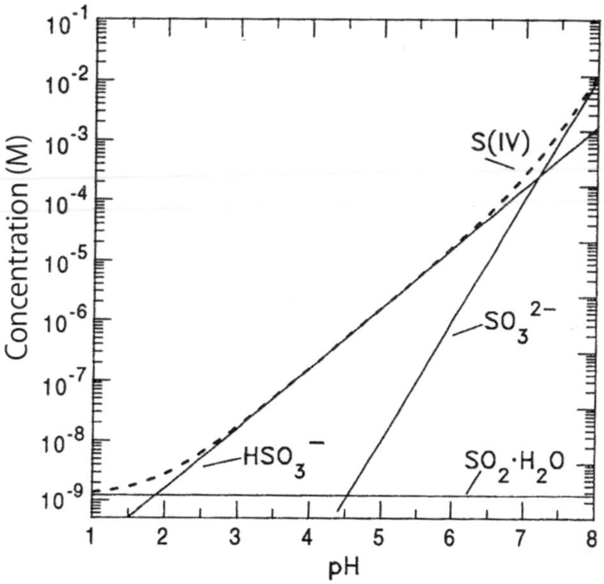

function of hydrogen ion exponent, pH (Seinfeld and Pandis 2006). As is well known, pH is defined by,

$$pH = -\log_{10}[H^+] \tag{7.183}$$

As seen in Fig. 7.28, the concentration of S(IV) in the aqueous solution increases rapidly with the increase of pH. In the aqueous solution of $pH = 2–7$, most of S (IV) is present as HSO_3^-. The concentrations of SO_3^{2-} and $SO_2 \bullet H_2O$ are dominant for $pH > 7$ and $pH < 2$, respectively, but the concentration of $SO_2 \bullet H_2O$ is kept constant by the Henry's equilibrium being independent on pH. The reason for the concentrations of HSO_3^- and SO_3^{2-} increase with pH is that H^+ is included in the right side of reactions (7.171), (7.172), and the equilibrium proceeds to the right side as the concentration of H^+ decreases with pH. This follows the Le Chatelier's principle, which states that in general when either of the concentration, temperature or pressure relevant to reactions is changed, the equilibrium shift to the direction to compensate the change. In Fig. 7.29, the pH dependence of the effective Henry's law constant for water defined by Eq. (7.181) is shown (Seinfeld and Pandis 2006). Reflecting the pH dependence of S(IV) concentration on pH as shown in Fig. 7.28, also increases rapidly with pH. For $pH \approx 1$ where only $SO_2 \bullet H_2O$ exists in the aqueous solution, the value of effective Henry's law constant is almost the same as given in Table 2.6, 1.36 M atm^{-1}, while it is as large as ~10^3 M atm^{-1} at $pH = 5$, and ~10^5 M atm^{-1} at $pH = 7$.

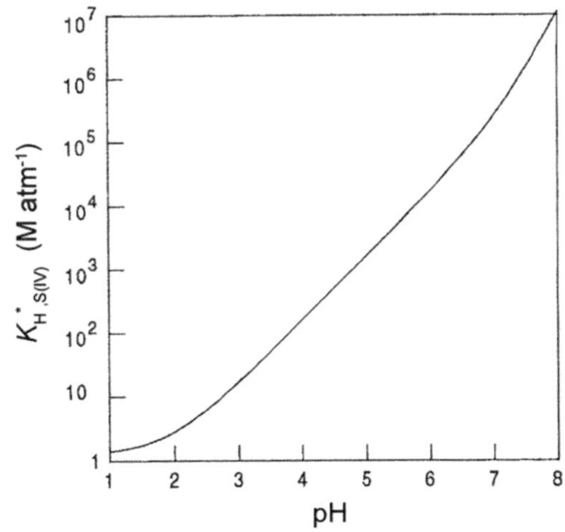

Fig. 7.29 Effective Henry's law constant $K_H{}^*{}_{,S(IV)}$ of SO_2 to water as a function of solution pH at 298 K(Seinfeld and Pandis 2006)

7.7.2.3 Oxidation Reaction of S(IV) in Aqueous Solution

The S(IV) taken up to water droplet is oxidized to form sulfate ion $SO_4{}^{2-}$, and the most important reactions for this oxidation process are the liquid phase reactions with H_2O_2 and O_3. The typical mixing ratio of H_2O_2 in the atmosphere is the order of ~1 ppbv which is much lower than O_3, but the Henry's law constant of H_2O_2 is 8.44×10^4 M atm^{-1} at 298 K, which is nearly 7 orders of magnitude lager than the value, 1.03×10^{-2} M atm^{-1}, for O_3 (see 2.6). Therefore, the H_2O_2 concentration in the aqueous solution is ~1×10^{-4} M, which is 5 orders of magnitude larger than O_3.

The dissociation constant for the ionic dissociation reaction of H_2O_2 dissolved in water,

$$H_2O_2 \quad H^+ + HO_2{}^- \tag{7.184}$$

is extremely small, 2.2×10^{-12} M atm^{-1} at 298 K (Pandis and Seinfeld 1989), so that H_2O_2 exists mainly as a molecule in the aqueous solution. It reacts with the main component of S(IV) in the solution, $HSO_3{}^-$,

$$HSO_3{}^- + H_2O_2 \quad SO_2OOH^- + H_2O \tag{7.185}$$

$$SO_2OOH^- + H^+ \rightarrow HSO_4{}^- + H^+ \tag{7.186}$$

to form hydrogen sulfate ion $HSO_4{}^-$ through peroxy sulfite ion, SO_2OOH^- (McArdle and Hoffmann 1983). Expressing $HSO_4{}^-$ as S(VI) since the valence of sulfur in $HSO_4{}^-$ is 6, the rate constant of the overall reaction of S(IV) and H_2O_2 can be defined as,

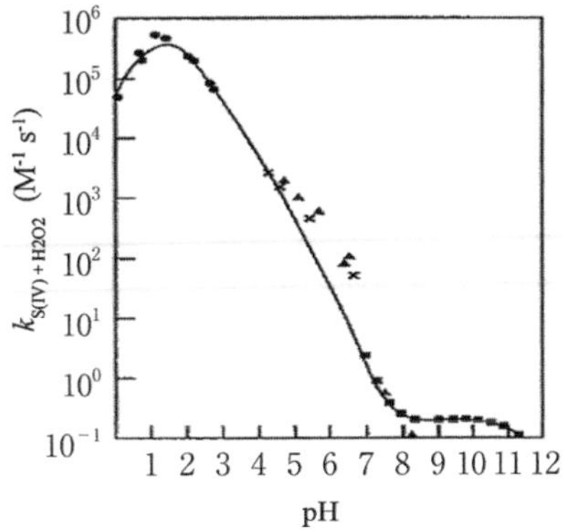

Fig. 7.30 Second order rate constant of the reaction of S (IV) and H_2O_2 in aqueous solution as a function of pH at 298 K: $d[S(IV)]/dt = k_{S(IV)+H2O2}\ [S(IV)][H_2O_2(aq)]$ (Martin and Damschen 1981) (Refer to the source for the literature of individual data in the figure)

$$d[S(VI)]/dt = k_{S(IV)+H2O2}[S(IV)][H_2O_2]. \tag{7.187}$$

Here, the value of $k_{S(IV)+H2O2}$ has been confirmed to be strongly dependent on pH, and the pH dependence of the experimental value of $k_{S(IV)+H2O2}$ is illustrated in Fig. 7.30 (Martin and Damschen 1981). As seen in Fig. 7.30, the rate constant of this reaction has a maximum value of ~5×10^5 M s^{-1} at around pH = 1.5, and decreases with the increase of pH. As seen in Fig. 7.28, the concentration of HSO_3^- increases with pH, which is in contrast to the pH dependence of $k_{S(IV)+H2O2}$ shown in Fig. 7.30. For this reason the oxidation rate of S(IV) by H_2O_2, a product of these two values, have a characteristic of almost constant being independent on pH.

The reaction of S(IV) and O_3 in aqueous solution is known to occur for either of the species, $SO_2{\bullet}H_2O$, HSO_3^- and SO_3^{2-},

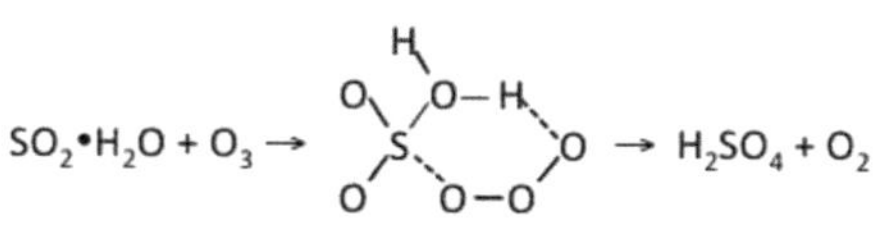

(7.188)

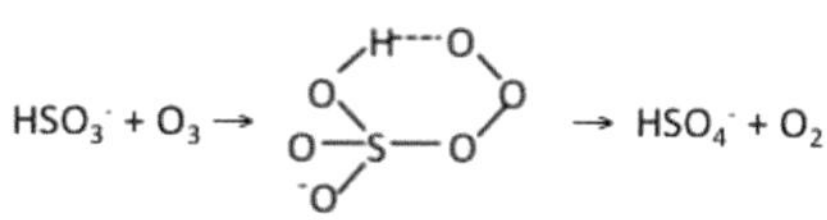

(7.189)

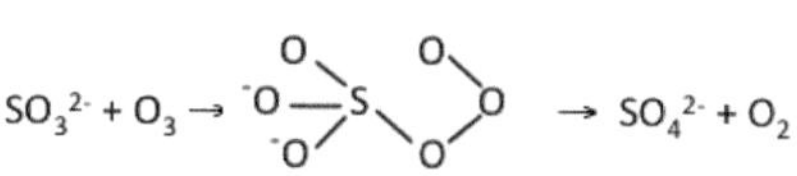

(7.190)

and forms sulfuric acid H_2SO_4, bisulfate ion HSO_4^-, and sulfate ion SO_4^{2-}, respectively, through each reaction complex (Hoffmann 1986). When we put each reaction rate constant as k_O, k_1, k_2, they have been obtained as $k_O = (2.4 \pm 1.1) \times 10^4$, $k_1 = (3.7 \pm 0.7) \times 10^5$ and $k_2 = (1.5 \pm 0.6) \times 10^9$ M s^{-1} (Hoffmann 1986). Since the reaction products, H_2SO_4, HSO_4^- and SO_4^{2-} are all in the oxidation state of 6, they are expressed as S(VI) in total,

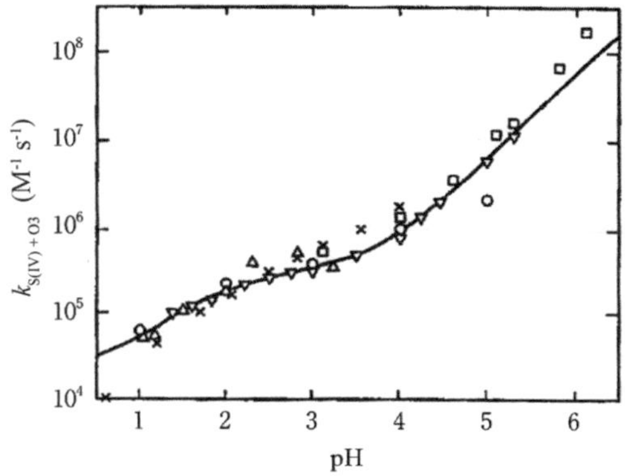

Fig. 7.31 Second order rate constant for the reaction of S(IV) and O_3 reaction as a function of solution pH at 298 K: $d[S(IV)]/dt = k_{S(IV)+O3}\ [S(IV)][O_3(aq)]$ (Seinfeld and Pandis 2006) (Refer to the source for the literature of individual data in the figure)

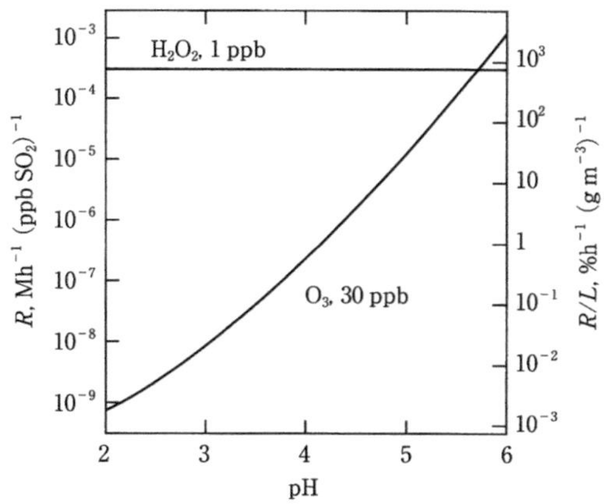

Fig. 7.32 Rate of aqueous phase oxidation of S(IV) by 1 ppb of H_2O_2 and 30 ppbv of O_3 as a function of solution pH at 298 K. R represents the aqueous phase reaction rate per ppb of gas phase SO_2. The R/L on the right axis represents reaction rate referred to gas phase SO_2 per unit cloud liquid water content (g m^{-3}) (Adapted from Seinfeld and Pandis 2006)

$$S(VI) = H_2SO_4 + HSO_4^- + SO_4^{2-} \quad (7.191)$$

and the reaction rate constant of overall reaction of S(IV) and O_3 can be defined by,

$$d[S(VI)]/dt = k_{S(IV)+O3}[S(IV)][O_3] \quad (7.192)$$

Figure 7.31 shows the pH dependence of $k_{S(IV)+O3}$ thus defined (Seinfeld and Pandis 2006). As shown in the figure, the rate constant of the reaction of S(IV) and O_3 increases with pH. Referring to the pH dependence of k_0, k_1 and k_2, the individual rate constant for reactions (7.188), (7.189) and (7.190), respectively, given by Hoffmann (1986), k_0 is independent on pH whereas k_1 and k_2 increases with pH so that the dependence of $k_{S(IV)+O3}$ shown in Fig. 7.30 reflects this feature. As seen in Fig. 7.28, since the concentrations of HSO_3^- and SO_3^- increase rapidly with pH, the oxidation reaction rate of O_3 by S(IV) multiplying them with the rate constants is thought to increase rapidly with pH.

Figure 7.32 depicts the pH dependence of the oxidation rate of SO_2 in aqueous solution per 1 ppbv of gaseous SO_2 using the values of above $k_{S(IV)+H2O2}$ and $k_{S(IV)}$

$_{+O3}$, taking the mixing ratios of H_2O_2 1 ppbv and O_3 30 ppbv in the atmosphere. As shown in the figure, the oxidation of SO_2 is mainly due to H_2O_2 for pH < 5, and the oxidation by O_3 predominates only for pH > 6.

7.7.2.4 pH of Clouds and Fog

The clouds and fog droplets are acidified by the uptake of SO_2 and other acidic species. Before discussing this, let us discuss the pH of water droplets by CO_2 in the atmosphere. The CO_2 molecule dissolved into water dissociates to bicarbonate ion, HCO_3^-, and carbonate ion, CO_3^{2-}, after forming a molecular complex between H_2O molecule, $CO_2 \bullet H_2O$, by two step dissociation,

$$CO_2(g) + H_2O \; CO_2 \cdot H_2O \tag{7.193}$$

$$CO_2 \cdot H_2O \; H^+ + HCO_3^- \tag{7.194}$$

$$HCO_3^- H^+ + CO_3^{2-}. \tag{7.195}$$

These processes are in parallel with the dissociation of SO_2 in the aqueous solution. Similar to the case of SO_2, the concentrations of $CO_2 \bullet H_2O$, HCO_3^-, and CO_3^{2-} can be written as,

$$[CO_2 \cdot H_2O] = K_{H,CO_2} p_{CO_2} \tag{7.196}$$

$$[HCO_3^-] = \frac{K_{C1}[CO_2 \cdot H_2O]}{[H^+]} = \frac{K_{H,CO_2} K_{C1} p_{CO_2}}{[H^+]} \tag{7.197}$$

$$[CO_3^{2-}] = \frac{K_{C2}[HCO_3^{2-}]}{[H^+]} = \frac{K_{H,CO_2} K_{C1} K_{C2} p_{CO_2}}{[H^+]^2} \tag{7.198}$$

by using Henry's law constants and dissociation constants. Here, $K_{H,CO2}$ is the Henry's law constant of CO_2 for water, K_{C1} and K_{C2} are the dissociation constants for reactions (7.194) and (7.195), respectively, and is the partial pressure (atm) of CO_2 in the atmosphere. Each value are given as $K_{H,CO2} = 3.4 \times 10^{-2}$ M atm^{-1} (Table 2.6), $K_{C1} = 4.3 \times 10^{-7}$ M, and $K_{C2} = 4.7 \times 10^{-11}$ M at 298 K (Seinfeld and Pandis 2006). The Henry's law constant of CO_2 is about one fourth of SO_2 and twice of NO_2. Meanwhile, K_{C1} and K_{C2} are 4 and 3 orders of magnitude smaller than K_{S1} and K_{S2}, respectively, showing that carboxylic acid is much weaker acid than sulfurous acid.

Assuming the partial pressure of CO_2 in the present day atmosphere is 4.0×10^{-4} atm (400 ppmv), $[H^+] = 2.42 \times 10^{-6}$ M is obtained from Eqs. (7.196) and (7.197) by putting $[CO_2 \bullet H_2O] = 1.36 \times 10^{-5}$ M and $[H^+] = [HCO_3^-]$, which implies pH = 5.62. Thus, the pH of water droplet equilibrated with atmospheric CO_2 is about 5.6. However, the pH of actual natural clouds, fog and rain droplet, is as low as pH = 4 over the open ocean due to the influence of H_2SO_4 due to DMS, and as high as pH = 7 when influenced by yellow sand or Saharan dust.

Assuming that the partial pressure of atmospheric SO_2 is 1.0×10^{-9} atm (1 ppbv), and only sulfurous acid is present in a water droplet, $[H^+] = 4.3 \times 10^{-6}$ M and pH = 5.4 is obtained by putting $[SO_2{\bullet}H_2O] = 1.4 \times 10^{-9}$ M and $[H^+] = [HSO_3^-]$ similarly to the case of CO_2. As seen in Fig. 7.29, under this condition, the concentration of SO_3^{2-} is two orders of magnitude lower than HSO_3^-, and can be neglected. Incidentally, pH = 4.9 is obtained for the SO_2 partial pressure is 1.0×10^{-8} atm (10 ppbv). On the other hand, if all of S(IV) in the droplet is converted to H_2SO_4, H_2SO_4 is strong acid and thus,

$$H_2SO_4 \ H^+ + HSO_4^- \qquad (7.199)$$

$$HSO_4^- \ H^+ + SO_4^{2-}. \qquad (7.200)$$

Since the dissociation constants of reactions (7.199) and (7.200) at 298 K are very large, 1000 M, and 1.0×10^{-2} M, respectively (Pandis and Seinfeld 1989), one molecule of S(IV) gives two H^+ ions. If we assume the partial pressure of SO_2 is 10^{-9} atm (1 ppbv), the molar concentration of SO_2 in the atmosphere is,

$$[SO_2(g)] = p_{SO2}/RT = 4.1 \times 10^{-11} \text{moles } l^{-1},$$

where $R = 0.082$ atm l K^{-1} mol^{-1} and $T = 298$ K. Assuming the cloud water content to be $L = 0.5 \times 10^{-6}$ l(aq)/l(gas), the concentration of H_2SO_4 in the water droplets when all of SO_2 is oxidized to H_2SO_4 is,

$$[H_2SO_4(aq)] = [SO_2(g)]/\left(0.5 \times 10^{-6}\right) = 0.82 \times 10^{-4} \text{mole } 1^{-1},$$

which gives $[H^+] = 1.6 \times 10^{-4}$ mole 1^{-1} and pH = 3.8. If the atmospheric mixing ratio of SO_2 is 10 ppbv, cloud water pH decreases to 2.8. In the actual field, clouds with average pH of 3.4–3.9 have been observed (Aneja and Kim 1993). As for fog droplets, taking $L = 0.05 \times 10^{-6}$ l(aq)/l(gas) (0.05 g m^{-3}), pH can be as low as 2.8 and 1.8 for 1 and 10 ppbv of SO_2., which is even lower than in the cloud droplets. Actually, fog with pH 2.2 has been observed in the field (Munger et al. 1983), and they are called "acid fog". In the case of rainwater, the water content is larger than clouds but has wide ranges. In general rainwater with pH-4–5 is often observed in the polluted atmosphere and they are called "acid rain". In the actual rainwater the process of uptake of sulfuric acid aerosols formed by the gas phase oxidation reaction of SO_2 also occurs in parallel.

Here, acidity of clouds, fog and rainwater have been discussed only from the viewpoint of sulfur chemistry. In actual clouds, fog, and rainwater further acidification by HNO_3 formed from NO_x, neutralization by NH_4^+ by the uptake of gaseous NH_3 and Ca^{2+} from soil dust occurs simultaneously, their pH is determined by the overall ion balance of these species.

References

Abbatt, J.P.D., Thomas, J.L., Abrahamsson, K., Boxe, C., Granfors, A., Jones, A.E., King, M.D., Saiz-Lopez, A., Shepson, P.B., Sodeau, J., Toohey, D.W., Toubin, C., Von Glasow, R., Wren, S.N., Yang, X.: Halogen activation via interactions with environmental ice and snow in the polar lower troposphere and other regions. Atmos. Chem. Phys. **12**, 6237–6271 (2012)

Adams, J.W., Holmes, N.S., Crowley, J.N.: Uptake and reaction of HOBr on frozen and dry NaCl/NaBr surfaces between 253 and 233 K. Atmos. Chem. Phys. **2**, 79–91 (2002)

Aiuppa, A., Baker, D.R., Webster, J.D.: Halogens in volcanic systems. Chem. Geol. **263**, 1–18 (2009)

Akimoto, H., Sakamaki, F.: Correlation of the ozone formation rates with hydroxyl radical concentrations in the propylene-nitrogen dioxide-dry air system: effective ozone formation rate constant. Environ. Sci. Technol. **17**, 94–99 (1983)

Akimoto, H., Hoshino, M., Inoue, G., Okuda, M.: Reaction mechanism of the photooxidation of the toluene-NO_2-O_2-N_2 system in the gas phase. Bull. Chem. Soc. Jpn. **51**, 2496–2502 (1978)

Akimoto, H., Hoshino, M., Inoue, G., Sakamaki, F., Washida, N., Okuda, M.: Design and characterization of the evacuable and bakable photochemical smog chamber. Environ. Sci. Technol. **13**, 471–475 (1979a)

Akimoto, H., Sakamaki, F., Hoshino, M., Inoue, G., Okuda, M.: Photochemical ozone formation in propylene-nitrogen oxide-dry air system. Environ. Sci. Technol. **13**, 53–58 (1979b)

Akimoto, H., Sakamaki, F., Inoue, G., Okuda, M.: Estimation of hydroxyl radical concentration in a propylene-NOx-dry air system. Environ. Sci. Technol. **14**, 93–97 (1980)

Akimoto, H., Kimitaka, K., Nakazawa, T., Washida, N. (eds.): Chemistry of the Tropospheric Atmosphere and Global Environment, Gakkai Shuppan Center, Toyko (2002). (in Japanese)

Albu, M., Barnes, I., Becker, K.H., Patroescu-Klotz, I., Mocanu, R., Benter, T.: Rate coefficients for the gas-phase reaction of OH radicals with dimethyl sulfide: Temperature and O_2 partial pressure dependence. Phys. Chem. Chem. Phys. **8**, 728–736 (2006)

Alicke, B., Hebestreit, K., Stutz, J., Platt, U.: Iodine oxide in the marine boundary layer. Nature **397**, 572–573 (1999)

Anastasio, C., Mozurkewich, M.: Laboratory studies of bromide oxidation in the presence of ozone: evidence for glass-surface mediated reaction. J. Atmos. Chem. **41**, 135–162 (2002)

Andino, J.M., Smith, J.N., Flagan, R.C., Goddard, W.A., Seinfeld, J.H.: Mechanism of atmospheric photooxidation of aromatics: a theoretical study. J. Phys. Chem. **100**, 10967–10980 (1996)

Aneja, V.P., Kim, D.-S.: Chemical dynamics of clouds at Mt. Mitchell, North Carolina. Air. Waste **43**, 1074–1083 (1993)

Arey, J., Aschmann, S.M., Kwok, E.S.C., Atkinson, R.: Alkyl nitrate, hydroxyalkyl nitrate, and hydroxycarbonyl formation from the NOx – air photooxidations of $C_5 - C_8$ n-alkanes. J. Phys. Chem. A **105**, 1020–1027 (2001)

Arnold, S.R., Chipperfield, M.P., Blitz, M.A., Heard, D.E., Pilling, M.J.: Photodissociation of acetone: atmospheric implications of temperature-dependent quantum yields. Geophys. Res. Lett. **31**, L07110 (2004). doi:10.1029/2003GL019099

Atkinson, R.: Kinetics and mechanisms of the gas-phase reactions of the hydroxyl radical with organic compounds. J. Phys. Chem. Ref. Data **1**, 1–246 (1989)

Atkinson, R.: Gas-phase tropospheric chemistry of volatile organic compounds: 1. Alkanes and alkenes. J. Phys. Chem. Ref. Data **26**, 215–290 (1997a)

Atkinson, R.: Atmospheric reactions of alkoxy and β-hydroxyalkoxy radicals. Int. J. Chem. Kinet. **29**, 99–111 (1997b)

Atkinson, R., Arey, J.: Atmospheric degradation of volatile organic compounds. Chem. Rev. **103**, 4605–4638 (2003)

Atkinson, R., Baulch, D.L., Cox, R.A., Hampson, R.F., Kerr, J.A., Rossi, M.J., Troe, J.: Evaluated kinetic, photochemical and heterogeneous data for atmospheric chemistry: supplement

V. IUPAC Subcommittee on Gas Kinetic Data Evaluation for Atmospheric Chemistry. J. Phys. Chem. Ref. Data **26**, 521–1011 (1997)

Atkinson, R., Baulch, D.L., Cox, R.A., Crowley, J.N., Hampson, R.F., Hynes, R.G., Jenkin, M.E., Rossi, M.J., Troe, J.: Evaluated kinetic and photochemical data for atmospheric chemistry: volume I – gas phase reactions of Ox, HOx, NOx, and SOx species. Atmos. Chem. Phys. **4**, 1461–1738 (2004)

Atkinson, R., Baulch, D.L., Cox, R.A., Crowley, J.N., Hampson, R.F., Hynes, R.G., Jenkin, M.E., Rossi, M.J., Troe, J.: Evaluated kinetic and photochemical data for atmospheric chemistry: volume II – gas phase reactions of organic species. Atmos. Chem. Phys. **6**, 3625–4055 (2006)

Atkinson, R., Baulch, D.L., Cox, R.A., Crowley, J.N., Hampson, R.F., Hynes, R.G., Jenkin, M.E., Rossi, M.J., Troe, J.: Evaluated kinetic and photochemical data for atmospheric chemistry: volume III – gas phase reactions of inorganic halogens. Atmos. Chem. Phys. **7**, 981–1191 (2007)

Baltaretu, C.O., Lichtman, E.I., Hadler, A.B., Elrod, M.J.: Primary atmospheric oxidation mechanism for toluene. J. Phys. Chem. A **113**, 221–230 (2009)

Bandow, H., Okuda, M., Akimoto, H.: Mechanism of the gas-phase reactions of C_3H_6 and NO_3 radicals. J. Phys. Chem. **84**, 3604–3608 (1980)

Barnes, I., Becker, K.H. Mahalopoulos, N.: An FTIR product study of the photooxidation of dimethyl disulfide. J. Atmos. Chem. **18**, 267–289 (1994)

Barnes, I., Bastian, V., Becker, K.H., Tong, Z.: Kinetics and products of the reactions of NO_3 with monoalkenes, dialkenes, and monoterpenes. J. Phys. Chem. **94**, 2413–2419 (1990)

Barnes, I., Hjorth, J., Mihalopoulos, N.: Dimethyl Sulfide and dimethyl sulfoxide and their oxidation in the atmosphere. Chem. Rev. **106**, 940–975 (2006)

Barrie, L.A., Bottenheim, J.W., Schnell, R.C., Crutzen, P.J., Rasmussen, R.A.: Ozone destruction and photochemical reactions at polar sunrise in the lower Arctic atmosphere. Nature **334**, 138–141 (1988)

Bartolotti, L.J., Edney, E.O.: Density functional theory derived intermediates from the OH initiated atmospheric oxidation of toluene. Chem. Phys. Lett. **245**, 119–122 (1995)

Bates, T.S., Lamb, B.K., Guenther, A.B., Dignon, J., Stoiber, R.E.: Sulfur emissions to the atmosphere from natural sources. J. Atmos. Chem. **14**, 315–337 (1992)

Beaver, M.R., Clair, J.M.S., Paulot, F., Spencer, K.M., Crounse, J.D., LaFranchi, B.W., Min, K.E., Pusede, S.E., Wooldridge, P.J., Schade, G.W., Park, C., Cohen, R.C., Wennberg, P.O.: Importance of biogenic precursors to the budget of organic nitrates: observations of multifunctional organic nitrates by CIMS and TD-LIF during BEARPEX 2009. Atmos. Chem. Phys. **12**, 5773–5785 (2012)

Becker, K.H.: The European Photoreactor EUPHORE: Design and Technical Development of the European Photoreactor and First Experimental Results. Final Report of the EC-Project Contract EV5V-CT92-0059, BUGH Wuppertal, Wuppertal (1996)

Birdsall, A.W., Dreoni, J., Elrod, M.J.: Investigation of the role of bicyclic peroxy radicals in the oxidation mechanism of toluene. J. Phys. Chem. A **114**, 10655–10663 (2010)

Blitz, M.A., Heard, D.E., Pilling, M.J., Arnold, S.R., Chipperfield, M.P.: Pressure and temperature-dependent quantum yields for the photodissociation of acetone between 279 and 327.5 nm. Geophys. Res. Lett. **31**, L06111 (2004). doi:10.1029/2003GL018793

Bloss, C., Wagner, V., Jenkin, M.E., Volkamer, R., Bloss, W.J., Lee, J.D., Heard, D.E., Wirtz, K., Martin-Reviejo, M., Rea, G., Wenger, J.C., Pilling, M.J.: Development of a detailed chemical mechanism (MCMv3.1) for the atmospheric oxidation of aromatic hydrocarbons. Atmos. Chem. Phys. **5**, 641–664 (2005)

Bobrowski, N., Hönninger, G., Galle, B., Platt, U.: Detection of bromine monoxide in a volcanic plume. Nature **423**, 273–276 (2003)

Bobrowski, N., von Glasow, R., Aiuppa, A., Inguaggiato, S., Louban, I., Ibrahim, O.W., Platt, U.: Reactive halogen chemistry in volcanic plumes. J. Geophys. Res. **112**, D06311 (2007). doi:10.1029/2006JD007206

Bohn, B.: Formation of peroxy radicals from OH – toluene adducts and O_2. J. Phys. Chem. A **105**, 6092–6101 (2001)

Bohn, B., Zetzsch, C.: Gas-phase reaction of the OH–benzene adduct with O_2: reversibility and secondary formation of HO_2. Phys. Chem. Chem. Phys. **1**, 5097–5107 (1999)

Brasseur, G.P., Orlanddo, J.L., Tyndall, G.S. (eds.): Atmospheric Chemistry and Global Change. Oxford University Press, New York (1999)

Brasseur, G.P., Prinn, R.G., Pszenny, A.P. (eds.): Atmospheric chemistry in a changing world, an integration and synthesis of a decade of tropospheric chemistry research. The International Global Atmospheric Chemistry Project of the International Geosphere-Biosphere Programme. Springer-Verlag, New York (2003)

Brune, W.: OH and HO_2: sources, interactions with nitrogen oxides, and ozone production. IGAC newsletter, No. 21. http://www.igacproject.org/sites/all/themes/bluemasters/images/NewsletterArchives/Issue_21_Sep_2000.pdf (2000)

Buys, Z., Brough, N., Huey, L.G., Tanner, D.J., von Glasow, R., Jones, A.E.: High temporal resolution Br_2, BrCl and BrO observations in coastal Antarctica. Atmos. Chem. Phys. **13**, 1329–1343 (2013)

Calvert, J.G., Atkinson, R., Kerr, J.A., Madronich, S., Moortgat, G.K., Wallington, T.J., Yarwood, G.: The Mechanisms of Atmospheric Oxidation of the Alkenes. Oxford University Press, New York (2000)

Calvert, J.G., Atkinson, R., Becker, K.H., Kamens, R.M., Seinfeld, J.H., Wallington, T.J., Yarwood, G.: The Mechanisms of Atmospheric Oxidation of the Aromatic Hydrocarbons. Oxford University Press, New York (2002)

Calvert, J.G., Derwent, R.G., Orlando, J.J., Tyndall, G.S., Wallington, T.J.: The Mechanisms of Atmospheric Oxidation of the Alkanes. Oxford University Press, New York (2008)

Calvert, J., Mellouki, A., Orlando, J., Pilling, M.J., Wallington, T.J.: The Mechanisms of Atmospheric Oxidation of the Oxygenates. Oxford University Press, New York (2011)

Carpenter, L.J.: Iodine in the marine boundary layer. Chem. Rev. **103**, 4953–4962 (2003)

Cartas-Rosado, R., Castro, M.: Theoretical study of reaction mechanisms of OH radical with toluene 1,2-epoxide/2-methyloxepin. J. Phys. Chem. A **111**, 13088–13098 (2007)

Cartas-Rosado, R., Alvarez-Idaboy, J.R., Galano-Jimenez, A., Vivier-Bunge, A.: A theoretical inveistigation of the mecjanism of the NO_3 addition to alkenes. J. Mol. Struct (Theochem) **684**, 51–59 (2004)

Chapman, S.: A theory of upper atmospheric ozone. Mem. R. Meteorol. Soc. **3**, 103–125 (1930)

Charlson, R.J., Lovelock, J.E., Andreae, M.O., Warren, S.G.: Oceanic phytoplankton, atmospheric sulphur, cloud albedo and climate. Nature **326**, 655–661 (1987)

Chen, S., Ren, X., Mao, J., Chen, Z., Brune, W.H., Lefer, B., Rappenglueck, B., Flynn, J., Olson, J., Crawford, J.H.: A comparison of chemical mechanisms based on TRAMP-2006 field data. Atmos. Environ. **44**, 4116–4125 (2010)

Chin, M., Davis, D.D.: Global sources and sinks of OCS and CS2 and their distributions. Glob. Biogeochem. Cycles **7**, 321–337 (1993)

Choi, S., Wang, Y., Salawitch, R.J., Canty, T., Joiner, J., Zeng, T., Kurosu, T.P., Chance, K., Richter, A., Huey, L.G., Liao, J., Neuman, J.A., Nowak, J.B., Dibb, J.E., Weinheimer, A.J., Diskin, G., Ryerson, T.B., da Silva, A., Curry, J., Kinnison, D., Tilmes, S., Levelt, P.F.: Analysis of satellite-derived Arctic tropospheric BrO columns in conjunction with aircraft measurements during ARCTAS and ARCPAC. Atmos. Chem. Phys. **12**, 1255–1285 (2012)

Claeys, M., Graham, B., Vas, G., Wang, W., Vermeylen, R., Pashynska, V., Cafmeyer, J., Guyon, P., Andreae, M.O., Artaxo, P., Maenhaut, W.: Formation of secondary organic aerosols through photooxidation of isoprene. Science **303**, 1173–1176 (2004)

Cowling, E.B.: Acid precipitation in historical perspective. Environ. Sci. Technol. **16**, 110A–123A (1982)

Crutzen, P.: A discussion of the chemistry of some minor constituents in the stratosphere and troposphere. Pure Appl. Geophys. **106–108**, 1385–1399 (1973)

Di Carlo, P., Brune, W.H., Martinez, M., Harder, H., Lesher, R., Ren, X.R., Thornberry, T., Carroll, M.A., Young, V., Shepson, P.B., Riemer, D., Apel, E., Campbell, C.: Missing OH reactivity in a forest: evidence for unknown reactive biogenic VOCs. Science **304**, 722–725 (2004)

Dickerson, R.R., Rhoads, K.P., Carsey, T.P., Oltmans, S.J., Burrows, J.P., Crutzen, P.J.: Ozone in the remote marine boundary layer: a possible role for halogens. J. Geophys. Res. **104**, 21385–21395 (1999)

Dillon, T.J., Crowley, J.N.: Direct detection of OH formation in the reactions of HO_2 with $CH_3C(O)O_2$ and other substituted peroxy radicals. Atmos. Chem. Phys. **8**, 4877–4889 (2008)

Dodge, M.C.: Combined use of modeling techniques and smog chamber data to derive ozone precursor relationships. In: Dimitriades, B. (ed.) Proceedings of the International Conference on Photochemical Oxidant Pollution and Its Control, EPA-600/3-77-001b, vol. II, pp. 881–889, US Environmental Protection Agency, Research Triangle Park, NC (1977)

Donahue, N.M., Drozd, G.T., Epstein, S.A., Prestoa, A.A., Kroll, J.H.: Adventures in ozoneland: down the rabbit-hole. Phys. Chem. Chem. Phys. **13**, 10848–10857 (2011)

Dusanter, S., Vimal, D., Stevens, P.S., Volkamer, R., Molina, L.T., Baker, A., Meinardi, S., Blake, D., Sheehy, P., Merten, A., Zhang, R., Zheng, J., Fortner, E.C., Junkermann, W., Dubey, M., Rahn, T., Eichinger, B., Lewandowski, P., Prueger, J., Holder, H.: Measurements of OH and HO_2 concentrations during the MCMA-2006 field campaign – part 2: model comparison and radical budget. Atmos. Chem. Phys. **9**, 6655 (2009)

Emmerson, K.M., Carslaw, N.: Night-time radical chemistry during the TORCH campaign. Atmos. Environ. **43**, 3220–3226 (2009)

Faloona, I., Tan, D., Brune, W., Hurst, J., Barket Jr., D., Couch, T.L., Shepson, P., Apel, E., Riemer, D., Thornberry, T., Carroll, M.A., Sillman, S., Keeler, G.J., Sagady, J., Hooper, D., Paterson, K.: Nighttime observations of anomalously high levels of hydroxyl radicals above a deciduous forest canopy. J. Geophys. Res. **106**, 24315–24333 (2001)

Fan, J., Zhang, R.: Atmospheric oxidation mechanism of isoprene. Environ. Chem. **1**, 140–149 (2004)

Finlayson-Pitts, B.J.: The tropospheric chemistry of sea salt: a molecular-level view of the chemistry of NaCl and NaBr. Chem. Rev. **103**, 4801–4822 (2003)

Finlayson-Pitts, B.J., Pitts Jr., J.N.: Chemistry of the Upper and Lower Atmosphere. Academic Press, San Diego (2000)

Foster, K.L., Plastridge, R.A., Bottenheim, J.W., Shepson, P.B., Finlayson-Pitts, B.J., Spicer, C. W.: The role of Br_2 and BrCl in surface ozone destruction at polar sunrise. Science **291**, 471–474 (2001)

Fox, D.L., Sickles, J.E., Kuhlman, M.R., Reist, P.C., Wilson, W.E.: Dedign and operating parameters for a large ambient aerosol chamber. J. Air Pollut. Control Assoc. **25**, 1049–1053 (1975)

Frieß, U., Wagner, T., Pundt, I., Pfeilsticker, K., Platt, U.: Spectroscopic measurements of tropospheric iodine oxide at Neumayer Station, Antarctica. Geophys. Res. Lett. **28**, 1941–1944 (2001)

Fuchs, H., Bohn, B., Hofzumahaus, A., Holland, F., Lu, K.D., Nehr, S., Rohrer, F., Wahner, A.: Detection of HO_2 by laser-induced fluorescence: calibration and interferences from RO_2 radicals. Atmos. Meas. Tech. **4**, 1209–1225 (2011)

Galano, A., Ruiz-Suárez, L.G., Vivier-Bunge, A.: On the mechanism of the OH initiated oxidation of acetylene in the presence of O_2 and NOx. Theor. Chem. Accounts **121**, 219–225 (2008)

García-Cruz, I., Castro, M., Vivier–Bunge, A.: DFT and MP2 molecular orbital determination of OH–toluene–O_2 isomeric structures in the atmospheric oxidation of toluene. J. Comput. Chem. **21**, 716–730 (2000)

Geyer, A., Bachmann, K., Hofzumahaus, A., Holland, F., Konrad, S., Klupfel, T., Patz, H.W., Perner, D., Mihelcic, D., Schafer, H.J., Volz-Thomas, A., Platt, U.: Nighttime formation of peroxy and hydroxyl radicals during the BERLIOZ campaign: observations and modeling studies. J. Geophys. Res. **108**, 8249 (2003). doi:10.1029/2001JD000656

Gillotay, D., Jenouvrier, A., Coquart, B., Merienne, M.F., Simon, P.C.: Ultraviolet absorption cross-sections of bromoform in the temperature range 295-240K. Planet. Space. Sci. **37**, 1127–1140 (1989)

Goldstein, A.H., Galbally, A.E.: Known and unexplored organic constituents in the earth's atmosphere. Environ. Sci. Technol. **41**, 1514–1521 (2007)

Gómez Martín, J.C., Ashworth, S.H., Mahajan, A.S., Plane, J.M.C.: Photochemistry of OIO: laboratory study and atmospheric implications. Geophys. Res. Lett. **36**, L09802 (2009). doi:10.1029/2009GL037642

Gong, H., Matsunaga, A., Ziemann, P.J.: Products and mechanism of secondary organic aerosol formation from reactions of linear alkenes with NO_3 radicals. J. Phys. Chem. A **109**, 4312–4324 (2005)

Grebenkin, S.Y., Krasnoperov, L.N.: Kinetics and thermochemistry of the hydroxycyclohexadienyl radical reaction with O_2 : $C_6H_6OH + O_2 \rightleftharpoons C_6H_6(OH)OO$. J. Phys. Chem. A **108**, 1953–1963 (2004)

Gross, A., Barnes, I., Sørensen, R.M., Kongsted, J., Mikkelsen, K.V.: A theoretical study of the reaction between $CH_3S(OH)CH_3$ and O_2. J. Phys. Chem. A **108**, 8659–8671 (2004)

Guenther, A., Karl, T., Harley, P., Wiedinmyer, C., Palmer, P.I., Geron, C.: Estimates of global terrestrial isoprene emissions using MEGAN (Model of Emissions of Gases and Aerosols from Nature). Atmos. Chem. Phys. **6**, 3181–3210 (2006)

Hatakeyama, S., Okuda, M., Akimoto, H.: Formation of sulfur dioxide and methanesulfonic acid in the photooxidation of dimethyl sulfide in the air. Geophys. Res. Lett. **9**, 583–586 (1982)

Hatakeyama, S., Washida, N., Akimoto, H.: Rate constants and mechanisms for the reaction of hydroxyl (OD) radicals with acetylene, propyne, and 2-butyne in air at 297 ± 2 K. J. Phys. Chem. **90**, 173–178 (1986)

Hatakeyama, S., Lai, H., Murano, K.: Formation of 2-hydroxyethyl hydroperoxide in an OH-initiated reaction of ethylene in air in the absence of NO. Environ. Sci. Technol. **29**, 833–835 (1995)

Hausmann, M., Platt, U.: Spectroscopic measurement of bromine oxide and ozone in the high Arctic during Polar Sunrise Experiment 1992. J. Geophys. Res. **99**, 25399–25413 (1994)

Haynes, W.M. (ed.): CRC Handbook of Chemistry and Physics, 93th ed. CRC Press, Boca Raton (2012–2013)

Heicklen, J.: In: Tuesday, C.S. (ed.) Chemical Reactions in Urban Atmospheres, pp. 55–59, American Elsevier, New York (1971)

Heicklen, J., Westberg, K., Cohen, N.: The Conversion of NO to NO_2 in Polluted Atmospheres, Publication No. 115–69. Center for Air Environment Studies, University Park, Pennsylvania (1969)

Hirokawa, J., Onaka, K., Kajii, Y., Akimoto, H.: Heterogeneous processes involving sodium halide particles and ozone: molecular bromine release in the marine boundary layer in the absence of nitrogen oxides. Geophys. Res. Lett. **25**, 2449–2452 (1998)

Hoffmann, M.R.: On the kinetics and mechanism of oxidation of aquated sulfur dioxide by ozone. Atmos. Environ. **20**, 1145–1154 (1986)

Hofzumahaus, A., Rohrer, F., Lu, K.D., Bohn, B., Brauers, T., Chang, C.C., Fuchs, H., Holland, F., Kita, K., Kondo, Y., Li, X., Lou, S.R., Shao, M., Zeng, L.M., Wahner, A., Zhang, Y.H.: Amplified trace gas removal in the troposphere. Science **324**, 1702–1704 (2009)

Hönninger, G., Platt, U.: Observations of BrO and its vertical distribution during surface ozone depletion at Alert. Atmos. Environ. **36**, 2481–2489 (2002)

Hönninger, G., Bobrowski, N., Palenque, E.R., Torrez, R., Platt, U.: Reactive bromine and sulfur emissions at Salar de Uyuni, Bolivia. Geophys. Res. Lett. **31**, L04101 (2004). doi:10.1029/2003GL018818

Ingham, T., Goddard, A., Whalley, L.K., Furneaux, K.L., Edwards, P.M., Seal, C.P., Self, D.E., Johnson, G.P., Read, K.A., Lee, J.D., Heard, D.E.: A flow-tube based laser-induced fluorescence instrument to measure OH reactivity in the troposphere. Atmos. Meas. Tech. **2**, 465–477 (2009)

IPCC, Climate Change: The Physical Science Basis, Contribution of working Group I to the Fifth Assessment Report of the Intergovernmental Oanel on Climate Change. Web page, http://www.ipcc.ch/teport/ar5/wg1/ (2013)

Jacob, D.: Introduction to Atmospheric Chemistry. Princeton University Press, Princeton (1999)

Jenkin, M.E., Saunders, S.M., Wagner, V., Pilling, M.J.: Protocol for the development of the Master Chemical Mechanism, MCM v3 (Part B): tropospheric degradation of aromatic volatile organic compounds. Atmos. Chem. Phys. **3**, 181–193 (2003)

Jobson, B.T., Niki, H., Yokouchi, Y., Bottenheim, J., Hopper, F., Leaitch, R.: Measurements of C2–C6 hydrocarbons during the Polar Sunrise 1992 Experiment: evidence for Cl atom and Br atom chemistry. J. Geophys. Res. **99**, 25355–25368 (1994)

Johnson, D., Marston, G.: The gas-phase ozonolysis of unsaturated volatile organic compounds in the troposphere. Chem. Soc. Rev. **37**, 699–716 (2008)

Johnson, D., Raoult, S., Rayez, M.-T., Rayez, J.-C., Lesclaux, R.: An experimental and theoretical investigation of the gas-phase benzene–OH radical adduct + O_2 reaction. Phys. Chem. Chem. Phys. **4**, 4678–4686 (2002)

Kanaya, Y., Akimoto, H.: Direct measurement of HOx radicals in the marine boundary layer: testing the current tropospheric chemistry mechanism. Chem. Rec. **2**, 199–211 (2002)

Kanaya, Y., Sadanaga, Y., Matsumoto, J., Sharma, U., Hirokawa, J., Kajii, Y., Akimoto, H.: Daytime HO_2 concentrations at Oki Island, Japan, in summer 1998: comparison between measurement and theory. J. Geophys. Res. **105**, 24205–24222 (2000)

Kanaya, Y., Yokouchi, Y., Matsumoto, J., Nakamura, K., Kato, S., Tanimoto, H., Furutani, H., Toyota, K., Akimoto, H.: Implication of iodine chemistry for daytime HO_2 levels at Rishiri Island. Geophys. Res. Lett. **29**(8) (2002). doi:10.1029/2001GL014061

Kanaya, Y., Cao, R., Kato, S., Miyakawa, Y., Kajii, Y., Tanimoto, H., Yokouchi, Y., Mochida, M., Kawamura, K., Akimoto, H.: Chemistry of OH and HO_2 radicals observed at Rishiri Island, Japan, in September 2003: missing daytime sink of HO_2 and positive nighttime correlations with monoterpenes. J. Geophys. Res. **112**, D11308 (2007a). doi:10.1029/2006JD007987

Kanaya, Y., Cao, R., Akimoto, H., Fukuda, M., Komazaki, Y., Yokouchi, Y., Koike, M., Tanimoto, H., Takegawa, N., Kondo, Y.: Urban photochemistry in central Tokyo: 1. Observed and modeled OH and HO_2 radical concentrations during the winter and summer of 2004. J. Geophys. Res. **112**, D21312 (2007b)

Kanaya, Y., Fukuda, M., Akimoto, H., Takegawa, N., Komazaki, Y., Yokouchi, Y., Koike, M., Kondo, Y.: Urban photochemistry in central Tokyo: 2. Rates and regimes of oxidant ($O_3 + NO_2$) production. J. Geophys. Res. **113**, D06301 (2008)

Karl, M., Brauers, T., Dorn, H.-P., Holland, F., Komenda, M., Poppe, D., Rohrer, F., Rupp, L., Schaub, A., Wahner, A.: Kinetic study of the OH-isoprene and O_3-isoprene reaction in the atmosphere simulation chamber, SAPHIR. Geophys. Res. Lett. **31**, L05117 (2004). doi:10.1029/2003GL019189

Karl, M., Dorn, H.-P., Holland, F., Koppmann, R., Poppe, D., Rupp, L., Schaub, A., Wahner, A.: Product study of the reaction of OH radicals with isoprene in the atmosphere simulation chamber SAPHIR. J. Atmos. Chem. **55**, 167–187 (2006)

King, M.D., Canosa-Mas, C.E., Wayne, R.P.: Frontier molecular orbital correlations for predicting rate constants between alkenes and the tropospheric oxidants NO_3, OH and O_3. Phys. Chem. Chem. Phys. **1**, 2231–2238 (1999)

Klotz, B., Barnes, I., Becker, K.H., Golding, B.T.: Atmospheric chemistry of benzeneoxide/oxepin. J. Chem. Soc. Faraday Trans. **93**, 1507–1516 (1997)

Klotz, B., Sørensen, S., Barnes, I., Becker, K.H., Etzkorn, T., Volkamer, R., Platt, U., Wirtz, K., Martín-Reviejo, M.: Atmospheric oxidation of toluene in a large-volume outdoor photoreactor: in situ determination of ring-retaining product yields. J. Phys. Chem. A **102**, 10289–10299 (1998)

Klotz, B., Barne, I., Golding, B.T., Becker, K.H.: Atmospheric chemistry of toluene-1,2-oxide/2-methyloxepin. Phys. Chem. Chem. Phys. **2**, 227–235 (2000)

Knipping, E.M., Lakin, M.J., Foster, K.L., Jungwirth, P., Tobias, D.J., Gerber, R.B., Dabdub, D., Finlayson-Pitts, B.J.: Experiments and simulations of ion-enhanced interfacial chemistry on aqueous NaCl aerosols. Science **288**, 301–306 (2000)

Knispel, R., Koch, R., Siese, M., Zetzsch, C.: Adduct formation of OH radicals with benzene, toluene and phenol and consecutive reactions of the adducts with nitrogen oxide and oxygen. Ber. Bunsenges. Phys. Chem. **94**, 1375–1379 (1990)

Kreher, K., Johnston, P.V., Wood, S.W., Nardi, B., Platt, U.: Ground-based measurements of tropospheric and stratospheric BrO at Arrival Heights, Antarctica. Geophys. Res. Lett. **23**, 3021–3024 (1997)

Kroll, J.H., Clarke, J.S., Donahue, N.M., Anderson, J.G., Demerjian, K.L.: Mechanism of HOx formation in the gas-phase ozone – alkene reaction. 1. Direct, pressure-dependent measurements of prompt OH yields. J. Phys. Chem. A **105**, 1554–1560 (2001a)

Kroll, J.H., Sahay, S.R., Anderson, J.G.: Mechanism of HOx formation in the gas-phase ozone-alkene reaction. 2. Prompt versus thermal dissociation of carbonyl oxides to form OH. J. Phys. Chem. A **105**, 4446–4457 (2001b)

Kroll, J.H., Hanisco, T.F., Donahue, N.M., Demerjian, K.L., Anderson, J.G.: Accurate, direct measurements of OH yields from gas-phase ozone-alkene reactions using an in situ LIF instrument. Geophys. Res. Lett. **28**, 3863–3866 (2001c)

Kroll, J.H., Ng, N.L., Murphy, S.M., Flagan, R.C., Seinfeld, J.H.: Secondary organic aerosol formation from isoprene photooxidation. Environ. Sci. Technol. **40**, 1869–1877 (2006)

Kwok, E.S.C., Aschmann, S.M., Arey, J., Atkinson, R.: Product formation from the reaction of the NO_3 radical with isoprene and rate constants for the reactions of methacrolein and methyl vinyl ketone with the NO_3 radical. Int. J. Chem. Kinet. **28**, 925–934 (1996a)

Kwok, E.S.C., Atkinson, R., Arey, J.: Isomerization of β-hydroxyalkoxy radicals formed from the OH radical-initiated reactions of $C_4 - C_8$ 1-alkenes. Environ. Sci. Technol. **30**, 1048–1052 (1996b)

Lawler, M.J., Sander, R., Carpenter, L.J., Lee, J.D., von Glasow, R., Sommariva, R., Saltzman, E. S.: HOCl and Cl_2 observations in marine air. Atmos. Chem. Phys. **11**, 7617–7628 (2011)

Lee, C., Kim, Y.J., Tanimoto, H., Bobrowski, N., Platt, U., Mori, T., Yamamoto, K., Hong, C.S.: High ClO and ozone depletion observed in the plume of Sakurajima volcano, Japan. Geophys. Res. Lett. **32**, L21809 (2005). doi:10.1029/2005GL023785

Lehrer, E., Wagenbach, D., Platt, U.: Aerosol chemical composition during tropospheric ozone depletion at Ny Ålesund/Svalbard. Tellus B **49**, 486–495 (1997)

Lelieveld, J., Butler, T.M., Crowley, J.N., Dillon, T.J., Fischer, H., Ganzeveld, L., Harder, H., Lawrence, M.G., Martinez, M., Taraborrelli, D., Williams, J.: Atmospheric oxidation capacity sustained by a tropical forest. Nature **452**, 737–740 (2008)

Levy II, H.: Normal atmosphere: large radical and formaldehyde concentrations predicted. Science **173**, 141–143 (1971)

Lewis, A.C., Carslaw, N., Marriott, P.J., Kinghorn, R.M., Morriso, P., Lee, A.L., Bartle, K.D., Pilling, M.J.: A larger pool of ozone-forming carbon compounds in urban atmospheres. Nature **405**, 778–781 (2000)

Liao, J., Huey, L.G., Tanner, D.J., Flocke, F.M., Orlando, J.J., Neuman, J.A., Nowak, J.B., Weinheimer, A.J., Hall, S.R., Smith, J.N., Fried, A., Staebler, R.M., Wang, Y., Koo, J.-H., Cantrell, C.A., Weibring, P., Walega, J., Knapp, D.J., Shepson, P.B., Stephens, C.R.: Observations of inorganic bromine (HOBr, BrO, and Br_2) speciation at Barrow, Alaska, in spring 2009. J. Geophys. Res. **117**, D00R16 (2012). doi:10.1029/2011JD016641

Lightfoot, P.D., Cox, R.A., Crowley, J.N., Destriau, M., Hayman, G.D., Jenkin, M.E., Moortgat, G.K., Zabel, F.: Organic peroxy radicals: Kinetics, spectroscopy and tropospheric chemistry. Atmos. Enviro **26A**, 1805–1961 (1992)

Liu, S.C., Trainer, M., Carriolli, M.A., Huber, G., Montzaka, D.D., Norton, R.B., Ridley, B.A., Walega, J.G., Atlas, E.L., Heikes, B.G., Huebert, B.J., Warreb, W.: A study of the photochemistry and ozone budget during the Mauna Loa Observatory photochemistry experiment. J. Geophys. Res. **97**, 10463–10471 (1992)

Lopez-Hilfiker, F.D., Constantin, K., Kercher, J.P., Thornton, J.A.: Temperature dependent halogen activation by N_2O_5 reactions on halide-doped ice surfaces. Atmos. Chem. Phys. **12**, 5237–5247 (2012)

Lovejoy, E.R., Hanson, D.R., Huey, L.G.: Kinetics and products of the gas-phase reaction of SO_3 with water. J. Phys. Chem. **100**, 19911–19916 (1996)

Lu, K.D., Rohrer, F., Holland, F., Fuchs, H., Bohn, B., Brauers, T., Chang, C.C., Häseler, R., Hu, M., Kita, K., Kondo, Y., Li, X., Lou, S.R., Nehr, S., Shao, M., Zeng, L.M., Wahner, A., Zhang, Y.H., Hofzumahaus, A.: Observation and modelling of OH and HO_2 concentrations in the Pearl River Delta 2006: a missing OH source in a VOC rich atmosphere. Atmos. Chem. Phys. **12**, 1541–1569 (2012)

Mahajan, A.S., Plane, J.M.C., Oetjen, H., Mendes, L., Saunders, R.W., Saiz-Lopez, A., Jones, C. E., Carpenter, L.J., McFiggans, G.B.: Measurement and modelling of tropospheric reactive halogen species over the tropical Atlantic Ocean. Atmos. Chem. Phys. **10**, 4611–4624 (2010)

Malkin, T.L., Goddard, A., Heard, D.E., Seakins, P.W.: Measurements of OH and HO_2 yields from the gas phase ozonolysis of isoprene. Atmos. Chem. Phys. **10**, 1441–1459 (2010)

Martin, L.R., Damschen, D.E.: Aqueous oxidation of sulfur dioxide by hydrogen peroxide at low pH. Atmos. Environ. **15**, 1615–1621 (1981)

Martino, M., Mills, G.P., Woeltjen, J., Liss, P.S.: A new source of volatile organoiodine compounds in surface seawater. Geophys. Res. Lett. **36**, L01609 (2009). doi:10.1029/2008GL036334

Matveev, V., Peleg, M., Rosen, D., Tov-Alper, D.S., Hebestreit, K., Stutz, J., Platt, U., Blake, D., Luria, M.: Bromine oxide—ozone interaction over the Dead Sea. J. Geophys. Res. **106**, 10375–10387 (2001)

McArdle, J.V., Hoffmann, M.R.: Kinetic and mechanism of oxidation of aquated sulfur dioxide by hydrogen peroxide at low pH. J. Phys. Chem. **87**, 5425–5429 (1983)

McElroy, M.B.: The Atmospheric Environment: Effects of Human Activities, 326 pp. Princeton University Press, Princeton (2002)

McFiggans, G., Plane, J.M.C., Allan, B.J., Carpenter, L.J., Coe, H., O'Dowd, C.: A modeling study of iodine chemistry in the marine boundary layer. J. Geophys. Res. **105**, 14371–14385 (2000)

McKee, M.L., Wine, P.H.: Ab initio study of the atmospheric oxidation of CS_2. J. Am. Chem. Soc. **123**, 2344–2353 (2001)

Mellouki, A., Le Bras, G., Sidebottom, H.: Kinetics and mechanisms of the oxidation of oxygenated organic compounds in the gas phase. Chem. Rev. **103**, 5077–5096 (2003)

Mielke, L.H., Furgeson, A., Osthoff, H.D.: Observation of $ClNO_2$ in a mid-continental urban environment. Environ. Sci. Technol. **45**, 8889–8896 (2011)

Milford, J.B., Russell, A.G., McRae, G.J.: A new approach to photochemical pollution control: implications of spatial patterns in pollutant responses to reductions in nitrogen oxides and reactive organic gas emissions. Environ. Sci. Technol. **23**, 1290–1301 (1989)

Mochida, M., Hirokawa, J., Akimoto, H.: Unexpected large uptake of O_3 on sea salts and the observed Br_2 formation. Geophys. Res. Lett. **27**, 2629–2632 (2000)

Mora-Diez, N., Boyd, R.J.: A computational study of the kinetics of the NO3 hydrogen-abstraction reaction from a series of aldehydes (XCHO: X = F, Cl, H, CH3). J. Phys. Chem. A **106**, 384–394 (2002)

Mozukewich, M.: Mechanisms for the release of halogens from sea-salt particles by free radical reactions. J. Geophys. Res. **100**, 14199–14207 (1995)

Munger, J.W., Jacob, D.J., Waldman, J.M., Hoffmann, M.R.: Fogwater chemistry in an urban atmosphere. J. Geophys. Res. **88**, 5109–5121 (1983)

Nehr, S., Bohn, B., Fuchs, H., Hofzumahaus, A.: HO_2 formation from the OH + benzene reaction in the presence of O_2. Phys. Chem. Chem. Phys. **13**, 10699–10708 (2011)

Ng, N.L., Kwan, A.J., Surratt, J.D., Chan, A.W.H., Chhabra, P.S., Sorooshian, A., Pye, H.O.T., Crounse, J.D., Wennberg, P.O., Flagan, R.C., Seinfeld, J.H.: Secondary organic aerosol (SOA) formation from reaction of isoprene with nitrate radicals (NO_3). Atmos. Chem. Phys. **8**, 4117–4140 (2008)

Nguyen, T.L., Park, J.H., Lee, K.J., Song, K.Y., Barker, J.R.: Mechanism and kinetics of the reaction $NO_3 + C_2H_4$. J. Phys. Chem. A **115**, 4894–4901 (2011)

Niki, H., Davy, E., Weinstock, B.: Mechanism of smog reactions. In: Photochemical Smog and Ozone Reactions, Advances in chemistry, vol. 113, pp. 16–57 (1972)

O'Brien, J.M., Czuba, E., Hastie, D.R., Francisco, J.S., Shepson, P.B.: Determination of the hydroxy nitrate yields from the reaction of $C_2 - C_6$ alkenes with OH in the presence of NO. J. Phys. Chem. A **102**, 8903–8908 (1998)

Oltmans, S.J., Komhyr, W.: Surface ozone distributions and variations from 1973–1984 measurements at the NOAA geophysical monitoring for climate change baseline observatories. J. Geophys. Res. **91**, 5229–5236 (1986)

Osthoff, H.D., Roberts, J.M., Ravishankara, A.R., Williams, E.J., Lerner, B.M., Sommariva, R., Bates, T.S., Coffman, D., Quinn, P.K., Dibb, J.E., Stark, H., Burkholder, J.B., Talukdar, R.K., Meagher, J., Fehsenfeld, F.C., Brown, S.S.: High levels of nitryl chloride in the polluted subtropical marine boundary layer. Nat. Geosci. **1**, 324–328 (2008)

Oum, K.W., Lakin, M.J., DeHaan, D.O., Brauers, T., Finlayson-Pitts, B.: Formation of molecular chlorine from the photolysis of ozone and aqueous sea-salt particles. Science **279**, 74–76 (1998a)

Oum, K.W., Lakin, M.J., Finlayson-Pitts, B.J.: Bromine activation in the troposphere by the dark reaction of O_3 with seawater ice. Geophys. Res. Lett. **25**, 3923–3926 (1998b)

Pandis, S.N., Seinfeld, J.H.: Sensitivity analysis of a chemical mechanism for aqueous-phase atmospheric chemistry. J. Geophys. Res. **94**, 1105–1126 (1989)

Peeters, J., Müller, J.-F.: HO_x radical regeneration in isoprene oxidation via peroxy radical isomerisations. II: experimental evidence and global impact. Phys. Chem. Chem. Phys. **12**, 14227–14235 (2010)

Pérez-Casany, M.P., Nebot-Gil, I., Sánchez-Marín, J.: Ab initio study on the mechanism of tropospheric reactions of the nitrate radical with alkenes: propene. J. Phys. Chem. A **104**, 6277–6286 (2000)

Perring, A.E., Wisthaler, A., Graus, M., Wooldridge, P.J., Lockwood, A.L., Mielke, L.H., Shepson, P.B., Hansel, A., Cohen, R.C.: A product study of the isoprene + NO_3 reaction. Atmos. Chem. Phys. **9**, 4945–4956 (2009)

Pfrang, C., King, M., Canosa-Mas, C.E., Wayne, R.P.: Correlations for gas-phase reactions of NO_3, OH and O_3 with alkenes: an update. Atmos. Environ. **40**, 1170–1179 (2006)

Platt, U., Wagner, T.: Satellite mapping of enhanced BrO concentrations in the troposphere. Nature **395**, 486–490 (1998)

Ramírez-Anguita, J.M., González-Lafont, À., Lluch, J.M.: Formation pathways of CH_3SOH from $CH_3S(OH)CH_3$ in the presence of O_2: a theoretical study. Theor. Chem. Acc. **123**, 93–103 (2009)

Raoult, S., Rayez, M.-T., Rayezand, J.-C., Lesclaux, R.: Gas phase oxidation of benzene: kinetics, thermochemistry and mechanism of initial steps. Phys. Chem. Chem. Phys. **6**, 2245–2253 (2004)

Read, K.A., Mahajan, A.S., Carpenter, L.J., Evans, M.J., Faria, B.V.E., Heard, D.E., Hopkins, J.R., Lee, J.D., Moller, S.J., Lewis, A.C., Mendes, L., McQuaid, J.B., Oetjen, H., Saiz-Lopez, A., Pilling, M.J., Plane, J.M.C.: Extensive halogen-mediated ozone destruction over the tropical Atlantic Ocean. Nature **453**, 1232–1235 (2008)

Regelin, E., Harder, H., Martinez, M., Kubistin, D., Ernest, C.T., Bozem, H., Klippel, T., Hosaynali-Beygi, Z., Fischer, H., Sander, R., Jöckel, P., Königstedt, R., Lelieveld, J.: HOx measurements in the summertime upper troposphere over Europe: a comparison of observations to a box model and a 3-D model. Atmos. Chem. Phys. Discuss. **12**, 30619–30660 (2012)

Ren, X., Brune, W.H., Mao, J., Mitchell, M.J., Lesher, R., Simpas, J.B., Metcalf, A.R., Schwab, J. J., Cai, C., Li, Y., Demerjian, K.L., Felton, H.D., Boynton, G., Adams, A., Perry, J., He, Y., Zhou, X., Hou, J.: Behavior of OH and HO_2 in the winter atmosphere in New York City. Atmos. Environ. **40**(Supplement 2), 252–263 (2006)

Ren, X., Olson, J.R., Crawford, J., Brune, W.H., Mao, J., Long, R.B., Chen, Z., Chen, G., Avery, M.A., Sachse, G.W., Barrick, J.D., Diskin, G.S., Huey, G., Fried, A., Cohen, R.C., Heikes, B., Wennberg, P.O., Singh, H.B., Blake, D., Shetter, R.: HOx chemistry during INTEX-A 2004: observation, model calculation, and comparison with previous studies. J. Geophys. Res. **113**, D05310 (2008). doi:10.1029/2007JD009166

Riedel, T.P., Bertram, T.H., Crisp, T.A., Williams, E.J., Lerner, B.M., Vlasenko, A., Li, S.-M., Gilman, J., de Gouw, J., Bon, D.M., Wagner, N.L., Brown, S.S., Thornton, J.A.: Nitryl chloride and molecular chlorine in the coastal marine boundary layer. Environ. Sci. Technol. **46**, 10463–10470 (2012)

Rollins, A.W., Kiendler-Scharr, A., Fry, J.L., Brauers, T., Brown, S.S., Dorn, H.-P., Dub'e, W.P., Fuchs, H., Mensah, A., Mentel, T.F., Rohrer, F., Tillmann, R., Wegener, R., Wooldridge, P.J., Cohen, R.C.: Isoprene oxidation by nitrate radical: alkyl nitrate and secondary organic aerosol yields. Atmos. Chem. Phys. **9**, 6685–6703 (2009)

Rose, A.H., Brandt, C.S.: Environmental irradiation test facility. Air. Pollut. Control Assoc **10**, 331–335 (1960)

Rossi, M.J.: Heterogeneous reactions on salts. Chem. Rev. **103**, 4823–4882 (2003)

Sadanaga, Y., Hirokawa, J., Akimoto, H.: Formation of molecular chlorine in dark condition: heterogeneous reaction of ozone with sea salt in the presence of ferric ion. Geophys. Res. Lett. **28**, 4433–4436 (2001)

Sadanaga, Y., Yoshino, A., Watanabe, K., Yoshioka, A., Wakazono, Y., Kanaya, Y., Kajii, Y.: Development of a measurement system of OH reactivity in the atmosphere by using a laser-induced pump and probe technique. Rev. Sci. Instrum. **75**, 2648–2655 (2004)

Saiz-Lopez, A., von Glasow, R.: Reactive halogen chemistry in the troposphere. Chem. Soc. Rev. **41**, 6448–6472 (2012)

Saiz-Lopez, A., Plane, J.M.C., Shillito, J.A.: Bromine oxide in the mid-latitude marine boundary layer. Geophys. Res. Lett. **31**, L03111 (2004). doi:10.1029/2003GL018956

Saiz-Lopez, A., Shillito, J.A., Coe, H., Plane, J.M.C.: Measurements and modelling of I_2, IO, OIO, BrO and NO_3 in the mid-latitude marine boundary layer. Atmos. Chem. Phys. **6**, 1513–1528 (2006)

Saiz-Lopez, A., Plane, J.M.C., Baker, A.R., Carpenter, L.J., von Glasow, R., Gómez Martín, J.C., McFiggans, G., Saunders, R.W.: Atmospheric chemistry of iodine. Chem. Rev. **112**, 1773–1804 (2012)

Sakamaki, F., Okuda, M., Akimoto, H., Yamazaki, H.: Computer modeling study of photochemical ozone formation in the propene-nitrogen oxides-dry air system. Generalized maximum ozone isopleth. Environ. Sci. Technol. **16**, 45–52 (1982)

Sander, R., Crutzen, P.J.: Model study indicating halogen activation and ozone destruction in polluted air masses transported to the sea. J. Geophys. Res. **101**, 9121–9138 (1996)

Sander, R., Rudich, Y., von Glasow, R., Crutzen, P.J.: The role of $BrNO_3$ in marine tropospheric chemistry: a model study. Geophys. Res. Lett. **26**, 2857–2860 (1999)

Sander, R., Keene, W.C., Pszenny, A.A.P., Arimoto, R., Ayers, G.P., Baboukas, E., Cainey, J.M., Crutzen, P.J., Duce, R.A., Hoenninger, G., Huebert, B.J., Maenhaut, W., Mihalopoulos, N., Turekian, V.C., Van Dingenen, R.: Inorganic bromine in the marine boundary layer: a critical review. Atmos. Chem. Phys. **3**, 1301–1336 (2003)

Sander, R., Kerkweg, A., Jockel, P., Lelieveld, J.: Technical note: the new comprehensive atmospheric chemistry module MECCA. Atmos. Chem. Phys. **5**, 445–450 (2005)

Sander, S.P., Baker, R., Golden, D.M., Kurylo, M.J., Wine, P.H., Abatt, J.P.D., Burkholder, J.B., Kolb, C.E., Moortgat, G.K., Huie, R.E., Orkin, V.L.: Chemical Kinetics and Photochemical Data for Use in Atmospheric Studies, Evaluation Number 17. JPL Publication 10-6, Pasadena, California (2011). Website:http://jpldataeval.jpl.nasa.gov/

Saunders, S.M., Jenkin, M.E., Derwent, R.G., Pilling, M.J.: Protocol for the development of the Master Chemical Mechanism, MCM v3 (Part A): tropospheric degradation of non-aromaticvolatile organic compounds. Atmos. Chem. Phys. **3**, 161–180 (2003)

Schmidt, V., Zhu, G.Y., Becker, K.H., Fink, E.H.: Study of OH reactions at high pressures by excimer laser photolysis-dye laser fluorescence. Ber. Bunsenges. Phys. Chem. **89**, 321 (1985)

Schneider, M., Luxenhofer, O., Deissler, A., Ballschmiter, K.: C1–C15 alkyl nitrates, benzyl nitrate, and bifunctional nitrates: measurements in California and south Atlantic air and global comparison using C2Cl4 and CHBr3 as marker. Environ. Sci. Technol. **32**, 3055–3062 (1998)

Schönhardt, A., Richter, A., Wittrock, F., Kirk, H., Oetjen, H., Roscoe, H.K., Burrows, J.P.: Observations of iodine monoxide columns from satellite. Atmos. Chem. Phys. **8**, 637–653 (2008)

Schwartz, S.E., Freiberg, J.E.: Mass-transport limitation to the rate of reaction of gases in liquid droplets: application to oxidation of SO_2 in aqueous solutions. Atmos. Environ. **15**, 1129–1144 (1981)

Seinfeld, J.H., Pandis, S.N.: Atmospheric Chemistry and Physics: Air Pollution to Climate Change, 2nd edn. John Wiley and Sons, Hoboken, NJ (2006)

Sheehy, P.M., Volkamer, R., Molina, L.T., Molina, M.J.: Oxidative capacity of the Mexico City atmosphere – Part 2: a RO_x radical cycling perspective. Atmos. Chem. Phys. **10**, 6993–7008 (2010)

Sillman, S., Logan, J.A., Wofsy, S.C.: The sensitivity of ozone to nitorgen oxides and hydrocarbons in regional ozone episodes. J. Geophys. Res. **95**, 1837–1851 (1990)

Simpson, W.R., von Glasow, R., Riedel, K., Anderson, P., Ariya, P., Bottenheim, J., Burrows, J., Carpenter, L.J., Frieß, U., Goodsite, M.E., Heard, D., Hutterli, M., Jacobi, H.-W., Kaleschke, L., Neff, B., Plane, J., Platt, U., Richter, A., Roscoe, H., Sander, R., Shepson, P., Sodeau, J., Steffen, A., Wagner, T., Wolff, E.: Halogens and their role in polar boundary-layer ozone depletion. Atmos. Chem. Phys. **7**, 4375–4418 (2007)

Sinha, V., Williams, J., Crowley, J.N., Lelieveld, J.: The comparative reactivity method: a new tool to measure total OH reactivity in ambient air. Atmos. Chem. Phys. **8**, 2213–2227 (2008)

Sinha, V., Williams, J., Lelieveld, J., Ruuskanen, T.M., Kajos, M.K., Patokoski, J., Hellen, H., Hakola, H., Mogensen, D., Boy, M., Rinne, J., Kulmala, M.: OH reactivity measurements within a boreal forest: evidence for unknown reactive emissions. Environ. Sci. Technol. **44**, 6614–6620 (2010)

Skov, H., Hjorth, J., Lohse, C., Jensen, N.R., Restelli, G.: Products and mechanisms of the reactions of the nitrate radical (NO_3) with isoprene, 1,3-butadiene and 2,3-dimethyl-1,3-butadiene in air. Atmos. Environ. Part A **26**, 2771–2783 (1992)

Smith, S.C., Lee, J.D., Bloss, W.J., Johnson, G.P., Ingham, T., Heard, D.E.: Concentrations of OH and HO_2 radicals during NAMBLEX: measurements and steady state analysis. Atmos. Chem. Phys. **6**, 1435–1453 (2006)

Smoydzin, L., von Glasow, R.: Modelling chemistry over the Dead Sea: bromine and ozone chemistry. Atmos. Chem. Phys. **9**, 5057–5072 (2009)

Sommariva, R., von Glasow, R.: Multiphase halogen chemistry in the tropical Atlantic ocean. Environ. Sci. Technol. **46**, 10429–10437 (2012)

Sommariva, R., Bloss, W.J., Brough, N., Carslaw, N., Flynn, M., Haggerstone, A.-L., Heard, D.E., Hopkins, J.R., Lee, J.D., Lewis, A.C., McFiggans, G., Monks, P.S., Penkett, S.A., Pilling, M.J., Plane, J.M.C., Read, K.A., Saiz-Lopez, A., Rickard, A.R., Williams, P.I.: OH and HO_2 chemistry during NAMBLEX: roles of oxygenates, halogen oxides and heterogeneous uptake. Atmos. Chem. Phys. **6**, 1135–1153 (2006)

Spicer, C.W., Chapman, E.G., Finlayson-Pitts, B.J., Plastridge, R.A., Hubbe, J.M., Fast, J.D., Berkowit, C.M.: Unexpectedly high concentrations of molecular chlorine in coastal air. Nature **394**, 353–356 (1998)

Spicer, C.W., Plastridge, R.A., Foster, K.L., Finlayson-Pitts, B.J., Bottenheim, J.W., Grannasd, A. M., Shepson, P.B.: Molecular halogens before and during ozone depletion events in the Arctic at polar sunrise: concentrations and sources. Atmos. Environ. **36**, 2721–2731 (2002)

Stickel, R.E., Chin, M., Daykin, E.P., Hynes, A.J., Wine, P.H., Wallington, T.J.: Mechanistic studies of the hydroxyl-initiated oxidation of carbon disulfide in the presence of oxygen. J. Phys. Chem. **97**, 13653–13661 (1993)

Stockwell, W.R., Calvert, J.G.: The mechanism of the HO-SO_2 reaction. Atmos. Environ. **17**, 2231–2235 (1983)
Stockwell, W.R., Kirchner, F., Kuhn, M., Seefeld, S.: A new mechanism for regional atmospheric chemistry modeling. J. Geophys. Res. **102**, 25847–25879 (1997)
Stockwell, W.R., Lawson, C.V., Saunders, E., Goliff, W.S.: A review of tropospheric atmospheric chemistry and gas-phase chemical mechanisms for air quality modeling. Atmosphere **3**, 1–32 (2012)
Stone, D., Whalley, L.K., Heard, D.E.: Tropospheric OH and HO_2 radicals: field measurements and model comparisons. Chem. Soc. Rev. **41**, 6348–6404 (2012)
Stutz, J., Ackermann, R., Fast, J.D., Barrie, L.: Atmospheric reactive chlorine and bromine at the Great Salt Lake, Utah. Geophys. Res. Lett. **29**(10), (2002). doi:10.1029/2002GL014812
Sudo, K., Takahashi, M., Akimoto, H.: CHASER: a global chemical model of the troposphere 2. Model results and evaluation. J. Geophys. Res. **107**(D21), 4586 (2002). doi:10.1029/2001JD001114
Suh, I., Zhang, R., Molina, L.T., Molina, M.J.: Oxidation mechanism of aromatic peroxy and bicyclic radicals from OH-toluene reactions. J. Am. Chem. Soc. **125**, 12655–12665 (2003)
Suh, I., Zhao, J., Zhang, R.: Unimolecular decomposition of aromatic bicyclic alkoxy radicals and their acyclic radicals. Chem. Phys. Lett. **432**, 313–320 (2006)
Taketani, F., Kanaya, Y., Pochanart, P., Liu, Y., Li, J., Okuzawa, K., Kawamura, K., Wang, Z., Akimoto, H.: Measurement of overall uptake coefficients for HO_2 radicals by aerosol particles sampled from ambient air at Mts. Tai and Mang (China). Atmos. Chem. Phys. **12**, 11907–11916 (2012)
Tan, D., Faloona, I., Simpas, J.B., Brune, W., Shepson, P.B., Couch, T.L., Sumner, A.L., Carroll, M.A., Thornberry, T., Apel, E., Riemer, D., Stockwell, W.: HOx budgets in a deciduous forest: results from the PROPHET summer 1998 campaign. J. Geophys. Res. **106**, 24407–24427 (2001a)
Tan, D., Faloona, I., Simpas, J.B., Brune, W., Olson, J., Crawford, J., Avery, M., Sachse, G., Vay, S., Sandholm, S., Guan, H.W., Vaughn, T., Mastromarino, J., Heikes, B., Snow, J., Podolske, J., Singh, H.: OH and HO_2 in the tropical Pacific: results from PEM-Tropics B. J. Geophys. Res. **106**, 32667–32681 (2001b)
Tang, T., McConnell, J.C.: Autocatalytic release of bromine from Arctic snow pack during polar sunrise. Geophys. Res. Lett. **23**, 2633–2636 (1996)
Tas, E., Peleg, M., Matveev, V., Zingler, J., Luria, M.: Frequency and extent of bromine oxide formation over the Dead Sea. J. Geophys. Res. **110**, D11304 (2005). doi:10.1029/2004JD005665
Tas, E., Peleg, M., Pedersen, D.U., Matveev, V., Pour Biazar, A., Luria, M.: Measurement -based modeling of bromine chemistry in the boundary layer: 1. Bromine chemistry at the Dead Sea. Atmos. Chem. Phys. **6**, 5589–5604 (2006)
Thornton, J.A., Kercher, J.P., Riedel, T.P., Wagner, N.L., Cozic, J., Holloway, J.S., Dubé, W.P., Wolfe, G.M., Quinn, P.K., Middlebrook, A.M., Alexander, B., Brown, S.S.: A large atomic chlorine source inferred from mid-continental reactive nitrogen chemistry. Nature **464**, 271–274 (2010)
Toyota, K., McConnell, J.C., Lupu, A., Neary, L., McLinden, C.A., Richter, A., Kwok, R., Semeniuk, K., Kaminski, J.W., Gong, S.-L., Jarosz, J., Chipperfield, M.P., Sioris, C.E.: Analysis of reactive bromine production and ozone depletion in the Arctic boundary layer using 3-D simulations with GEM-AQ: inference from synoptic-scale patterns. Atmos. Chem. Phys. **11**, 3949–3979 (2011)
Tuazon, E.C., Aschmann, S.M., Arey, J., Atkinson, R.: Products of the gas-phase reactions of a series of methyl-substituted ethenes with the OH radical. Environ. Sci. Technol. **32**, 2106–2112 (1998)
Tuckermann, M., Ackermann, R., Gölz, C., Lorenzen-Schmidt, H., Senne, T., Stutz, J., Trost, B., Unold, W., Platt, U.: DOAS-observation of halogen radical-catalyzed arctic boundary layer

ozone destruction during the ARCTOC-campaigns 1995 and 1996 in Ny-Ålesund, Spitsbergen. Tellus B **49**, 533–555 (1997)

Vance, A., McGonigle, A.J.S., Aiuppa, A., Stith, J.L., Turnbull, K., von Glasow, R.: Ozone depletion in tropospheric volcanic plumes. Geophys. Res. Lett. **37**, L22802 (2010). doi:10.1029/2010GL044997

Vogt, R., Crutzen, P.J., Sander, R.: A mechanism for halogen release from sea-salt aerosol in the remote marine boundary layer. Nature **383**, 327–330 (1996)

Volkamer, R., Platt, U., Wirtz, K.: Primary and secondary glyoxal formation from aromatics: experimental evidence for the bicycloalkyl – radical pathway from benzene, toluene, and p-xylene. J. Phys. Chem. A **105**, 7865–7874 (2001)

Volkamer, R., Sheehy, P., Molina, L.T., Molina, M.J.: Oxidative capacity of the Mexico City atmosphere – Part 1: a radical source perspective. Atmos. Chem. Phys. **10**, 6969–6991 (2010)

von Glasow, R.: Atmospheric chemistry in volcanic plumes. PNAS **107**, 6594–6599 (2010)

Wagner, V., Jenkin, M.E., Saunders, S.M., Stanton, J., Wirtz, K., Pilling, M.J.: Modelling of the photooxidation of toluene: conceptual ideas for validating detailed mechanisms. Atmos. Chem. Phys. **3**, 89–106 (2003)

Wagner, T., Ibrahim, O., Sinreich, R., Frieß, U., von Glasow, R., Platt, U.: Enhanced tropospheric BrO over Antarctic sea ice in mid winter observed by MAX-DOAS on board the research vessel Polarstern. Atmos. Chem. Phys. **7**, 3129–3142 (2007)

Wallace, J.M., Hobbs, P.V.: Atmospheric Science: An Introductory Survey, 2nd edn. Academic Press, Burlington, MA (2006)

Wallington, T., Armmann, M., Atkinson, R., Cox, R.A., Crowley, J.N., Hynes, R., Jenkin, M.E., Mellouki, W., Rossi, M.J., Troe, J.: IUPAC Subcommittee for Gas Kinetic Data Evaluation for Atmospheric Chemistry, Evaluated Kinetic Data, Gas-phase Reactions (2012). http://www.iupac-kinetic.ch.cam.ac.uk/

Wang, J., Doussin, J.F., Perrier, S., Perraudin, E., Katrib, Y., Pangui, E., Picquet-Varrault, B.: Design of a new multi-phase experimental simulation chamber for atmospheric photosmog, aerosol and cloud chemistry research. Atmos. Meas. Tech. **4**, 2465–2494 (2011)

Warneck, P.: Chemistry of the Natural Atmosphere. Academic Press, San Diego (1988)

Wayne, R.P.: Chemistry of Atmospheres, 3rd edn. Oxford University Press, New York (2000)

Weinstock, B.: In Tuesday, C.S. (ed.) Chemical Reactions in Urban Atmospheres, pp. 54–55, American Elsevier, New York (1971)

Wennberg, P.: Atmospheric chemistry: bromine explosion. Nature **397**, 299–301 (1999)

Whalley, L.K., Furneaux, K.L., Goddard, A., Lee, J.D., Mahajan, A., Oetjen, H., Read, K.A., Kaaden, N., Carpenter, L.J., Lewis, A.C., Plane, J.M.C., Saltzman, E.S., Wiedensohler, A., Heard, D.E.: The chemistry of OH and HO_2 radicals in the boundary layer over the tropical Atlantic Ocean. Atmos. Chem. Phys. **10**, 1555–1576 (2010)

Whalley, L.K., Edwards, P.M., Furneaux, K.L., Goddard, A., Ingham, T., Evans, M.J., Stone, D., Hopkins, J.R., Jones, C.E., Karunaharan, A., Lee, J.D., Lewis, A.C., Monks, P.S., Moller, S., Heard, D.E.: Quantifying the magnitude of a missing hydroxyl radical source in a tropical rainforest. Atmos. Chem. Phys. **11**, 7223–7233 (2011)

Winer, A.M., Graham, R.A., Doyle, G.J., Bekowies, P.J., McAfee, J.M., Pitts Jr., J.N.: An evacuable environmental chamber and solar simulator facility for the study of atmospheric photochemistry. Adv. Environ. Sci. Technol. **10**, 461–511 (1980)

Yoshino, Y., Nakashima, Y., Miyazaki, K., Kato, S., Suthawaree, J., Shimo, N., Matsunaga, S., Chatani, S., Apel, E., Greenberg, J., Guenther, A., Ueno, H., Sasaki, H., Hoshi, J., Yokota, H., Ishii, K., Kajii, Y.: Air quality diagnosis from comprehensive observations of total OH reactivity and reactive trace species in urban central Tokyo. Atmos. Environ. **49**, 51–59 (2012)

Yu, J., Jeffries, E.: Atmospheric photooxidation of alkylbenzenes—II. Evidence of formation of epoxide intermediates. Atmos. Environ. **31**, 2281–2287 (1997)

Zhang, L., Qin, Q.-Z.: Theoretical studies on CS_2OH-O_2: a possible intermediate in the OH initiated oxidation of CS_2 by O_2. J. Mol. Struct. (Theochem) **531**, 375–379 (2000)

Zhang, D., Zhang, R.: Mechanism of OH formation from ozonolysis of isoprene: a quantum-chemical study. J. Am. Chem. Soc. **124**, 2692–2703 (2002)

Zhang, D., Lei, W., Zhang, R.: Mechanism of OH formation from ozonolysis of isoprene: kinetics and product yields. Chem. Phys. Lett. **358**, 171–179 (2002)

Zhang, D., Zhang, R.: Ozonolysis of α-pinene and β-pinene: kinetics and mechanism. J. Chem. Phys. **122**, 114308–114319 (12 pages) (2005)

Zhang, Y., Huang, J.-P., Henze, D.K., Seinfeld, J.H.: Role of isoprene in secondary organic aerosol formation on a regional scale. J. Geophys. Res. **112**, D20207 (2007). doi:10.1029/2007JD008675

Zhang, L., Jacob, D.J., Boersma, K.F., Jaffe, D.A., Olson, J.R., Bowman, K.W., Worden, J.R., Thompson, A.M., Avery, M.A., Cohen, R.C., Dibb, J.E., Flock, F.M., Fuelberg, H.E., Huey, L. G., McMillan, W.W., Singh, H.B., Weinheimer, A.J.: Transpacific transport of ozone pollution and the effect of recent Asian emission increases on air quality in North America: an integrated analysis using satellite, aircraft, ozonesonde, and surface observations. Atmos. Chem. Phys. **8**, 6117–6136 (2008). doi:10.5194/acp-8-6117-2008

Zhao, T.L., Gong, S.L., Bottenheim, J.W., McConnell, J.C., Sander, R., Kaleschke, L., Richter, A., Kerkweg, A., Toyota, K., Barrie, L.A.: A three-dimensional model study on the production of BrO and Arctic boundary layer ozone depletion. J. Geophys. Res. **113**, D24304 (2008). doi:10.1029/2008JD010631

Zingler, J., Platt, U.: Iodine oxide in the Dead Sea Valley: evidence for inorganic sources of boundary layer IO. J. Geophys. Res. **110**, D07307 (2005). doi:10.1029/2004JD004993

Chapter 8
Stratospheric Reaction Chemistry

One of the most peculiar chemical characteristics of the earth's atmosphere is the formation of O_3 through the photolysis of O_2, one of the major constituents of the atmosphere, by the ultraviolet radiation in the stratosphere. Photochemically produced O_3 in the stratosphere forms the ozone layer, which absorbs nearly 100 % of the ultraviolet radiation with wavelength shorter than 300 nm preventing it to reach to the earth's surface. Since the DNA molecules constituting the cells of creatures on the earth are destructed photochemically, when they absorb the light with wavelengths shorter than 300 nm, organisms cannot survive under such ultraviolet radiation. From the point of view of the earth's history, a story is told that the ozone layer formed due to the increase of O_2 in the atmosphere by photosynthesis of primitive organisms in the ocean, and it prevents harmful ultraviolet radiation from reaching the earth's surface, therefore enabling the migration of the earth's creatures to the land and bringing about the biosphere as seen in the present day (Berkner and Marshall 1965). This implies inversely that the decrease of stratospheric ozone for any reason would jeopardize the existence of terrestrial organisms, and the depletion of the ozone layer by human activities has been a big issue since the 1970s (Dotto et al. 1978; Middleton and Tolbert 2000; Finlayson-Pitts and Pitts 2000). Among them, the depletion of ozone layer by chlorofluorocarbons (CFCs) has been actualized, and atmospheric chemists have played a central role in raising the issue, understanding the phenomena, and bringing about a solution.

Almost all species related to the chemical reactions in the stratosphere except O_3 have their origin on the earth's surface, and are brought to the stratosphere from the troposphere. Since it takes about 1–2 years for the transport and mixing between the troposphere and stratosphere, compounds with sufficiently long lifetimes to reach to the stratosphere are rather limited among the numerous chemical species present in the troposphere, as seen in the previous chapter. For this reason, reaction chemistry in the stratosphere is far more simple compared to that in the troposphere, and is expected to be described more precisely by a chemical reaction model than for the troposphere.

H. Akimoto, *Atmospheric Reaction Chemistry*, Springer Atmospheric Sciences,
DOI 10.1007/978-4-431-55870-5_8

Stratospheric chemistry is described in detail by Brasseur and Solomon (2005), including the mesosphere, and is also given in the textbooks by Warneck (1988), Brasseur et al. (1999), Finlayson-Pitts and Pitts (2000), Wayne (2000), McElroy (2002), Seinfeld and Pandis(2006). A review on the reactions of halogen radicals in the stratosphere has been given by Bedjanian and Pullet (2003), and an updated review on stratospheric ozone depletion has been provided periodically by WMO (2011). In the present chapter, chemical reaction system is described exclusively among the stratospheric chemistry in which transport and reaction are coupled together.

8.1 Pure Oxygen Atmosphere and Ozone Layer

Chapman (1930a, b) showed successfully that the characteristics of the height of the ozone layer and ozone density in the earth's atmosphere can be described by assuming only oxygen is present as a reactive species in the atmosphere, and this reaction scheme is called pure oxygen theory or Chapman mechanism. Photolysis of O_2 occurs by the solar radiation of wavelength shorter than 242 nm (see Sect. 4.3.1). Only the photolytic process,

$$O_2 + h\nu(\lambda < 242\ \text{nm}) \rightarrow O(^3P) + O(^3P) \tag{8.1}$$

is energetically possible in the stratosphere where the solar radiation of wavelength longer than 180 nm reaches. From the ground state oxygen atoms $O(^3P)$, O_3 is formed by the reaction,

$$O(^3P) + O_2 + M \rightarrow O_3 + M, \tag{8.2}$$

and this is the sole reaction in the atmosphere to directly produce O_3. The O_3 molecule produced returns to O_2 by the reaction of $O(^3P)$ atoms (Sect. 5.1.2) or by photolytic reactions (Sect. 4.3.2) such as,

$$O_3 + O(^3P) \rightarrow 2O_2 \tag{8.3}$$

$$O_3 + h\nu \rightarrow O(^3P) + O_2 \tag{8.4}$$

$$\rightarrow O(^1D) + O_2. \tag{8.5}$$

Although most of the excited oxygen atoms $O(^1D)$ thus formed are quenched (see Sect. 7.1 and Table 5.1) in the atmosphere,

$$O(^1D) + N_2 \rightarrow O(^3P) + N_2 \tag{8.6}$$

$$O(^1D) + O_2 \rightarrow O(^3P) + O_2, \tag{8.7}$$

a part of them react with trace gases in the stratosphere, and give a large impact on stratospheric O_3 concentration as described in the next section. Thus, the mixing ratios of O and O_3 in the stratosphere are determined by the photo-equilibrium by reactions (8.1), (8.2), (8.3), (8.4), and (8.5), and photochemical reactions with trace species which will be described later.

Since the concentration of $O(^1D)$ is much smaller than $O(^3P)$, by simply expressing $O(^3P)$ as O neglecting $O(^1D)$, the rate equations for O and O_3 can be expressed,

$$\frac{d[O]}{dt} = 2j_{8.1} - k_{8.2}[O][O_2][M] - k_{8.3}[O][O_3] - j_{8.4+8.5}[O_3] \tag{8.8}$$

$$\frac{d[O_3]}{dt} = k_{8.2}[O][O_2][M] - k_{8.3}[O][O_3] - j_{8.4+8.5}[O_3] \tag{8.9}$$

Summing these equations and defining the sum of O and O_3 as odd oxygen O_x,

$$[O_x] = [O] + [O_3],$$

then,

$$\frac{d[O_x]}{dt} = \frac{d[O]}{dt} + \frac{d[O_3]}{dt} = 2j_{8.1}[O_2] - 2k_{8.3}[O][O_3], \tag{8.10}$$

which shows simply O_x is produced by the photolysis of O_2 and dissipated by the reaction of O and O_3 (reaction 8.3). Here, assuming the steady state, $d[O_3]/dt = 0$ and $d[O_x]/dt = 0$, Eqs. (8.9) and (8.10) give,

$$k_{8.2}[O][O_2][M] = k_{8.3}[O][O_3] + j_{8.4+8.5}[O_3] \tag{8.11}$$

and

$$j_{8.1}[O_2] = k_{8.3}[O][O_3], \tag{8.12}$$

respectively. From Eqs. (8.11) and (8.12), the quadratic equation with stationary concentration of O_3,

$$j_{8.4+8.5}k_{8.3}[O_3]^2 + j_{8.1}k_{8.3}[O_2][O_3] - j_{8.1}k_{8.2}[O_2]^2[M] = 0 \tag{8.13}$$

is obtained, and the solution is

$$[O_3]_{ss} = \frac{-j_{8.1} + \sqrt{j_{8.1}^2 + 4j_{8.1}j_{8.4+8.5}k_{8.2}[M]/k_{8.3}}}{2j_{8.4}}[O_2]. \tag{8.14}$$

The vertical profile of the stationary state concentration of O_3 in the stratosphere can be obtained by using the values of $j_{8.1}$, $j_{8.4}$, $[O_2]$, and $M(=[O_2]+[N_2])$ at each altitude. In approximation, assuming $2\,j_{8.1}\,[O_2] << j_{8.4}[O_3]$ and $k_{8.2}\,[O_2]\,[M] >> k_{8.3}[O_3]$ in the stratosphere, Eq. (8.8) under the steady state of $d[O]/dt = 0$ gives,

$$[O]_{ss} = \frac{2j_{8.1}[O_2] + j_{8.4+8.5}[O_3]}{k_{8.2}[O_2][M] + k_{8.3}[O_3]} \approx \frac{j_{8.4+8.5}[O_3]}{k_{8.2}[O_2][M]}. \tag{8.15}$$

Substituting this to Eq. (8.10),

$$\frac{d[O_x]}{dt} = 2j_{8.1}[O_2] - \frac{2k_{8.3}j_{8.4+8.5}[O_3]^2}{k_{8.2}[O_2][M]}. \tag{8.16}$$

Here, since most of O_x is O_3 in the stratosphere, putting $d[O_x]/dt = d[O_3]/dt = 0$ in the steady state, from Eq. (8.16),

$$[O_3]_{ss} \approx \sqrt{\frac{j_{8.1}k_{8.2}[M]}{j_{8.4+8.5}}[O_2]} \tag{8.17}$$

is obtained. Equation (8.17) is the approximation of Eq. (8.14). Since the altitude dependence of j_4 is relatively small among the parameters, $[O_3]_{ss}$ is mainly determined by the dependence of $j_{8.1}$, $[O_2]$ and $[M]$ on altitude. Since $j_{8.1}$ is proportional to the actinic flux, $j_{8.1}$ increases with altitude, while $[O_2]$ and $[M]$ decrease with altitude. Therefore, $[O_3]_{ss}$ has maximum at a certain altitude, which is the ozone layer deduced from the Chapman mechanism.

Figure 8.1 shows the comparison of ozone density profile in the mid-latitude between the calculated and observed values. The horizontal lines with the observation curve represent the observed range of ozone density. It can be seen that the Chapman mechanism predicts the maximum of the ozone layer at around 25 km and its ozone density almost properly. However, there are two points of deviation of the theoretical curve from the observation. One is the overestimation of the density by a factor of two above the altitude of ozone maximum and the theoretical value of the altitude of the maximum is higher than observation by a few km. Another is that in the lower stratosphere to the troposphere, the theoretical value of ozone decreases rapidly whereas the observed value keeps nearly a constant value.

The first point is due to the incomplete treatment of the pure oxygen theory, ignoring trace constituents other than O, O_2 and O_3, which will be described in detail below. The second point is due to the transport of ozone within the stratosphere and into the troposphere. Photochemical lifetime of stratospheric O_3 is the order of 10 min at the altitude of 45 km and the diurnal cycle is observed. On the

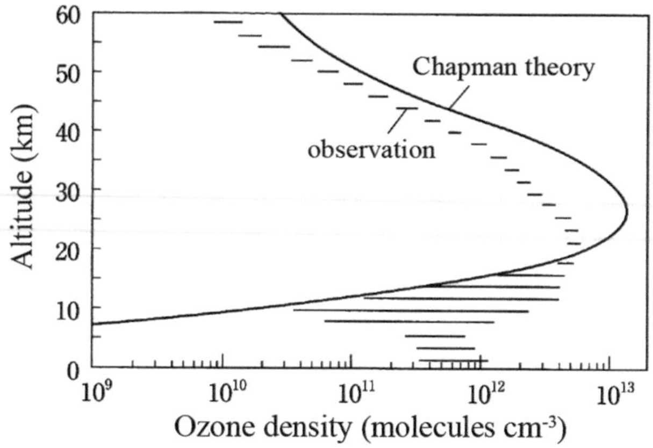

Fig. 8.1 Comparison of vertical ozone profiles calculated by the pure oxygen atmosphere theory (Chapman mechanism) and by observation (Adapted from Shimazaki et al. 1987)

other hand, the lifetime is more than a few days at the altitude of 20 km, so that it is necessary to consider the effect of transport in order to reproduce the O_3 distribution at the lower altitude than this. In the lower stratosphere, the decrease rate of O_3 density with the decrease of altitude is much smaller than calculated by the photo-equilibrium due to the descending O_3 from the ozone layer at the higher altitude. The deviation of O_3 density at lower altitude than the lower stratosphere in Fig. 8.1 is due to this reason.

8.2 Ozone Loss Cycles by Trace Constituents

Although the vertical profile of stratospheric ozone given by the pure oxygen theory had been thought to be correct for nearly 30 years since advocated by Chapman in 1930, a more accurate rate constant of the reaction (8.3) was obtained in 1960s, and by using the updated value, it was revealed that the profile by the pure oxygen theory over-predicted by about a factor of two as seen in Fig. 8.1. Since the other rate constants related to the pure oxygen theory, $j_{8.1}$, $k_{8.2}$, $j_{8.4}$ were thought to be accurate, it implied that the there is a chemical reaction other than reaction (8.3) to destruct O_3. As described in sect. 7.1, a chain reaction is a prerequisite process for the trace species to affect the higher concentration of O_3. The possibility of such a chain reaction was first suggested regarding the effect of water vapor on the upper atmosphere by Bates and Nicolet(1950), but it was at the end of 1960s to the beginning of 1970s when it were investigated more in detail. The importance of nitrogen oxides for stratospheric ozone was then proposed by Crutzen (1970), and further the relevance of chlorine was pointed by Stolarski and Cicerone (1974) and Wofsy and McElroy (1974). The stratospheric ozone chemistry research in those days was reviewed by Nicolet (1975). Among the research on ozone dissipation by trace gases in those days, the advocate based on atmospheric chemistry by Johnston (1971) on the possibility of ozone destruction by oxides of nitrogen emitted from supersonic stratospheric transport (SST) is particularly important, since it opened human's eyes to the impact of anthropogenic activity on the stratosphere for the first time. Although the flight of SST was not realized, the followed advocacy of the possibility of ozone destruction by chlorofluorocarbons (CFCs) by Molina and Rowland (1974) was actually realized, as will be seen in sects. 8.3 and 8.4, which has had a large impact on the real world by resulting the prohibition of manufacturing of CFCs by an international agreement called the Montreal Protocol.

As for the chain reactions contributing the dissipation of ozone in the stratosphere, there are three processes called HO_x, NO_x, and ClO_x cycles, which can be written formally as,

$$X + O_3 \rightarrow XO + O_2 \quad (8.18)$$

$$XO + O \rightarrow X + O_2 \quad (8.19)$$

$$\text{net} \quad O + O_3 \rightarrow O_2 + O_2.$$

In the HO_x, NO_x, and ClO_x cycles, OH, NO, and Cl mainly act as X, respectively, and play a role in causing O_3 dissipation as a net reaction, the same as reaction (8.3).

8.2.1 Hydrogen Containing Species and HO_x Cycle

The sources of active species containing H atom like OH, HO_2, etc. in the stratosphere are H_2O, H_2, and CH_4. Although water vapor (H_2O) exists at 0.1–1 % in the troposphere, the vapor pressure drops when it passes through the cold tropopause (at 195 K, above tropics), and the mixing ratio of water vapor in the stratosphere is typically 5–6 ppmv. It should be noted, however, the half of H_2O present in the stratosphere is that produced in situ by the oxidation of CH_4. Stratospheric H_2O is a source of OH by the reaction with $O(^1D)$ formed in the photolysis of O_3 just the same as in the troposphere (Sect. 5.1.4).

$$O(^1D) + H_2O \rightarrow 2OH \quad (8.20)$$

Meanwhile, the mixing ratio of troposphere-origin H_2, is about the same as in the troposphere, ca. 0.55 ppmv in the lower stratosphere, and it decreases to ca. 0.4 ppmv above 30 km in the upper stratosphere (Brasseur and Solomon 2005). Similarly, the mixing ratio of troposphere-origin CH_4 is ca. 1.8 ppmv at the tropopause, nearly the same as in the troposphere, and it decreases rapidly within the stratosphere to ca. 0.3 ppmv in the upper stratosphere (Brasseur and Solomon 2005). Neither of H_2 nor CH_4 is photolyzed in the stratosphere but they reacts to form OH by the reaction with $O(^1D)$ similar to H_2O (see Sect. 5.1.6 for CH_4).

$$O(^1D) + H_2 \rightarrow OH + H \quad (8.21)$$

$$O(^1D) + CH_4 \rightarrow OH + CH_3 \tag{8.22}$$

From CH_3 formed in reaction (8.22), H, HCO, H_2 and CO are produced via HCHO in the similar reaction process as in the troposphere mentioned in Chap. 7, sect. 7.1. The H and HCO are then transformed to HO_2 by the reaction with O_2, and they constitute ozone dissipation cycle in the stratosphere.

$$OH + O_3 \rightarrow HO_2 + O_2 \tag{8.23}$$

$$HO_2 + O \rightarrow OH + O_2 \tag{8.24}$$

$$\text{net} \quad O + O_3 \rightarrow O_2 + O_2$$

In the lower stratosphere (<30 km) where the O atom concentration is low, the chain reaction,

$$OH + O_3 \rightarrow HO_2 + O_2 \tag{8.25}$$

$$HO_2 + O_3 \rightarrow OH + 2O_2 \quad (8.25)$$

$$\text{net} \quad O_3 + O_3 \rightarrow 3O_2$$

is more important, while in the upper stratosphere (>40 km) where the O atom concentration is high, the chain reaction,

$$OH + O \rightarrow H + O_2 \tag{8.26}$$

$$H + O_2 + M \rightarrow HO_2 + M \tag{8.27}$$

$$HO_2 + O \rightarrow OH + O_2 \quad (8.24)$$

$$\text{net} \quad O + O + M \rightarrow O_2 + M$$

also operates for dissipating the odd oxygen O_x. The sum of H, OH and HO_2 is called odd hydrogen family, HO_x, and the above chain cycles are collectively called HO_x cycle.

He main termination reaction of the HO_x cycle (Sect. 5.3.5) is,

$$OH + HO_2 \rightarrow H_2O + O_2 \tag{8.28}$$

while the reaction,

$$HO_2 + HO_2 + M \rightarrow H_2O_2 + M \tag{8.29}$$

is not effective since the formed H_2O_2 is photolyzed (Sect. 4.2.8) rapidly in the stratosphere,

$$H_2O_2 + h\nu \rightarrow 2OH. \tag{8.30}$$

In the lower stratosphere, the reaction of HO_2 with NO (Sect. 5.3.2),

$$HO_2 + NO \rightarrow OH + NO_2 \tag{8.31}$$

is also important competing with the above reaction (8.25) with O_3. While this reaction is the most important reaction to produce O_3 in the troposphere (Sects. 7.3.1 and 7.3.2), it forms a chain cycle combing with the HO_x cycle in the stratosphere as,

$$OH + O_3 \rightarrow HO_2 + O_2 \tag{8.23}$$

$$HO_2 + NO \rightarrow OH + NO_2 \tag{8.31}$$

$$NO_2 + h\nu \rightarrow NO + O \tag{8.32}$$

$$O + O_2 + M \rightarrow O_3 + M. \tag{8.2}$$

When the both sides of the equations are summed up, they are cleared to leave nothing, which means nothing happens. This kind of chain reaction is called "null cycle", where null means zero in German. In the null cycle neither net production nor destruction of ozone occurs.

Furthermore in the lower stratosphere, the reactions,

$$OH + H_2O_2 \rightarrow H_2O + CO_2 \tag{8.33}$$

$$OH + CO \rightarrow H + CO_2 \tag{8.34}$$

$$OH + HNO_3 \rightarrow H_2O + NO_3 \tag{8.35}$$

$$OH + HO_2NO_2 \rightarrow H_2O + O_2 + NO_2 \tag{8.36}$$

$$OH + HCl \rightarrow H_2O + Cl \tag{8.37}$$

are also important as loss processes of OH in addition to reactions with O, O_3 and HO_2.

Figure 8.2 shows a comparison of satellite observation and model calculation for the vertical profiles of OH and HO_2 above Fairbanks, Alaska (65°N) in 1997 (Jucks et al. 1998). As shown in the figure, the maximum densities of [OH] and [HO_2] are 2×10^7 and 1.8×10^7 molecules cm^{-3} at the altitude of 45 km and 35–40 km, respectively, and [HO_x] as their sum has maximum value of 3.4×10^7 molecules cm^{-3} in the upper stratosphere at 40–45 km. The model calculation using the rate constants given by the NASA/JPL evaluation No. 12 (DeMore et al. 1997) agrees well with the vertical profile of OH density in the whole stratosphere, while that for HO_2 agrees well in the upper stratosphere, but overestimate its density about 30 % in the lower stratosphere. The observed ratio of [HO_2]/[OH] is <1 in the upper stratosphere, and ~2 at the altitude of 30 km.

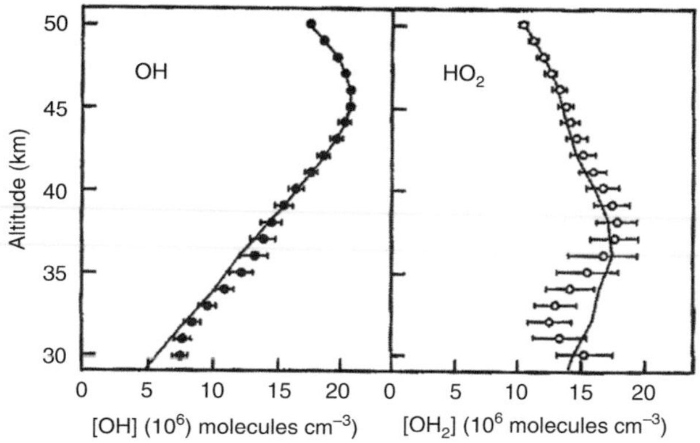

Fig. 8.2 Comparison of satellite observation in 1977 (*symbols*) and model calculation (*solid lines*) for the vertical profiles of the OH and HO_2 mixing ratios in the stratosphere over Fairbanks, Alaska (Adapted from Jucks et al. 1998)

8.2.2 *Nitrogen Containing Species and NO_x Cycle*

The source of active nitrogen, NO, NO_2, etc. in the stratosphere is N_2O of tropospheric origin. The mixing ratio of N_2O in the stratosphere is kept as about 320 ppbv of the tropospheric value up to the altitude of 20 km, and it decreases rapidly to about 20 ppbv at 40 km (Brasseur and Solomon 2005; Seinfeld and Pandis 2006). The 90 % of this rapid decrease is due to the photolysis (Sect. 4.3.4),

$$N_2O + h\nu \rightarrow N_2 + O(^1D), \qquad (8.38)$$

and the rest is mainly due to the reaction with $O(^1D)$ (Sect. 5.1.5) formed in the photolysis of O_3,

$$O(^1D) + N_2O \rightarrow N_2 + O_2 \quad (0.39) \qquad (8.39)$$

$$\rightarrow 2NO \quad (0.61). \qquad (8.40)$$

The NO formed in reaction (8.40) is the main source of NO_x in the stratosphere. The recommended value for the branching ratio of reaction (8.40) is 0.61 by the NASA/JPL panel evaluation No. 17 (Sander et al. 2011) (see Sect. 5.1.5).

The NO formed in the stratosphere reacts with O_3 to be converted to NO_2, and NO_2 returns to NO by reacting with O atoms, which results in net O_3 loss as,

$$NO + O_3 \rightarrow NO_2 + O_2 \qquad (8.41)$$

$$NO_2 + O \rightarrow NO + O_2 \qquad (8.42)$$

$$\text{net} \quad O + O_3 \rightarrow 3O_2.$$

Typical conversion time for the NO to NO_2 by reaction (8.41) in the stratosphere is about 1 min, and that of NO_2 to NO by reaction (8.42) is about 10 min. In stratospheric chemistry the nitrogeneous species including N_2O_5, HNO_3, $ClONO_2$, etc. in addition to NO and NO_2, are called odd nitrogen family, NO_x, and chain reaction including these species is called NO_x cycle. It should be noted the definition of NO_x here is different from in tropospheric chemistry where NO_x is defied as $NO + NO_2$. The chain reaction consisting of reaction (8.41) and (8.42) occurs in high efficiency in the upper stratosphere where the concentration of O atoms are high. When the photolysis of NO_2 competes reaction (8.42),

$$NO_2 + h\nu \rightarrow NO + O, \tag{8.32}$$

a series of the processes,

$$NO + O_3 \rightarrow NO_2 + O_2 \tag{8.41}$$
$$NO_2 + h\nu \rightarrow NO + O \tag{8.32}$$
$$O + O_2 + M \rightarrow O_3 + M \tag{8.2}$$

forms null cycle, and net loss of O_3 does not occur.

Another reaction between NO_x and O_x,

$$NO_2 + O_3 \rightarrow NO_3 + O_2 \tag{8.41}$$

produces NO_3, but NO_3 is photolyzed very rapidly during the daytime (Sect. 4.2.4),

$$NO_3 + h\nu \rightarrow NO_2 + O \quad (0.89) \tag{8.44}$$
$$\rightarrow NO + O_2, \quad (0.11) \tag{8.45}$$

and dissipates within a few seconds. The branching ratio of photolytic reaction (8.44) and (8.45) is known to depend on wavelength and temperature, it is deduced to be about 9:1 under the stratospheric conditions (Sander et al. 2011). The net sequence of reaction (8.41) and (8.44) corresponds to the photolysis of O_3. In the nighttime, NO_3 is stable, and further react with NO_2 to form N_2O_5 (Sect. 5.5.2).

$$NO_3 + NO_2 + M \rightarrow N_2O_5 + M \tag{8.46}$$

The produced N_2O_5 is dissipated by the photolysis in the daytime (Sect. 4.2.4),

$$N_2O_5 + h\nu \rightarrow NO_2 + NO_3 \tag{8.47}$$
$$\rightarrow NO_3 + NO + O. \tag{8.48}$$

As for the photolytic process, reaction (8.47) is known to be the main rout but reaction (8.48) also proceeds partly under the stratospheric conditions (Atkinson et al. 2004; Sander et al. 2011). The heterogeneous reactions of N_2O_5 on ice particle of polar stratospheric clouds (PSC) (Sects. 6.1.5 and 6.5.1),

$$N_2O_5(g) + H_2O(s) \rightarrow 2\ HNO_3(g) \tag{8.49}$$

will be discussed in the next section. In contrast the thermal decomposition reaction of N_2O_5 (Sect. 5.5.2),

$$N_2O_5 + M \rightarrow NO_3 + NO_2 + M \tag{8.50}$$

is not important in general in the stratosphere since the temperature low.

As for the loss process of NO_2 in the stratosphere, cross reactions with HO_x cycle mentioned in the previous section,

$$NO_2 + OH + M \rightarrow HONO_2 + M \tag{8.51}$$

$$NO_2 + HO_2 + M \rightarrow HO_2NO_2 + M \tag{8.52}$$

are the most important (Sects. 5.2.4 and 5.3.4). Particularly, $HONO_2$ (=HNO_3) is an important reservoir of NO_x in the lower stratosphere, and reaction (8.51) is important as a termination reaction of the HO_x and NO_x cycles. Also, HO_2NO_2 formed in reaction (8.52) exists as an important reservoir of NO_x since it is thermally stable in the lower and middle stratosphere where the temperature is low. The chemical loss processes of $HONO_2$ and HO_2NO_2 are the photolysis and reaction with OH (Sect. 5.2.5 for $HONO_2$).

$$HONO_2 + h\nu \rightarrow OH + NO_2 \tag{8.53}$$

$$HONO_2 + OH \rightarrow NO_3 + H_2O \tag{8.54}$$

$$HO_2NO_2 + h\nu \rightarrow HO_2 + NO_2 \quad (0.80) \tag{8.55}$$

$$\rightarrow OH + NO_3 \quad (0.20) \tag{8.56}$$

$$HO_2NO_2 + OH \rightarrow NO_2 + H_2O \tag{8.57}$$

The thermal decomposition of HO_2NO_2 (Sect. 5.3.4),

$$HO_2NO_2 + M \rightarrow HO_2 + NO_2 + M \tag{8.58}$$

is not important in general due to the low temperature.

Other than these processes, the production of $ClONO_2$ by the cross reaction of NO_2 with the main carrier ClO (Sect. 5.3.4) of ClO_x cycle mentioned later, is important as the reservoir of NO_x.

$$ClO + NO_2 + M \rightarrow ClONO_2 + M \tag{8.59}$$

The dissipating reaction of $ClONO_2$ is mainly photolysis (Sect. 4.4.2),

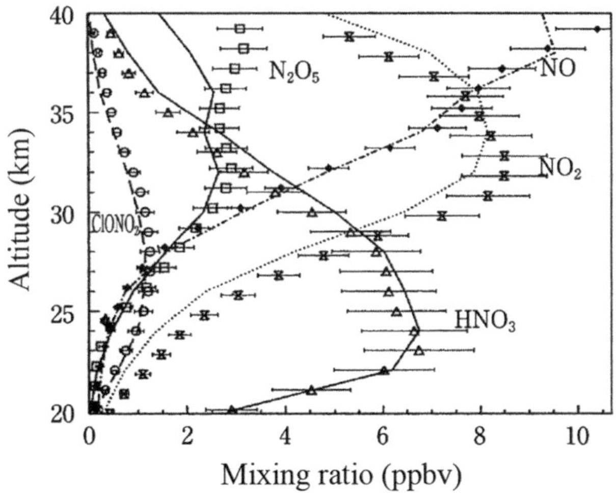

Fig. 8.3 Comparison of balloon observation (*symbols*) and model calculation (*lines*) for the vertical profiles of odd nitrogen (NO, NO_2, N_2O_5, HNO_3, $ClONO_2$) mixing ratio in the stratosphere at dawn over New Mexico, US (Adapted from Seinfeld and Pandis 2006 based on Sen et al. 1998)

$$ClONO_2 + h\nu \rightarrow ClO + NO_2 \quad (0.4) \tag{8.60}$$

$$\rightarrow Cl + NO_3 \quad (0.6). \tag{8.61}$$

The heterogeneous reaction of $ClONO_2$ in the polar region will be described in sect. 8.4.

Figure 8.3 depicts the comparison between the observed vertical profiles of odd nitrogen compounds (NO, NO_2, N_2O_5, HNO_3, $ClONO_2$) in the stratosphere obtained by an FTIR (Fourier transform infrared spectrometer) loaded on a balloon in New Mexico, United States, in autumn 1993, and 24 h-averaged calculated values assuming photochemical equilibrium (Sen et al. 1998). As shown in the figure, the mixing ratio of NO in the mid-latitude increases from ca. 0.1 ppbv at 20 km to ca. 10 ppbv at 40 km by a factor of 100. On the other hand, the mixing ratio of NO_2 increases from ca. 1 ppbv at 20 km to the maximum of ca. 8 ppbv at 30–35 km, and then decreases at the higher altitude to ca. 5 ppbv at 40 km. In the lower and mid-latitude, the most part of NO_x exists as HNO_3 and its mixing ratio is ca. 3 ppbv at 20 km, maximizes to ca. 6 ppbv at around 23 km, and then decreases rapidly at the higher altitude than 30 km. The mixing ratio of N_2O_5 is the highest at higher than 30 km, but shows a large diurnal cycle, high in the nighttime and low in the daytime (Brasseur and Solomon 2005). The largest mixing ratio at dawn shown in Fig. 8.3 is ca. 3 ppbv over 30 km, which is about 20 % of NO_x and decrease to 5–10 % in the daytime average. The mixing ratio of $ClONO_2$ shows the maximum of ca. 1 ppbv at 25–30 km. The total amount of odd nitrogen is almost constant at ca. 18 ppbv at higher altitude than 30 km, and decreases to ca. 10 and 3 ppbv at 25 and 20 km, respectively (Sen et al. 1998). As shown in Fig. 8.3, the calculation assuming the photo-equilibrium reproduces the absolute values and vertical profiles of mixing ratios of NO_x well, suggesting that the reaction system of NO_x chemistry in the stratosphere is understood fully. Within the ozone hole where PSC develops, N_2O_5 and HNO_3 are removed almost completely from the gas phase by heterogeneous reactions, and it should be noted that the profile of NO_x is largely different from Fig. 8.3. The stratospheric reaction chemistry relevant to PSC within the ozone hole is described in sect. 8.4.

8.2.3 Chlorine Containing Species and ClO_x Cycle

Among the halogen compounds emitted in the troposphere, marine origin CH_3Cl has the relatively long atmospheric lifetime of 1.5 years (Brasseur and Solomon 2005), and partly reaches to the stratosphere to becomes a Cl atom source in the natural stratosphere. The mixing ratio of CH_3Cl in the clean troposphere is ca. 550 ppbv, and it is thought to intrude into the stratosphere with this mixing ratio. From CH_3Cl molecules reached to the stratosphere, Cl atoms and chloromethyl radical are formed by photolysis (Sect. 4.3.8),

$$CH_3Cl + h\nu \rightarrow CH_3 + Cl \tag{8.62}$$

and reaction with OH,

$$CH_3Cl + OH \rightarrow CH_2Cl + H_2O, \tag{8.63}$$

respectively. The CH_2Cl formed in reaction (8.63) is the alkyl-type radical, and it proceeds through the following reaction processes,

$$CH_2Cl + O_2 + M \rightarrow CH_2ClO_2 + M \tag{8.64}$$

$$CH_2ClO_2 + NO \rightarrow CH_2ClO + NO_2 \tag{8.65}$$

$$CH_2ClO + O_2 \rightarrow ClCHO + M \tag{8.66}$$

to produce formyl chloride (ClCHO). From formyl chloride, Cl atoms are released by the similar type of photolysis as formaldehyde,

$$ClCHO + h\nu \rightarrow HCO + Cl \tag{8.67}$$

The Cl atoms released by reactions (8.62) and (8.67) react mainly with O_3,

$$Cl + O_3 \rightarrow ClO + O_2, \tag{8.68}$$

to form ClO radicals (Sect. 5.6.1). As for the reaction of ClO, the reactions with O atoms (Sect. 5.1.3) and NO in the upper and lower stratosphere are important, respectively.

$$ClO + O \rightarrow Cl + O_2 \tag{8.69}$$

$$ClO + NO \rightarrow Cl + NO_2 \tag{8.70}$$

Reactions (8.68) and (8.69) forms an O_3 dissipation cycle,

$$\begin{array}{lll} & Cl + O_3 \rightarrow ClO + O_2, & (8.68) \\ & ClO + O \rightarrow Cl + O_2 & (8.69) \\ \hline \text{net} & O + O_3 \rightarrow 3O_2, & \end{array}$$

and reactions (8.68) and (8.70) forms a null cycle,

$$Cl + O_3 \rightarrow ClO + O_2 \tag{8.68}$$

$$ClO + NO \rightarrow Cl + NO_2 \tag{8.70}$$

$$NO_2 + h\nu \rightarrow NO + O \tag{8.32}$$

$$O + O_2 + M \rightarrow O_3 + M \tag{8.2}$$

Here, the active chlorine family, Cl, ClO, HOCl and $ClONO_2$, is called ClO_x, and the chain reaction consisting of them is called the ClO_x cycle. The rate determining steps of the above two ClO_x cycles are ClO + O and ClO + NO, respectively, and the contributions of these two cycles at the altitude of 40 km in the upper stratosphere are calculated to be nearly equal, based on the mixing ratios of [O] and [NO], and their rate constants (Seinfeld and Pandis 2006).

The Cl atoms, a chain carrier of ClO_x cycles, react with species other than O_3 to form HCl,

$$Cl + CH_4 \rightarrow CH_3 + HCl \tag{8.71}$$

$$Cl + H_2 \rightarrow H + HCl \tag{8.72}$$

$$Cl + HO_2 \rightarrow O_2 + HCl. \tag{8.73}$$

The main loss processes of HCl are photolysis (Sect. 4.4.3) and reaction with OH and $O(^1D)$,

$$HCl + h\nu \rightarrow H + Cl \tag{8.74}$$

$$HCl + OH \rightarrow H_2O + Cl \tag{8.75}$$

$$HCl + O(^1D) \rightarrow OH + Cl \tag{8.76}$$

$$\rightarrow H + ClO. \tag{8.77}$$

However, since the rate constants of these reactions are not very large, reactions (8.71), (8.72) and (8.73) act as the effective chain termination reactions, and HCl becomes the most important reservoir of ClO_x in the stratosphere.

As for the reaction of ClO, another chain carrier of the ClO_x cycle, the reaction with NO_2 (Sect. 5.6.5) is the most important, except the reactions with O and NO mentioned above.

$$ClO + NO_2 + M \rightarrow ClONO_2 + M \tag{8.78}$$

Although a major loss process of $ClONO_2$ is the photolysis (Sect. 4.4.2),

$$ClONO_2 + h\nu \rightarrow ClO + NO_2 \tag{8.79}$$

$$\rightarrow Cl + NO_3 \tag{8.80}$$

the photolytic rate is not very large, so that $ClONO_2$ becomes an important reservoir in the mid-stratosphere. As shown later, the amount of $ClONO_2$ is about 50 % of HCl at 25–30 km as a reservoir of ClO_x, and ClO_x exists almost exclusively as HCl at $\geq$30 km.

Other than these reactions, the reaction of ClO with HO_2 (Sect. 5.6.4),

$$ClO + HO_2 \rightarrow HOCl + O_2 \tag{8.81}$$

is important as an HOCl forming reaction. Since HOCl is easily photolyzed by visible light (Sect. 4.4.4),

$$HOCl + h\nu \rightarrow OH + Cl, \tag{8.82}$$

it cannot be an effective reservoir. However, the mixing ratio of HOCl gets higher in the stratosphere of polar night, and its heterogeneous reactions are important together with those of $ClONO_2$. Other than these reactions, photolysis of ClO itself (Sect. 4.4.5) and reaction with OH (Sect. 5.6.3) have to be considered.

$$1ClO + h\nu \rightarrow Cl + O \tag{8.83}$$

$$ClO + OH \rightarrow Cl + HO_2 \quad (0.94) \tag{8.84}$$

$$\rightarrow HCl + O_2 \quad (0.06) \tag{8.85}$$

Although the ratio of reaction (8.85) is small, it affect to decrease the [ClO]/[HCl] ratio in the upper stratosphere.

Figure 8.4 depicts the observed vertical profiles of chlorine species in the mid-latitude (30–60°N) stratosphere obtained by satellite observation in 2004 (WMO 2007). As shown in the figure, the mixing ratio of total Cl is ca. 3.6 ppbv. Since the mixing ratio of the natural source Cl is ca. 550 pptv, the rest can be ascribed to the anthropogenic chlorine compounds described in sect. 8.3, which amounts to a few times the natural emission. Among the chlorine compounds, HCl is the most abundant, and the mixing ratio increases from ca. 1.5 ppbv at 20 km to ca. 3 ppbv at 40 km. Next abundant species is $ClONO_2$ which amounts to the maximum mixing ratio of ca. 1 ppbv at 20–30 km. The mixing ratio of ClO maximizes at ca. 0.5 ppbv at 35–40 km, and other chlorine species such as HOCl, etc. are less than 0.2 ppbv at all altitude.

Accompanying to the increase of the mixing ratio of ClO_x due to the increase of anthropogenic chlorine containing compounds, self-reaction between ClO_x is getting important, which will be described in sect. 8.3.

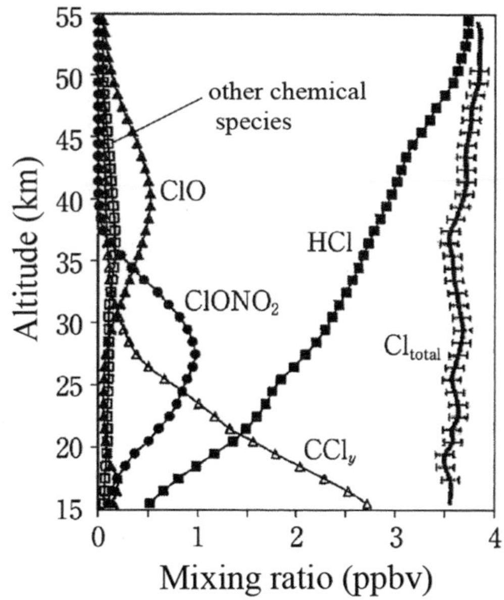

Fig. 8.4 Vertical profiles of active chlorine compounds by the satellite observation in 2004 in the stratosphere at northern hemisphere mid-latitude (30–60°N). CCl_y is the total organic chlorine compounds (Adapted from WMO 2007)

8.2.4 Reactions of Other Halogen (Bromine, Iodine, Fluorine) Compounds

Other than chlorine, halogen compounds containing bromine, iodine and fluorine are present in the stratosphere. Among these, the most important from the point of ozone depletion is bromine. Although iodine also forms an ozone-destructing chain cycle, its impact is limited since its mixing ratio is low. In contrast, fluorine does not constitute such a chain cycle due to its reactivity, and does not contribute to ozone destruction.

Bromine The most important bromine compound of natural origin is CH_3Br from marine organisms as with CH_3Cl. The mixing ratio of CH_3Br in the clean troposphere is ca. 10 pptv, and since the atmospheric lifetime is ca. 1.5 years, nearly the same as CH_3Cl, bromine is thought to be present in the natural stratosphere with this mixing ratio (Brasseur and Solomon 2005). In spite of the fact that the mixing ratio of Br is only one 50th of Cl in the natural stratosphere, the impact of Br on ozone dissipation is thought to be comparable to Cl. The reason is ascribed to the very efficient ozone destructing chain cycle of Br as compared to Cl.

After reaching to the stratosphere, CH_3Br releases Br atoms by photolysis (Sect. 4.3.8) and reaction with OH.

$$CH_3Br + h\nu \rightarrow CH_3 + Br \tag{8.86}$$

$$CH_3Br + OH \rightarrow CH_2Br + H_2O \tag{8.87}$$

The rate constant of the OH reaction (8.87) is about the same as for CH_3Cl (Table 5.2), but the absorption cross section of CH_3Br at the solar ultraviolet region in the stratosphere (around 200 nm) is about a few tens of times larger than CH_3Cl (Fig. 4.32), the release of Br by th0065 photolysis occurs at much faster rate than CH_3Cl. This is one of the reasons that the chain cycle of Br is efficient for the ozone destruction. Since the Br atoms released into the stratosphere react with O_3 and the formed BrO reacts with O atoms to regenerate Br, the BrO_x cycle is formed to dissipate O_3 just like the ClO_x cycle.

$$Br + O_3 \longrightarrow BrO + O_2 \tag{8.97}$$

$$\begin{array}{ll} & BrO + O \longrightarrow Br + O_2 \\ \hline \text{net} & O + O_3 \longrightarrow 3O_2 \end{array} \tag{8.89}$$

Other than the reaction with O atoms, BrO reacts with NO and OH to regenerate Br,

$$BrO + NO \rightarrow Br + NO_2 \tag{8.90}$$

$$BrO + OH \rightarrow Br + HO_2. \tag{8.91}$$

However, the more important Br regenerating process is the photolysis of BrO itself,

$$BrO + h\nu \rightarrow Br + O. \tag{8.92}$$

Since BrO has a large absorption cross section in wavelength regions longer than 300 nm (Fig. 4.40), the conversion efficiency from BrO to Br by photolysis is much higher than in the case of ClO photolysis to give Cl.

On the other hand, the formation of HBr by the reaction of Br and HO_2,

$$Br + HO_2 \rightarrow HBr + O_2 \tag{8.93}$$

is important as the chain termination reaction of the BrO_x cycle. Since Br does not react with CH_4 and H_2, the above reaction is the major process to from HBr. The produced HBr returns to Br rather efficiently by the photolysis and reactions with OH and $O(^1D)$,

$$HBr + h\nu \rightarrow H + Br \tag{8.94}$$

$$HBr + OH \rightarrow Br + H_2O \tag{8.95}$$

$$HBr + O(^1D) \rightarrow Br + OH. \tag{8.96}$$

Therefore, HBr cannot be an important reservoir, and the efficiency of reaction (8.93) for the chain termination of the BrO_x cycle is not very high.

Another chain termination reaction of BrO operates with NO_2 and HO_2 similar to the case of ClO,

$$BrO + NO_2 + M \rightarrow BrONO_2 + M \tag{8.97}$$

$$BrO + HO_2 \rightarrow HOBr + O_2, \tag{8.98}$$

to form $BrONO_2$ and HOBr. However, both of $BrONO_2$ and HOBr are photolyzed (Sects. 4.4.2 and 4.44.4) with high rates,

$$BrONO_2 + h\nu \rightarrow BrO + NO_2 \tag{8.99}$$

$$HOBr + h\nu \rightarrow OH + Br, \tag{8.100}$$

they cannot be effective reservoirs and the efficiency of reactions (8.88) and (8.89) as chain termination are also low. Thus, the low effectiveness of the chain termination reactions is another reason for the high efficiency of the BrO_x cycle and enhances the importance of Br in the ozone destruction reaction even though the mixing ratio is relatively low. The $BrONO_2$ and HOBr are stable in the absence of solar radiation, and play an important role in the multiphase chain reaction system including heterogeneous reactions on PSC together with $ClONO_2$ and HOCl in the stratospheric polar night, which will be mentioned in the next section.

The mixing ratio of bromine compounds in the present day stratosphere is ca. 20 pptv, which is less than one 150th of chlorine compounds whose mixing ratio is ca. 3.5 ppbv. Since the mixing ratio of Br from natural origins is ca. 10 pptv, this means that anthropogenic Br compounds intrude into the stratosphere in about the same amount as CH_3Br from the natural source. With the increase of the emissions of anthropogenic halocarbons, the cross reactions between ClO_x and BrO_x are getting important, and will be described in sect. 8.3.

Figure 8.5 shows the vertical distribution of active bromine compounds in 24-h average at 30°N obtained by the model calculation (McElroy 2002). Summing up all these compounds, the total mixing ratio of bromine in the stratosphere is ca. 20 pptv, as mentioned earlier. Among BrO_x, BrO and $BrONO_2$ are the most abundant, at ca. 10 pptv each, at the altitudes higher and lower than 35 km, respectively. Meanwhile, the mixing ratio of Br atoms increases with the altitude monotonically, and reaches to ca. 10 pptv, nearly the same as BrO at the stratopause. The mixing ratio of HBr is very low, less than 1 pptv, implying that it does not act as an effective Br reservoir, being different from HCl in the case of Cl. The mixing ratio of HOBr is less than 1 pptv in the upper stratosphere, and even decreases further at altitudes of less than 30 km.

Iodine Although several biogenic iodine compounds are emitted from natural sources, their photolytic lifetime in the troposphere is short, e.g. only a few days even for the relatively long-life species like CH_3I (Brasseur and Solomon 2005). Therefore, in general, it is thought that tropospheric CH_3I does not intrude into the stratosphere. However, it has been suggested that strong updraft in the tropics could

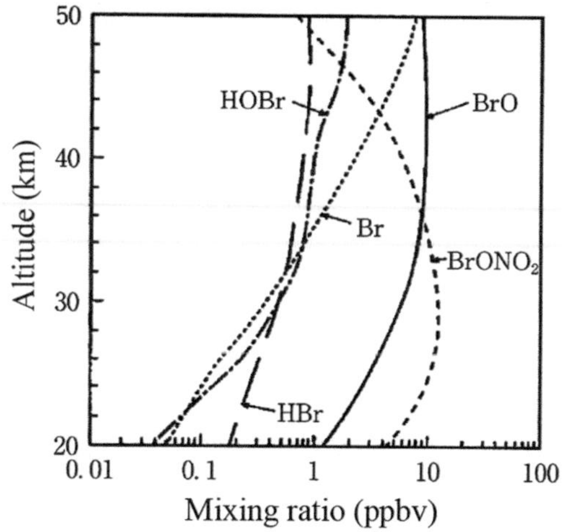

Fig. 8.5 Vertical profiles of active bromine compounds in the stratosphere by model calculation in 24-h average at 30°N (Adapted from McElroy 2002)

rapidly transport CH_3I near the sea-surface to the stratosphere, and may further cause dissipation of O_3 (Solomon et al. 1994).

Iodine atoms are released from CH_3I and reach to the stratosphere by rapid photolysis (Sect. 4.3.8).

$$CH_3I + h\nu \rightarrow CH_3 + I \tag{8.101}$$

The reactions of I atoms are similar to Br, e.g. with O_3,

$$I + O_3 \rightarrow IO + O_2. \tag{8.102}$$

However, the photolysis rate of formed IO (Sect. 4.3.4) is very large,

$$IO + h\nu \rightarrow I + O \tag{8.103}$$

and together with reactions,

$$IO + NO \rightarrow I + NO_2 \tag{8.104}$$

$$IO + O \rightarrow I + O_2 \tag{8.105}$$

they compose the IO_x cycle resulting from O_3 destruction. Since the intermediates such as $IONO_2$ and HOI formed in the reactions of IO with NO_2 and HO_2 are photolyzed more rapidly than $BrONO_2$ and HOBr, they do not work as termination reactions, and the chain reaction is characterized by very high efficiency.

Fluorine: Among halogens, fluorine is very different from Cl, Br and I in respect that it does not cause O_3 dissipation. Fluorine is transported to the stratosphere in the form of anthropogenic CFCs and HCFCs as described in the following section, and also as CF_4 and hydrofluorocarbons (HFCs). From the decomposition processes of these compounds (see sect. 8.3), COF_2, COFCl, etc. are formed, and their photolysis and the succeeding reactions give F atoms and other reactive species,

$$COF_2 + h\nu \rightarrow FCO + F \tag{8.106}$$

$$COFCl + h\nu \rightarrow FCO + Cl \tag{8.107}$$

$$FCO + O_2 + M \rightarrow FC(O)O_2 + M \tag{8.108}$$

$$FC(O)O_2 + NO \rightarrow FCO_2 + NO_2 \tag{8.109}$$

$$FCO_2 + NO \rightarrow FNO + CO_2. \tag{8.110}$$

The rate constants of reactions (8.108), (8.109) and (8.110) at 296 K have been obtained by Wallington et al. (1994) as $k_{8.108} = (1.2 \pm 0.2) \times 10^{-12}$, $k_{8.109} = (2.5 \pm 0.8) \times 10^{-11}$, $k_{8.110} = (1.3 \pm 0.7) \times 10^{-10}$ cm^3 molecule^{-1} s^{-1}, respectively, and they are all known to be fast reactions.

The rate constant of the reaction of F atoms and O_3,

$$F + O_3 \rightarrow FO + O_2 \tag{8.111}$$

is $k_{8.111}$(298 K) $= 1.0 \times 10^{-11}$ cm^3 molecule^{-1} s^{-1}, which is about the same as $k_{8.68}$(298 K) $= 1.2 \times 10^{-11}$ cm^3 molecule^{-1} s^{-1} for the reaction of $Cl + O_3$. However, in case of F, the reaction of F atom with the atmospheric major component, O_2,

$$F + O_2 + M \rightleftarrows FO_2 + M \tag{8.112}$$

has the equilibrium constant $K_{8.112}$ (298 K) $= 3.7 \times 10^{-16}$ cm^3 molecule^{-1}, which is much larger than that for Cl, 2.9×10^{-21} cm^3 molecule^{-1} (Sander et al. 2011), so that the equilibrium between F and FO_2 inclines to FO_2 to give $[FO_2]/[F] \sim 10^4$. Therefore, the reaction of $F + O_3$ is much less important than the reaction with O_2. Meanwhile, F atoms react with CH_4, H_2O and H_2 with rate constants of the order of $\sim 10^{-11}$ cm^3 molecule^{-1} s^{-1} (Sander et al. 2011) to form HF.

$$F + CH_4 \rightarrow HF + CH_3 \tag{8.113}$$

$$F + H_2O \rightarrow HF + OH \tag{8.114}$$

$$F + H_2 \rightarrow HF + H \tag{8.115}$$

Since the bond dissociation energy of H-F is very large ($D_{298} = 568$ kJ mol^{-1}), HF is photochemically stable and toward reactions becomes a very stable reservoir so it is eventually transported to the troposphere and washed away by precipitation.

The FO_2 formed by reaction (8.112) does not react with O_3, and its main reactions are thought to be,

$$FO_2 + NO \rightarrow FNO + O_2 \tag{8.116}$$

$$FNO + h\nu \rightarrow F + NO \tag{8.117}$$

(Wallington et al. 1995; Burley et al. 1993).

Although FO formed by the reaction of F atoms and O_3 can react with NO and O to regenerate F (Sander et al. 2011),

$$FO + NO \rightarrow F + NO_2 \tag{8.118}$$

$$FO + O \rightarrow F + O_2, \tag{8.119}$$

F atoms are mostly removed as HF from the reaction system, and O_3 destruction cycle by F does not have an effect.

The abundance of F compounds in the stratosphere has been obtained as 1.2 ppbv by satellite observations as of 1985, and most compounds are present as HF at the altitude of higher than 30 km (Zander et al. 1992).

8.3 Gas Phase Chain Reactions and Ozone Depletion by CFCs

While the HO_x, NO_x and ClO_x cycles, which tend to decrease O_3 mixing ratios in the stratosphere, are naturally occurring chemical reactions caused by trace species of natural origin, Molina and Rowland (1974) advocated the possibility of the depletion of the ozone layer by anthropogenic chlorofluorocarbons (CFCs) adding large amount of Cl to the stratosphere. The CFCs are the carbon compounds containing only chlorine and fluorine in a molecule and they are all anthropogenic. Since they undergo neither photolysis nor reactions with active species such as OH, etc., they have no sink in the troposphere, and all reached to the stratosphere where they are photolyzed to release Cl atoms. There are many other anthropogenic substances causing ozone destruction in the stratosphere, such as hydrochlorofluorocarbon (HCFCs) containing hydrogen in addition to chlorine d fluorine in a molecule, 1, 1, 1-trichloroethane (CH_3CCl_3), carbon tetrachloride (CCl_4), and bromochlorofluorocarbons (halon) containing bromine, chlorine and fluorine in a molecule (WMO 2011; Brasseur and Solomon 2005). They are collectively called halocarbons. The atmospheric lifetime of the species that do not have a hydrogen atom in a molecule, such as CFCs, halons, CCl_4, etc., are determined by the photolysis rate in the stratosphere and are in general very long, a few tens to more than 100 years. Since the hydrogen containing species like HCFCs and CH_3CCl_3 are consumed by the reaction with OH in the troposphere, their atmospheric lifetimes are shorter, a few to ten-odd years comparable to CH_4, but they also reach to the stratosphere and release Cl atoms to take part in the ozone depletion (Brasseur and Solomon 2005). Among CFCs, the mixing ratios of

CF_2Cl_2 (CFC-12) and $CFCl_3$ (CFC-11) are the highest, ca. 540 and 240 ppbv in 2008, respectively, followed by 190 and 90 ppbv of other anthropogenic chlorine-containing compounds, CHF_2Cl (HCFC-22) and CCl_4, respectively. The total amount of chlorine in the stratosphere is ca. 3.5 ppbv, including 550 ppbv of natural origin CH_3Cl (WMO 2011). Similarly, as for anthropogenic bromine compounds, CF_2ClBr (halon-1211) and CF_3Br (halon-1301) have the highest mixing ratios, ca. 4.3 and 3.2 pptv, respectively. The total bromine mixing ratio is ca. 20 pptv by adding these halons, and CH_3Br at ca. 10 pptv which is from both natural and anthropogenic origins (WMO 2011).

Ultraviolet radiation with 200–220 nm reaches to the altitude of 30–40 km in the middle stratosphere, which is called the "atmospheric window" of the stratosphere (see Chap. 4, sect. 4.1). The absorption spectra of CFCs, HCFCs and halons have large cross section around 200 nm (Figs. 4.33, 4.34, and 4.35) and match with the atmospheric window resulting efficient photolysis. For example, CFC-12 is photolyzed to give Cl atoms.

$$CF_2Cl_2 + h\nu \rightarrow CF_2Cl + Cl \quad (8.120)$$

Simultaneously formed CF_2Cl is an alkyl type radical, and the succeeding processes release another Cl atom,

$$CF_2Cl + O_2 + M \rightarrow CF_2ClO_2 + M \quad (8.121)$$

$$CF_2ClO_2 + NO \rightarrow CF_2ClO + NO_2 \quad (8.122)$$

$$CF_2ClO + M \rightarrow COF_2 + Cl, \quad (8.123)$$

and carbonyl fluoride COF_2 is produced. In the case of $CFCl_3$, including the photolysis of similarly formed carbonyl chlorofluoride (COFCl), all three Cl atoms are ultimately released. Other than photolysis, CFCs react with $O(^1D)$, and in the case of CF_2Cl_2, the main reaction is,

$$CF_2Cl_2 + O(^1D) \rightarrow CF_2Cl + ClO \quad (8.124)$$

to form CF_2Cl and ClO. The release of Br and BrO from halons are similar due to photolysis and the reaction with $O(^1D)$.

Figure 8.6 compares the vertical profiles of the O_3 destruction rates by each of the O_x, HO_x, NO_x, ClO_x and BrO_x cycles, calculated by a model considering all the reactions mentioned in Sects. 8.1, 8.2 and in this section, with those values calculated based on the observed values of radicals by a balloon (Osterman et al. 1997). As shown in the figure, these gas phase chain cycles cause large O_3 destruction above 30 km in the upper stratosphere, and in the middle and lower stratosphere below 30 km the dissipation rate is less than 2×10^6 molecules cm^{-3} s^{-1}. Figure 8.7 shows the relative contribution of each cycle to the O_3 destruction (Osterman et al. 1997). As shown in Fig. 8.7, the contribution of HO_x is the most important in the upper stratosphere above 40 km and also in the lower stratosphere

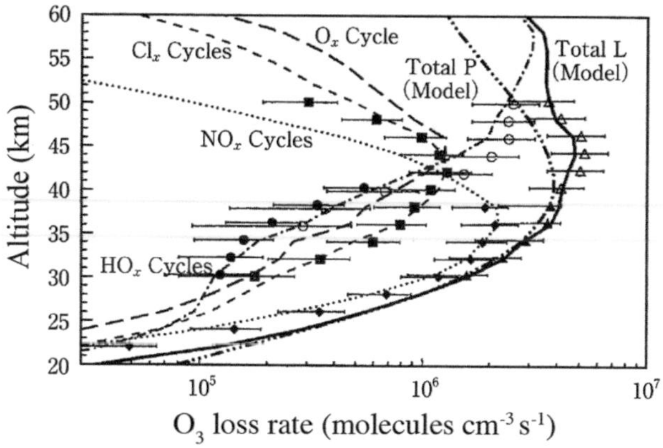

Fig. 8.6 Comparison of observed (*symbols*) and model calculated (*lines*) vertical profiles of ozone dissipation rate by each cycle of O_x, ClO_x, NO_x and HO_x (Adapted from Seinfeld and Pandis 2006 based on Osterman et al. 1997 updating the reaction rate constants). ●, ○ are the contribution of HO_x based on different observations, ■, ◆ are the contribution of ClO_x and NO_x based on observation, respectively, and ▲, △ are the total dissipation rate from observation and that partly using model calculation for unobserved NO_x, respectively. —— and (2点鎖線) are total dissipation and formation rate of O_3 by model calculation, respectively

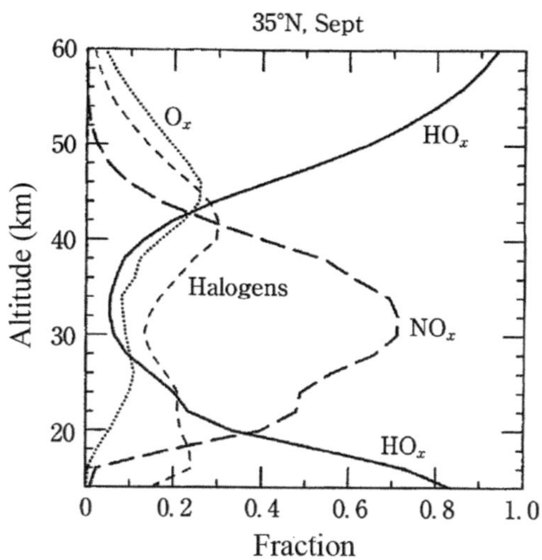

Fig. 8.7 Contributions of O_x, HO_x, NO_x and ClO_x cycles to the ozone depletion by altitude (Adapted from Seinfeld and Pandis 2006 based on Osterman et al. 1997 updating the reaction rate constants)

below 20 km, and NO_x cycle is the most important in between in the middle stratosphere. Meanwhile, the contribution of halogen cycles (ClO_x, BrO_x) maximizes at around 40 km.

From these considerations, it has been deduced theoretically that the ozone depletion due to the increase of anthropogenic CFCs and halons occurs in the upper stratosphere at 30–35 km at the mid- and low latitude, and the decreased amount of ozone from 1980 to 2004 is 5–10 % (WMO 2007). This value agrees well the observed value of ca. 7 % decrease obtained by satellites and sondes (WMO 2007). In contrast, the more distinct ozone depletion in the lower stratosphere in the polar region, known as the "ozone hole," is due to the multiphase chain reaction involving PSCs as discussed in the next section.

8.4 Heterogeneous Reactions on PSCs and Ozone Hole

As mentioned in the previous section, ozone depletion by the increase of CFCs and halons had been anticipated to occur at 5–10 % in the 33–35 km altitude region. In contrast, it was discovered surprisingly that the ozone column density in the Antarctic Continent in spring (September–October) decreased to ca. 100 DU (Dobson Unit, 100 DU is equivalent to 1 mm thickness of O_3 at 0 °C, 1 atm) from the normal value of ca. 300 DU (Farman et al. 1985; Chubachi 1985), and the ozone depletion was found to be happening in the lower stratosphere. This phenomenon was revealed by satellite observation to occur in the spatial scale to cover the whole of Antarctica, and is called Antarctic ozone hole. Figure 8.8 shows typical vertical profiles of the partial pressure of O_3 before and after the development of ozone hole (Hofmann et al. 1987). Although there were various views for the cause of this phenomenon shortly after the discovery, Anderson et al. (1989) found by aircraft observation that ClO and O_3 have distinct anti-correlation spatially and temporarily in the ozone hole as depicted in Fig. 8.9, and proved that the direct cause is due to the chemistry of the ClO_x cycle.

However, the springtime is the dawn of Antarctica, and enough UV radiation does not reach it. Therefore, O atom mixing ratio is low, and the phenomenon cannot be explained by the ClO_x cycle considering only the gas phase reactions

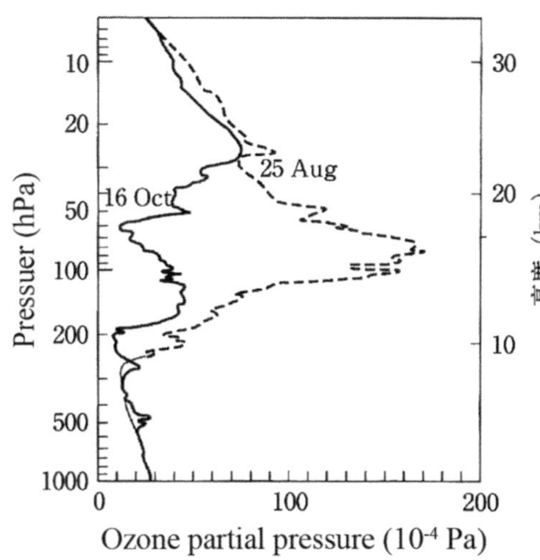

Fig. 8.8 Comparison of the ozone vertical profiles in 1986 before ozone hole formation (August 25) and after its development (October 26) (Adapted from Hofmann et al. 1987)

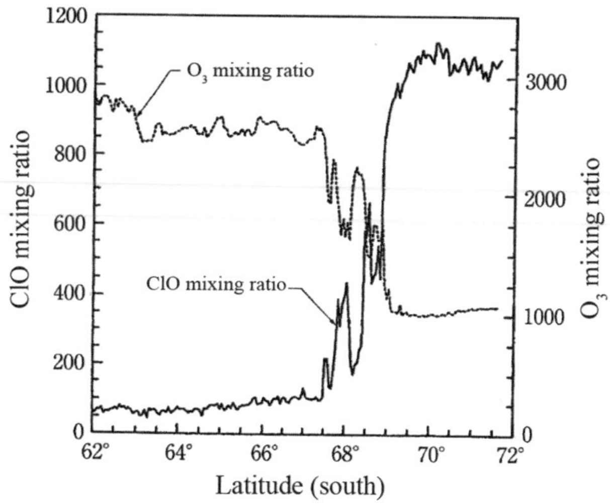

Fig. 8.9 Anti-correlation between O_3 and Cl mixing ratios in the ozone hole observed by an aircraft (Adapted from Anderson et al. 1989)

described in the previous section. Owing to the later studies, the ozone hole was found to be caused by the combination of a physical phenomenon called the polar vortex, due to the peculiar meteorology over Antarctica, and chemical phenomenon consisting of heterogeneous reactions on the surface of PSCs and succeeding gas phase chain reactions. A very stable polar vortex develops from winter to spring over Antarctica because of the cooling of the atmosphere by the cold continent, which provides the reaction field like a closed vessel blocking the outside air to mix with (Schoeberl et al. 1992; Brasseur and Solomon 2005). Since the stratospheric air is very dry, with only a few ppmv of water vapor, clouds are not formed in general. However, the temperature of the stratosphere in Antarctic winter is lower than 195 K, water vapor freezes to form clouds called nacreous clouds, or mother-of-pearl, and the clouds co-condensing HCl, HNO_3 etc. are called PSCs (see Chap. 6, sect. 6.5). The altitude of the appearance of PSCs is 10–25 km, which roughly agrees with the regime where O_3 is depleted extremely in the ozone hole.

Later, polar ozone depletion was found to occur not only over the Antarctica but also the Arctic circle (McElroy et al. 1986; Müller et al. 1997). However, in the Arctic, different from the Antarctic, a polar vortex does not sufficiently develop due to the larger inhomogeneity of the earth surface along the latitude, and higher temperatures than Antarctica, the size of ozone hole is smaller in general, but the similar ozone depletion due to CFCs etc. occurs by similar chemical reactions to those in the Antarctic. Particularly, ozone depletion comparable to the Antarctic ozone hole for density and spatial size was observed over the Arctic in 2011 (Manney et al. 2011).

The most important heterogeneous reaction with PSCs is the Cl_2 release into the gas phase by the reaction of $ClONO_2$(g) in the gas phase and HCl(s) on the solid surface (Sect. 6.5.5).

$$ClONO_2(g) + HCl(s) \rightarrow Cl_2(g) + HNO_3(s) \quad (8.125)$$

By this reaction, photolytically active Cl_2 is formed from HCl and $ClONO_2$ reserved in the polar vortex during polar night. In spring, when the polar stratosphere starts to receive the solar radiation, Cl_2 is rapidly photolyzed to release Cl atoms and ozone destruction proceeds. From Cl_2 formed in reaction (8.125), HOCl is formed,

$$Cl_2 + h\nu \rightarrow 2Cl \quad (8.126)$$

$$Cl + O_3 \rightarrow ClO + O_2 \quad (8.68)$$

$$ClO + HO_2 \rightarrow HOCl + O_2 \quad (8.81)$$

and ClO and HO_2 are regenerated by the photolysis of HOCl,

$$HOCl + h\nu \rightarrow OH + Cl \quad (8.82)$$

$$Cl + O_3 \rightarrow ClO + O_2 \quad (8.68)$$

$$OH + O_3 \rightarrow HO_2 + O_2 \quad (8.23)$$

so that a chain reaction is constituted and O_3 is dissipated by reactions (8.68) and (8.23). Since the ozone hole develops at altitudes lower than 20 km as seen in Fig. 8.8, the O atom mixing ratio is low, and it is a distinct feature that the reaction to regenerate Cl atoms mentioned in Sect. 8.2.3,

$$ClO + O \rightarrow Cl + O_2 \quad (8.69)$$

does not work effectively. Also in the polar vortex, the mixing ratio of NO_2 in the gas phase is very low since NO_2 is mostly taken into PSCs as HNO_3. Therefore, the chain termination reaction,

$$ClO + NO_2 + M \rightarrow ClONO_2 + M \quad (8.78)$$

is not effective, so that the above O_3, dissipation reactions are promoted effectively. Among the PSCs, NAT (nitric acid trihydrate) (sect. 6.5 Table 6.2) is particularly stable, grows to a large particle with the diameter of 1–20 μm, and is removed from the stratosphere to the troposphere by gravitational sedimentation. This process brings about ultimate removal of NO_x from the stratosphere resulting in the very low mixing ratio of NO_2 and acceleration of the O_3 dissipation cycle under the solar radiation in springtime.

Next to reaction (8.116), the hydrolysis reaction of $ClONO_2$ by $H_2O(s)$ on PSCs (Sect. 6.5.4),

$$ClONO_2(g) + H_2O(s) \rightarrow HOCl(g) + HNO_3(s) \tag{8.127}$$

is an important reaction to release photolytically active HOCl into the gas phase. While HOCl is photolyzed,

$$HOCl + h\nu \rightarrow OH + Cl \tag{8.82}$$

to give OH and Cl as seen above and facilitate the O_3 destruction by the gas phase homogeneous reaction, HOCl participates in a heterogeneous reaction with HCl (s) in PSCs (Sect. 6.5.3),

$$HOCl(g) + HCl(s) \rightarrow Cl_2(g) + H_2O(s) \tag{8.128}$$

to release Cl_2 into the gas phase.

As other heterogeneous reactions on PSCs, the reactions of N_2O_5,

$$N_2O_5(g) + H_2O(s) \rightarrow 2HNO_3(s) \tag{8.129}$$

$$N_2O_5(g) + HCl(s) \rightarrow ClNO_2(g) + HNO_3(s) \tag{8.130}$$

are also known (Sects. 6.5.1 and 6.5.2). Reaction (8.130) releases photochemically active $ClNO_2$ into the gas phase, and Cl atoms are formed by the photolysis of $ClNO_2$,

$$ClNO_2 + h\nu \rightarrow Cl + NO_2. \tag{8.131}$$

Meanwhile, the active nitrogen is taken into PSCs as HNO_3 by reactions (8.129) and (8.130).

A characteristic of the gas phase reaction in the polar vortex is that the importance of radical-radical reactions of ClO with ClO, and ClO with BrO since the mixing ratios of ClO and BrO become very high in spring as compared to the normal stratosphere. The formation of ClO dimer, ClOOCl (Sect. 5.6.6), and its photolysis can facilitate O_3 depletion as has been proposed by Molina and Molina (1987).

$$ClO + ClO + M \rightarrow ClOOCl + M \tag{8.132}$$

At present, ClOOCl is the most abundant Cl species in the gas phase at polar night in the lower stratosphere over the polar region (Brasseur and Solomon 2005). The ClOOCl is photolyzed (Sect. 4.4.6) in the polar vortex in spring,

$$ClOOCl + h\nu \rightarrow Cl + ClOO \tag{8.133}$$

$$\rightarrow 2Cl + O_2, \tag{8.134}$$

to yield Cl and ClOO, and ClOO is further decomposed photolytically or thermally to $Cl + O_2$,

$$ClOO + h\nu \rightarrow Cl + O_2 \quad (8.135)$$

$$ClOO + M \rightarrow Cl + O_2 + M. \quad (8.136)$$

The O_3 destruction reactions through the ClO dimer can be expressed as,

$$ClO + ClO + M \rightarrow ClOOCl + M \quad (8.132)$$

$$ClOOCl + h\nu \rightarrow 2\,Cl + O_2 \quad (8.134)$$

$$2\,[Cl + O_3 \rightarrow ClO + O_2] \quad (8.68)$$

$$\text{net} \quad 2\,O_3 \rightarrow 3\,O_2.$$

Similarly, the O_3 destruction via the cross reaction of ClO and BrO is important in the polar vortex in spring (McElroy et al. 1986).

$$BrO + ClO \rightarrow Br + OClO \quad (8.137)$$

$$\rightarrow Br + ClOO \quad (8.138)$$

$$\rightarrow BrCl + O_2 \quad (8.139)$$

Here, formed OClO is known to photolyse to give O atom and ClO (Sect. 4.4.7),

$$OClO + h\nu \rightarrow O + ClO. \quad (8.140)$$

BrCl formed in the reaction (8.139) is also photolyzed easily to give Br and Cl atoms (Sect. 4.4.1),

$$BrCl + h\nu \rightarrow Br + Cl. \quad (8.141)$$

Since the mutual reaction of ClO, and reaction of ClO with BrO regenerate Cl or Br except the O atom forming reaction (8.140), they all accelerate the O_3 depleting chain reaction. The ozone destruction by these reactions are expressed either by,

$$BrO + ClO \rightarrow Br + ClOO \quad (8.138)$$

$$ClOO + M \rightarrow Cl + O_2 + M \quad (8.136)$$

$$Cl + O_3 \rightarrow ClO + O_2 \quad (8.68)$$

$$Br + O_3 \rightarrow BrO + O_2 \quad (8.88)$$

$$\text{net} \quad 2\,O_3 \rightarrow 3\,O_2.$$

or,

$$BrO + ClO \longrightarrow BrCl + O_2 \quad (8.139)$$

$$BrCl + h\nu \longrightarrow Br + Cl \quad (8.142)$$

$$Cl + O_3 \longrightarrow ClO + O_2 \quad (8.68)$$

$$Br + O_3 \longrightarrow BrO + O_2$$

$$\text{net} \quad 2\,O_3 \longrightarrow 3\,O_2. \quad (8.88)$$

The relative contribution of ClO + ClO and BrO + ClO to the ozone depletion in the Antarctic ozone hole has been estimated as 60 and 40 %, respectively (Seinfeld and Pandis 2006).

8.5 Stratospheric Sulfur Chemistry

Most inorganic and organic sulfur compounds emitted from the earth's surface do not reach to the stratosphere since their atmospheric lifetime is shorter than 1 year. One exception, however, is the carbonyl sulfide COS emitted from marine organisms and volcanoes. COS is not photolyzed in the troposphere and the rate constant of the reaction with OH is as small as 2.0×10^{-15} cm^3 molecule^{-1} s^{-1} (298 K) (Table 5.2), its tropospheric lifetime is as long as a few years and reaches to the stratosphere. The mixing ratio of COS is ca. 500 pptv in the troposphere and enters into the stratosphere with this value. In the stratosphere it is easily photolyzed, and the mixing ration decreases to ca. 15 pptv at the altitude of 35 km (Brasseur and Solomon 2005). By the photolysis of COS, sulfuric acid (H_2SO_4) molecules are formed as gaseous molecules, and then they condense to form sulfuric acid aerosol since the vapor pressure of H_2SO_4 is very low. The sulfuric acid aerosol layer in the stratosphere spreads around the altitude of 20 km, and called Junge layer after the name of the discoverer. The particle diameter of stratospheric aerosol is 0.01–1 μm, and it affects the climate through changing the thermal budget of the earth by reflecting, absorbing and scattering the solar and terrestrial radiation (Brühl et al. 2012). Sulfur compounds in the stratosphere other than COS include SO_2, sporadically input by a large-scale volcano eruption, which increases the stratospheric aerosols temporarily, and tends to decreases the earth surface temperature for some years.

The COS reached to the stratosphere is photolyzed as described in Chap. 4, Sect. 4.3.6,

$$COS + h\nu \rightarrow CO + S \quad (8.143)$$

to release S atoms. The sulfur atoms in the atmosphere react with O_2 immediately,

$$S + O_2 \rightarrow SO + O \tag{8.144}$$

to form sulfur monoxide SO (Donovan and Little 1972). Also COS reacts with oxygen atom $O(^3P)$ to form SO,

$$O(^3P) + COS \rightarrow CO + SO. \tag{8.145}$$

However, the rate constant of this reaction is relatively small, 1.3×10^{-14} cm^3 molecule^{-1} s^{-1} (298 K) (Table 5.1), so that the photolysis is the major pathway of the source of SO. The SO formed in reaction (8.144) reacts with O_2 to form SO_2 (Black et al. 1982; Atkinson et al. 2004),

$$SO + O_2 \rightarrow SO_2 + O. \tag{8.146}$$

The SO_2 is subject to photolysis (Sect. 4.3.7) in the stratosphere,

$$SO_2 + h\nu \rightarrow SO + O, \tag{8.147}$$

and also reacts with OH (Sect. 5.2.6) and forms $HOSO_2$ radicals as seen in the troposphere.

$$OH + SO_2 + M \rightarrow HOSO_2 + M \tag{8.148}$$

The $HOSO_2$ reacts with O_2 in the atmosphere,

$$HOSO_2 + O_2 \rightarrow HO_2 + SO_3 \tag{8.149}$$

to form sulfur trioxide SO_3 (Stockwell and Calvert 1983). Sulfuric acid molecule H_2SO_4 is formed from SO_3 by the homogeneous or heterogeneous reaction with H_2O (Jayne et al. 1997; Atkinson et al. 2004),

$$SO_3 + H_2O \rightarrow H_2SO_4 \tag{8.150}$$

The formed H_2SO_4 is a gaseous molecule but condenses to form sulfuric acid aerosol due to its low vapor pressure. Observation of gaseous H_2SO_4 in the stratosphere has been made by a negative ion mass spectrometer board on a balloon (Krieger and Arnold 1994). Figures 8.10 and 8.11 depicts an example of the measurement (Reiner and Arnold 1997). Since the saturation vapor pressure of H_2SO_4 increases with temperature, the mixing ratio of H_2SO_4(g) increases with the altitude, maximizes at 35–38 km, and decreases rapidly at the higher altitude. This is due to the depletion of COS, and the formation of H_2SO_4 does not occur much in the upper stratosphere.

Figure 8.11 shows the vertical profiles of the mixing ratios of sulfur species by model calculation (Turco et al. 1979). As shown in the figure, COS is photolyzed within the lower stratosphere, and H_2SO_4 exists as a major sulfur species with the

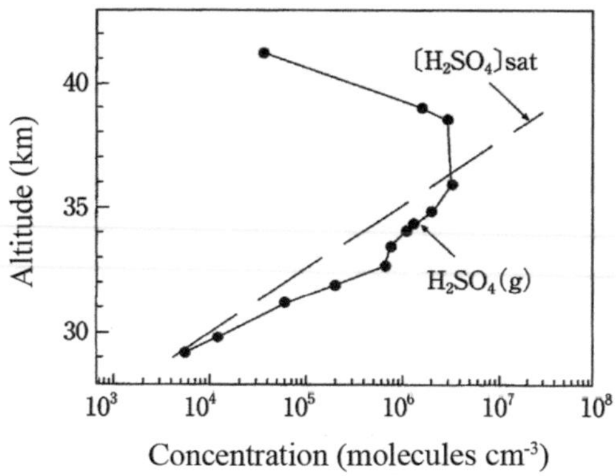

Fig. 8.10 Vertical profile of H_2SO_4 (g) in the stratosphere by observation (Adapted from Reiner and Arnold 1997). *Dashed line* is the equilibrium vapor pressure of H_2SO_4 on liquid sulfuric acid aerosols

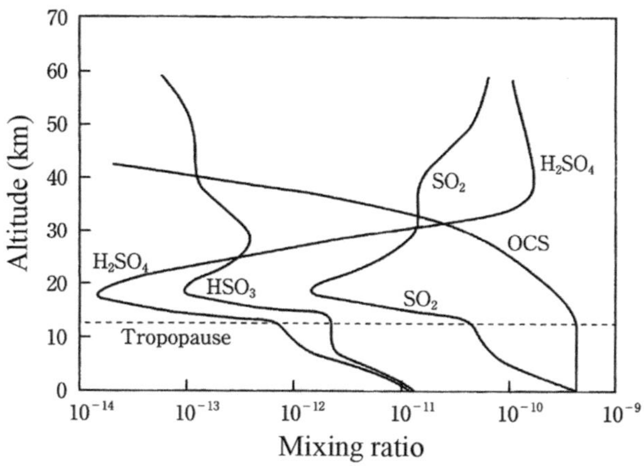

Fig. 8.11 Vertical profiles of sulfur compounds in the stratosphere by model calculation (Adapted from Turco et al. 1979)

mixing ratio of 100–200 pptv above 35 km. SO_2 exists at a few tens of pptv near the tropopause, decreases rapidly within a few km from the tropopause, and increases again in the upper stratosphere to 10–30 pptv.

References

Anderson, J.G., Brune, W.H., Proffitt, M.H.: Ozone destruction by chlorine radicals within the Antarctic vortex – the spatial and temporal evolution of ClO-O_3 anticorrelation based on in situ ER-2 data. J. Geophys. Res. **94**, 11465–11479 (1989)

Atkinson, R., Baulch, D.L., Cox, R.A., Crowley, J.N., Hampson, R.F., Hynes, R.G., Jenkin, M.E., Rossi, M.J., Troe, J.: Evaluated kinetic and photochemical data for atmospheric chemistry: volume I – gas phase reactions of Ox, HOx, NOx, and SOx species. Atmos. Chem. Phys. **4**, 1461–1738 (2004)

Bates, D.R., Nicolet, M.: The photochemistry of atmospheric water vapor. J. Geophys. Res. **55**, 301–327 (1950)

Bedjanian, Y., Poulet, G.: Kinetics of halogen oxide radicals in the stratosphere. Chem. Rev. **103**, 4639–4655 (2003)

Berkner, L.V., Marshall, L.C.: On the origin and rise of oxygen concentration in the Earth's atmosphere. J. Atmos. Sci. **22**, 225–261 (1965)

Black, G., Sharpless, R.L., Slanger, T.G.: Rate coefficients for SO reactions with O_2 and O_3 over the temperature range 230–420 K. Chem. Phys. Lett. **93**, 598–602 (1982)

Brasseur, G.P., Solomon, S.: Aeronomy of the Middle Atmosphere: Chemistry and Physics of the Stratosphere and Mesosphere, 3rd edn. Springer, Dordrecht (2005)

Brasseur, G.P., Orlando, J.J., Tyndall, G.S.: Atmospheric Chemistry and Global Change. Oxford University Press, New York (1999)

Brühl, C., Lelieveld, J., Crutzen, P.J., Tost, H.: The role of carbonyl sulphide as a source of stratospheric sulphate aerosol and its impact on climate. Atmos. Chem. Phys. **12**, 1239–1253 (2012)

Burley, J.D., Miller, C.E., Johnston, H.S.: Spectroscopy and photoabsorption cross sections of FNO. J. Mol. Spectrosc. **158**, 377–391 (1993)

Chapman, S.: A theory of upper atmospheric ozone. Mem. Roy. Meteorol. Soc. **3**, 103–125 (1930a)

Chapman, S.: On ozone and atomic oxygen in the upper atmosphere. Philos. Mag. **10**, 369–383 (1930b)

Chubachi, S.: A special ozone observation at Syowa Station, Antarctica from February 1982 to January 1983. In: Zerefos, C.S., Ghazi, A. (eds.) Atmospheric Ozone. pp. 606–610. Reidel, Dordrecht (1985)

Crutzen, P.J.: The influence of nitrogen oxides on atmospheric ozone content. Qurt. J. Roy. Meteorol. Soc. **96**, 320–325 (1970)

DeMore, W.M., Golden, D.M., Hampson, R.F., Kurylo, M.J., Howard, C.J., Ravishankara, A.R., Kolb, C.E., Molina, M.J.: Chemical Kinetics and Photochemical Data for Use in Atmospheric Studies, Evaluation Number 12, pp. 97–104. JPL Publication, Pasadena (1997)

Donovan, R.J., Little, D.J.: The rate of the reaction $S(3^3P_J) + O_2$. Chem. Phys. Lett. **13**, 488–490 (1972)

Dotto, L., Schiff, H., Schiff, H.: The Ozone War. Doubleday, Garden City (1978)

Farman, J.C., Gardiner, B.G., Shankin, J.D.: Large losses of total ozone in Antarctica reveal seasonal ClO_x/NO_x interaction. Nature **315**, 207–210 (1985)

Finlayson-Pitts, B.J., Pitts Jr., J.N.: Chemistry of the Upper and Lower Atmosphere. Academic, San Diego (2000)

Hofmann, D.J., Harder, J.W., Rolf, S.R., Rosen, J.M.: Balloon-borne observations of the development and vertical structure of the Antarctic ozone hole in 1986. Nature **326**, 59–62 (1987)

Jayne, J.T., Pöschl, U., Chen, Y., Dai, D., Molina, L.T., Worsnop, D.R., Kolb, C.E., Molina, M.J.: Pressure and temperature dependence of the gas-phase reaction of SO_3 with H_2O and the heterogeneous reaction of SO_3 with H_2O/H_2SO_4 surfaces. J. Phys. Chem. A **101**, 10000–10011 (1997)

Johnston, H.S.: Reduction of stratospheric ozone by nitrogen oxide catalysis from supersonic transport exhaust. Science **173**, 517–522 (1971)

Jucks, K.J., Johnson, D.G., Chance, K.V., Traub, W.A., Margitan, J.J., Osterman, G.B., Salawitch, R.J., Sasano, Y.: Observations of OH, HO_2, H_2O, and O_3 in the upper stratosphere; implications for HO_x photochemistry. Geophys. Res. Lett. **25**, 3935–3938 (1998)

Krieger, A., Arnold, F.: First composition measurements of stratospheric negative ions and inferred gaseous sulfuric acid in winter Arctic vortex: implications for aerosols and hydroxyl radical formation. Geophys. Res. Lett. **21**, 1259–1262 (1994)

Manney, G.L., Santee, M.L., Rex, M., Livesey, N.J., Pitts, M.C., Veefkind, P., Nash, E.R., Wohltmann, I., Lehmann, R., Froidevaux, L., Poole, L.R., Schoeberl, M.R., Haffner, D.P., Davies, J., Dorokhov, V., Gernandt, H., Johnson, B., Kivi, R., Kyrö, E., Larsen, N., Levelt, P. F., Makshtas, A., McElroy, C.T., Nakajima, H., Parrondo, M.C., et al.: Unprecedented Arctic ozone loss in 2011. Nature **478**, 469–475 (2011)

McElroy, M.B.: The Atmospheric Environment: Effects of Human Activities. Princeton University Press, Princeton (2002)

McElroy, M.B., Salawitch, R.J., Wofsy, S.C., Logan, J.A.: Reductions of Antarctic ozone due to synergistic interactions of chlorine and bromine. Nature **321**, 759–762 (1986)

Middleton, A.M., Tolbert, M.A.,.: Stratospheric Ozone Depletion. RSC Publishing, Sausalito (2000)

Molina, L.T., Molina, M.J.: Production of Cl_2O_2 from the self-reaction of the ClO radical. J. Phys. Chem. **91**, 433–436 (1987)

Molina, M.J., Rowland, F.S.: Stratospheric sink for chlorofluoromethanes: chlorine atom catalyzed destruction of ozone. Nature **249**, 810–812 (1974)

Müller, R., Crutzen, P.J., Groo, J.-U., Bürhl, C., Russell III, J.M., Gernandt, H., McKennal, D.S., Tuck, A.F.: Severe chemical ozone loss in the Arctic during the winter of 1995–96. Nature **389**, 709–712 (1997)

Nicolet, M.: Stratospheric ozone; an introduction to its study. Rev. Geophys. **13**, 593–636 (1975)

Osterman, G.B., Salawitch, R.J., Sen, B., Toon, G.C., Stachnik, R.A., Pickett, H.M., Margitan, J.J., Blavier, J.-F., Peterson, D.B.: Balloon-borne measurements of stratospheric radicals and their precursors: implications for the production and loss of ozone. Geophys. Res. Lett. **24**, 1107–1110 (1997)

Reiner, T., Arnold, F.: Stratospheric SO_3: upper limits inferred from ion composition measurements – implications for H_2SO_4 and aerosol formation. Geophys. Res. Lett. **24**, 1751–1754 (1997)

Sander, S.P., Baker, R., Golden, D.M., Kurylo, M.J., Wine, P.H., Abatt, J.P.D., Burkholder, J.B., Kolb, C.E., Moortgat, G.K., Huie, R.E., Orkin, V.L.: Chemical Kinetics and Photochemical Data for Use in Atmospheric Studies: Evaluation Number 17, pp. 10–16. JPL Publication, Pasadena (2011)

Schoeberl, M.R., Lait, L.R., Newman, P.A., Rosenfield, J.E.: The structure of the polar vortex. J. Geophys. Res.: Atmos. **97**, 7859–7882 (1992)

Seinfeld, J.H., Pandis, S.N.: Atmospheric Chemistry and Physics: Air Pollution to Climate Change, 2nd edn. Wiley, Hoboken, NJ (2006)

Sen, B., Toon, G.C., Osterman, G.B., Blavier, J.-F., Margitan, J.J., Salawitch, R.J., Yue, G.K.: Measurements of reactive nitrogen in the stratosphere. J. Geophys. Res. **103**, 3571–3585 (1998)

Shimazaki, T.: Stratospheric Ozone, 2nd ed., University of Tokyo Press (1987) (in Japanese)

Solomon, S., Garcia, R.R., Ravishankara, A.R.: On the role of iodine in ozone depletion. J. Geophys. Res. **99**, 20491–20499 (1994)

Stockwell, W.R., Calvert, J.G.: The mechanism of the HO-SO_2 reaction. Atmos. Environ. **17**, 2231–2235 (1983)

Stolarski, R.S., Cicerone, R.J.: Stratospheric chlorine: a possible sink for ozone. Can. J. Chem. **52**, 1610–1615 (1974)

Turco, R.P., Hamill, P., Toon, O.B., Whitten, R.C., Kiang, C.S.: A one-dimensional model describing aerosol formation and evolution in the stratosphere: I. Physical process and mathematical analogs. J. Atmos. Sci. **36**, 699–717 (1979)

Wallington, T.J., Ellermann, T., Nielsen, O.J., Sehested, J.: Atmospheric chemistry of FCO_x radicals: UV spectra and self-reaction kinetics of FCO and $FC(O)O_2$ and kinetics of some reactions of FCO_x with O_2, O_3, and NO at 296 K. J. Phys. Chem. **98**, 2346–2356 (1994)

Wallington, T.J., Schneider, W.F., Szente, J.J., Maricq, M.M., Nielsen, O.J., Sehested, J.: Atmospheric chemistry of FNO and FNO_2: reactions of FNO with O_3, $O^{(3}P)$, HO_2, and HCl and the reaction of FNO_2 with O_3. J. Phys. Chem. **99**, 984–989 (1995)

Warneck, P.: Chemistry of the Natural Atmosphere. Academic, San Diego (1988)

Wayne, R.P.: Chemistry of Atmospheres, 3rd edn. Oxford University Press, New York (2000)

WMO (World Meteorological Organization), Scientific Assessment of Ozone Depletion: 2006, Global Ozone Research and Monitoring Project – Report No. 50, Geneva, 2007

WMO (World Meteorological Organization), Scientific Assessment of Ozone Depletion: 2010, Global Ozone Research and Monitoring Project – Report No.52, Geneva, 2011

Wofsy, S.C., McElroy, M.B.: HOx, NOx, and ClOx: their role in atmospheric chemistry. Can. J. Chem. **52**, 1582–1591 (1974)

Zander, R., Gunson, M.R., Farmer, C.B., Rinsland, C.P., Irion, F.W., Mahieu, E.: The 1985 chlorine and fluorine inventories in the stratosphere based on ATMOS observations at 30° north latitude. J. Atmos. Chem. **15**, 171–186 (1992)

Index

H. Akimoto, *Atmospheric Reaction Chemistry*, Springer Atmospheric Sciences,
DOI 10.1007/978-4-431-55870-5

D

P

MIX
Papier aus verantwortungsvollen Quellen
Paper from responsible sources
FSC® C105338

If you have any concerns about our products,
you can contact us on
ProductSafety@springernature.com

In case Publisher is established outside the EU,
the EU authorized representative is:
Springer Nature Customer Service Center GmbH
Europaplatz 3, 69115 Heidelberg, Germany

Printed by Libri Plureos GmbH
in Hamburg, Germany